Springer INdAM Series

Volume 7

For further volumes:
http://www.springer.com/series/10283

Maria Gorelik • Paolo Papi
Editors

Advances in Lie Superalgebras

Springer

Editors
Maria Gorelik
Department of Mathematics
The Weizmann Institute of Science
Rehovot, Israel

Paolo Papi
Dipartimento di Matematica
Sapienza – Università di Roma
Roma, Italy

ISSN: 2281-518X
ISSN: 2281-5198 (electronic)
Springer INdAM Series
ISBN 978-3-319-34628-1
ISBN 978-3-319-02952-8 (eBook)
DOI 10.1007/978-3-319-02952-8
Springer Cham Heidelberg New York Dordrecht London

Cover Design: Raffaella Colombo, Giochi di Grafica, Milano, Italy
Typesetting with LaTeX: PTP-Berlin, Protago TeX-Production GmbH, Germany (www.ptp-berlin.de)

Springer is part of Springer Science+Business Media (www.springer.com)

Preface

The first examples of Lie superalgebras appear in algebraic topology in the late 40's (the Whitehead product on homotopy groups is a Lie superalgebra bracket) and in the context of deformation theory of complex structures (Nijenhuis, Frölicher and Nijenhuis) in the late 50's. Shortly after, Gerstenhaber, in a series of fundamental papers, shed new light on the role of Lie superalgebras in his theory of deformation of rings and algebras, while Spencer and his collaborators developed applications to pseudogroup structures on manifolds.

A renewed interest in Lie superalgebras came from Physics in the early 70's: many examples arise naturally as "supersymmetries" for quantum field theories, e.g. in the Wess-Zumino model[1]. It was however Kac's landmark Advances paper[2] which established the study of Lie superalgebras as a branch of Algebra in its own right.

Since then the subject has received dramatic developments, so that up to now more than 900 papers having "Lie superalgebra" in their title can be counted in the MathSciNet.

This volume originates from "Lie Superalgebras", held in Roma, Istituto Nazionale di Alta Matematica "Francesco Severi", December 14-19, 2012.

It consists of original papers and/or extended expositions of the talks delivered at the conference.

We believe that the contributions, kindly offered by the invited speakers, clearly illustrate one of the most remarkable features of the theory of Lie superalgebras which is, the astonishing range of its connections with other branches of Mathematics and Mathematical Physics.

It is our pleasure to thank Professor Vincenzo Ancona, President of Indam, the Scientific Committee and the entire staff of Indam, for allowing us the opportunity to gather so many specialists in such a highly stimulating meeting.

---

[1] A thorough discussion of the role of Lie superalgebras up to the mid 70's can be found in Corwin-Ne'eman-Sternberg, Rev. Mod. Ph., **47**, 575–603 (1975).

[2] Kac, V.G.: Lie superalgebras. Advances in Math. **26**(1) 8–96 (1977).

By happy coincidence, the publication of this volume coincides with Victor Kac's seventieth birthday. It would be hard to believe that the theory of Lie superalgebras would have progressed so far without his contribution in the field. With the consent of all contributing authors, we would like to dedicate this volume to him.

Rehovot and Roma
September 2013

Maria Gorelik
Paolo Papi

Contents

Superbosonisation, Riesz superdistributions, and highest weight modules

Alexander Alldridge and Zain Shaikh

Abstract Superbosonisation, introduced by Littelmann–Sommers–Zirnbauer, is a generalisation of bosonisation, with applications in Random Matrix Theory and Condensed Matter Physics. In this survey, we link the superbosonisation identity to Representation Theory and Harmonic Analysis and explain two new proofs, one *via* the Laplace transform and one based on a multiplicity freeness statement.

1 Introduction

Supersymmetry (SUSY) has its origins in Quantum Field Theory. It is usually associated with High Energy Physics, especially with SUGRA, where the fermionic fields correspond to physical quantities, the mathematical incarnation of a (as yet, hypothetical) fundamental phenomenon. However, beyond this fascinating and deep theory, and its independent mathematical interest, SUSY also has applications in quite different areas of physics, notably, in Condensed Matter.

Here, the generators of supersymmetry do not correspond to physical quantities. Rather, they appear as effective symmetries of models for low-temperature limits of the fundamental Quantum Field Theory. This idea goes under the name of the *Supersymmetry Method*, and was developed by Efetov and Wegner [9].

Its particular merit is the possibility to derive, by the use of Harmonic Analysis on certain symmetric superspaces, precise closed form expressions for statistical quantities – such as the moments of the conductance of a metal with impurities [29, 30] – in a regime where the system becomes critical, for instance, exhibiting a transition from localisation to diffusion, which is not tractable by other methods.

A. Alldridge (✉)
Mathematical Institute, University of Cologne, Weyertal 86–90, 50931 Köln, Germany
e-mail: alldridg@math.uni-koeln.de

Z. Shaikh
Mathematical Institute, University of Cologne, Weyertal 86–90, 50931 Köln, Germany
e-mail: zain@math.uni-koeln.de

M. Gorelik, P. Papi (eds.): *Advances in Lie Superalgebras.* Springer INdAM Series 7,
DOI 10.1007/978-3-319-02952-8_1,

In connection with the physics of thin wires, the subject was well studied in the 1990s; it has recently gained substantial new interest, since the 'symmetry classes' investigated in this context [4, 14, 31] have been found to occur as 'edge modes' of certain 2D systems dubbed 'topological insulators' (resp. superconductors) [12].

Mathematically, several aspects of the method beg justification. One both subtle and salient point is the transformation of certain integrals over flat superspace in high dimension $N \to \infty$, which occur as expressions for statistical Green's functions, into integrals over a curved superspace of fixed rank and dimension – the latter being more amenable to asymptotic analysis (by steepest descent or stationary phase). Traditionally, this step is performed by the use of the so-called Hubbard–Stratonovich transformation, which is based on a careful deformation of the integration contour.

This poses severe analytical problems, which to the present day have only been overcome in cases derived from random matrix ensembles that follow the normal distribution [15]. To extend the Supersymmetry Method's range beyond Gaussian disorder, for instance to establish universality for invariant random matrix ensembles, a complementary tool was introduced, based on ideas of Fyodorov [13]: the *Superbosonisation Identity* of Littelmann–Sommers–Zirnbauer [21]. (A more complete account of the history of superbosonisation is to be found in the introduction of [3].)

We now proceed to describe this identity. In general, it holds in the context of unitary, orthogonal, and unitary-symplectic symmetry. We restrict ourselves to the first case (of unitary symmetry), although our methods carry over to the other cases.

One considers the space $W := \mathbf{C}^{p|q\times p|q}$ of square super-matrices and a certain subsupermanifold Ω of purely even codimension, whose underlying (Riemannian symmetric) manifold Ω_0 is the product of the positive Hermitian $p \times p$ matrices with the unitary $q \times q$ matrices. Let f be a superfunction defined and holomorphic on the tube domain based on $\mathrm{Herm}^+(p) \times \mathrm{Herm}(q)$. The superbosonisation identity states

$$\int_{\mathbf{C}^{p|q\times n}\oplus\mathbf{C}^{n\times p|q}} |Dv|\, f(Q(v)) = C \int_{\Omega} |Dy|\, \mathrm{Ber}(y)^n f(y), \tag{1}$$

for some finite positive constant C, provided f has sufficient decay at infinity along the manifold Ω_0. Here, Q is the quadratic map $Q(v) = vv^*$, $|Dv|$ is the flat Berezinian density, and $|Dy|$ is a Berezinian density on Ω, invariant under a certain natural transitive supergroup action we will specify below.

Remark that any $\mathrm{GL}(n,\mathbf{C})$-invariant superfunction on $\mathbf{C}^{p|q\times n} \oplus \mathbf{C}^{n\times p|q}$ may be written in the form $f(Q(v))$. Thus, a notable feature of the formula is that it puts the 'hidden supersymmetries' (from $\mathrm{GL}(p|q,\mathbf{C})$) into evidence through the invariant integral over the homogeneous superspace Ω where 'manifest symmetries' (from $\mathrm{GL}(n,\mathbf{C})$) enter *via* some character (namely, $\mathrm{Ber}(y)^n$).

A remarkable special case occurs when $p = 0$. Then Eq. (1) reduces to

$$\int_{\mathbf{C}^{0|q\times n}\oplus\mathbf{C}^{n\times 0|q}} |Dv|\, f(Q(v)) = C \int_{\mathrm{U}(q)} |Dy|\, \det(y)^{-n} f(y),$$

which is known as the *Bosonisation Identity* in physics. Notice that the left-hand side is a purely fermionic Berezin integral, whereas the right-hand side is purely bosonic.

Formally, it turns fermions $\psi\bar{\psi}$ into bosons $e^{i\varphi}$. If in addition $q=1$, we obtain the *Cauchy Integral Formula*.

At the other extreme, if $q=0$, then $\Omega=\mathrm{Herm}^+(p)$, and Eq. (1) is a classical identity due to Ingham and Siegel [18, 26], well-known to harmonic analysts. It admits a far-reaching generalisation in the framework of Euclidean Jordan algebras [11]. The first one to use it in the physics context that inspired superbosonisation was Fyodorov [13]. Moreover, the right-hand side of the identity, *viz.*

$$\langle T_n, f\rangle := \int_{\mathrm{Herm}^+(p)} |Dy| \det(y)^n f(y)$$

is the so-called (unweighted) *Riesz distribution*. After suitable renormalisation, it becomes analytic in the parameter n, a fact that was exploited in the analytic continuation of holomorphic discrete series representations by Rossi and Vergne [10, 24].

This observation links the identity to equivariant geometry and Lie theoretic Representation Theory, and this was our motivation to re-investigate the identity.

In this survey, we explain two new proofs of the superbosonisation identity, which exploit these newly found connections. One of these proofs is based on Representation Theory. Namely, as it turns out, the two sides of the identity are given by certain special relatively invariant functionals on two highest weight modules of the Lie superalgebra $\mathfrak{g}' := \mathfrak{gl}(2p|2q,\mathbf{C})$, which are infinite dimensional for $p>0$. Their equality is an immediate consequence of a multiplicity one statement. We will explain the main ingredients of the proof; details shall be published elsewhere.

On the other hand, to actually identify the r.h.s. as a functional on the corresponding representation requires the construction of an intertwining operator in the form of a certain weighted Laplace transform $\mathscr{L}_n$. This leads to another proof, based on comparing Laplace transforms. The functional analytic details of this proof can be found in [3]. Here, we only explain the main ideas.

Combining both points of view leads to further developments. Indeed, the representations related to the superbosonisation identity depend on a parameter $n \geq p$. Using functional equations, one may show that the r.h.s. is analytic as a distribution-valued function of n. In a forthcoming paper, we shall use this fact to investigate the analytic continuation of the representations.

In what follows, we use the terminology of supergeometry freely. For the reader's convenience, some basics are summarised in an Appendix.

2 The superbosonisation module

In this section, we define, by the use of Howe duality, a particular supermodule, which will turn out to be intimately related to the superbosonisation identity.

2.1 The oscillator representation

We begin by reviewing some standard material on the Weyl–Clifford algebra in a form well suited for our purposes. Consider $V := U \oplus U'$ where $U := \underline{\mathrm{Hom}}(\mathbf{C}^n, \mathbf{C}^{p|q})$, $U' := \underline{\mathrm{Hom}}(\mathbf{C}^{p|q}, \mathbf{C}^n)$, and $\underline{\mathrm{Hom}}(\cdot,\cdot)$ denotes the set of all linear maps with its usual $\mathbf{Z}/2\mathbf{Z}$ grading. Then $V^* = U^* \oplus U'^*$. The supersymplectic form on $V \oplus V^*$ is

$$s(a+f, b+g) := f(b) - (-1)^{|a||g|} g(a),$$

where $a, b \in V$ and $f, g \in V^*$ are homogeneous. The trace and supertrace forms allow us to identify $V^* = \underline{\mathrm{Hom}}(\mathbf{C}^{p|q}, \mathbf{C}^n) \oplus \underline{\mathrm{Hom}}(\mathbf{C}^n, \mathbf{C}^{p|q})$.

Let $\mathfrak{h}_V$ be the central extension of the Abelian Lie superalgebra $V \oplus V^*$ by $\mathbf{C}$ determined by s. Denoting the generator of $\mathbf{C}$ by $\mathbf{1}$, the non-zero bracket relations of $\mathfrak{h}_V$ are $[v, w] = s(v, w)\mathbf{1}$ for $v, w \in V \oplus V^*$. By definition, the *Weyl–Clifford algebra* is

$$\mathrm{WCl}(V) := \mathfrak{U}(\mathfrak{h}_V)/(\mathbf{1} - 1),$$

where $\mathbf{1} \in \mathfrak{h}_V \subseteq \mathfrak{U}(\mathfrak{h}_V)$ and $1 \in \mathfrak{U}(\mathfrak{h}_V)$ is the unit of the universal enveloping algebra.

Canonical $\mathfrak{spo}$ subalgebra The Weyl–Clifford algebra inherits an ascending filtration $\mathrm{WCl}_n(V)$ from the tensor algebra $\bigotimes(V \oplus V^*)$. The PBW theorem implies that $\mathrm{gr}\,\mathrm{WCl}(V) = S(V \oplus V^*)$. Conversely, $\mathrm{WCl}(V)$ inherits a canonical augmentation from $\mathfrak{U}(\mathfrak{h}_V)$; moreover, the kernel of the canonical map $\bigotimes(V \oplus V^*) \to \mathrm{WCl}(V)$ is generated by quadratic relations without linear term, and it follows that there is a canonical splitting of the map $\mathrm{WCl}(V) \to S(V \oplus V^*)$ in degree two, the image of which we denote by $\mathfrak{s}$.

On general grounds, we have $[\mathrm{WCl}_a(V), \mathrm{WCl}_b(V)] \subseteq \mathrm{WCl}_{a+b-1}(V)$, so $\mathrm{gr}\,\mathfrak{s}$ preserves the filtration; since, moreover, $\mathbf{C}1$ is central in $\mathrm{WCl}(V)$, $\mathfrak{s}$ is a Lie superalgebra and the degree one part $\mathrm{gr}_1\,\mathrm{WCl}(V) \cong V \oplus V^* \subseteq \mathrm{WCl}(V)$ is an $\mathfrak{s}$-module. This sets up an isomorphism $\mathfrak{s} \cong \mathfrak{spo}(V \oplus V^*)$, where the latter is the Lie subsuperalgebra of $\mathfrak{gl}(V \oplus V^*)$ consisting of the endomorphisms leaving s infinitesimally invariant.

Oscillator module Let $V \oplus V^* = X \oplus Y$ be a complex polarisation of $V \oplus V^*$. That is, X and Y are maximal isotropic subspaces. The Weyl–Clifford algebra $\mathrm{WCl}(V)$ has a module

$$S_X := \mathrm{WCl}(V)/\mathrm{WCl}(V) \cdot X = \mathfrak{U}(\mathfrak{h}_V) \otimes_{\mathfrak{U}(X \oplus \mathbf{C}1)} \mathbf{C},$$

called the *oscillator* (or *spinor*) *representation*. Here, $\mathbf{C}$ is understood to be the module of the Abelian Lie subsuperalgebra $V \oplus \mathbf{C}1 \subseteq \mathfrak{h}_V$ on which $\mathbf{1}$ acts as the identity and X acts by zero.

We have $S_X \cong S(Y) = \mathbf{C}[X]$ as modules over the Abelian Lie subsuperalgebra Y of $\mathfrak{h}_V$, where the action of Y on $f \in \mathbf{C}[X]$ is by left multiplication and X acts by superderivations. Since any $f \in \mathbf{C}[X]$, which is annihilated by X, is necessarily constant, the maximal proper submodule of S_X is zero, and S_X is a simple module over $\mathrm{WCl}(V)$. Since $\mathfrak{s}$ is a Lie subsuperalgebra of $\mathrm{WCl}(V)$, S_X is also an $\mathfrak{s}$-module.

The isomorphism class of S_X as a $\mathrm{WCl}(V)$-module does not depend on the choice of the polarisation. For $X := V$, $Y := V^*$, we simply write $S := S_X = \mathbf{C}[V]$. However, we will need to choose other polarisations to accommodate the action of real forms.

2.2 An application of Howe duality

Let $G_{\mathbf{C}} := \mathrm{GL}(n, \mathbf{C})$ be the complexification of the unitary group $G = \mathrm{U}(n)$. Define an action of $G_{\mathbf{C}}$ on V by $g \cdot (u, \varphi) := (ug^{-1}, g\varphi)$ for $g \in G_{\mathbf{C}}$, $(u, \varphi) \in V$; on V^*, we have the contragredient $G_{\mathbf{C}}$-action. The induced action of $G_{\mathbf{C}}$ on $V \oplus V^*$ is faithful and preserves the form s; hence, it realises the Lie algebra $\mathfrak{g}$ of $G_{\mathbf{C}}$ as a subalgebra of $\mathfrak{s}$.

Let $\mathfrak{g}' := \mathfrak{z}_{\mathfrak{s}}(\mathfrak{g})$ be the centraliser of $\mathfrak{g}$ in $\mathfrak{s}$. Then $\mathfrak{g}' \cong \mathfrak{gl}(2p|2q, \mathbf{C})$ and $\mathfrak{z}_{\mathfrak{s}}(\mathfrak{g}') = \mathfrak{g}$. In other words, $(\mathfrak{g}, \mathfrak{g}')$ form a *dual pair*, see [17]. We shall henceforth decompose elements of $\mathfrak{g}'$ as 2×2 block matrices with blocks of size $p|q \times p|q$.

Howe's celebrated duality [17, Theorem 8] yields the following conclusion:

Proposition 1 *The $\mathfrak{g}'$-submodule $\mathbf{C}[V]^{G_{\mathbf{C}}} = \mathbf{C}[V]^G$ of $\mathbf{C}[V]$ formed by the $G_{\mathbf{C}}$-invariant superpolynomials is simple.*

In order to identify this representation, we introduce a triangular decomposition of $\mathfrak{s} = S^2(V \oplus V^*)$. Define

$$\mathfrak{s}_+ := S^2(V), \quad \mathfrak{s}_0 := V \otimes V^*, \quad \mathfrak{s}_- := S^2(V^*).$$

This defines a $\mathbf{Z}$-grading of $\mathfrak{s}$, *i.e.* $\mathfrak{s}_\pm$ are Abelian Lie subsuperalgebras, normalised by the Lie subsuperalgebra $\mathfrak{s}_0$, and $[\mathfrak{s}_+, \mathfrak{s}_-] \subseteq \mathfrak{s}_0$. We observe that $\mathfrak{g} \subseteq \mathfrak{s}_0$, so that it preserves the triangular decomposition. Hence, $\mathfrak{g}'$ inherits a decomposition

$$\mathfrak{p}^+ := \mathfrak{g}'_+ := \mathfrak{g}' \cap \mathfrak{s}_+, \quad \mathfrak{k} := \mathfrak{g}'_0 := \mathfrak{g}' \cap \mathfrak{s}_0, \quad \mathfrak{p}^- := \mathfrak{g}'_- := \mathfrak{g}' \cap \mathfrak{s}_-$$

from $\mathfrak{s}$. For $Y \in \mathfrak{g}'$ decomposed into blocks, labeled A, B, C, D from left to right and top to bottom, we have $(A, D) \in \mathfrak{k}$, $B \in \mathfrak{g}'_+$ and $C \in \mathfrak{g}'_-$. Both of $\mathfrak{p}_\pm = \mathfrak{g}'_\pm$ are Abelian Lie superalgebras isomorphic to $\mathfrak{gl}(p|q, \mathbf{C})$ as super-vector spaces. The subalgebra $\mathfrak{k}$ can also be characterised as largest subalgebra of $\mathfrak{g}' = \mathfrak{gl}(2p|2q, \mathbf{C})$ that preserves the decomposition of V as $U \oplus U'$, *cf.* [17].

Highest weight From the definitions, we see that the constants $\mathbf{C}1 \subseteq \mathbf{C}[V]$ are annihilated by $\mathfrak{s}_+$ and left invariant as a subspace by the action of $\mathfrak{s}_0$. We have the following result.

Proposition 2 *The simple $\mathfrak{g}'$-module $\mathbf{C}[V]^{G_{\mathbf{C}}}$ has a highest weight. For a Borel subsuperalgebra $\mathfrak{b} \subseteq \mathfrak{g}'$ contained in $\mathfrak{k} \oplus \mathfrak{p}^+$, it is given as the restriction to a Cartan subalgebra of the $\mathfrak{k}$-character λ defined by $\lambda(\mathrm{diag}(A, D)) := \frac{n}{2}(\mathrm{str}(D) - \mathrm{str}(A))$.*

Let $\mathfrak{h}$ be the Cartan subalgebra spanned by the elementary matrices E^A_{aa}, E^D_{dd}, $1 \leq a, d \leq p+q$, the superscripts A and D referring to the upper left and lower right block of $\mathfrak{g}'$, respectively. The basis $\delta_1, \ldots, \delta_{2p}, \varepsilon_1, \ldots, \varepsilon_{2q}$ of $\mathfrak{h}^*$ is defined to be dual to the ordered basis

$$E^A_{11}, \ldots, E^A_{pp}, E^D_{11}, \ldots, E^D_{pp}, E^A_{p+1,p+1}, \ldots, E^A_{p+q,p+q}, E^D_{p+1,p+1}, \ldots, E^D_{p+q,p+q}.$$

Let $\mathfrak{b} \subseteq \mathfrak{g}'$ be the Borel subsuperalgebra determined uniquely by $\mathfrak{b} \cap \mathfrak{k}$ being the direct product of the standard Borels for $\mathfrak{k}_j$, $j = 1, 2$, and $\mathfrak{b} \cap (\mathfrak{p}^+ \oplus \mathfrak{p}^-) = \mathfrak{p}^+$. Then $\mathfrak{b}$ is contained in the parabolic subalgebra $\mathfrak{k} \oplus \mathfrak{p}^+$, and its Dynkin diagram is:

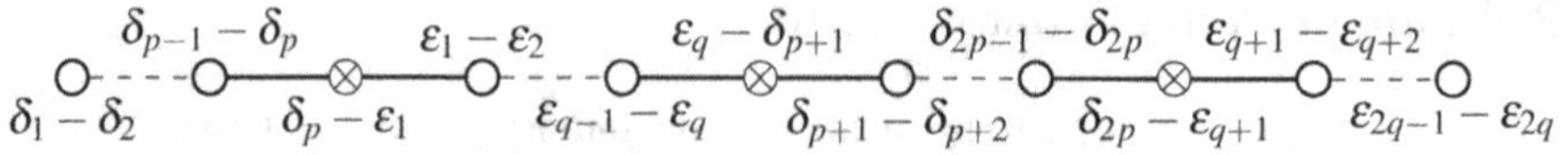

By standard facts [7], one obtains the following statement.

Proposition 3 *The simple $\mathfrak{g}'$-module $L(\lambda)$ has finite dimension if and only if $p=0$.*

Note that the highest weight $\lambda|_{\mathfrak{h}}$ is integral or half-integral, depending on whether n is even or odd. Moreover, $L(\lambda)$ is atypical whenever $pq>0$.

3 A relatively invariant functional

In this section, we show how to realise the left hand side of the superbosonisation identity as a relatively invariant functional on a globalisation of $L(\lambda)$.

In what follows, recall facts and definitions pertaining to supergroup pairs as summarised in the Appendix.

3.1 Globalisation of the oscillator representation

In order to globalise the oscillator representation $\mathbf{C}[V]$, we need to consider real forms. We shall use the following concept, *cf.* [5,8].

Definition 1 A real $\mathbf{Z}/2\mathbf{Z}$ graded vector space $U=U_{\bar{0}}\oplus U_{\bar{1}}$ with a fixed complex structure on $U_{\bar{1}}$ will be called a cs *vector space*. Given a complex super-vector space W, we call a *cs* vector subspace U a cs *form* of W if $W_{\bar{0}}=U_{\bar{0}}\oplus iU_{\bar{0}}$ and $W_{\bar{1}}=U_{\bar{1}}$.

We introduce a *cs* form $V_{\mathbf{R}}$ of V by positing

$$V_{\mathbf{R},\bar{0}}:=\mathscr{X}_{\bar{0}}:=V_{\bar{0}}\cap\operatorname{Herm}(n+p)=\big\{(L,L^*)\bigm| L\in\mathbf{C}^{p\times n}\big\},$$

where L^* denotes conjugate transpose. By $V^*_{\mathbf{R},\bar{0}}:=\mathscr{Y}_{\bar{0}}:=\big\{(K^*,K)\bigm| K\in\mathbf{C}^{p\times n}\big\}$, we define a *cs* form of V^*. Then $\mathscr{X}_{\bar{0}}\oplus\mathscr{Y}_{\bar{0}}$ is a totally real polarisation of $V_{\bar{0}}\oplus V^*_{\bar{0}}$.

Boson-boson sector Let $F^0:=L^2(\mathscr{X}_{\bar{0}})$, where we take the Lebesgue measure on $V_{\mathbf{R},\bar{0}}$ induced by the Euclidean form $\operatorname{tr}(LL^*)$. Consider the Heisenberg group H^0, *i.e.* the connected and simply connected real Lie group with Lie algebra $(V_{\mathbf{R},\bar{0}}\times V^*_{\mathbf{R},\bar{0}}\times i\mathbf{R})\cap\mathfrak{h}_{V_{\bar{0}}}$. The Schrödinger model F^0 of the oscillator representation of H^0 defines a representation of $\mathfrak{sp}(\mathscr{X}_{\bar{0}}\oplus\mathscr{Y}_{\bar{0}},\mathbf{R})$, which integrates to a unitary representation of the double cover $\mathrm{Mp}:=\mathrm{Mp}(\mathscr{X}_{\bar{0}}\oplus\mathscr{Y}_{\bar{0}},\mathbf{R})$ of $\mathrm{Sp}:=\mathrm{Sp}(\mathscr{X}_{\bar{0}}\oplus\mathscr{Y}_{\bar{0}},\mathbf{R})$ [23,27].

Let $\tilde{U}^0$ be the lift to Mp of the maximal compact subgroup of Sp. That is, we have $\tilde{U}^0=\mathrm{U}(np)\times_{\mathrm{U}(1)}\mathrm{U}(1)$, the $\sqrt{\det}$ double cover of $\mathrm{U}(np)$.

We consider $S_{\mathscr{X}_{\bar{0}}}$, the formal power series ring on $\mathscr{X}_{\bar{0}}$, and the Gaussian $\Gamma^0:=e^{-\operatorname{tr}LL^*/2}\in\widehat{S}_{\mathscr{X}_{\bar{0}}}$. The action of $\mathfrak{h}_{V_{\bar{0}}}$ extends to this space. Then Γ^0 is annihilated by $X_{\bar{0}}\subseteq\mathfrak{h}_{V_{\bar{0}}}$, where $X_{\bar{0}}\oplus Y_{\bar{0}}$ is the totally complex polarisation of $V_{\bar{0}}\oplus V^*_{\bar{0}}$ given by

$$X_{\bar{0}}:=\big\{(L,K,-K,-L)\bigm| L\in\mathbf{C}^{p\times n},K\in\mathbf{C}^{n\times p}\big\},$$
$$Y_{\bar{0}}:=\big\{(L,K,K,L)\bigm| L\in\mathbf{C}^{p\times n},K\in\mathbf{C}^{n\times p}\big\}.$$

Thus, we have that $\mathbf{C}[\mathscr{X}_{\bar{0}}]\Gamma^0 = \mathbf{C}[\mathscr{X}_{\bar{0}}]e^{-\operatorname{tr}(LL^*)/2} \subseteq L^2(\mathscr{X}_{\bar{0}}) = F^0$ is the space of $\tilde{U}^0$-finite vectors in F^0, and as an $\mathfrak{h}_{V_{\bar{0}}}$- and $(\mathfrak{s}^0, \tilde{U}^0)$-module, where we define $\mathfrak{s}^0 := \mathfrak{sp}(V_{\bar{0}} \oplus V_{\bar{0}}^*, \mathbf{C})$, it is isomorphic to $S_{X_{\bar{0}}} = \mathbf{C}[X_{\bar{0}}]$ [1, Lemma 4.1].

Fermion-fermion sector A similar argument applies to $V_{\bar{1}}$, the only difference being that real forms need not to be chosen. Indeed, setting $\mathscr{X}_{\bar{1}} := V_{\bar{1}}$, $\mathscr{Y}_{\bar{1}} := V_{\bar{1}}^*$, we have a complex polarisation of $V_{\bar{1}} \oplus V_{\bar{1}}^*$. Thus, $F^1 := S(V_{\bar{1}}^*) = \bigwedge(V_{\bar{1}}^*) = S_{\mathscr{X}_{\bar{1}}}$ as a module of the Clifford algebra $\mathrm{WCl}(V_{\bar{1}})$. It contains the Gaussian $\Gamma^1 := e^{\operatorname{tr}(K_1K_2)/2}$, where K_1 and K_2, respectively, denote the identity of $\mathbf{C}^{0|q\times n} = U_{\bar{1}}^{\prime *}$ and $\mathbf{C}^{0|n\times q} = U_{\bar{1}}^*$. Similar to the above, Γ^1 is annihilated by $X_{\bar{1}} \subseteq \mathfrak{h}_{V_{\bar{1}}}$, where the spaces

$$X_{\bar{1}} := \{(L,K,-K,-L) \mid L \in \mathbf{C}^{0|q\times n}, K \in \mathbf{C}^{n\times 0|q}\},$$
$$Y_{\bar{1}} := \{(L,K,K,L) \mid L \in \mathbf{C}^{0|q\times n}, K \in \mathbf{C}^{n\times 0|q}\},$$

form a complex polarisation of $V_{\bar{1}} \oplus V_{\bar{1}}^*$. Since Γ^1 is invertible, we have $\mathbf{C}[\mathscr{X}_{\bar{1}}]\Gamma^1 = \mathbf{C}[\mathscr{X}_{\bar{1}}]e^{\operatorname{tr}(K_1K_2)/2} = F^1$, and as an $\mathfrak{h}_{V_{\bar{1}}}$-module, it is isomorphic to $S_{X_{\bar{1}}} = \mathbf{C}[X_{\bar{1}}]$.

Full graded picture Let $X := X_{\bar{0}} \oplus X_{\bar{1}}$ and observe that $\mathscr{X}_{\bar{0}} \oplus \mathscr{X}_{\bar{1}} = V_{\mathbf{R}}$. By [23, Lemma 5.4], we obtain that $\mathbf{C}[V_{\mathbf{R}}]\Gamma \subseteq F := F^0 \otimes F^1$ is isomorphic to $\mathbf{C}[X]$ as a $\mathfrak{h}_V$-module, where

$$\Gamma := \Gamma^0 \cdot \Gamma^1 = e^{-\operatorname{str}(X^2)/4}, \quad X = \begin{pmatrix} 0 & L^* & K_2 \\ L & 0 & 0 \\ K_1 & 0 & 0 \end{pmatrix}.$$

Moreover, let $\mathfrak{s}^1 := \mathfrak{o}(V_{\bar{1}} \oplus V_{\bar{1}}^*, \mathbf{C})$ and $\tilde{U}_{\mathbf{C}}^1$ be the complex Lie group that the Lie algebra exponentiates to in the Clifford algebra $\mathrm{WCl}(V_{\bar{1}})$ (*i.e.* the the complex spin group $\mathrm{Spin}(nq, \mathbf{C})$, the simply connected double cover of $\mathrm{SO}(nq, \mathbf{C})$). Then $\mathbf{C}[V_{\mathbf{R}}]\Gamma$ is isomorphic to $\mathbf{C}[X]$ as a $\tilde{U}_{\mathbf{C}}$-module, where $\tilde{U}_{\mathbf{C}} := \tilde{U}_{\mathbf{C}}^0 \times \tilde{U}_{\mathbf{C}}^1$ and $\tilde{U}_{\mathbf{C}}^0$ is the complexification of $\tilde{U}^0$. That is, $\tilde{U}_{\mathbf{C}}^0 = \mathrm{GL}(nq, \mathbf{C}) \times_{\mathbf{C}^\times} \mathbf{C}^\times$, the $\sqrt{\det}$ double cover of $\mathrm{GL}(nq, \mathbf{C})$. In summary, $\mathbf{C}[V_{\mathbf{R}}]\Gamma \cong \mathbf{C}[X]$ as $(\mathfrak{s}, \tilde{U}_{\mathbf{C}})$-modules.

3.2 *Action on Schwartz superfunctions*

Recall the terminology summarised in the Appendix. We consider $V_{\mathbf{R}}$ as a *cs* manifold, namely, the *cs* affine superspace associated with the *cs* vector space $V_{\mathbf{R}}$.

We define $\mathscr{S}(V_{\mathbf{R}}) := \mathscr{S}(V_{\mathbf{R},\bar{0}}) \otimes \bigwedge V_{\bar{1}}^* \subseteq \Gamma(\mathscr{O}_{V_{\mathbf{R}}})$, where

$$\mathscr{S}(V_{\mathbf{R},\bar{0}}) := \left\{ f \in \mathscr{C}^\infty(V_{\mathbf{R},\bar{0}}) \,\middle|\, \forall p \in \mathbf{C}[V_{\mathbf{R},\bar{0}}] : \int_{V_{\mathbf{R},\bar{0}}} |p(L)f(L)|^2 \, dL \, d\bar{L} < \infty \right\}$$

is the Schwartz space of $V_{\mathbf{R},\bar{0}}$. We find that

$$\mathbf{C}[V_{\mathbf{R}}]\Gamma \subseteq \mathscr{S}(V_{\mathbf{R}}) \subseteq L^2(V_{\mathbf{R},0}) \otimes \bigwedge V_{\bar{1}}^* = F.$$

Since the leftmost of these is the space of $\tilde{U}^0 \times \tilde{U}_{\mathbf{C}}^1$-finite vectors of the $\mathrm{Mp} \times U_{\mathbf{C}}^1$-module F, this is a chain of dense inclusions. In fact, $\mathscr{S}(V_{\mathbf{R}})$ is the space of smooth vectors of F, considered as an $\mathrm{Mp}\times\tilde{U}^1_C=\mathrm{Mp}(np,\mathbf{R})\times\mathrm{Spin}(nq,\mathbf{C})$-module [16]. As one easily checks, the action of $\mathfrak{s}$ extends to $\mathscr{S}(V_{\mathbf{R}})$, and this gives a representation of the *cs* supergroup pair $(\mathfrak{s},\mathrm{Mp}\times\tilde{U}^1)$, for any real form $\tilde{U}^1$ of $\tilde{U}^1_{\mathbf{C}}$.

Let $|Dv|$ be the Berezinian density associated with the standard coordinate system on $V_{\mathbf{R}}$ (see the Appendix). Then we have the following fact.

Proposition 4 *The Berezin integral defines a functional $|Dv|$ on the space $\Gamma_c(\mathscr{O}_{V_{\mathbf{R}}})$ of compactly supported superfunctions by $\langle|Dv|,f\rangle:=\int_{V_{\mathbf{R}}}|Dv|\,f(v)$. It has a unique continuous extension to $\mathscr{S}(V_{\mathbf{R}})$.*

As $\mathfrak{s}$-modules, $S_V=\mathbf{C}[V]$ and $S_X=\mathbf{C}[X]$ are isomorphic. Since both Γ and U, U', U^*, U'^* are $G_{\mathbf{C}}$-invariant, we see that as $\mathfrak{g}'$-modules, we have

$$L(\lambda)=\mathbf{C}[V]^{G_{\mathbf{C}}}\cong\mathbf{C}[X]^{G_{\mathbf{C}}}\cong\mathbf{C}[V_{\mathbf{R}}]^{G_{\mathbf{C}}}\Gamma.$$

Thus, the latter is a copy of $L(\lambda)$ in $\mathscr{S}(V_{\mathbf{R}})^G\subseteq\mathscr{S}(V_{\mathbf{R}})$. Since $|Dv|$ is invariant under the compact group G, it is determined by its restriction to $\mathscr{S}(V_{\mathbf{R}})^G$, which is determined by its values on $\mathbf{C}[V_{\mathbf{R}}]^G\Gamma\cong L(\lambda)$, due to the density of $\mathbf{C}[V_{\mathbf{R}}]\Gamma\subseteq\mathscr{S}(V_{\mathbf{R}})$. In particular, the latter restriction is non-zero. Computing on compactly supported superfunctions $\Gamma_c(\mathscr{O}_{V_{\mathbf{R}}})\subseteq\mathscr{S}(V_{\mathbf{R}})$, the following proposition readily follows.

Proposition 5 *The functional $|Dv|$ is $\mathfrak{k}$-relatively invariant for the character $-\lambda$. Hence, its restriction to $L(\lambda)\cong\mathbf{C}[V_{\mathbf{R}}]^G\Gamma$ spans the space $\underline{\mathrm{Hom}}_{\mathfrak{k}}(L(\lambda),\mathbf{C}_{-\lambda})$.*

4 The Riesz superdistribution

In this section, we introduce a certain quadratic morphism Q, which pushes the Berezin integration functional $|Dv|$ forward to a superdistribution on (an integration cycle in) $\underline{\mathrm{End}}(\mathbf{C}^{p|q})$. Moreover, we define the Riesz superdistribution R_n and state the superbosonisation identity as a relation between these superdistributions.

In what follows, recall the notion of S-valued points from the Appendix.

4.1 The super-Grassmannian and the integration cycle Ω

Consider the complex super-Grassmannian $Y:=\mathrm{Gr}_{p|q,2p|2q}(\mathbf{C})$. Following [22] with minor modifications, it is given as follows: Given a subset $I\subseteq 2p|2q$ of $p|q$ homogeneous indices, let U_I be the superdomain with S-valued points

$$R_I=\overset{\;p|q}{\begin{matrix}K\\I\end{matrix}\begin{pmatrix}Z_I\\1\end{pmatrix}},\tag{2}$$

where I indexes the rows containing the identity and K indexes the other rows.

Given another set J of $p|q$ indices, let B_{JI} be the $p|q\times p|q$ submatrix of R_I formed by the rows indexed by J. Then U_{IJ} is defined to be the maximal open subdomain of U_I on which B_{IJ} is invertible. The equation $Z_J=Z_IB_{JI}^{-1}$ expresses the entries of Z_J as rational functions of the entries of Z_I. Then $Y=\mathrm{Gr}_{p|q,2p|2q}(\mathbf{C})$ is defined to be the complex supermanifold obtained by gluing these data. We identify $W:=\mathbf{C}^{p|q\times p|q}$ with the standard affine open patch U_{I_0}, $I_0=\{p+1,\dots,2p|q+1,\dots,2q\}$.

Supergroup actions Consider the complex Lie supergroup $G'_{\mathbf{C}} := \mathrm{GL}(2p|2q,\mathbf{C})$ whose Lie superalgebra is $\mathfrak{g}'$. In the sequel we will write $g \in_S G'_{\mathbf{C}}$ (for any *cs* manifold S) in the form

$$\begin{array}{c} \\ p|q \\ p|q \end{array}\begin{array}{c} \begin{matrix} p|q & p|q \end{matrix} \\ \begin{pmatrix} A & B \\ C & D \end{pmatrix} \end{array}. \tag{3}$$

Then $G'_{\mathbf{C}}$ acts transitively on $\mathrm{Gr}_{p|q,2p|2q}(\mathbf{C})$ by left multiplication. For $g \in_S G'_{\mathbf{C}}$ given in the form above and $Z \in_S W$, we have

$$g \cdot Z = (AZ+B)(CZ+D)^{-1} \in_S W, \tag{4}$$

whenever $CZ+D \in_S \mathrm{GL}(p|q,\mathbf{C})$. In particular, the action of the complex supergroup $K_{\mathbf{C}} := \mathrm{GL}(p|q,\mathbf{C}) \times \mathrm{GL}(p|q,\mathbf{C})$, realised as a closed subsupergroup of $G'_{\mathbf{C}}$ by requiring $B = C = 0$, leaves the affine patch $W \subseteq Y$ invariant. This also holds for the closed subsupergroup P^+ of $G'_{\mathbf{C}}$ given by $A = D = 1$, $C = 0$.

Consider the closed Lie subsupergroup P^- of $G'_{\mathbf{C}}$ given by $A = D = 1$, $B = 0$. Then for $o_0 := 0 \in W \subseteq Y$, the isotropy subsupergroup is $G'_{\mathbf{C},o_0} = K_{\mathbf{C}}P^-$, which intersects trivially with P^+. In particular, P^+ acts simply transitively on W.

We define a *cs* form H of $K_{\mathbf{C}}$ by specifying the real form $H_0 \subseteq K_{\mathbf{C},0}$ to be $\mathrm{GL}(p,\mathbf{C}) \times \mathrm{U}(q)$, embedded into $K_{\mathbf{C},0}$ as the matrices $\mathrm{diag}(A,D,(A^*)^{-1},D')$ with $A \in \mathrm{GL}(p,\mathbf{C})$, $D,D' \in \mathrm{U}(q)$. Since H is a closed subsupergroup of $K_{\mathbf{C},cs}$, the orbit $\Omega := H.o_1$, where o_1 is the identity matrix in W, is a closed *cs* submanifold of W_{cs}. The isotropy supergroup H_o is the intersection (fibre product) of the diagonal subsupergroup $\mathrm{GL}(p|q,\mathbf{C})_{cs}$ with H. In particular, we have $H_{o_1,0} = \mathrm{U}(p) \times \mathrm{U}(q)$, embedded diagonally into H, and $\Omega_0 \cong \mathrm{Herm}^+(p) \times \mathrm{U}(q)$.

4.2 The Q morphism

We let a quadratic map $Q : V \to W$ be defined as

$$Q(L,L') := LL', \quad L \in \mathbf{C}^{p|q\times n}, L' \in \mathbf{C}^{n\times p|q}. \tag{5}$$

It is clearly $G_{\mathbf{C}}$-invariant. It gives rise to a corresponding morphism of complex supermanifolds. Defining a *cs* form $W_{\mathbf{R}}$ of W by setting $W_{\mathbf{R},\bar{0}} := \mathrm{Herm}(p) \times \mathrm{Herm}(q)$, Q descends to a morphism $V_{\mathbf{R}} \to W_{\mathbf{R}}$ of *cs* manifolds.

Proposition 6 *The pullback along the morphism Q induces a continuous linear map $Q^\sharp : \mathscr{S}(W_{\mathbf{R}}) \to \mathscr{S}(V_{\mathbf{R}})$. In fact, for $Q^\sharp(f)$ to lie in $\mathscr{S}(V_{\mathbf{R}})$, it is sufficient for $f \in \Gamma(\mathscr{O}_{W_{\mathbf{R}}})$ to have rapid decay at infinity along* $\mathrm{Herm}^+(p)$, i.e.

$$\sup_{z\in\mathrm{Herm}^+(p)} \left|(1+\|z\|)^N (Df)(z,w)\right| < \infty$$

for all $w \in \mathrm{Herm}(q)$, $N \in \mathbf{N}$, and $D \in S(W)$. Here, Df denotes the natural action of $S(W)$ by constant coefficient differential operators. In particular, $Q_\#(|Dv|)$, defined by $\langle Q_\sharp(|Dv|), f\rangle := \langle |Dv|, Q^\sharp(f)\rangle$ for $f \in \mathscr{S}(W_{\mathbf{R}})$, is a continuous linear functional on $\mathscr{S}(W_{\mathbf{R}})$ with support in the closure of $\mathrm{Herm}^+(p)$. *Thus, it extends to a continuous functional on the space of all $f \in \Gamma(\mathscr{O}_{W_{\mathbf{R}}})$ with rapid decay along* $\mathrm{Herm}^+(p)$.

Moreover, note that we have

$$Q^\sharp(\mathbf{C}[W_\mathbf{R}]e^{-\operatorname{str}}) = \mathbf{C}[V_\mathbf{R}]^{G_\mathbf{C}}\Gamma, \tag{6}$$

since the $G_\mathbf{C}$-invariants of $\mathbf{C}[V_\mathbf{R}]$ are generated in degree two, *cf.* [21].

There is an action of a suitable twofold cover $\tilde{H}$ of the *cs* form H of $K_\mathbf{C}$ on the space $\mathscr{S}(V_\mathbf{R})$. Explicitly, it can be written for any $\tilde{h} \in_S \tilde{H}$ lying above $h = \operatorname{diag}(A,D) \in_S H$, $f \in \mathscr{S}(V_\mathbf{R})$, and $(L,L') \in_S V_\mathbf{R}$, as

$$(\tilde{h}\cdot f)(L,L') = \operatorname{Ber}(A)^{n/2}\operatorname{Ber}(D)^{-n/2}f(D^{-1}L,L'A).$$

Under the pullback $Q^\sharp$, this corresponds to the *twisted action* $\cdot_\lambda$ of $\tilde{H}$ on $\mathscr{S}(W_\mathbf{R})$, $(\tilde{h}\cdot_\lambda f)(w) = \operatorname{Ber}(A)^{n/2}\operatorname{Ber}(D)^{-n/2}(h\cdot f)(w)$, where the *untwisted action* of H is

$$(h\cdot f)(w) := f(D^{-1}wA). \tag{7}$$

The subspaces $\mathbf{C}[V_\mathbf{R}]\Gamma$ and $\mathbf{C}[W_\mathbf{R}]e^{-\operatorname{str}/2}$ are invariant for the induced $\mathfrak{k}$-action.

4.3 Statement of the theorem

In what follows, recall the facts on Berezin integration summarised in the Appendix.

The homogeneous *cs* manifold $\Omega = H.o_1 = H/H_{o_1}$ is a locally closed *cs* submanifold of W_{cs}. It admits a non-zero H-invariant Berezinian density $|Dy|$ [2]. It can given quite explicitly, see [3], and we follow the normalisation introduced there.

Observe that Ω has purely even codimension in W_{cs}. Thus, we have a canonical splitting $\Omega \cong \Omega_0 \times W_{\bar{1}}$, defining a retraction r of Ω, which we call *standard.*

Riesz superdistribution When $n \geq p$, define the functional T_n, called the *Riesz superdistribution*, by

$$\langle T_n, f\rangle := \int_\Omega |Dy|\operatorname{Ber}(y)^n f(y)$$

for any entire superfunction $f \in \Gamma(\mathscr{O}_W)$ that satisfies Paley–Wiener type estimates along the tube $T_{bb} := \operatorname{Herm}^+(p) + i\operatorname{Herm}(p)$, *i.e.*

$$\sup_{z\in T_{bb}}\left|e^{-R\|\Im z\|}(1+\|z\|)^N(Df)(z,w)\right| < \infty \tag{8}$$

for any $D \in S(W)$, $N \in \mathbf{N}$, $w \in \mathbf{C}^{q\times q}$, and some $R > 0$. Here, we write $\Im z$ for $\frac{1}{2i}(z-z^*)$. The integral is taken with respect to the standard retraction, and its convergence is proved in [3].

Our terminology is explained by the fact that for $q = 0$, T_n coincides with the unweighted *Riesz distribution* for the parameter n, see [11]. Using the super Laplace transform and some Functional Analysis, one proves the following [3].

Proposition 7 *Let $n \geq p$. Then the functional T_n extends continuously to the space of all superfunctions of rapid decay along* $\operatorname{Herm}^+(p)$.

Conical superfunctions and Gindikin Γ function To state the superbosonisation identity, we introduce the following set of rational superfunctions on W. For any $Z = (Z_{ij}) \in_S W = \mathfrak{gl}(p|q, \mathbf{C})$ and $1 \leq k \leq p+q$, we consider the kth principal minor $[Z]_k := (Z_{ij})_{1\leq i,j\leq k}$.

Whenever $[Z]_k$ is invertible, we set $\Delta_k(Z) := \mathrm{Ber}([Z]_k)$, and whenever all principal minors of Z are invertible and $\mathbf{m} = (m_1, \dots, m_{p+q}) \in \mathbf{Z}^{p+q}$, we define

$$\Delta_{\mathbf{m}} := \Delta_1^{m_1-m_2} \cdots \Delta_{p+q-1}^{m_{p+q-1}-m_{p+q}} \Delta_{p+q}^{m_{p+q}}. \tag{9}$$

These functions are called *conical superfunctions*. They are characterised as the unique rational superfunctions that are eigenfunctions of a suitable Borel [3].

Fix a superfunction $f \in \Gamma(\mathscr{O}_\Omega)$ and $x \in_S W_{cs}$. Whenever the integral converges, we define the *Laplace transform* of f at x by

$$\mathscr{L}(f)(x) := \int_\Omega |Dy|\, e^{-\operatorname{str}(xy)} f(y),$$

where we write $|Dy|$ for the invariant Berezinian on Ω. All integrals will be taken with respect to the standard retraction on Ω. Provided the integral exists, we define

$$\Gamma_\Omega(\mathbf{m}) := \mathscr{L}(\Delta_{\mathbf{m}})(1) = \int_\Omega |Dy| e^{-\operatorname{str}(y)} \Delta_{\mathbf{m}}(y), \tag{10}$$

and call this the *Gindikin Γ function* of Ω. The following is proved in [3].

Proposition 8 *Let $O \subseteq W_{cs}$ be the open* cs *submanifold on which all principal minors of Z are invertible. For $x \in_S O$, the integral $\mathscr{L}(\Delta_{\mathbf{m}})(x^{-1})$ converges absolutely if and only if $m_j > j-1$ for $j = 1, \dots, p$. In this case, $\Gamma_\Omega(\mathbf{m})$ exists, and we have*

$$\mathscr{L}(\Delta_{\mathbf{m}})(x^{-1}) = \Gamma_\Omega(\mathbf{m}) \Delta_{\mathbf{m}}(x).$$

In fact, the function $\Gamma_\Omega(\mathbf{m})$ can be determined explicitly, as follows, *cf.* [3].

Theorem 1 *Let $m_j > j-1$ for all $j = 1, \dots, p$. We have*

$$\Gamma_\Omega(\mathbf{m}) = (2\pi)^{\frac{p(p-1)}{2}} \prod_{j=1}^{p} \Gamma(m_j - j + 1) \prod_{k=1}^{q} \frac{\Gamma(q-k+1)}{\Gamma(m_{p+k}+q-k+1)} \frac{\Gamma(m_{p+k}+k)}{\Gamma(m_{p+k}-p+k)}.$$

In particular, $\Gamma_\Omega(\mathbf{m})$ extends uniquely as a meromorphic function of $\mathbf{m} \in \mathbf{C}^{p+q}$.

Superbosonisation identity We can finally state the superbosonisation identity. To that end, denote for $n \geq p$: $\Gamma_\Omega(n) := \Gamma_\Omega(n, \dots, n) > 0$, and let $R_n := \Gamma_\Omega(n)^{-1} T_n$ be the *normalised Riesz superdistribution*. Then we have the following theorem [3,21].

Theorem 2 *Let $n \geq p$. Then we have $Q_\sharp(|Dv|) = \sqrt{\pi}^{np} R_n$. For any holomorphic superfunction f on the open subspace of W over $T_{bb} + \mathbf{C}^{q\times q}$, satisfying the estimate in Eq.* (8) *for some $R > 0$ and any $D \in S(W)$, $w \in \mathbf{C}^{q\times q}$, and $N \in \mathbf{N}$, we have*

$$\int_{V_\mathbf{R}} |Dv|\, f(Q(v)) = \frac{\sqrt{\pi}^{np}}{\Gamma_\Omega(n)} \int_\Omega |Dy| \operatorname{Ber}(y)^n f(y).$$

In particular, this applies to any f in the space $\mathbf{C}[W_\mathbf{R}]e^{-\operatorname{str}/2}$ from Eq. (6).

5 Proofs of the superbosonisation identity

We end this survey by explaining two proofs of the superbosonisation identity. The first proof, which we only sketch briefly, makes heavy use of Functional Analysis to reduce everything to a trivial computation. The second proof equates the Riesz superdistribution R_n to a relatively invariant functional on a suitable representation of $\mathfrak{g}'$ and uses some basic representation theory to prove the identity.

5.1 Analytic proof

One can prove the superbosonisation identity (*i.e.* Theorem 2) by computing the super version of the Euclidean Laplace transform of both sides of the identity and comparing the results. We give a brief sketch of the procedure; for a more detailed discussion, in particular, of the relevant topologies, see [3, Appendix C].

The space of continuous linear functionals on $\mathscr{S}(W_{\mathbf{R}})$ is denoted by $\mathscr{S}'(W_{\mathbf{R}})$. Its elements are called *tempered superdistributions*. Clearly, $\mathscr{S}'(W_{\mathbf{R}})$ embeds continuously as a subspace into $\Gamma_c(\mathscr{O}_{W_{\mathbf{R}}})'$. The elements in the image are characterised as those functionals on $\Gamma_c(\mathscr{O}_{W_{\mathbf{R}}})$ that are continuous for the topology induced by $\mathscr{S}(W_{\mathbf{R}})$. Let $\mu \in \mathscr{S}'(W_{\mathbf{R}})$. One can show the following [3].

Proposition 9 *There exists a largest open subspace $\Upsilon_{\mathscr{S}'}(\mu) \subseteq W_{\mathbf{R}}$ such that for every* cs *manifold S and any $w \in_S \Upsilon_{\mathscr{S}}(\mu)$, we have $e^{-\operatorname{str}(w\cdot)}\mu \in \Gamma(\mathscr{O}_S)\widehat{\otimes}\mathscr{S}'(W_{\mathbf{R}})$, where $\widehat{\otimes}$ denotes the completed tensor product (w.r.t. the injective or, equivalently, the projective tensor product topology).*

Let $z = x + iy \in_S W_{cs}$ where $y \in_S W_{\mathbf{R}}$ and $x \in_S \Upsilon_{\mathscr{S}}(\mu)$ (this does *not* determine x, y uniquely). Then we define the *Laplace transform* of μ by $\mathscr{L}(\mu)(z) := \mathscr{F}(e^{-\operatorname{str}(x\cdot)}\mu)(y)$, where $\mathscr{F}$ denotes the Fourier transform. This definition makes sense, since by a straightforward extension of Schwartz's classical theory of the Laplace transform, we have $e^{-\operatorname{str}(x\cdot)}\mu \in \Gamma(\mathscr{O}_S)\widehat{\otimes}\mathscr{S}(W_{\mathbf{R}})$, and hence the Fourier transform (w.r.t. $W_{\mathbf{R}}$) is contained in the same space.

The following is a special case of results from [3].

Proposition 10 *There is a holomorphic superfunction on the tube $\Upsilon_{\mathscr{S}}(\mu) + iW_{\mathbf{R}}$ whose value at z is $\mathscr{L}(\mu)(z)$. The superdistribution μ is determined by $\mathscr{L}(\mu)$.*

We now give an account of the analytic proof of the superbosonisation identity.

Proof. (of Theorem 2, analytic version) Let $T \subseteq W$ be the open subspace whose underlying open set is $T_{bb} + \mathbf{C}^{q\times q}$, where we recall that $T_{bb} = \operatorname{Herm}^+(p) + i\operatorname{Herm}(p)$. For any $z \in_S T$ such that all principal minors of z are invertible, we have

$$\mathscr{L}(T_n)(z) = \mathscr{L}(\Delta_{n,\dots,n})(z) = \Gamma_\Omega(n)\Delta_{n,\dots,n}(z^{-1}) = \Gamma_\Omega(n)\operatorname{Ber}(z)^{-n},$$

in view of Proposition 8. Computing Gaussian Berezin integrals over affine superspace gives $\mathscr{L}(Q_\sharp(|Dv|))(z) = \sqrt{\pi}^{np}\operatorname{Ber}(z)^{-n}$, proving the claim. □

5.2 Representation theoretic proof

We end our survey with a representation theoretic proof of the superbosonisation identity. Let G' be the *cs* form $G'_{\mathbf{C}}$ with underlying Lie group $G'_0 := \mathrm{U}(p,p) \times \mathrm{U}(2q)$.

Line bundle The $\mathfrak{k}$-character 2λ integrates to a character $\chi_{2\lambda}(\mathrm{diag}(A,D)) = \mathrm{Ber}(D)^n \mathrm{Ber}(A)^{-n}$ of $K_{\mathbf{C}}$. Extending $\chi_{2\lambda}$ trivially to $K_{\mathbf{C}}P^-$, we may define a holomorphic line bundle on the complex homogeneous superspace $G'_{\mathbf{C}}/K_{\mathbf{C}}P^- = Y$ by $L_{2\lambda} := G'_{\mathbf{C}} \times^{K_{\mathbf{C}}P^-_{\mathbf{C}}} \chi_{2\lambda}$. We will also consider its restriction to $D := G'.o_0$, where $o_0 \in W_0$ is the zero matrix. Observe that the underlying space of D is

$$D_0 = \mathrm{U}(p,p)/(\mathrm{U}(p) \times \mathrm{U}(p)) \times \mathrm{U}(2q)/(\mathrm{U}(q) \times \mathrm{U}(q)).$$

Then $G'_{\mathbf{C}}$ resp. G' acts on sections of $L_{2\lambda}$ resp. $L_{2\lambda}|_D$. Denote the latter by $\pi_{2\lambda}$.

On general grounds, the cocycle defining $L_{2\lambda}$ is $\chi_{2\lambda}(s_J^{-1}s_I)$ where $s_I : U_I \to G'_{\mathbf{C}}$ is a local section of the $K_{\mathbf{C}}P^-_{\mathbf{C}}$-principal bundle $G'_{\mathbf{C}} \to Y$. The transition matrix for the super-Grassmannian Y on U_{IJ} is given by $Z_J = Z_I B_{JI}^{-1}$. A short calculation shows

$$s_J^{-1}s_I = \begin{pmatrix} 1 & 0 \\ 0 & B_{JI}^{-1} \end{pmatrix} \begin{pmatrix} 1 & Z \\ 0 & 1 \end{pmatrix} \in_{U_{IJ}} K_{\mathbf{C}}P^-_{\mathbf{C}},$$

so that the defining cocycle of $L_{2\lambda}$ is $\chi_{2\lambda}(s_J^{-1}s_I) = \mathrm{Ber}(B_{JI})^n$.

Highest weight section We construct a global section $|0\rangle$ of the line bundle $L_{2\lambda}$ as follows: In the trivialisation on U_I given by s_I, it is defined by $|0\rangle_I(Z_I) = \mathrm{Ber}(s_I(Z_I))^{-n}$. It is not difficult to see that this indeed defines a global section of $L_{2\lambda}$, which on the standard affine patch $W = U_{I_0}$ is just the constant function 1.

Let $Z \in_S D \cap W$ and $g \in_S G'$ be such that $g \cdot Z \in_S D \cap W$, *i.e.* the action remains in the affine patch. Then the action of g on a section f of $L_{2\lambda}|_D$ is given at Z by

$$\pi_{2\lambda}(g)f(Z) = \chi_{2\lambda}(k(g,Z))f(g^{-1} \cdot Z),$$

where

$$k(g,Z) := \begin{matrix} \\ {\scriptstyle p|q} \\ {\scriptstyle p|q} \end{matrix} \!\!\begin{pmatrix} \overset{p|q}{(A - BD^{-1}C)(1 + ZD^{-1}C)^{-1}} & \overset{p|q}{0} \\ 0 & CZ + D \end{pmatrix} \in_S K_{\mathbf{C}}.$$

Since holomorphic sections of $L_{2\lambda}$ are determined by their restriction to $W = U_{I_0}$, one computes that $|0\rangle$ is fixed by the action of P^- and transforms under $K_{\mathbf{C}}$ by the character $\chi_{2\lambda}$. Therefore, there is a $\mathfrak{g}'$-equivariant map

$$M(2\lambda) := \mathfrak{U}(\mathfrak{g}') \otimes_{\mathfrak{k} \oplus \mathfrak{p}^+} \mathbf{C}_\lambda \to \Gamma(D, L_{2\lambda}|_D)$$

from the parabolic Verma module, which maps the highest weight vector to $|0\rangle$.

Construction of an intertwiner We define the *weighted Laplace transform* $\mathscr{L}_n$ by

$$\mathscr{L}_n(f)(z) := \Gamma_\Omega(n)^{-1}\,\mathscr{L}(f\Delta_{n,\dots,n})(z/2) = \Gamma_\Omega(n)^{-1}\int_\Omega |Dy|e^{-\operatorname{str}(zy)/2}f(y)\operatorname{Ber}(y)^n,$$

whenever this makes sense. For $h \in_S H$, we have

$$\mathscr{L}_n(h\cdot f)(z) = \chi_{2\lambda}(h)\,\mathscr{L}_n(f)(h^{-1}\cdot z), \tag{11}$$

where $h\cdot f$ denotes the untwisted H-action introduced in Eq. (7).

The *Cayley transform* is $\gamma(Z) := (1+Z)(1-Z)^{-1}$, and the *weighted Cayley transform* γ_n is defined by $\gamma_n(F)(Z) := \operatorname{Ber}(1-Z)^{-n}F(\gamma(Z))$, when this makes sense. For the diagonal subsupergroup $K'_{\mathbf{C}} := \operatorname{diag}\operatorname{GL}(p|q,\mathbf{C})$ of $K_{\mathbf{C}}$, γ is $K'_{\mathbf{C}}$-equivariant. Thus, $\gamma_n\circ\mathscr{L}_n$ intertwines the $\mathfrak{k}'$-action by π_n on $\Gamma(D,L_{2\lambda}|_D)$ and the untwisted $\mathfrak{k}'$-action on $\mathbf{C}[W_{\mathbf{R}}]e^{-\operatorname{str}/2}$.

By Proposition 8, we have $(\gamma_n\circ\mathscr{L}_n)(e^{-\operatorname{str}/2}) = 1 = |0\rangle|_{D\cap W}$, and more generally,

$$(\gamma_n\circ\mathscr{L}_n)(\Delta_{\mathbf{m}}e^{-\operatorname{str}/2})(z) = (n)_{\mathbf{m}}\,\Delta_{\mathbf{m}+n}(1-z) \tag{12}$$

for $\mathbf{m}\in\mathbf{Z}^{p+q}$ with $m_j > j-1$ for $j\leq p$. Here, $\mathbf{m}+n := (m_1+n,\dots,m_{p+q}+n)$ and $(n)_{\mathbf{m}} := \Gamma_\Omega(n)^{-1}\Gamma_\Omega(\mathbf{m}+n)$. Then $\Delta_{\mathbf{m}}$ is contained in $\mathbf{C}[W_{\mathbf{R}}]$ if and only if

$$m_1\geq m_2\geq\cdots\geq m_p\geq 0,\quad m_{p+1}\leq m_{p+2}\leq\cdots m_{p+q}\leq 0. \tag{13}$$

For these $\mathbf{m}$, $\Gamma_\Omega(\mathbf{m}+n)\neq 0$ if and only if in addition $m_{p+1}\geq p-n$, by Theorem 1.

Those $\Delta_{\mathbf{m}}$ that are polynomial are exactly the lowest weight vectors for the $K_{\mathbf{C}}$-action on $\mathbf{C}[W_{\mathbf{R}}]$ [3]. Moreover, according to [25, Proposition 3.1], $\mathbf{C}[W_{\mathbf{R}}]$ is semisimple (and multiplicity free). Thus, we have a $(\mathfrak{k},H)$-module decomposition

$$\mathbf{C}[W_{\mathbf{R}}] = \bigoplus\nolimits_{\mathbf{m}}\underline{\operatorname{End}}(L^{p|q}(\mu_{\mathbf{m}})) = \bigoplus\nolimits_{\mathbf{m}}\mathbf{C}[W_{\mathbf{R}}]_{\mathbf{m}},$$

where $\mathbf{m}$ runs over all multi-indices satisfying the assumptions of Eq. (13), $\mathbf{C}[W_{\mathbf{R}}]_{\mathbf{m}} := \mathfrak{U}(\mathfrak{k})\Delta_{\mathbf{m}}$, and $L^{p|q}(\mu_{\mathbf{m}})$ is the simple finite-dimensional $\mathfrak{gl}(p|q,\mathbf{C})$-module of highest weight $\mu_{\mathbf{m}} := -\sum_{j=1}^{p}m_j\delta_j + \sum_{j=p+1}^{p+q}m_j\varepsilon_{j-p}$. By Eqs. (11) and (12), and $\mathscr{L}_n(fe^{-\operatorname{str}/2})(z) = \mathscr{L}_n(f)(1-z)$, the summand $\mathbf{C}[W_{\mathbf{R}}]_{\mathbf{m}}$ is annihilated by $\mathscr{L}_n$ if $m_{p+1} < n-p$ and mapped injectively otherwise. Since $e^{-\operatorname{str}/2}$ is $K'_{\mathbf{C}}$-invariant, $\gamma_n\circ\mathscr{L}_n$ is a split epimorphism of $\mathfrak{k}'$-modules from $\mathbf{C}[W_{\mathbf{R}}]e^{-\operatorname{str}/2}$ onto its image.

We now give a second proof of the superbosonisation identity.

Proof. (of Theorem 2, representation theoretic version) We consider the untwisted $\mathfrak{k}$-action on $\mathbf{C}[W_{\mathbf{R}}]e^{-\operatorname{str}/2}$. By the invariance of $|Dy|$, the restriction of R_n defines an element of $\underline{\operatorname{Hom}}_{\mathfrak{k}}(\mathbf{C}[W_{\mathbf{R}}]e^{-\operatorname{str}/2},\mathbf{C}_{-2\lambda})$. Since $Q^\sharp : \mathbf{C}[W_{\mathbf{R}}]e^{-\operatorname{str}/2}\otimes\mathbf{C}_\lambda\to\mathbf{C}[V_{\mathbf{R}}]\Gamma$ is a surjective $\mathfrak{k}$-equivariant map, we obtain an injection

$$Q_\sharp : \underline{\operatorname{Hom}}_{\mathfrak{k}}(\mathbf{C}[V_{\mathbf{R}}]\Gamma,\mathbf{C}_{-\lambda})\to\underline{\operatorname{Hom}}_{\mathfrak{k}}(\mathbf{C}[W_{\mathbf{R}}]e^{-\operatorname{str}/2}\otimes\mathbf{C}_\lambda,\mathbf{C}_{-\lambda})$$

by left exactness of the hom functor $\underline{\operatorname{Hom}}_{\mathfrak{k}}(\cdot,\cdot)$. The latter of these hom spaces is $\underline{\operatorname{Hom}}_{\mathfrak{k}}(\mathbf{C}[W_{\mathbf{R}}]e^{-\operatorname{str}/2},\mathbf{C}_{-2\lambda})$. Since $\gamma_n\circ\mathscr{L}_n(e^{-\operatorname{str}/2}) = |0\rangle$, after Cayley transformation, we obtain a parabolic of $\mathfrak{g}'$ whose nilradical annihilates $\mathscr{L}_n(e^{-\operatorname{str}/2})$.

In particular, the dimension of $\underline{\operatorname{Hom}}_{\mathfrak{k}}(\mathbf{C}[W_{\mathbf{R}}]e^{-\operatorname{str}/2},\mathbf{C}_{-2\lambda})$ is at most one, and it is spanned by both $Q_\sharp(|Dv|)$ and R_n. Computing constants, the claim follows. □

Appendix

Supergeometry We summarise some basic definitions from supergeometry.

Definition 2 A **C**-*superspace* is a pair $X = (X_0, \mathscr{O}_X)$ where X_0 is a topological space and $\mathscr{O}_X$ is a sheaf of supercommutative superalgebras over **C** with local stalks. A *morphism* $f : X \to Y$ of **C**-superspaces is a pair $(f_0, f^\sharp)$ comprising a continuous map $f_0 : X_0 \to Y_0$ and a sheaf map $f^\sharp : f_0^{-1}\mathscr{O}_Y \to O_X$, which is local in the sense that $f^\sharp(\mathfrak{m}_{Y,f_0(x)}) \subseteq \mathfrak{m}_{X,x}$ for any x, where $\mathfrak{m}_{X,x}$ is the maximal ideal of $\mathscr{O}_{X,x}$.

Global sections $f \in \Gamma(\mathscr{O}_X)$ of $\mathscr{O}_X$ are called *superfunctions*. Due to the locality condition, the *value* $f(x) := f + \mathfrak{m}_{X,x} \in \mathscr{O}_{X,x}/\mathfrak{m}_{X,x}$ is defined for any x. *Open subspaces* of a **C**-superspace X are given by $(U, \mathscr{O}_X|_U)$, for any open subset $U \subseteq X_0$.

We consider two types of model spaces.

Definition 3 For a complex super-vector space V, we define $\mathscr{O}_V := \mathscr{H}_{V_{\bar{0}}} \otimes \bigwedge(V_{\bar{1}})^*$ where $\mathscr{H}$ denotes the sheaf of holomorphic functions. The space $(V_{\bar{0}}, \mathscr{O}_V)$ is called the *complex affine superspace* associated with V, and denoted by V.

If instead, V is a *cs* vector space (see Definition 1), then we define the sheaf $\mathscr{O}_V := \mathscr{C}^\infty_{V_{\bar{0}}} \otimes \bigwedge(V_{\bar{1}})^*$, where $\mathscr{C}^\infty$ denotes the sheaf of complex-valued smooth functions. The space $(V_{\bar{0}}, \mathscr{O}_V)$ is called the cs *affine superspace* associated with V, and denoted by V. (The *cs* terminology is due to J. Bernstein.)

In turn, this gives two flavours of supermanifolds.

Definition 4 Let X be a **C**-superspace, whose underlying topological space X_0 is Hausdorff, and which admits a cover by open subspaces isomorphic to open subspaces of some complex resp. *cs* affine superspace V, where V may vary. Then X is called a *complex supermanifold* resp. a cs *manifold*.

Complex supermanifolds and *cs* manifolds form full subcategories of the category of **C**-superspaces that admit finite products. The assignment sending the complex affine superspace V to the *cs* affine superspace obtained by forgetting the complex structure on $V_{\bar{0}}$ extends to a product-preserving cs-*ification* functor from complex supermanifolds to *cs* manifolds; the *cs*-ification of X is denoted by X_{cs}.

This point of view is also espoused by Witten in recent work [28].

Supergroups and supergroup pairs We give some basic definitions on supergroups. Details can be found in [6, 8, 22].

Definition 5 A *complex Lie supergroup* (resp. a cs *Lie supergroup*) is a group object in the category of complex supermanifolds (resp. *cs* manifolds). A *morphism* of (complex or *cs*) Lie supergroups is a morphism of group objects in the category of complex supermanifolds (resp. *cs* manifolds). The *cs*-ification functor maps complex Lie supergroups to *cs* Lie supergroups and morphisms of complex Lie supergroups to morphisms of *cs* Lie supergroups.

Definition 6 A complex (resp. *cs*) *supergroup pair* $(\mathfrak{g}, G_0)$ is given by a complex (resp. real) Lie group G_0 and a complex Lie superalgebra $\mathfrak{g}$, together with a morphism $\mathrm{Ad} : G_0 \to \mathrm{Aut}(\mathfrak{g})$ of complex (resp. real) Lie groups such that $\mathfrak{g}_{\bar{0}}$ is the Lie algebra of G_0 (resp. its complexification), Ad extends the adjoint action of G_0 on $\mathfrak{g}_{\bar{0}}$, and $[\cdot,\cdot]$ extends $d\,\mathrm{Ad}$. A *morphism of supergroup pairs* $(d\phi, \phi_0)$ consists of a morphism ϕ_0 of complex (resp. real) Lie groups and a morphism $d\phi$ of Lie superalgebras that is ϕ_0-equivariant for the Ad-actions, such that $d\phi$ extends $d(\phi_0)$.

The following is well-known, *cf.* [6, 19, 20].

Proposition 11 *There is an equivalence of the categories of complex (resp.* cs*) Lie supergroups and of complex (resp.* cs*) supergroup pairs. It maps any Lie supergroup to the pair consisting of its Lie superalgebra and its underlying Lie group.*

Definition 7 A closed embedding of (complex resp. *cs*) Lie supergroups is called a closed (complex resp. *cs*) *subsupergroup*. A *closed supergroup subpair* $(\mathfrak{h}, H_0) \subseteq (\mathfrak{g}, G_0)$ consists of a Lie subsuperalgebra $\mathfrak{h} \subseteq \mathfrak{g}$ and a closed subgroup $H_0 \subseteq G_0$, such that $(\mathfrak{h}, H_0)$ is a supergroup pair. Given a complex Lie supergroup G, a cs *form* of G is a closed subsupergroup H of G_{cs} such that in the supergroup pairs $(\mathfrak{h}, H_0)$ and $(\mathfrak{g}, G_0)$ of H resp. G, one has $\mathfrak{h} = \mathfrak{g}$. In this case, H_0 is a real form of G_0.

If G is a complex Lie supergroup with associated supergroup pair $(\mathfrak{g}, G_0)$, then $(\mathfrak{g}, H_0)$, for a closed subgroup $H_0 \subseteq G_0$, is the supergroup pair of a *cs* form of G if and only if H_0 is a real form of G_0, or equivalently, if $(\mathfrak{g}, H_0)$ is a *cs* supergroup pair. To define a *cs* form H of G, it thus suffices to specify a real form $H_0 \subseteq G_0$.

Points If $\mathbf{C}$ is any category, and X is an object of $\mathbf{C}$, then an *S-valued point* (where S is another object of $\mathbf{C}$) is defined to be a morphism $x : S \to X$. Suggestively, one writes $x \in_S X$ in this case, and denotes the set of all $x \in_S X$ by $X(S)$.

For any morphism $f : X \to Y$, one may define a set-map $f_S : X(S) \to Y(S)$ by

$$f_S(x) := f(x) := f \circ x \in_S Y, \quad x \in_S X.$$

Taking the *generic point* $x = \mathrm{id}_X \in_X X$, the values $f(x)$ completely determine f. The following statement is known as Yoneda's Lemma: For any collection of maps $f_S : X(S) \to Y(S)$, there exists a morphism $f : X \to Y$ such that $f_S(x) = f(x)$ for all $x \in_S X$ if and only if $f_T(x(t)) = f_S(x)(t)$, for all $t : T \to S$. Here, $x(t)$ are called *specialisations of* x, so the condition states that (f_S) is invariant under specialisation.

Thus, the *Yoneda embedding* functor $X \mapsto X(-)$ from $\mathbf{C}$ to $[\mathbf{C}^{op}, \mathbf{Sets}]$ is fully faithful. It preserves products, so if $\mathbf{C}$ admits finite products, it induces a fully faithful embedding of the category of group objects in $\mathbf{C}$ into the category $[\mathbf{C}^{op}, \mathbf{Grp}]$ of group-valued functors. In other words, an object X of $\mathbf{C}$ is a group object if and only if for any S, $X(S)$ admits a group law that is invariant under specialisation.

Berezin integrals Let X be a *cs* manifold and $\mathscr{B}er_X$ to the Berezinian sheaf. The sheaf of *Berezinian densities* $|\mathscr{B}er|_X$ is the twist by the orientation sheaf. Given local coordinates $(x^a) = (x, \xi)$ on U, one may consider the distinguished basis $|D(x^a)|$ of the module of Berezinian densities $|\mathscr{B}er|_X$ [22].

A *retraction* is a morphism $r : X \to X_0$, left inverse to the canonical embedding $j : X_0 \to X$. A system of coordinates (x,ξ) of X is called *adapted* to r if $x = r^\sharp(x_0)$. Given an adapted system, we write $\omega = |D(x,\xi)|\, f$ and $f = \sum_{I\subseteq\{1,\dots,q\}} r^\sharp(f_I)\,\xi^I$ for unique $f_I \in \Gamma(\mathscr{O}_{X_0})$, where $\dim X = *|q$. Then one defines $\int_{X/X_0} \omega := |dx_0|\, f_{\{1,\dots,q\}}$, which is an ordinary density on X_0. This quantity only depends on r, and not on the choice of an adapted system of coordinates. If this density is absolutely integrable on X_0, then we say that ω is *absolutely integrable* with respect to r, and define $\int_X \omega := \int_{X_0} \int_{X/X_0} \omega$. Unless $\operatorname{supp}\omega$ is compact, this quantity depends heavily on r.

Acknowledgements This research was funded by the grants no. DFG ZI 513/2-1 and SFB TR/12, provided by Deutsche Forschungsgemeinschaft (DFG). We wish to thank Martin Zirnbauer for extensive discussions, detailed comments, and for bringing this topic to our attention. We thank INdAM for its hospitality. The first named author wishes to thank Jacques Faraut for his interest, and the second named author wishes to thank Bent Ørsted for some useful comments.

References

1. Adams, J.: The theta correspondence over **R**. In: Harmonic Analysis, Group Representations, Automorphic Forms, and Invariant Theory, pp. 1–39. World Scientific, Hackensack, NJ (2007)
2. Alldridge, A., Hilgert, J.: Invariant Berezin integration on homogeneous supermanifolds. J. Lie Theory **20**, 65–91 (2010)
3. Alldridge, A., Shaikh, Z.: Superbosonisation via Riesz superdistributions. Under revision for Forum of Math., Sigma. Available on arXiv. http://arxiv.org/abs/1301.6569v2. Cited 30 Oct 2013
4. Altland, A., Zirnbauer, M.R.: Nonstandard symmetry classes in mesoscopic normal-superconducting hybrid structures. Phys. Rev. B **55**, 1142–1161 (1997)
5. Bouarroudj, S., Grozman, P.Y., Leites, D.A., Shchepochkina, I.M.: Minkowski superspaces and superstrings as almost real-complex supermanifolds. Theor. Math. Phys. **173**, 1687–1708 (2012)
6. Carmeli, C., Caston, L., Fioresi, R.: Mathematical Foundations of Supersymmetry. European Mathematical Society, Zürich (2011)
7. Cheng, S.-J., Wang, W.: Dualities and Representations of Lie Superalgebras. American Mathematical Society, Providence, RI (2012)
8. Deligne, P., Morgan, J.W.: Notes on supersymmetry (following Joseph Bernstein). In: Quantum Fields and Strings: A Course for Mathematicians, Vol. 1, 2. Princeton, NJ, 1996/1997, pp. 41–97. American Mathematical Society, Providence, RI (1999)
9. Efetov, K.: Supersymmetry in Disorder and Chaos. Cambridge University Press, Cambridge (1997)
10. Faraut, J., Korányi, A.: Function spaces and reproducing kernels on bounded symmetric domains. J. Funct. Anal. **88**, 64–89 (1990)
11. Faraut, J., Korányi, A.: Analysis on Symmetric Cones. Oxford University Press, New York (1994)
12. Freed, D.S., Moore G.W.: Twisted equivariant matter. Ann. Henri Poincaré Online First, 1–97 (2013). Available on-line at http://dx.doi.org/10.1007/s00023-013-0236-x. Cited 3 Sep 2013
13. Fyodorov, Y.V.: Negative moments of characteristic polynomials of random matrices: Ingham–Siegel integral as an alternative to Hubbard–Stratonovich transformation. Nuclear Phys. B **621**, 643–674 (2002)
14. Heinzner, P., Huckleberry, A., Zirnbauer, M.R.: Symmetry classes of disordered fermions. Commun. Math. Phys. **257**, 725–771 (2005)

15. Müller-Hill, J., Zirnbauer, M.R.: Equivalence of domains for hyperbolic Hubbard–Stratonovich transformations. J. Math. Phys. **52**, 053506, 25 (2011)
16. Howe, R.: Quantum mechanics and partial differential equations. J. Funct. Anal. **38**, 188–254 (1980)
17. Howe, R.: Remarks on classical invariant theory. Trans. Amer. Math. Soc. **313**, 539–570 (1989)
18. Ingham, A. E.: An integral which occurs in statistics. Proc. Cambridge Philos. Soc. **29**, 271–276 (1933)
19. Kostant, B.: Graded manifolds, graded Lie theory, and prequantization. In: Differential Geometrical Methods in Mathematical Physics (Proc. Sympos., Univ. Bonn, Bonn, 1975), pp. 177–306. Lecture Notes in Math. vol. 570. Springer-Verlag, Berlin Heidelberg New York (1977)
20. Koszul, J.-L.: Graded manifolds and graded Lie algebras. In: Proceedings of the International Meeting on Geometry and Physics (Florence, 1982), pp. 71–84. Pitagora, Bologna (1983)
21. Littelmann, P., Sommers, H.-J., Zirnbauer, M.R.: Superbosonization of invariant random matrix ensembles. Commun. Math. Phys. **283**, 343–395 (2008)
22. Manin, Y.I.: Gauge Field Theory and Complex Geometry. 2nd ed. Springer-Verlag, Berlin Heidelberg New York (1997)
23. Nishiyama, K.: Oscillator representations for orthosymplectic algebras. J. Algebra **129** 231–262 (1990)
24. Rossi, H., Vergne, M.: Analytic continuation of the holomorphic discrete series of a semisimple Lie group. Acta Math. **136**, 1–59 (1976)
25. Scheunert, M., Zhang, R.B.: The general linear supergroup and its Hopf superalgebra of regular functions. J. Algebra **254**, 44–83 (2002)
26. Siegel, C.L.: Über die analytische Theorie der quadratischen Formen. Ann. of Math. (2) **36**, 527–606 (1935)
27. Weil, A.: Sur certains groupes d'opérateurs unitaires. Acta Math. **111**, 143–211 (1964)
28. Witten, E.: Notes on supermanifolds and integration. Available on arXiv. http://arxiv.org/abs/1209.2199. Cited 3 Sep 2013
29. Zirnbauer, M.R.: Fourier analysis on a hyperbolic supermanifold with constant curvature. Commun. Math. Phys. **141**, 503–522 (1991)
30. Zirnbauer, M.R.: Super Fourier analysis and localization in disordered wires. Phys. Rev. Lett. **69**, 1584–1587 (1992)
31. Zirnbauer, M.R.: Riemannian symmetric superspaces and their origin in random-matrix theory. J. Math. Phys. **37**, 4986–5018 (1996)

Homological algebra for osp(1/2n)

Kevin Coulembier

Abstract We discuss several topics of homological algebra for the Lie superalgebra $\mathfrak{osp}(1|2n)$. First we focus on Bott-Kostant cohomology, which yields classical results although the cohomology is not given by the kernel of the Kostant Laplace operator. Based on this cohomology we can derive strong Bernstein-Gelfand-Gelfand resolutions for finite dimensional $\mathfrak{osp}(1|2n)$-modules. Then we state the Bott-Borel-Weil theorem which follows immediately from the Bott-Kostant cohomology by using the Peter-Weyl theorem for $\mathfrak{osp}(1|2n)$. Finally we calculate the projective dimension of irreducible and Verma modules in the category $\mathcal{O}$.

1 Introduction

The Lie superalgebra $\mathfrak{osp}(1|2n)$ plays an exceptional role in the theory of Lie superalgebras, see [12]. Contrary to the other simple finite dimensional Lie superalgebras the Harish-Chandra map yields an isomorphism $Z(\mathfrak{g}) \cong S(\mathfrak{h})^W$. Closely related to this observation is the fact that the category of finite dimensional representations is semisimple. In other words, all integral dominant highest weights are typical and every finite dimensional representation is completely reducible. As a consequence the algebra of regular functions on a Lie supergroup with superalgebra $\mathfrak{osp}(1|2n)$ satisfies a Peter-Weyl decomposition.

Because of these extraordinary properties, the algebra $\mathfrak{osp}(1|2n)$ and its representation theory is relatively well-understood, see e.g. [9, 16, 20]. In this paper we prove that certain standard topics of homological algebra for $\mathfrak{osp}(1|2n)$ allow elegant conclusions of the classical type. In particular the remarkable connection with the Lie algebra $\mathfrak{so}(2n+1)$, see e.g. [20], is confirmed. Another approach to the results in the current paper would be to make use of the equivalence of categories proved by Gorelik, since all regular integral blocks of $\mathfrak{osp}(1|2n)$ are strongly typical.

K. Coulembier (✉)
Department of Mathematical Analysis, Ghent University, Krijgslaan 281, 9000 Gent, Belgium
e-mail: coulembier@cage.ugent.be

M. Gorelik, P. Papi (eds.): *Advances in Lie Superalgebras.* Springer INdAM Series 7,
DOI 10.1007/978-3-319-02952-8_2,

First we focus on cohomology of the nilradical $\mathfrak{n}$ of the Borel subalgebra $\mathfrak{b}$ with values in finite dimensional $\mathfrak{osp}(1|2n)$-representations. Since the coboundary operator commutes with the Cartan subalgebra $\mathfrak{h}$ these cohomology groups are $\mathfrak{h}$-modules. For *Lie algebras* it can be proved that the cohomology is isomorphic to the kernel of the Kostant Laplace operator, see [13]. This operator is equivalent to an element of $S(\mathfrak{h})^W$. From the results in [1, 2, 13] it then follows that every weight in the kernel of the Laplace operator (or equivalently in the cohomology) only appears with multiplicity one in the space of cochains.

For *Lie superalgebras* in general the kernel of the Laplace operator is larger than the cohomology groups, see [7], even for $\mathfrak{osp}(1|2n)$ as we will see. We will also find that the weights appearing in the cohomology groups appear inside the space of cochains with multiplicities greater than one. We compute the cohomology by quotienting out an exact subcomplex, such that the resulting complex is isomorphic to that of $\mathfrak{so}(2n+1)$.

We use this result to obtain Bott-Borel-Weil (BBW) theory for $\mathfrak{osp}(1|2n)$. The classical BBW result in [2] computes the sheaf cohomology on line bundles over the flag manifold of a semisimple Lie group. In general it is a difficult task to compute these cohomology groups for supergroups. BBW theory for the typical blocks was obtained in [18]. Important further insight was gained in [10, 19, 22].

Since all blocks are typical for $\mathfrak{osp}(1|2n)$ the BBW theorem for $\mathfrak{osp}(1|2n)$ is included in the results in [18]. The $\mathfrak{n}$-cohomology results mentioned above could then be derived from the BBW result. Here we take the inverse approach because, despite being more computational, it clearly reveals the mechanism that makes the kernel of the Laplace operator larger than the cohomology groups, here caused by non-isotropic odd roots. When the kernel of the Laplace operator coincides with the cohomology it was proved in [7] that the irreducible modules of basic classical Lie superalgebras have a strong Bernstein-Gelfand-Gelfand (BGG) resolution (see [1]).

In this paper we prove that finite dimensional modules of $\mathfrak{osp}(1|2n)$ always possess a strong BGG resolution. As can be expected from [7] the main difficulty is dealing with the property that the kernel of the Kostant Laplace operator is larger than the cohomology. By making extensive use of the BGG theorem for $\mathfrak{osp}(1|2n)$ of [16] and our result on $\mathfrak{n}$-cohomology we can overcome this difficulty. Other results on BGG resolutions for basic classical Lie superalgebras were obtained in [4,5,7,8].

Finally we focus on the projective dimension of structural modules in the category $\mathcal{O}$ for $\mathfrak{osp}(1|2n)$. The main result is that the projective dimension of irreducible and Verma modules with a regular highest weight is given in terms of the length of the element of Weyl group making them dominant. In particular we obtain that the global dimension of a regular block in $\mathcal{O}$ is $2n^2$.

The remainder of the paper is organised as follows. In Sect. 2 we introduce some notations and conventions. The cohomology groups $H^k(\mathfrak{n},-)$ are calculated in Sect. 3. This result is then used in Sect. 4 to derive BGG resolutions. In Sect. 5 the $\mathfrak{n}$-cohomology result is translated into a BBW theorem. In Sec. 6 the projective dimensions in the category $\mathcal{O}$ are calculated. Finally there are two appendices. In Appendix 1 the technical details of the computation of the $\mathfrak{n}$-cohomology are given.

In Appendix 2 we state some facts about the BGG category $\mathscr{O}$ for basic classical Lie superalgebras.

2 Preliminaries

For the complex basic classical Lie superalgebra $\mathfrak{g} = \mathfrak{osp}(1|2n)$ we consider the simple positive roots

$$\delta_1 - \delta_2, \delta_2 - \delta_3, \cdots, \delta_{n-1} - \delta_n, \delta_n$$

corresponding to the standard system of positive roots, see [12]. For this system, the set of even positive roots is given by

$$\Delta_{\bar{0}}^+ = \{\delta_i - \delta_j | 1 \leq i < j \leq n\} \cup \{\delta_i + \delta_j | 1 \leq i \leq j \leq n\}$$

and the set of odd positive roots by

$$\Delta_{\bar{1}}^+ = \{\delta_i | 1 \leq i \leq n\}.$$

This leads to the value $\rho = \sum_{j=1}^n (n + \frac{1}{2} - j)\delta_j$ for half the difference between the sum of even roots and the sum of odd roots.

The Cartan subalgebra of $\mathfrak{osp}(1|2n)$ is denoted by $\mathfrak{h}$. The subalgebra consisting of positive (negative) root vectors is denoted by $\mathfrak{n}$ ($\overline{\mathfrak{n}}$). The corresponding triangular decomposition is given by $\mathfrak{osp}(1|2n) = \overline{\mathfrak{n}} + \mathfrak{h} + \mathfrak{n}$. The Borel subalgebra is denoted by $\mathfrak{b} = \mathfrak{h} + \mathfrak{n}$.

The Weyl group W of $\mathfrak{osp}(1|2n)$ is the same as for the underlying Lie algebra $\mathfrak{sp}(2n)$ (and isomorphic to the Weyl group of $\mathfrak{so}(2n+1)$), where the action is naturally extended to include the odd roots of $\mathfrak{osp}(1|2n)$. By the dotted action of $w \in W$ on elements $\lambda \in \mathfrak{h}^*$ we mean the ρ-shifted action: $w \cdot \lambda = w(\lambda + \rho) - \rho$. Since the Weyl group is the same as for the underlying Lie algebra, the notion of the Chevallay-Bruhat ordering and the length $|w| = l(w)$ of an element $w \in W$, remains unchanged. However, the notion of strongly linked weights, see Sect. 5.1 in [11] or Sect. 10.4 in [17], should be interpreted with respect to ρ and not $\rho_{\bar{0}}$ (half the sum of the positive roots of $\mathfrak{sp}(2n)$). Through the identification of the Weyl groups and root lattices of $\mathfrak{osp}(1|2n)$ and $\mathfrak{so}(2n+1)$, this shifted action coincides. In particular the characters of irreducible highest weights modules of $\mathfrak{osp}(1|2n)$ and $\mathfrak{so}(2n+1)$ coincide, see e.g. [20].

The set of integral dominant weights is denoted by $\mathscr{P}^+ \subset \mathfrak{h}^*$. For each $\lambda \in \mathfrak{h}^*$ the corresponding Verma module is denoted by $M(\lambda) = \mathrm{U}(\mathfrak{g}) \otimes_{\mathrm{U}(\mathfrak{b})} \mathbb{C}_\lambda$. Where $\mathbb{C}_\lambda$ is the one dimensional $\mathfrak{b}$-module with properties $H\mathbb{C}_\lambda = \lambda(H)\mathbb{C}_\lambda$ for all $H \in \mathfrak{h}$ and $\mathfrak{n}\mathbb{C}_\lambda = 0$. The quotient of $M(\lambda)$ with respect to its unique maximal submodule is irreducible and denoted by $L(\lambda)$. The module $L(\lambda)$ is finite dimensional if and only if $\lambda \in \mathscr{P}^+$. For each $\mu \in \mathfrak{h}^*$ we denote the central character associated with it by $\chi_\mu : Z(\mathfrak{g}) \to \mathbb{C}$.

The spaces of k-chains for $\overline{\mathfrak{n}}$-homology in an $\mathfrak{osp}(1|2n)$-module V are denoted by $C_k(\overline{\mathfrak{n}}, V) = \Lambda^k \overline{\mathfrak{n}} \otimes V$. These spaces are naturally $\mathfrak{h} + \overline{\mathfrak{n}}$-modules where the action is the tensor product of the adjoint action and the restricted action on the $\mathfrak{osp}(1|2n)$-module V. The boundary operator $\delta_k^* : C_k(\bar{\mathfrak{n}}, V) \to C_{k-1}(\bar{\mathfrak{n}}, V)$ is defined by

$$\delta_k^*(Y \wedge f) = -Y \cdot f - Y \wedge \delta_{k-1}^*(f) \quad \text{and} \quad \delta_0^* = 0,$$

for $Y \in \bar{\mathfrak{n}}$ and $f \in C_{k-1}(\bar{\mathfrak{n}}, V))$, see e.g. [7]. This operator is an $\mathfrak{h}$-module morphism and satisfies $\delta_k^* \circ \delta_{k+1}^* = 0$. The homology groups are defined as $H_k(\bar{\mathfrak{n}}, V) = \ker \delta_k^* / \mathrm{im}\delta_{k+1}^*$ and are naturally $\mathfrak{h}$-modules. A general approach to the concept of Lie superalgebra cohomology can e.g. be found in Chap. 16 in [16].

For an abelian category $\mathscr{A}$, the right derived functors (see [21]) of the left exact functor given by $\mathrm{Hom}_{\mathscr{A}}(A, -)$, for A an object of $\mathscr{A}$, are denoted by $\mathrm{Ext}_{\mathscr{A}}^k(A, -)$, where $\mathrm{Ext}_{\mathscr{A}}^1(A, -)$ is also written as $\mathrm{Ext}_{\mathscr{A}}(A, -)$. When the category of finitely generated $\mathfrak{a}$-modules is considered, for some algebra $\mathfrak{a}$, the name of the category is replaced by $\mathfrak{a}$.

3 Bott–Kostant cohomology

The main result of this section is the following description of the homology and cohomology of the nilradical of the Borel subalgebra of $\mathfrak{osp}(1|2n)$ or its dual, with values in irreducible representations of $\mathfrak{osp}(1|2n)$.

Theorem 1 *The (co)homology of $\mathfrak{n}$ and $\bar{\mathfrak{n}}$ in the irreducible finite dimensional $\mathfrak{osp}(1|2n)$-representation $L(\lambda)$ is given by*

$$H^k(\mathfrak{n}, L(\lambda)) = \bigoplus_{w \in W(k)} \mathbb{C}_{w \cdot \lambda} \qquad H_k(\mathfrak{n}, L(\lambda)) = \bigoplus_{w \in W(k)} \mathbb{C}_{-w \cdot \lambda}$$

$$H^k(\bar{\mathfrak{n}}, L(\lambda)) = \bigoplus_{w \in W(k)} \mathbb{C}_{-w \cdot \lambda} \qquad H_k(\bar{\mathfrak{n}}, L(\lambda)) = \bigoplus_{w \in W(k)} \mathbb{C}_{w \cdot \lambda},$$

with $W(k)$ the set of elements of the Weyl group with length k, see Sect. 0.3 in [11].

One of these results implies the other three according to Lemma 6.22 in [6], Lemma 4.6 in [7] or Theorem 17.6.1 in [17]. The remainder of this section is devoted to proving the property for the $\bar{\mathfrak{n}}$-homology, where the more technical steps in the proof are given in Appendix 1.

For each root the corresponding space of root vectors is one dimensional. For each positive root $\alpha \in \Delta^+$, we fix one root vector with weight $-\alpha$ and denote it by $Y_\alpha \in \bar{\mathfrak{n}}$. We choose the normalisation such that $[Y_{\delta_i}, Y_{\delta_i}] = Y_{2\delta_i}$ holds. Each element $f \in C_d(\bar{\mathfrak{n}}, V)$ of the form $f = Y_{\alpha_1} \wedge \cdots \wedge Y_{\alpha_d} \otimes v$ for certain positive roots $\alpha_1, \cdots, \alpha_d$ and $v \in V$ is called a monomial. For convenience v will often be considered to be a weight vector. We say that $f = Y_{\alpha_1} \wedge \cdots \wedge Y_{\alpha_d} \otimes v$ contains a monomial $Y_{\beta_1} \wedge \cdots \wedge Y_{\beta_k} \in \Lambda^k \bar{\mathfrak{n}}$ if $\{\beta_1, \cdots, \beta_k\} \subset \{\alpha_1, \cdots, \alpha_d\}$.

Definition 1 The $\mathfrak{h}$-submodule of $C_\bullet(\bar{\mathfrak{n}}, V)$ spanned by all monomials that do not contain any $Y_{2\delta_i}$ or $Y_{\delta_i}^{\wedge 2}$ for $i \in \{1, \cdots, n\}$ is denoted by $R_\bullet(\bar{\mathfrak{n}}, V)$ and the subvectorspace spanned by all monomials that do contain a $Y_{2\delta_i}$ or $Y_{\delta_i}^{\wedge 2}$ is denoted by

$W_\bullet(\overline{\mathfrak{n}},V)$, then

$$C_\bullet(\overline{\mathfrak{n}},V)=R_\bullet(\overline{\mathfrak{n}},V)\oplus W_\bullet(\overline{\mathfrak{n}},V).$$

The subspaces $A_\bullet^{(j)}$ and $B_\bullet^{(j)}$ of $W_\bullet(\overline{\mathfrak{n}},V)$ are defined as

$$A_\bullet^{(j)} = \mathrm{Span}\{Y_{\delta_j}^{\wedge 2}\wedge f|\ \mathrm{f}\in C_\bullet(\overline{\mathfrak{n}},V) \text{ contains no } Y_{2\delta_j}, Y_{2\delta_i} \text{ or } Y_{\delta_i}^{\wedge 2} \text{ for } i<j\}$$

$$B_\bullet^{(j)} = \mathrm{Span}\{Y_{2\delta_j}\wedge f|\ \mathrm{f}\in C_\bullet(\overline{\mathfrak{n}},V) \text{ contains no } Y_{2\delta_i} \text{ or } Y_{\delta_i}^{\wedge 2} \text{ for } i<j\}.$$

The subspace $R_\bullet(\overline{\mathfrak{n}},L(\lambda))$ for $\lambda\in\mathscr{P}^+\subset\mathfrak{h}^*$ is isomorphic as an $\mathfrak{h}$-module to the the corresponding full spaces of chains for the nilradical of $\mathfrak{so}(2n+1)$ and the corresponding representation of $\mathfrak{so}(2n+1)$ with the same highest weight λ. In particular, $R_k(\overline{\mathfrak{n}},L(\lambda))=0$ for $k>n^2$.

Using the results in Appendix 1 we can prove that the homology of $C_\bullet(\overline{\mathfrak{n}},V)$ can essentially be described in terms of $R_\bullet(\overline{\mathfrak{n}},V)$. This is based on the fact that the homology of a complex does not change after quotienting out an exact complex:

Proposition 1 *Let $S_\bullet\subset C_\bullet(\overline{\mathfrak{n}},V)$ be an exact subcomplex (and $\mathfrak{h}$-submodule). The operator $d: C_\bullet(\overline{\mathfrak{n}},V)/S_\bullet\to C_\bullet(\overline{\mathfrak{n}},V)/S_\bullet$ canonically induced from δ^* satisfies*

$$H_\bullet(C/S)\cong \ker d_k/\mathrm{im} d_{k+1}\cong \ker\delta_k^*/\mathrm{im}\delta_{k+1}^*\cong H_\bullet(C)$$

as $\mathfrak{h}$-modules.

Proof. The operator d is defined as $d(f+S)=\delta^*(f)+S$ for $f\in C_\bullet(\overline{\mathfrak{n}},V)$. The morphism

$$\eta:\ker\delta^*\to\ker d\quad \eta(f)=f+S$$

is well-defined. Since $\eta(\mathrm{im}\delta^*)\subset\mathrm{im} d$ this descends to a morphism $\widetilde{\eta}$: $\ker\delta^*/\mathrm{im}\delta^*\to\ker d/\mathrm{im} d$.

We prove that this is injective. Assume that $f\in\ker\delta^*\backslash\mathrm{im}\delta^*$, we have to prove that f is not of the form $\delta^*(g)+s$ for $s\in S_\bullet$. If f were of this form it immediately would follow that $s\in\ker\delta^*\cap S_\bullet=\mathrm{im}\delta^*\cap S_\bullet$ and therefore $f\in\mathrm{im}\delta^*$, which is a contradiction.

Finally we prove that $\widetilde{\eta}$ is also surjective. Every element in $\ker d_k/\mathrm{im} d_{k+1}$ is represented by some $a\in C_\bullet(\overline{\mathfrak{n}},V)$ such that $\delta^*a=s\in S_\bullet$. Since $\delta^*s=0$ and $S_\bullet$ forms an exact complex, there is a certain $s_1\in S_\bullet$ such that $s=\delta^*s_1$. The element $a-s_1$ is clearly inside $\ker\delta^*$, so the proposition follows from $\eta(a-s_1)=a+S_\bullet$.

Theorem 2 *For any $\mathfrak{osp}(1|2n)$-module V, the subspace $A_\bullet\subset C_\bullet(\overline{\mathfrak{n}},V)$ satisfies $A_\bullet\cap\ker\delta^*=\{0\}$,*

$$(A_\bullet\oplus\delta^*A_\bullet)\cap R_\bullet(\overline{\mathfrak{n}},V)=\{0\} \text{ and } A_\bullet\oplus\delta^*A_\bullet\oplus R_\bullet(\overline{\mathfrak{n}},V)=C_\bullet(\overline{\mathfrak{n}},V).$$

Proof. The proof makes use of the results in Lemma 3 and Theorem 7 in Appendix 1.

The property $A_\bullet\cap\ker\delta^*=\{0\}$ follows immediately from Theorem 7. This implies that $A_\bullet\oplus\delta^*A_\bullet$ is in fact a direct sum, since $\delta^*A_\bullet\subset\ker\delta^*$.

If $r = f + g$ with $r \in R_\bullet(\overline{\mathfrak{n}}, V)$, $f \in A_\bullet$ and $g \in \delta^* A_\bullet$, then $\phi(h) = 0$ for any $h \in A_\bullet$ such that $\delta^* h = g$, with ϕ the isomorphism defined in Theorem 7, and therefore $g = 0$. Since $A_\bullet \subset W_\bullet(\overline{\mathfrak{n}}, V)$, $r = f$ implies $r = 0 = f$ according to Definition 1 and we obtain $(A_\bullet \oplus \delta^* A_\bullet) \cap R_\bullet(\overline{\mathfrak{n}}, V) = \{0\}$.

The last property follows from the previous one and dimensional considerations. The first property in the theorem 7 implies that $\dim A_{k+1} = \dim \delta^* A_{k+1} = \dim(\delta^* A)_k$, together with Lemma 3 this yields $\dim(\delta^* A)_k = \dim B_k$. Therefore $\dim A_k + \dim(\delta^* A)_k + \dim R_k(\mathfrak{n}, V) = \dim C_k(\overline{\mathfrak{n}}, V)$ according to Definition 1 and Lemma 3.

Remark 1 Thus far the fact that V is not just an $\mathfrak{h} + \overline{\mathfrak{n}}$-module but also an $\mathfrak{osp}(1|2n)$-module has not been used. Theorem 2 could therefore be used to calculate $\overline{\mathfrak{n}}$-homology with values in arbitrary finite dimensional $\mathfrak{h} + \overline{\mathfrak{n}}$-modules.

Now we can give the proof of Theorem 1.

Proof. We calculate the Euler characteristic of the homology: $\sum_{i=0}^{\infty}(-1)^i \mathrm{chH_i}(\overline{\mathfrak{n}}, \mathrm{L}(\lambda))$

$$\begin{aligned} &= \sum_{i=0}^{\infty}(-1)^i \mathrm{ch}(\Lambda^i \overline{\mathfrak{n}}) \mathrm{chL}(\lambda) \\ &= \frac{\prod_{\alpha \in \Delta_{\overline{0}}^+}(1 - e^{-\alpha})}{\prod_{\gamma \in \Delta_{\overline{1}}^+}(1 + e^{\gamma})} \frac{\prod_{\gamma \in \Delta_{\overline{1}}^+}(e^{\gamma/2} + e^{-\gamma/2})}{\prod_{\alpha \in \Delta_{\overline{0}}^+}(e^{\alpha/2} - e^{-\alpha/2})} \sum_{w \in W}(-1)^{|w|} e^{w(\lambda+\rho)} \\ &= \sum_{w \in W}(-1)^{|w|} e^{w \cdot \lambda}, \end{aligned}$$

which is the technique through which Kostant obtained the Weyl character formula from this type of cohomology in [13].

Now from Sect. 4 in [7] it follows that $H_k(\overline{\mathfrak{n}}, L(\Lambda)) \subset \ker \Box$ with $\Box$ the Kostant Laplacian $\Box$ on $C_k(\overline{\mathfrak{n}}, V)$. This operator $\Box$ is a quadratic element of $\mathrm{U}(\mathfrak{h})$. From Proposition 1 and Theorem 2 it follows that this property can be made stronger to $H_k(\overline{\mathfrak{n}}, L(\Lambda)) \subset \ker \Box_{R_k}$. The $\mathfrak{h}$-module R_k is isomorphic to the chains for Bott-Kostant cohomology for $\mathfrak{so}(2n+1)$. There the cohomology is well-known and equal to the kernel of the Kostant Laplace operator, which takes the same form as for $\mathfrak{osp}(1|2n)$. Therefore the result in [13] and the observation of the connection between $\mathfrak{osp}(1|2n)$ and $\mathfrak{so}(2n+1)$ in Sect. 2 yields

$$H_k(\overline{\mathfrak{n}}, L(\lambda)) \subset \bigoplus_{w \in W(k)} \mathbb{C}_{w \cdot \lambda}.$$

The Euler characteristic then implies that these inclusions must be equalities.

The results on cohomology of $\mathfrak{n}$ can be reinterpreted in terms of Ext-functors in the category $\mathcal{O}$ as defined in Appendix 2.

Corollary 1 *For* $\mathfrak{g} = \mathfrak{osp}(1|2n)$, $\lambda \in \mathcal{P}^+$ *and* $\mu \in \mathfrak{h}^*$, *the property*

$$\mathrm{Ext}^k_{\mathcal{O}}(M(\mu), L(\lambda)) = \begin{cases} 1 & \text{if } \mu = w \cdot \lambda \text{ with } |w| = k \\ 0 & \text{otherwise} \end{cases}$$

holds.

Proof. As in the classical case the Frobenius reciprocity $\mathrm{Hom}_{\mathcal{O}}(\mathrm{U}(\mathfrak{g})\otimes_{\mathrm{U}(\mathfrak{b})}\mathbb{C}_\mu, V) = \mathrm{Hom}_{\mathfrak{b}}(\mathbb{C}_\mu, \mathrm{Res}^{\mathfrak{g}}_{\mathfrak{b}} V)$ holds for all $V \in \mathcal{O}$. This gives an equality of functors $\mathcal{O} \to$ Sets and since the functor $\mathrm{Res}^{\mathfrak{g}}_{\mathfrak{b}}$ is exact we can take the right derived functor of both left exact functors above to obtain

$$\mathrm{Ext}^k_{\mathcal{O}}(M(\mu), V) = \mathrm{Ext}^k_{\mathfrak{b}}(\mathbb{C}_\mu, \mathrm{Res}^{\mathfrak{g}}_{\mathfrak{b}} V)$$

If we use $\mathrm{Hom}_{\mathfrak{b}}(\mathbb{C}_\mu, -) = \mathrm{Hom}_{\mathfrak{h}}(\mathbb{C}_\mu, -) \circ \mathrm{Hom}_{\mathfrak{n}}(\mathbb{C}, -)$, the fact that $\mathrm{Hom}_{\mathfrak{h}}(\mathbb{C}_\mu, -)$ is exact and $\mathrm{Ext}^k(\mathfrak{n}, -) = H^k(\mathfrak{n}, -)$, see Lemma 4.7 in [7], we obtain

$$\mathrm{Ext}^k_{\mathcal{O}}(M(\mu), V) = \mathrm{Hom}_{\mathfrak{h}}\left(\mathbb{C}_\mu, H^k(\mathfrak{n}, V)\right).$$

The corollary then follows from Theorem 1.

4 Bernstein–Gelfand–Gelfand resolutions

The main result of this section is that all finite dimensional modules of $\mathfrak{osp}(1|2n)$ can be resolved in terms of direct sums of Verma modules. Such resolutions are known as (strong) BGG resolutions and were discovered first for semisimple Lie algebras in [1].

Theorem 3 *Every finite dimensional representation $L(\lambda)$ of $\mathfrak{osp}(1|2n)$ has a resolution in terms of Verma modules of the form*

$$0 \to \bigoplus_{w \in W(n^2)} M(w\cdot\lambda) \to \cdots \to \bigoplus_{w \in W(j)} M(w\cdot\lambda) \to \cdots$$
$$\to \bigoplus_{w \in W(1)} M(w\cdot\lambda) \to M(\lambda) \to L(\lambda) \to 0.$$

In the remainder of this section we provide the results needed to prove Theorem 3. We will make extensive use of the notions and results on the category $\mathcal{O}$ in Appendix 2.

First, we state the BGG theorem for $\mathfrak{osp}(1|2n)$, which was proved by Musson in Theorem 2.7 in [16]:

Theorem 4 (BGG theorem) *For $\mathfrak{g} = \mathfrak{osp}(1|2n)$ and $\lambda, \mu \in \mathfrak{h}^*$ it holds that $[M(\lambda) : L(\mu)] \neq 0$ if and only if $\mu \uparrow \lambda$ (μ is strongly linked to λ).*

Using this we obtain the following corollary.

Corollary 2 *Consider $\mathfrak{g} = \mathfrak{osp}(1|2n)$ and $\mu, \lambda \in \mathfrak{h}^*$. If $\mathrm{Ext}_{\mathcal{O}}(M(\mu), M(\lambda)) \neq 0$ then $\mu \uparrow \lambda$ but $\mu \neq \lambda$.*

Proof. The property $\mathrm{Ext}_{\mathcal{O}}(M(\mu), M(\lambda)) \neq 0$ holds if and only if there is a short exact non-split sequence of the form $M(\lambda) \hookrightarrow M \twoheadrightarrow M(\mu)$ for an $M \in \mathcal{O}$. That $\mu \neq \lambda$ must hold follows immediately from the fact that otherwise M would contain two highest weight vectors of weight λ, which both generate a Verma module.

The remainder of the proof is then equivalent with the proof of Theorem 6.5 in [11]. We consider the projective cover $P(\mu)$ of $M(\mu)$, which exists and has a standard filtration by Lemma 6. This filtration $0 = P_0 \cdots \subset P_1 \subset \cdots P_n = P$ satisfies $P_i/P_{i-1} \cong M(\mu_i)$ with $\mu \uparrow \mu_i$ by the combination of Theorem 4 and Lemma 7.

The canonical map $P(\mu) \to M(\mu)$ extends to $\phi : P(\mu) \to M$ and since the exact sequence does not split we obtain that for some i, $\phi(P_i) \cap M(\lambda) \neq 0$ while $\phi(P_{i-1}) \cap M(\lambda) = 0$. This implies that $M(\lambda)$ has a nonzero submodule which is a homomorphic image of $M(\mu_i)$ and therefore $[M(\lambda) : L(\mu_i)] \neq 0$. Applying Theorem 4 again yields $\mu_i \uparrow \lambda$.

These two results lead to $\mu \uparrow \lambda$.

Now we can prove the following consequence of this corollary.

Lemma 1 *Consider $w \in W$, $\lambda \in \mathscr{P}^+$ and a module M with a standard filtration where the occurring Verma modules are of the form $M(w' \cdot \lambda)$ with $l(w') \geq l(w)$, then*

$$\mathrm{Ext}_{\mathscr{O}}(M(w \cdot \lambda), M) = 0.$$

Furthermore, any module S in $\mathscr{O}_{\chi_\lambda}$ with standard filtration has a filtration of the form $S = S^{(0)} \supseteq S^{(1)} \supseteq \cdots S^{(n^2)} \supseteq S^{(n^2+1)} = 0$, where $S^{(j)}/S^{(j+1)}$ is isomorphic to the direct sum of Verma modules with highest weights $u \cdot \lambda$ with $u \in W(n^2 - j)$.

Proof. The first statement is an immediate application of Corollary 2 if M is a Verma module. The remainder can then be proved by induction on the filtration length. Assume it is true for filtration length $p-1$ and M has filtration length p. Then there is a short exact sequence

$$0 \to N \to M \to M(w_p \cdot \lambda) \to 0$$

for N having a standard filtration of the prescribed kind of length $p-1$ and $l(w_p) \geq l(w)$. Applying the functor $\mathrm{Hom}_{\mathscr{O}}(M(w \cdot \lambda), -)$ and its right derived functors gives a long exact sequence

$$\begin{aligned}0 &\to \mathrm{Hom}_{\mathscr{O}}(M(w \cdot \lambda), N) \to \mathrm{Hom}_{\mathscr{O}}(M(w \cdot \lambda), M) \to \mathrm{Hom}_{\mathscr{O}}(M(w \cdot \lambda), M(w_p \cdot \lambda)) \\ &\to \mathrm{Ext}_{\mathscr{O}}(M(w \cdot \lambda), N) \to \mathrm{Ext}_{\mathscr{O}}(M(w \cdot \lambda), M) \to \mathrm{Ext}_{\mathscr{O}}(M(w \cdot \lambda), M(w_p \cdot \lambda)) \to \cdots.\end{aligned}$$

Since $\mathrm{Ext}_{\mathscr{O}}(M(w \cdot \lambda), N) = 0 = \mathrm{Ext}_{\mathscr{O}}(M(w \cdot \lambda), M(w_p \cdot \lambda))$ by the induction step we obtain $\mathrm{Ext}_{\mathscr{O}}(M(w \cdot \lambda), M) = 0$.

In order to prove the second claim we consider an arbitrary module K in $\mathscr{O}_{\chi_\lambda}$ with a standard filtration,

$$K = K_0 \supset K_1 \supset \cdots \supset K_d = 0 \qquad \text{with} \qquad K_i/K_{i+1} \cong M(w_{(i)} \cdot \lambda).$$

Consider an arbitrary i such that $w_{(i)}$ has the minimal length appearing in the set $\{w_{(j)}, j = 0, \cdots, d-1\}$, since $\mathrm{Ext}_{\mathscr{O}}(M(w_{(i)} \cdot \lambda), K_{i+1}) = 0$ by the first part of the lemma it follows that $M(w_{(i)}) \subset K_i \subset K$. Therefore the direct sum of all these Verma modules are isomorphic to a submodule of K. This submodule can be quotiented out and the statement follows by iteration.

As in [1] we start by constructing a resolution of $L(\lambda)$ in terms of modules induced by the spaces of chains

$$C_\bullet(\overline{\mathfrak{n}},L(\lambda)) \cong \Lambda^\bullet\overline{\mathfrak{n}}\otimes L(\lambda) \cong \Lambda^\bullet(\mathfrak{g}/\mathfrak{b})\otimes L(\lambda),$$

which will possess standard filtrations by construction. For the classical case, restricting to the block in $\mathcal{O}$ which $L(\lambda)$ belongs to, exactly reduces from $C_k(\overline{\mathfrak{n}},V)$ to $H_k(\overline{\mathfrak{n}},V)$. Corollary 2 then already yields the BGG resolutions. In fact, according to the results in [13] only one Casimir operator is needed for this reduction, the quadratic one. Applying this procedure in the case of Lie superalgebras would however lead to a resolution in terms of the kernel of the Laplace operator, which is still larger than the homology groups, as discussed in Sect. 3. In case the kernel of the Laplace operator agrees with the cohomology, strong BGG resolutions for basic classical Lie superalgebras always exist, according to the result in [7].

Lemma 2 *For each finite dimensional representation $L(\lambda)$ of $\mathfrak{osp}(1|2n)$, there is a finite resolution of the form*

$$\cdots \to D_k \to \cdots \to D_1 \to D_0 \to L(\lambda) \to 0,$$

where each D_k has a standard filtration. Moreover D_k has a filtration $D_k = S_k^{(0)} \supseteq S_k^{(1)} \supseteq \cdots S_k^{(n^2)} \supseteq S_k^{(n^2+1)} = 0$, where $S_k^{(j)}/S_k^{(j+1)}$ is isomorphic to the direct sum of Verma modules with highest weights $w\cdot\lambda$ with $l(w) = n^2 - j$.

Proof. The first step of the construction is parallel to the classical case. We can define an exact complex of $\mathfrak{g}$-modules of the form

$$\cdots \to \mathrm{U}(\mathfrak{g})\otimes_{\mathrm{U}(\mathfrak{b})}(\Lambda^k\mathfrak{g}/\mathfrak{b}\otimes L(\lambda)) \to \cdots \to$$
$$\mathrm{U}(\mathfrak{g})\otimes_{\mathrm{U}(\mathfrak{b})}(\Lambda^1\mathfrak{g}/\mathfrak{b}\otimes L(\lambda)) \to \mathrm{U}(\mathfrak{g})\otimes_{\mathrm{U}(\mathfrak{b})}L(\lambda) \to L(\lambda) \to 0$$

where the maps are given by the direct analogs of those in [1], or Sect. 6.3 in [11], see also Eq. (4.1) in [5]. In fact it suffices to do this for $L(\lambda)$ trivial, since a straightforward tensor product can be taken afterwards.

Now we can restrict the resolution to the block of the category $\mathcal{O}$ corresponding to the central character χ_λ, which still yields an exact complex. Lemma 1 then implies that the modules which appear must be of the proposed form.

It remains to be proved that the resolution is finite. This follows from the observation that for k large enough all the weights appearing in $C_k(\overline{\mathfrak{n}},L(\Lambda))$ are lower than those in the set $\{w(\lambda+\rho)-\rho | w\in W\}$.

Now we can prove Theorem 3. Contrary to the classical case in [1], where the BGG resolutions are constructed to obtain an alternative derivation for the Bott-Kostant cohomology groups, we will need our result on the $\overline{\mathfrak{n}}$-homology to derive the BGG resolutions.

Proof. Since the modules appearing in the resolution in Lemma 2 have a filtration in terms of Verma modules, this corresponds to a projective resolution in the category of $\overline{\mathfrak{n}}$-modules. This can therefore be applied to calculate the right derived functors of the left exact contravariant functor $\mathrm{Hom}_{\overline{\mathfrak{n}}}(-,\mathbb{C})$ acting on $L(\lambda)$, see [21]. These functors satisfy $\mathrm{Ext}^k_{\overline{\mathfrak{n}}}(L(\lambda),\mathbb{C}) = H_k(\overline{\mathfrak{n}},L(\lambda))^*$, see Lemma 4.6 and Lemma 4.7 in [7]. By applying this we obtain that the homology $H_k(\overline{\mathfrak{n}},L(\lambda))$ is equal to the homology of the finite complex of $\mathfrak{h}$-modules

$$\cdots \to D_k/(\overline{\mathfrak{n}}D_k) \to \cdots \to D_1/(\overline{\mathfrak{n}}D_1) \to D_0/(\overline{\mathfrak{n}}D_0) \to 0,$$

where the maps are naturally induced from the ones in Lemma 2.

The $\mathfrak{h}$-modules $D_k/(\overline{\mathfrak{n}}D_k)$ are exactly given by all the highest weights of the Verma modules appearing in the standard filtration of D_k. We take the largest k such that the filtration of D_k contains a Verma module of highest weight $w\cdot\lambda$ with $l(w) < k$. Since such a weight can not be in $H_k(\overline{\mathfrak{n}},L(\lambda))$ by Theorem 1, it is not in the kernel of the mapping $D_k/(\overline{\mathfrak{n}}D_k) \to D_{k-1}/(\overline{\mathfrak{n}}D_{k-1})$ (it is not in the image of $D_{k+1}/(\overline{\mathfrak{n}}D_{k+1}) \to D_k/(\overline{\mathfrak{n}}D_k)$ since we chose k maximal). We fix such a $w\cdot\lambda$ for D_k with minimal $l(w)$. According to Lemma 2 $M(w\cdot\lambda)$ is actually a submodule of D_k. Under the $\mathfrak{g}$-module morphism in Lemma 2 this submodule is mapped to a submodule in D_{k-1}. The highest weight vector of $M(w\cdot\lambda)$ is mapped to a highest weight vector in D_{k-1}. Since the projection onto $D_{k-1}/(\overline{\mathfrak{n}}D_{k-1})$ is not zero this highest weight vector is not inside another Verma module. This implies that the quotient of D_{k-1} with respect to the image of $M(w\cdot\lambda)$ still has a standard filtration. The appearance of $M(w\cdot\lambda)$ in D_k and D_{k-1} forms an exact subcomplex which can be quotiented out and according to Proposition 1 the resulting complex is still exact.

This procedure can be iterated until the resolution in Lemma 2 is reduced to a resolution of the form of Lemma 2 for which we use the same notations and where it holds that $S_k^{(j)} = 0$ if $j > n^2-k$. In a similar step we can quotient out the Verma submodules of $S_k^{(n^2-k)}$ that do not contribute to $H_k(\overline{\mathfrak{n}},L(\lambda))$.

Then we can focus on the submodules $S_k^{(n^2-k)} \subset D_k$ of the resulting complex. Because of the link with the $\overline{\mathfrak{n}}$-homology each of the highest weight vectors of the Verma modules is not mapped to the highest weight vector of a Verma module in the filtration of D_{k-1}. Theorem 4 implies that the image of a Verma module in $S_k^{(n^2-k)}$ under the composition of the map in Lemma 2 with the projection onto $D_{k-1}/S_{k-1}^{(n^2-k+1)}$ must be zero since the filtration of $D_{k-1}/S_{k-1}^{(n^2-k+1)}$ contains only Verma modules with highest weight $u\cdot\lambda$ with $l(u)\geq k$. So $S_k^{(n^2-k)}$ gets mapped to $S_{k-1}^{(n^2-k+1)} \subset D_{k-1}$, and thus there is a subcomplex of the desired form in Theorem 3. The complex originating from quotienting out this subcomplex is exact, which can again be seen from the connection with $\overline{\mathfrak{n}}$-homology. Therefore we obtain that the subcomplex of the modules $S_k^{(n^2-k)}$ must also be exact and Theorem 3 is proven.

5 Bott–Borel–Weil theory

In this section we use the algebraic reformulation of the result of Bott, Borel and Weil in [2] for algebraic groups to describe the Bott–Borel–Weil theorem for the algebraic supergroup $OSp(1|2n)$. This rederives the result for $\mathfrak{osp}(1|2n)$ in Theorem 1 in [18].

Theorem 5 *Consider* $\mathfrak{g} = \mathfrak{osp}(1|2n)$ *and* $\mathbb{C}_\lambda$ *the irreducible* $\mathfrak{b}$*-module with* $\mathfrak{h}\mathbb{C}_{-\lambda} = -\lambda(\mathfrak{h})\mathbb{C}_{-\lambda}$.

- *If* λ *is regular, there exists a unique element of the Weyl group* W *rendering* $\Lambda := w(\lambda+\rho)-\rho$ *dominant. In this case*

$$H^k(G/B, G\times_B \mathbb{C}_{-\lambda}) = \begin{cases} L(\Lambda) & \text{if } |w| = k \\ 0 & \text{if } |w| \neq k \end{cases}.$$

- *If* λ *is not regular,* $H^k(G/B, G\times_B \mathbb{C}_{-\lambda}) = 0$.

Proof. For any $\mathfrak{b}$-module the holomorphic sections of the flag manifold satisfy $H^0(G/B, G\times_B V) = \mathrm{Hom}_{\mathfrak{b}}(\mathbb{C}, V\otimes\mathscr{R})$ with the $\mathfrak{g}\times\mathfrak{g}$-module $\mathscr{R}$ given by the algebra of regular functions (the finite dual of the Hopf algebra U($\mathfrak{g}$)) on $OSp(1|2n)$, see the proof of Lemma 2 in [10]. This algebra corresponds to the finite dual of the super Hopf algebra U($\mathfrak{g}$). The derived functors therefore satisfy $H^k(G/B, G\times_B V) = \mathrm{Ext}^k_{\mathfrak{b}}(\mathbb{C}, V\otimes\mathscr{R})$. Since $\mathscr{R}$ corresponds to the algebra of matrix elements, see Lemma 3.1 in [22] and the category of finite dimensional $\mathfrak{osp}(1|2n)$-representations is semisimple the Peter-Weyl type theorem

$$\mathscr{R} = \bigoplus_{\Lambda\in\mathscr{P}^+} L(\Lambda)\times L(\Lambda)$$

follows immediately. Therefore BBW theory is expressed as

$$H^k(G/B, G\times_B \mathbb{C}_{-\lambda}) = \bigoplus_{\Lambda\in\mathscr{P}^+} \mathrm{Hom}_{\mathfrak{h}}(\mathbb{C}_\lambda, H^k(\mathfrak{n}, L(\Lambda)))L(\Lambda).$$

The result then follows from Theorem 1.

6 Projective dimension in $\mathscr{O}$ of simple and Verma modules

In this section we calculate projective dimensions of simple and Verma modules in $\mathscr{O}$, which also gives the global dimension of the category $\mathscr{O}$. For semisimple Lie algebras this was obtained by Mazorchuk in a general framework to calculate projective dimensions of structural modules in [14]. Part of this approach extends immediately to $\mathfrak{osp}(1|2n)$, where the global dimension can actually be calculated from the BGG resolutions in Theorem 3. However here, we follow an approach similar to the classical one sketched in Sect. 6.9 in [11].

Theorem 6 *For* $\mathfrak{g} = \mathfrak{osp}(1|2n)$ *and* $\lambda \in \mathscr{P}^+$*, the following equalities on the projective dimensions hold:*

(i) $p.d.M(w\cdot\lambda) = l(w)$;
(ii) $p.d.L(w\cdot\lambda) = 2n^2 - l(w)$;
(iii) $gl.d.\mathscr{O}_{\chi_\lambda} = 2n^2$.

Proof. By Lemma 8, statement (i) is true for $w = 1$, or $l(w) = 0$. Then we proceed by induction on the length of w.

We use the general fact that if there is a short exact sequence of the form $A \hookrightarrow B \twoheadrightarrow C$ then

$$p.d.A \leq \max\{p.d.B, p.d.C - 1\} \quad \text{and} \quad p.d.C \leq \max\{p.d.A + 1, p.d.B\},$$

see [11,21].

Assume (i) holds for all w such that $l(w) < k$, then we take some $w \in W(k)$ and denote the kernel of the canonical morphism $P(w \cdot \lambda) \twoheadrightarrow M(w \cdot \lambda)$ by N. The module N has a standard filtration and the components can be obtained from the combination of Theorem 4 and Lemma 7. Therefore we obtain $p.d.N = l(w) - 1$. The short exact sequence $N \hookrightarrow P(w \cdot \lambda) \twoheadrightarrow M(w \cdot \lambda)$ implies $p.d.N \leq p.d.M(w \cdot \lambda) - 1$ and $p.d.M(w \cdot \lambda) \leq p.d.N + 1$ and we obtain $p.d.M(w \cdot \lambda) = k$.

This proves (i). The result of (i) implies (ii) for $l(w) = n^2$ since then $M(w \cdot \lambda) = L(w \cdot \lambda)$ by Theorem 4. From this point on statement (ii) can also be proved by induction, now using the short exact sequences of the form $N' \hookrightarrow M(w \cdot \lambda) \twoheadrightarrow L(w \cdot \lambda)$ with N' the unique maximal submodule of $M(w \cdot \lambda)$.

The result of (ii) immediately implies (iii).

Remark 2 Since projective modules in the category $\mathcal{O}$ have a standard filtration, see Lemma 6, a projective resolution of V provides a complex with homology $H_k(\bar{\mathfrak{n}}, V)$, for any basic classical Lie superalgebra. In particular it follows that the projective dimension of V in the category $\mathcal{O}$ is larger than or equal to the projective dimension as an $\bar{\mathfrak{n}}$-module. In fact, for $\mathfrak{osp}(1|2n)$, using the technique from the proof of Proposition 2 in [14], and the result in Theorem 3, one obtains that the projective dimension in $\mathcal{O}$ is at least twice the projective dimension as an $\bar{\mathfrak{n}}$-module. The result for $\mathfrak{g} = \mathfrak{osp}(1|2n)$ in Theorem 6 exactly states that this bound is actually an equality.

Appendix 1: Structure of the space of chains $C_\bullet(\bar{\mathfrak{n}}, V)$

In this appendix we obtain some technical results about the spaces $C_\bullet(\bar{\mathfrak{n}}, V)$, $R_\bullet(\bar{\mathfrak{n}}, V)$, $W_\bullet(\bar{\mathfrak{n}}, V)$, $A_\bullet$ and $B_\bullet$ as introduced in Sect. 3. Here V is a finite dimensional $\mathfrak{osp}(1|2n)$-module, although the same results would hold for an arbitrary finite dimensional $\mathfrak{h} + \bar{\mathfrak{n}}$-module.

Lemma 3 *The spaces $\{A_\bullet^{(j)}\}$ and $\{B_\bullet^{(k)}\}$ of Definition 1 are linearly independent. For $A_\bullet = \bigoplus_{j=1}^n A_\bullet^{(j)}$ and $B_\bullet = \bigoplus_{j=1}^n B_\bullet^{(j)}$ it holds that*

$$W_\bullet(\bar{\mathfrak{n}}, V) = A_\bullet \oplus B_\bullet \text{ and } A_k \cong B_{k-1}$$

as $\mathfrak{h}$-modules for $k \in \mathbb{N}$.

Proof. The monomials in the spaces $W_\bullet(\bar{\mathfrak{n}}, V)$, $A_\bullet$ and $B_\bullet$ form bases of these spaces. Therefore the proof can be written in terms of these monomials.

For every monomial in the span of the spaces $\{A_\bullet^{(j)}\}$ there is a certain k, such that it contains $Y_{\delta_k}^{\wedge 2}$ but no $Y_{2\delta_i}$ for $i \leq k$, which separates this space from the span of the spaces $\{B_\bullet^{(j)}\}$. If for $a_j \in A_\bullet^{(j)}$, the element $\sum_{j=1}^n a_j$ is zero we can prove that every a_j must be zero. If k is the lowest number such that a_k is not zero, then a_k contains $Y_{\delta_k}^{\wedge 2}$ while none of the other terms contain this, therefore $a_k = 0$.

Every monomial in $W_\bullet(\overline{\mathfrak{n}}, V)$ contains some term $Y_{2\delta_i}$ or some term $Y_{\delta_j}^{\wedge 2}$. If the lowest such i is strictly lower than the lowest such j, this monomial is inside $A_\bullet$, if the lowest such i is higher or equal to the lowest such j the monomial is inside $B_\bullet$. This proves $W_\bullet(\overline{\mathfrak{n}}, V) = A_\bullet \oplus B_\bullet$.

Finally the morphism $A_k^{(j)} \to B_{k-1}^{(j)}$ defined by mapping $Y_{\delta_j}^{\wedge 2} \wedge f \to Y_{2\delta_j} \wedge f$ is clearly well-defined and bijective for every j.

Definition 2 We introduce two subsets of the even positive roots $\Delta_{\overline{0}}^+$ of $\mathfrak{osp}(1|2n)$

$$M = \{\delta_i - \delta_j | \forall i < j\} \text{ and } P = \{\delta_i + \delta_j | \forall i < j\}.$$

The grading D on a monomial in $C_\bullet(\overline{\mathfrak{n}}, V)$ is defined as

$$D(Y_{\alpha_1} \wedge \cdots \wedge Y_{\alpha_d} \otimes v) = \sharp\{\alpha_k \in M, k = 1, \ldots, d\} - \sharp\{\alpha_k \in P, k = 1, \ldots, d\} + D(v),$$

where $D(v) = \sum_{i=1}^n \mu_i$ for v a weight vector of weight $\sum_{i=1}^n \mu_i \delta_i$.

Since the root vectors corresponding to the roots in M and P are even and V is finite dimensional, the grading is finite and we define $(C_\bullet(\overline{\mathfrak{n}}, V))\,[i]$ as the span of all the monomials f that satisfy $D(f) = i$. Also for subspaces $L_\bullet \subset C_\bullet(\overline{\mathfrak{n}}, V)$ we set

$$(L_\bullet)\,[i] = (C_\bullet(\overline{\mathfrak{n}}, V))\,[i] \cap L_\bullet \tag{1}$$

The following lemma follows immediately from the definition of the boundary operator.

Lemma 4 *The boundary operator $\delta^* : C_\bullet(\overline{\mathfrak{n}}, V) \to C_\bullet(\overline{\mathfrak{n}}, V)$ acting on a monomial f with $D(f) = p$ yields $\delta^* f = \sum_j f_j$ for monomials f_j that satisfy $D(f_j) \leq p$.*

The following calculation will be crucial for computing the cohomology.

Lemma 5 *For $Y_{\delta_j}^{\wedge k} \wedge f \in C_\bullet(\overline{\mathfrak{n}}, V)$ the boundary operator acts as*

$$\begin{aligned}\delta^*(Y_{\delta_j}^{\wedge k} \wedge f) &= -\frac{1}{2}k(k-1)Y_{2\delta_j} \wedge Y_{\delta_j}^{\wedge k-2} \wedge f \\ &\quad + k(-1)^k Y_{\delta_j}^{\wedge k-1} \wedge Y_{\delta_j} \cdot f + (-1)^k Y_{\delta_j}^{\wedge k} \wedge \delta^* f.\end{aligned}$$

Proof. From the immediate calculation

$$\begin{aligned}\delta^*(Y_{\delta_j}^{\wedge k} \wedge f) &= -(k-1)Y_{2\delta_j} \wedge Y_{\delta_j}^{\wedge k-2} \wedge f \\ &\quad + (-1)^k Y_{\delta_j}^{\wedge k-1} \wedge Y_{\delta_j} \cdot f - Y_{\delta_j} \wedge \delta^*(Y_{\delta_j}^{\wedge k-1} \wedge f)\end{aligned}$$

the statement can be proven by induction on k.

The previous results can now be brought together to come to the main conclusion of this appendix. The following result states that the coboundary operator maps the subspaces $A_\bullet$ bijectively to spaces isomorphic with $B_\bullet$.

Theorem 7 *The morphism*

$$\phi : A_\bullet \to C_\bullet(\overline{\mathfrak{n}}, V)/(A_\bullet \oplus R_\bullet(\overline{\mathfrak{n}}, V)) \cong B_\bullet$$

given by the composition of the boundary operator $\delta^* : A_\bullet \to C_\bullet(\overline{\mathfrak{n}}, V)$ *with the canonical projection onto* $C_\bullet(\overline{\mathfrak{n}}, V)/(A_\bullet \oplus R_\bullet(\overline{\mathfrak{n}}, V))$ *is an isomorphism.*

Proof. First we prove that the morphism $\phi^{(l)}$ given by ϕ acting on the restriction to $(A_\bullet)[l]$ (as defined in Eq. (1)) composed with the restriction

$$C_\bullet(\overline{\mathfrak{n}}, V)/(A_\bullet \oplus R_\bullet(\overline{\mathfrak{n}}, V)) \to (C_\bullet(\overline{\mathfrak{n}}, V)/(A_\bullet \oplus R_\bullet(\overline{\mathfrak{n}}, V)))\,[l]$$

is an isomorphism. We take a general element of $(A_\bullet)[l]$ and expand it according to the decomposition $A_\bullet = \bigoplus_{j=1}^n A_\bullet^{(j)}$ in Definition 1:

$$h = \sum_{j=1}^{n} \sum_{k=2}^{N_j} Y_{\delta_j}^{\wedge k} \wedge h_k^{(j)}$$

where $h_k^{(j)}$ does not contain Y_{δ_j}, $Y_{2\delta_j}$ or $Y_{\delta_i}^{\wedge 2}$ and $Y_{2\delta_i}$ for $i < j$ and $D(h_k^{(j)}) = l$. According to Lemma 5 the action of δ^* combined with projection onto $C_\bullet(\overline{\mathfrak{n}}, V)[l]$ is given by

$$\begin{aligned}(\delta^* h)\,[l] = & -\frac{1}{2} \sum_{j=1}^{n} \sum_{k=2}^{N_j} k(k-1) Y_{2\delta_j} \wedge Y_{\delta_j}^{\wedge k-2} \wedge h_k^{(j)} \\ & + \sum_{j=1}^{n} \sum_{k=2}^{N_j} (-1)^k Y_{\delta_j}^{\wedge k} \wedge \left(\delta^* h_k^{(j)}\right)[l]\end{aligned}$$

since degree of the monomials in the terms $Y_{\delta_j}^{\wedge k-1} \wedge Y_{\delta_j} \cdot h_k^{(j)}$ is strictly lower than l. Assume p is the smallest number for which $h_p^{(n)}$ is different from zero and assume that $h \in \ker \phi^{(l)}$. The term $Y_{2\delta_n} \wedge Y_{\delta_n}^{\wedge p-2} \wedge h_p^{(n)}$ is not inside $A_\bullet \oplus R_\bullet(\overline{\mathfrak{n}}, V)$ and $h_p^{(n)}$ does not contain Y_{δ_n} or any $Y_{2\delta_i}$ or $Y_{\delta_i}^{\wedge 2}$. Therefore there is no other term appearing in $(\delta^* h)\,[l]$ to compensate this one and we obtain $h_k^{(n)} \equiv 0$ for every k. Then from similar arguments we obtain by induction that $h_k^{(j)} \equiv 0$ must hold for every j and k, so $\phi^{(l)}$ is injective. The isomorphism $A_k \cong B_{k-1}$ from Lemma 3, which can clearly be refined to $(A_k)\,[l] \cong (B_{k-1})\,[l]$, then shows that injectivity implies surjectivity.

Lemma 4 implies that δ^* never raises degree, a property that is immediately inherited by ϕ. The combination of this with the fact that the grading is finite, leads to the conclusion that ϕ is bijective since the $\{\phi^{(l)}\}$ are.

Appendix 2: Category $\mathscr{O}$ for basic classical Lie superalgebras

The BGG category $\mathscr{O}$ for a basic classical Lie superalgebra $\mathfrak{g}$ is the full subcategory of the category of $\mathfrak{g}$-modules of modules M that satisfy the conditions:

- M is a finitely generated $U(\mathfrak{g})$-module.
- M is $\mathfrak{h}$-semisimple.
- M is locally $U(\mathfrak{n})$-finite.

In this appendix we mention some properties of this category which are needed in Sect. 4 and Sect. 6. For more details on category $\mathscr{O}$ for Lie (super)algebras, see [1, 3, 11, 14, 15]. We use notations similar to the rest of the paper, but now for arbitrary basic classical Lie superalgebras.

The following results are due to Mazorchuk, see Proposition 1 and Theorem 2 in [15], or Brundan, see Theorem 4.4 in [3].

Lemma 6 *In the category $\mathscr{O}$ for basic classical Lie superalgebras each irreducible representation $L(\mu)$ has a projective cover and each projective module in $\mathscr{O}$ has a standard filtration.*

The projective cover of $L(\lambda)$ is denoted by $P(\lambda)$ and is also the projective cover of $M(\lambda)$.

Lemma 7 (BGG reciprocity) *For a basic classical Lie superalgebra $\mathfrak{g}$ the following relation holds between the standard filtration of the projective module $P(\lambda)$ and the Jordan-Hölder series of the Verma module $M(\mu)$:*

$$(P(\lambda):M(\mu)) = [M(\mu):L(\lambda)].$$

Proof. This is a special case of Corollary 4.5 in [3], but can also easily be proved directly. Firstly, we have $[M(\mu):L(\lambda)] = [M(\mu)^\vee : L(\lambda)]$. For any module $M \in \mathscr{O}$ it holds that $[M:L(\lambda)] = \dim \mathrm{Hom}_{\mathscr{O}}(P(\lambda), M)$, since this is true for M irreducible and $\mathrm{Hom}_{\mathscr{O}}(P(\lambda), -)$ is an exact functor and thus preserves short exact sequences. Together this yields

$$[M(\mu):L(\lambda)] = \dim \mathrm{Hom}_{\mathscr{O}}\left(P(\lambda), M(\mu)^\vee\right).$$

The statement then follows from $\dim \mathrm{Hom}_{\mathscr{O}}(P(\lambda), M(\mu)^\vee) = (P(\lambda):M(\mu))$, which can be proved similarly as Theorem 3.7 in [11].

If an integral dominant weight is the highest one inside the class of weights corresponding to a central character (which is always true for typical highest weights) we obtain the classical result that the corresponding Verma module is projective.

Lemma 8 *Suppose $\Lambda \in \mathscr{P}^+$ is the highest weight inside the set $\{\mu \in \mathfrak{h}^* | \chi_\mu = \chi_\Lambda\}$, then $M(\Lambda)$ is a projective module in $\mathscr{O}$.*

Proof. The proof does not change from the proof of Proposition 3.8 in [11] because of the extra condition on χ_Λ.

Acknowledgements The author is a Postdoctoral Fellow of the Research Foundation - Flanders (FWO). The author wishes to thank Ruibin Zhang for fruitful discussions on this topic.

References

1. Bernstein, I.N., Gel'fand, I.M., Gel'fand, S.I.: Differential operators on the base affine space and a study of g-modules. In: Lie groups and their representations, pp. 21–64. Halsted, New York (1975)
2. Bott, R.: Homogeneous vector bundles. Ann. of Math. **66**, 203–248 (1957)
3. Brundan, J.: Tilting modules for Lie superalgebras. Comm. Algebra **32**, 2251–2268 (2004)
4. Brundan, J., Stroppel, C.: Highest weight categories arising from Khovanov's diagram algebra. II. Koszulity. Transform. Groups. **15**, 1–45 (2010)
5. Cheng, S-J., Kwon, J-H., Lam, N.: A BGG-type resolution for tensor modules over general linear superalgebra. Lett. Math. Phys. **84**, 75–87 (2008)
6. Cheng, S-J., Wang, W.: Dualities and representations of Lie superalgebras. Graduate Studies in Mathematics, 144. American Mathematical Society, Providence, RI (2012)
7. Coulembier, K.: Bernstein-Gelfand-Gelfand resolutions for basic classical Lie superalgebras. Accepted in J. Algebra
8. Coulembier, K., Zhang, R.B.: Bott-Borel-Weil theory and algebra of functions for Lie superalgebras. In preparation
9. Gorelik, M., Lanzmann, E.: The minimal primitive spectrum of the enveloping algebra of the Lie superalgebra osp(1,2l). Adv. Math. **154**, 333–366 (2000)
10. Gruson, C., Serganova, V.: Cohomology of generalized supergrassmannians and character formulae for basic classical Lie superalgebras. Proc. Lond. Math. Soc. **101**, 852–892 (2010)
11. Humphreys, J.E.: Representations of semisimple Lie algebras in the BGG category O. Graduate Studies in Mathematics, 94. American Mathematical Society, Providence, RI (2008)
12. Kac, V.G.: Lie superalgebras. Adv. Math. **26**, 8–96 (1977)
13. Kostant, B.: Lie algebra cohomology and the generalized Borel-Weil theorem. Ann. of Math. (2) **74**, 329–387 (1961)
14. Mazorchuk, V.: Some homological properties of the category $\mathcal{O}$. Pacific J. Math. **232**, 313–341 (2007)
15. Mazorchuk, V.: Parabolic category $\mathcal{O}$ for classical Lie superalgebras. In: Gorelik, M., Papi. P. (eds.) Springer INdAM Series 7. Springer International Publishing, Cham (2014)
16. Musson, I.M.: The enveloping algebra of the Lie superalgebra osp(1,2r). Represent. Theory **1**, 405–423 (1997)
17. Musson, I.M.: Lie superalgebras and enveloping algebras. Graduate Studies in Mathematics, vol. 131. American Mathematical Society, Providence, RI (2012)
18. Penkov, I.: Borel-Weil-Bott theory for classical Lie supergroups. J. Soviet Math. **51**, 2108–2140 (1990)
19. Penkov, I., Serganova, V.: Cohomology of G/P for classical complex Lie supergroups G and characters of some atypical G-modules. Ann. Inst. Fourier (Grenoble) **39**, 845–873 (1989)
20. Rittenberg, V., Scheunert, M.: A remarkable connection between the representations of the Lie superalgebras osp(1,2n) and the Lie algebras o(2n+1). Comm. Math. Phys. **83**, 1–9 (1982)
21. Weibel, C.A.: An introduction to homological algebra. Cambridge University Press, Cambridge (1994)
22. Zhang, R.B.: Quantum superalgebra representations on cohomology groups of non-commutative bundles. J. Pure Appl. Algebra **191**, 285–314 (2004)

Finiteness and orbifold vertex operator algebras

Alessandro D'Andrea

Abstract In this paper, I investigate the ascending chain condition of right ideals in the case of vertex operator algebras satisfying a finiteness and/or a simplicity condition. Possible applications to the study of finiteness of orbifold VOAs are discussed.

1 Introduction

It has been observed in many instances, see [10] and references therein, that a strong finiteness condition on a (simple) vertex operator algebra, or VOA, is inherited by subalgebras of invariant elements under the action of a reductive (possibly finite) group of automorphisms. This amounts to a quantum version of Hilbert's basis theorem for finitely generated commutative algebras, but is typically dealt with, in the relevant examples, by means of invariant theory.

A big issue that needs to be addressed in all attempts towards proving the above statement in a general setting is its failure in the trivial commutative case. A commutative vertex algebra is nothing but a commutative differential algebra, and it has long been known that both the noetherianity claim contained in Hilbert's basis theorem, and the finiteness property of invariant subalgebras, cannot hold for differential commutative algebras. Counterexamples are easy to construct, and a great effort has been spent over the years into finding the appropriate generalization of differential noetherianity. Every investigation of finiteness of vertex algebras must first explain the role played by noncommutativity and its algebraic consequences.

In this paper, I announce some results in this direction, and claim that every strongly finitely generated simple vertex operator algebra satisfies the ascending chain condition on its right ideals. Here, a VOA is simple if it has no nontrivial quotient VOAs,

A. D'Andrea (✉)
Dipartimento di Matematica, Università di Roma "La Sapienza", P. le Aldo Moro 5, 00185 Roma, Italy
e-mail: dandrea@mat.uniroma1.it

M. Gorelik, P. Papi (eds.): *Advances in Lie Superalgebras.* Springer INdAM Series 7,
DOI 10.1007/978-3-319-02952-8_3, © Springer International Publishing Switzerland 2014

whereas the right ideals involved in the ascending chain conditions are subspaces that are stable under both derivation and right multiplication with respect to the normally ordered product; even a simple VOA may have very many ideals of this sort, and they are better suited for addressing finiteness conditions. Right noetherianity of simple VOAs is the first algebraic property, as far as I know, that can be proved on a general level, and explains an important difference between the commutative and noncommutative situation.

The paper is structured as follows: in Sects. 2 and 3, I rephrase the vertex algebra structure in the context of left-symmetric algebras, and describe how the normally ordered product and the singular part of the Operator Product Expansion relate to each other. In Sects. 4 and 5, I recall the concept of strong generators for a VOA, and explain its interaction with Li's filtration [9], and its generalization to structures that are weaker than proper VOAs. Section 6 explains the role of what I call *full ideals* into proving some version of noetherianity for a VOA. Speculations on how to use noetherianity in order to address strong finiteness of invariant subalgebras of a strongly finitely generated VOA are given in Sect. 7. I thankfully acknowledge Victor Kac for his suggestion that Lemma 1 might be useful in the study of finiteness of orbifold VOAs.

2 What is a vertex operator algebra?

2.1 Left-symmetric algebras

A *left-symmetric algebra* is a (complex) vector space A endowed with a bilinear product $\cdot : A \otimes A \to A$ which is in general neither commutative nor associative. The associator $(a,b,c) = (ab)c - a(bc)$ must however satisfy the following left-symmetry axiom:

$$(a,b,c) = (b,a,c),$$

for every choice of $a,b,c \in A$. One may similarly define *right-symmetric algebras* by requiring that $(a,b,c) = (a,c,b)$. Clearly, an associative algebra is both left- and right-symmetric. If A is any (non-commutative, non-associative) algebra, the commutator $[a,b] = ab - ba$ satisfies

$$[a,[b,c]] =$$
$$[[a,b],c] + [b,[a,c]] + (b,a,c) - (a,b,c) - (c,a,b) + (a,c,b) + (c,b,a) - (b,c,a),$$

for all $a,b,c \in A$. When A is either left- or right-symmetric, this reduces to the ordinary Jacobi identity

$$[a,[b,c]] = [[a,b],c] + [b,[a,c]],$$

and the commutator thus defines a Lie bracket on A. In a left-symmetric algebra, commutativity implies associativity, as

$$(a,b,c) = [c,a]b + a[c,b] - [c,ab]. \tag{1}$$

A similar identity holds in the right-symmetric case.

2.2 *Differential graded left-symmetric algebras*

A *differential graded left-symmetric algebra* (henceforth, a *DGLsA*) is a non-negatively graded vector space $A = \oplus_{n\geq 0} A^n$, endowed with a unital left-symmetric product $\cdot : A \otimes A \to A$, and a derivation $\partial : A \to A$, satisfying:

- $\mathbf{1} \in A^0$;
- $A^m \cdot A^n \subset A^{m+n}$;
- $\partial A^n \subset A^{n+1}$.

Throughout the paper, we will assume all A^n to be finite-dimensional vector spaces.

Example 1 Let $A = \mathbb{C}[x]$, and set $\partial = x^2 d/dx$. If we choose x to have degree 1, then A is a differential graded commutative algebra, hence also a DGLsA.

2.3 *Lie conformal algebras*

A *Lie conformal algebra* is a $\mathbb{C}[\partial]$-module L endowed with a λ-bracket

$$R \otimes R \ni a \otimes b \mapsto [a_\lambda b] \in R[\lambda]$$

satisfying

- $[\partial a_\lambda b] = -\lambda [a_\lambda b], \qquad [a_\lambda \partial b] = (\partial + \lambda)[a_\lambda b];$
- $[a_\lambda b] = -[b_{-\partial-\lambda} a];$
- $[a_\lambda [b_\mu c]] - [b_\mu [a_\lambda c]] = [[a_\lambda b]_{\lambda+\mu} c],$

whenever $a, b, c \in R$. Lie conformal algebras have been introduced in [8] and studied in [5] in order to investigate algebraic properties of local families of formal distributions. This notion, and its multi-variable generalizations [1], are deeply related to linearly compact infinite-dimensional Lie algebras and their representation theory.

2.4 *Vertex algebras*

Let V be a complex vector space. A *field* on V is defined as a formal power series $\phi(z) \in (\operatorname{End} V)[[z, z^{-1}]]$ with the property that $\phi(z)v \in V((z)) = V[[z]][z^{-1}]$, for every $v \in V$. In other words, if

$$\phi(z) = \sum_{i \in \mathbb{Z}} \phi_i z^{-i-1}$$

then $\phi_n(v) = 0$ for sufficiently large n.

A *vertex algebra* is a (complex) vector space V endowed with a linear *state-field correspondence* $Y : V \to (\operatorname{End} V)[[z, z^{-1}]]$, a *vacuum element* $\mathbf{1}$ and a linear *(infinitesimal) translation operator* $\partial \in \operatorname{End} V$ satisfying the following properties:

- **Field axiom.** $Y(v, z)$ is a field for all $v \in V$.
- **Locality.** For every $a, b \in V$ one has

  $$(z-w)^N [Y(a,z), Y(b,w)] = 0$$

  for sufficiently large N.

- **Vacuum axiom.** The vacuum element **1** is such that

$$Y(\mathbf{1},z) = \mathrm{id}_V, \qquad Y(a,z)\mathbf{1} \equiv a \mod zV[[z]],$$

for all $a \in V$.
- **Translation invariance.** ∂ satisfies

$$[\partial, Y(a,z)] = Y(\partial a, z) = \frac{d}{dz}Y(a,z),$$

for all $a \in V$.

One usually writes

$$Y(a,z) = \sum_{j\in\mathbb{Z}} a_{(j)}z^{-j-1}.$$

and views the $\mathbb{C}$-bilinear maps $a \otimes b \mapsto a_{(j)}b, j \in \mathbb{Z}$, as products describing the vertex algebra structure. The normally ordered product $ab = :ab: = a_{(-1)}b$ determines all negatively labeled products as

$$j!a_{(-j-1)}b = (\partial^j a)_{(-1)}b.$$

Non-negatively labeled products can be grouped in a generating series

$$[a_\lambda b] = \sum_{n\geq 0} \frac{\lambda^n}{n!}a_{(n)}b,$$

which is showed to define a Lie conformal algebra structure. The compatibility conditions between the normally ordered product and the λ-bracket are well understood [2, 6], and amount to imposing quasi-commutativity

$$[a,b] = \int_{-\partial}^{0} d\lambda\, [a_\lambda b], \tag{2}$$

and the noncommutative Wick formula

$$[a_\lambda bc] = [a_\lambda b]c + b[a_\lambda c] + \int_0^\lambda d\mu\, [[a_\lambda b]_\mu c]. \tag{3}$$

As a consequence, the normally ordered product may fail to be associative. The associator $(a,b,c) := (ab)c - a(bc)$ can be expressed in the form

$$(a,b,c) = \left(\int_0^\partial d\lambda\, a\right)[b_\lambda c] + \left(\int_0^\partial d\lambda\, b\right)[a_\lambda c], \tag{4}$$

hence it satisfies $(a,b,c) = (b,a,c)$. V is therefore a left-symmetric algebra with respect to its normally ordered product. Because of (1) and (2), one obtains commutativity and associativity of the normally ordered product as soon as the λ-bracket vanishes. The operator ∂ is a derivation of all products. As the normally ordered product is non-associative, we will denote by $: a_1a_2\ldots a_n :$ the product $a_1(a_2(\ldots(a_{n-1}a_n)\ldots))$ obtained by associating on the right.

2.5 Vertex operator algebras

In this paper, a *vertex operator algebra* (henceforth, a *VOA*) is a non-negatively graded vector space $V = \bigoplus_{n\geq 0} V^n$, endowed with a vertex algebra structure such that

- the normally ordered product and translation operator ∂ make V into a DGLsA;
- $\mathrm{Tor}(V) = V^0 = \mathbb{C}\mathbf{1}$;
- there exists a *Virasoro element* — i.e., an element $\omega \in V^2$ satisfying
$$[\omega_\lambda \omega] = (\partial + 2\lambda)\omega + \frac{c}{12}\lambda^3 \mathbf{1},$$
for some $c \in \mathbb{C}$ — such that $[\omega_\lambda a] = (\partial + n\lambda)a + O(\lambda^2)$, for all $a \in V^n$.

As a consequence, $V^i{}_{(n)}V^j \subset V^{i+j-n-1}$, $\partial V^i \subset V^{i+1}$. By $\mathrm{Tor}\,V$, I mean the torsion of V when viewed as a $\mathbb{C}[\partial]$-module.

3 Interaction between normally ordered product and λ-bracket

As the structure of a vertex algebra is described by the normally ordered product, along with the λ-bracket, it is interesting to figure out how much either product determines the other.

3.1 The normally ordered product of a VOA determines the λ-bracket

We know that the λ-bracket of elements in a vertex algebra V is polynomial in λ, and determines the commutator as in (2). If we choose elements $c_j \in V$ so that

$$[a_\lambda b] = \sum_{j=0}^{n} \lambda^j c_j,$$

then we may compute

$$[\partial^i a, b] = (-1)^i \cdot \sum_{j=0}^{n} \int_{-\partial}^{0} \lambda^{i+j} c_j d\lambda = \sum_{j=0}^{n} (-1)^j \frac{\partial^{i+j+1} c_j}{i+j+1},$$

hence

$$\partial^{n-i}[\partial^i a, b] = \sum_{j=0}^{n} \frac{(-1)^j}{i+j+1} \cdot \partial^{n+j+1} c_j. \tag{5}$$

As soon as we are knowledgeable about the normally ordered product of the vertex algebra V, we are able to compute the left-hand side of (5) for every $i = 0, \ldots, n$; as coefficients of the right-hand sides form a non-degenerate matrix, we can then solve (5) as a system of linear equations, and recover uniquely the values of $\partial^{n+j+1} c_j$, $j = 0, \ldots, n$. In other words, the normally ordered product determines each coefficient c_j up to terms killed by ∂^{n+j+1}.

We have already seen that every VOA is a DGLsA with respect to its normally ordered product.

Theorem 1 *A DGLsA structure may be lifted to a VOA structure in at most one way.*

Proof. It is enough to show that the normally ordered product uniquely determines the λ-bracket. Let $a \in V^h, b \in V^k$. Then $[a_\lambda b]$ is a polynomial in λ of degree at most $n = h+k-1$. Proceeding as above, we may determine all of its coefficients up to terms killed by some power of ∂. However, $\operatorname{Tor} V = \mathbb{C}\mathbf{1}$, so $[a_\lambda b]$ is determined up to multiples of $\mathbf{1}$. By (4), we have

$$(a,u,b) = \left(\int_0^\partial d\lambda\, a\right)[u_\lambda b] + \left(\int_0^\partial d\lambda\, u\right)[a_\lambda b].$$

If we choose u so that u, a are $\mathbb{C}[\partial]$-linearly independent, we may now determine unknown central terms in $[a_\lambda b]$.

Such a choice of u is always possible, as we may assume without loss of generality that $a \notin \operatorname{Tor} V$, otherwise $[a_\lambda b] = 0$ [5]; we may also assume that V has rank at least two, otherwise, if a is non-torsion, unknown central terms in $[a_\lambda a]$ can be computed using

$$(a,a,a) = 2\left(\int_0^\partial d\lambda\, a\right)[a_\lambda a].$$

The value of $[a_\lambda a]$ now uniquely determines the Lie conformal algebra structure.

3.2 The λ-bracket determines vertex algebra ideals

If A and B are subsets of V, define products

$$\begin{aligned} A \cdot B &= \operatorname{span}_{\mathbb{C}}\langle a_{(n)} b \,|\, a \in A, b \in B, n \in \mathbb{Z}\}, \\ [\![A,B]\!] &= \operatorname{span}_{\mathbb{C}}\langle a_{(n)} b \,|\, a \in A, b \in B, n \geq 0\}. \end{aligned}$$

If B is a $\mathbb{C}[\partial]$-submodule of V, then $A \cdot B$, $[\![A,B]\!]$ are also $\mathbb{C}[\partial]$-submodules. If A, B are both $\mathbb{C}[\partial]$-submodules of V, then $A \cdot B = B \cdot A$, $[\![A,B]\!] = [\![B,A]\!]$. A $\mathbb{C}[\partial]$-submodule $I \subset V$ is a *vertex algebra ideal* if $I \cdot V \subset I$; it is a *Lie conformal algebra ideal* if $[\![I,V]\!] \subset I$. An element $a \in V$ is *central* if $[\![a,V]\!] = 0$.

Lemma 1 ([3,4]) *If $B, C \subset V$ are $\mathbb{C}[\partial]$-submodules, then $[\![A,B]\!] \cdot C \subset [\![A, B \cdot C]\!]$. In particular, if X is a subset of V, then $[\![X,V]\!]$ is an ideal of V.*

This observation has an immediate drawback: every vertex algebra V is in particular a Lie conformal algebra. If I is an ideal of this Lie conformal algebra structure, then $J = [\![I,V]\!] \subset I$ is an ideal of the vertex algebra V, which is certainly contained in I. The induced λ-bracket on the quotient I/J is trivial. We may rephrase this by saying that every Lie conformal algebra ideal of V sits centrally on a vertex algebra ideal. In the case of a VOA, a stronger statement holds:

Theorem 2 *Let V be a VOA. A subspace $\mathbf{1} \notin I \subset V$ is an ideal for the vertex algebra structure if and only if it is an ideal of the underlying Lie conformal algebra.*

Proof. The grading of V is induced by the Virasoro element ω. We know that $[\![I,\omega]\!] \subset [\![I,V]\!] \subset I$, hence I must contain all homogeneous components of each of its elements. However, if $a \in V$ is a homogeneous element (of nonzero degree), then $a \in [\![a,\omega]\!]$. This forces I to equal $[\![I,V]\!]$, which is a vertex algebra ideal.

Remark 1 Notice that $\mathbb{C}\mathbf{1}$ is always a Lie conformal algebra ideal of V, but is never an ideal of the vertex operator algebra structure.

3.3 Different notions of ideal in a vertex algebra

A vertex algebra structure is made up of many ingredients, that may stand by themselves to provide meaningful concepts. In particular, a vertex algebra is naturally endowed with both a (differential) left-symmetric product, and a Lie conformal algebra structure, and we may consider ideals with respect to each of the above structures. To sum it up, we have

- *vertex algebra ideals*: ideals of the vertex algebra structure — closed under ∂, $:ab:$, $[a_\lambda b]$;
- *Lie conformal algebra ideals:* ideals of the Lie conformal algebra structure — closed under $\partial, [a_\lambda b]$;
- *DLs ideals:* ideals of the differential left-symmetric structure — closed under ∂, $:ab:$.

When V is a VOA, we have seen that the first two notions (more or less) coincide. In what follows, we will mostly be concerned with simple VOAs, i.e., VOAs with no nontrivial vertex ideals. Notice that even a simple VOA does possess many DLs ideals. Both the normally ordered product and the differential ∂ increase the grading, so that if $a \in V^h$, then the DLs ideal generated by a is contained in $\oplus_{n\geq h} V^n$.

We conclude that the only nontrivial concept in a simple VOA is that of DLs ideal; thus, the term *ideal* will henceforth refer to DLs ideals alone. Notice that we may distinguish between left, right and two-sided ideals, whereas vertex algebra and Lie conformal algebra ideals are always two-sided.

4 Finiteness of VOAs

4.1 Strong generators of a VOA

When dealing with finiteness of vertex algebras, the notion that has naturally emerged in the (both mathematical and physical) literature depends only on the (differential) left-symmetric algebra structure. A vertex algebra V is called *strongly finitely generated* if there exists a finite set of generators such that normally ordered products of derivatives of the generators $\mathbb{C}$-linearly span V; this is equivalent to being able to choose finitely many quantum fields so that every element of V can be obtained from the vacuum state by applying a suitable polynomial expression in the corresponding creation operators. This definition makes no reference whatsoever to the λ-bracket; when dealing with finiteness phenomena it is natural to only resort to concepts that are independent of the Lie conformal algebra structure.

4.2 Hilbert's Basissatz and the fundamental theorem of invariant theory

If $A = \oplus_{n\geq 0} A^n$ is a finitely generated commutative associative unital graded algebra, and G is a reductive group acting on A by graded automorphisms, then the subalgebra A^G of G-invariants is also finitely generated. Hilbert's celebrated proof of this fact uses noetherianity of A in an essential way: if I is the ideal of A generated by the positive degree part A^G_+ of A^G, then any finite subset of A^G_+ generating I as an ideal is also a finite set of generators of A^G as an algebra.

4.3 Does the orbifold construction preserve finiteness of a VOA?

It is natural to ask whether Hilbert's strategy can be extended to the wider setting of VOAs. Indeed, the mathematical and physical literature provide scattered example of strongly finitely generated (simple) VOAs for which the invariant subalgebra relative to the action of a reductive group of graded automorphisms stays strongly finitely generated. However, no general argument is known that applies to all examples.

A major difficulty in understanding the general algebraic aspect of the above phenomena depend on its failure in commutative examples. We have seen that a commutative VOA is nothing but a differential commutative associative algebra. However, it is not difficult to provide examples of differentially finitely generated commutative associative algebras whose invariant part with respect to the action of a finite group of graded automorphisms does not stay finitely generated.

The strongly finite generatedness of invariant subalgebra does therefore depend on noncommutative quantum features, and any attempt to provide a general proof must address the problem of understanding why the commutative case behaves so differently.

4.4 Failure of noetherianity in the differential commutative setting and non-finiteness of invariant subalgebras

Consider the commutative ring $A = \mathbb{C}[u^{(n)}, n \in \mathbb{N}]$ of polynomials in the countably many indeterminates $u^{(n)}$. Setting $\partial u^{(n)} = u^{(n+1)}$ uniquely extends to a derivation of A, thus making it into a differential commutative algebra.

Consider now the unique differential automorphism σ of A satisfying $\sigma(u) = -u$. Then clearly $\sigma(u^{(n)}) = -u^{(n)}$ and $\sigma(u^{(n_1)} \dots u^{(n_h)}) = (-1)^h u^{(n_1)} \dots u^{(n_h)}$. It is not difficult to see that $A^{\langle\sigma\rangle} = \mathbb{C}[u^{(i)}u^{(j)}, i, j \in \mathbb{N}]$. However, $A^{\langle\sigma\rangle}$ admits no finite set of differential algebra generators.

Remark 2 If we endow A with a trivial λ-bracket, then A is an example of a commutative vertex algebra. Notice that setting $\deg(u^{(n)}) = n+1$ provides A with a grading compatible with the vertex algebra structure. However, A is not a VOA as there is no Virasoro element inducing this grading.

It is easy to adapt Hilbert's argument to the differential commutative setting *once* noetherianity is established. An inevitable consequence of the above counterexample is that the differential commutative algebra A must fail to satisfy the ascending chain condition on differential ideals. This fact has long been known [11], and effort has been put into providing some weaker statement replacing and generalizing noetherianity. We recall the following classical result:

Theorem 3 (Ritt) *Let A be finitely generated as a differential commutative K-algebra, where K is a field of characteristic zero. Then A satisfies the ascending chain condition on its* **radical** *differential ideals.*

In Ritt's language, radical differential ideals are *perfect*, and generators of a perfect ideal as an ideal (resp. as a perfect ideal) are called strong (resp. weak) generators. The above statement claims that all perfect ideals have a finite set of weak generators, but they may well fail to have a finite set of strong generators.

Under a different meaning of weak vs. strong generators, this difference of finiteness property shows up again in the context of VOAs.

5 An abelianizing filtration for VOAs

The problem of finding strong generators for a VOA can be addressed by using a decreasing abelianizing filtration introduced[1] in [9]. We recall here (a slight variant of) its definition and some of its main properties. In what follows, if $X, Y \subset A$ are subsets, we will set $AB = \text{span}_{\mathbb{C}}\langle ab | a \in A, b \in B\rangle$. Notice that $AB \neq A \cdot B$, in general.

5.1 Li's filtration

If A is a DGLsA, set $E_n(A), n \in \mathbb{N}$, to be the linear span of all products (with respect to all possible parenthesizations)

$$\partial^{d_1} a_1 \, \partial^{d_2} a_2 \dots \partial^{d_h} a_h,$$

where $a_i \in A$ are homogeneous elements, and $d_1 + \dots d_h \geq n$. Also set $E_n(A) = A$ if $n < 0$. The $E_i(A), i \in \mathbb{N}$, form a decreasing sequence

$$A = E_0(A) \supset E_1(A) \supset \dots \supset E_n(A) \supset \dots$$

of subspaces of A, and clearly satisfy

$$E_i(A)E_j(A) \subset E_{i+j}(A); \tag{6}$$
$$\partial E_i(A) \subset E_{i+1}(A). \tag{7}$$

[1] Li's setting is more general than ours, as the grading is only assumed to be bounded from below.

In particular, each $E_i(A)$ is an ideal of A. If $a \in E_i(A) \setminus E_{i+1}(A)$, then we will say that a has rank i, and will denote by $[a]$ the element $a + E_{i+1}(A) \in E_i(A)/E_{i+1}(A)$.

Lemma 2 *If V is a VOA, then $[E_i(V), E_j(V)] \subset E_{i+j+1}(V)$ for all i, j.*

Proof. Follows immediately from (2).

Proposition 1 *Let V be a VOA. Then $[a][b] = [ab], \partial[a] = [\partial a]$ make*

$$grV = \oplus_{i \geq 0} E_i(V)/E_{i+1}(V)$$

into a graded commutative (associative) differential algebra.

Proof. Well definedness of the product is clear. Its commutativity follows from Lemma 2. By (1), associativity follows from commutativity and left-symmetry of the product in V. Finally, ∂ is well-defined, and its derivation property descends to the quotient.

Remark 3 Li proves that, if V is a VOA, then $\mathrm{gr}V$ can be endowed with a Poisson vertex algebra structure [7]. However, we will not need this fact.

Theorem 4 (Li) *Let X be a subset of homogeneous elements of a VOA V. Then X strongly generates V if and only if elements $[x], x \in X$, generate grV as a differential commutative algebra.*

In other words, a VOA V is strongly finitely generated if and only if $\mathrm{gr}V$ is finitely generated as a differential commutative algebra.

5.2 Strong generators of ideals

The problem of finding strong generators for a VOA is closely connected to that of finding nice sets of generators for its ideals.

Recall that, if A is a DGLsA, $I \subset A$ is a *(two-sided, right) ideal* of A if it is a (two-sided, right) homogeneous differential ideal. We denote by $(X))$ the smallest right ideal of A containing a given subset $X \subset A$, and similarly, by $((X))$, the smallest two-sided ideal containing X. A subspace $U \subset A$ is *strongly generated* by $X \subset U$ if $U = (\mathbb{C}[\partial]X)A$. When dealing with strongly generated ideals, we will henceforth abuse notation and write XA for $(\mathbb{C}[\partial]X)A$.

We rephrase another of Li's results as follows

Theorem 5 *Let I be a right ideal of a VOA V. Then grI is a (differential) ideal of grV, and $X \subset V$ strongly generates I if and only if $[x], x \in X$, generate grI as a differential ideal of grV.*

We can easily apply this statement to elements of Li's filtration.

Proposition 2 *Let X be a set of homogeneous generators of a VOA V. Then $E_d(V)$ is strongly generated by monomials*

$$:(\partial^{d_1}x_1)(\partial^{d_2}x_2)\dots(\partial^{d_{h-1}}x_{h-1})(\partial^{d_h}x_h):,$$

where $x_i \in X$, and $d_i > 0$ satisfy $d_1 + \cdots + d_h = d$. In particular, if V is finitely generated, then $E_d(V)$ is a strongly finitely generated ideal.

Proof. It follows immediately by noticing that $E_n(V)/E_{n+1}(V)$ is linearly generated by classes of monomials

$$:(\partial^{d_1}x_1)(\partial^{d_2}x_2)\dots(\partial^{d_{h-1}}x_{h-1})(\partial^{d_h}x_h):,$$

where $x_i \in X$, and $d_i \geq 0$ satisfy $d_1+\dots+d_h = n$.

5.3 Weak vertex algebras

In order to construct and use Li's filtration, we do not need the full power of VOAs. Indeed, the $E_i(A)$ always constitute a decreasing filtration of the DGLsA A and satisfy (6), (7). In order to show that $\mathrm{gr}A$ is commutative and associative, we also need

$$[E_i(A), E_j(A)] \subset E_{i+j+1}(A). \tag{8}$$

This certainly holds in VOAs, but stays true under weaker conditions.

Definition 1 A *weak VOA* is a DGLsA $A = \oplus_{i\geq 0}A^i$ satisfying (8).

Example 2

- Every non-negatively graded differential commutative (associative) algebra is a weak vertex operator algebra.
- Every VOA is a weak vertex operator algebra.
- Let V be a VOA, $I \subset V$ a two-sided ideal. Then V/I is a weak vertex operator algebra: indeed, V/I is a DGLsA and constructing Li's filtration commutes with the canonical projection. Notice that V/I fails to be a VOA, unless I is a vertex algebra ideal.

If A is a weak VOA, then $\partial^d A \subset \oplus_{i\geq d}A^i$, hence $E_n(A) \subset \oplus_{i\geq n}A^i$. Consequently, $\cap_n E_n(A) = (0)$, and $E_i(A)\cap A^j = (0)$ as soon as $i > j$. Propositions 1, 2 and Theorems 4, 5 easily generalize to the weak VOA setting.

Chains of inclusions between ideals in a weak VOA also behave nicely, due to the following observation:

Lemma 3 *Let $I \subset J$ be right ideals of a weak VOA A satisfying $grI = grJ$. Then $I = J$.*

Proof. If $X \subset I$ generates $\mathrm{gr}I$ as an ideal of $\mathrm{gr}A$, then it also generates $\mathrm{gr}J$, hence $I = J = XA$.

6 The ascending chain condition in a VOA

6.1 Full ideals

Definition 2 Let I be a right ideal of a VOA V. Then I is *full* if $E_N(V) \subset I$ for sufficiently large values of N.

Full ideals are important because of the following key observation.

Theorem 6 *Let V be a strongly finitely generated VOA, $I \subset V$ a full right ideal. Then I is a strongly finitely generated ideal.*

Proof. As I is full, it contains $E_N(V)$ for some $N \geq 0$. Then $\bar{I} = I/E_N(V)$ is an ideal of the quotient weak VOA $\bar{V} = V/E_N(V)$.

Notice that if $u_1, \ldots, u_n$ are (strong) generators of V, then $\bar{u}_1, \ldots, \bar{u}_n$ generate $\bar{V}$, hence elements $[\bar{u}_i]$ generate $\mathrm{gr}\,\bar{V}$ as a differential commutative associative algebra. However, only finitely many derivatives of each $[\bar{u}_i]$ are nonzero. Therefore, $\mathrm{gr}\,\bar{V}$ is a finitely generated, and not just differentially finitely generated, commutative algebra. By Hilbert's basis theorem, the ideal $\mathrm{gr}\,\bar{I}$ is finitely generated, and we may apply the weak VOA version of Theorem 5 to show that I is strongly finitely generated modulo some $E_N(V)$. However, Proposition 2 shows that all ideals $E_N(V)$ are strongly finitely generated, hence I is so too.

By using a variant of the argument in Sect. 3.1, one is able to prove the following statement.

Lemma 4 *Let I be a right ideal of the VOA V. Then I is full as soon as any one of the following properties is satisfied*

- *I is nonzero and V is a simple VOA;*
- *I contains some derivative of the Virasoro element ω, provided that the central charge is nonzero;*
- *I is two-sided, and contains some derivative of the Virasoro element ω.*

6.2 Noetherianity

Proposition 3 *Let V be a finitely generated VOA. Then V satisfies the ascending chain condition on its full right ideals.*

Proof. If

$$I_1 \subset I_2 \subset \ldots \subset I_n \subset I_{n+1} \subset \ldots$$

is an ascending sequence of full right ideals, set $I = \cup_n I_n$. Then I is a full ideal, and we may use Theorem 6 to locate a finite $X \subset I$ such that $I = XV$. Due to finiteness of X, one may find $N \geq 0$ such that $X \subset I_N$. Then $I = XV \subset I_N$.

All the following statements are now of immediate proof.

Theorem 7 *Every simple VOA satisfies the ascending chain condition on its right ideals.*

Theorem 8 *Let V be a VOA, $X \subset V$ a subset containing $\partial^i \omega$ for some $i \geq 0$. Then there exists a finite subset $X_0 \subset X$ such that $((X)) = ((X_0))$.*

Theorem 9 *Let V be a simple VOA, $X \subset V$. Then there exists a finite subset $X_0 \subset X$ such that $(X)) = (X_0))$.*

We may rephrase Theorem 7 by saying that every simple finitely generated VOA is right-noetherian.

Remark 4 Notice that, unless V is associative (e.g., when V is commutative), subspaces of the form XV may fail to be right ideals, so the above reasoning **does not** prove that if

$$X_1 \subset X_2 \subset \ldots \subset X_n \subset X_{n+1} \subset \ldots$$

is an increasing family of subsets, then the corresponding sequence

$$X_1V \subset X_2V \subset \ldots \subset X_nV \subset X_{n+1}V \subset \ldots$$

stabilizes. In other words, we do not know whether a simple fintiely generated VOA must satisfy the ascending chain condition **also** on its subspaces of the form XV.

Remark 5 Finite generation of every right ideal I in a simple finitely generated VOA V is a strong claim. However, one often needs a stronger statements which may easily fail.

Say that $I = (X))$ or even $I = XV$. Then it is true that one may find a finite subset $X_0 \subset I$ such that $I = X_0V$, but there is no clear way to force $X_0 \subset X$. The standard proof of this fact would require the ascending chain condition in the stronger form stated above.

7 Speculations on Hilbert's approach to finiteness in the VOA orbifold setting

7.1 Subspaces of the form XV

Let a, b be elements of a VOA V. Then (4) shows that $(a,b,c) \in aV + bV$ for every $c \in V$. However, $(a,b,c) = (ab)c - a(bc)$; as $a(bc) \in aV$, then $(ab)c \in aV + bV$ for all $c \in V$. We can summarize this in the following statement:

Lemma 5 *Let V be a VOA, $X \subset V$ a collection of homogeneous elements not containing* **1**. *Then $XV = \langle X \rangle_+ V$.*

Proof. It is enough to show that if u is a product of (derivatives) of elements from X, then $uV \subset XV$. This follows from the previous lemma and an easy induction on the number of terms in the product.

Proposition 4 *Let $U \subset V$ be VOAs, $X \subset V$ a collection of homogeneous elements not containing* **1**. *Then*

$$X \text{ strongly generates } U \implies U_+ \subset XV \implies U_+ \subset XV + VX.$$

The above implications can be reversed for certain classes of subalgebras.

7.2 Split subalgebras

Let $U \subset V$ be VOAs.

Definition 3 U is a *split subalgebra* of V if there exists a graded $\mathbb{C}[\partial]$-submodule decomposition $V = U \oplus M$ such that $UM \subset M$.

Whenever U is a split subalgebra of V, there exists a $\mathbb{C}[\partial]$-linear splitting $R : V \to U$ which is a homomorphism of U-modules. The splitting clearly satisfies $R^2 = R$, and $R(uv) = uR(v), R(vu) = R(v)u$ for every $u \in U, v \in V$.

Example 3 If G is a reductive group acting on the finitely generated VOA V by graded automorphisms, then V^G is a split subalgebra of V.

Theorem 10 *Let U be a split subalgebra of the VOA V, $X \subset U$ a collection of homogeneous elements not containing* **1**. *Then*

$$U_+ \subset XV + VX \implies X \text{ strongly generates } U.$$

Proof. Let $u \in U$ be a homogeneous element of positive degree. As we know that $u \in U_+ \subset XV + VX$, then there exist finitely many nonzero elements $r_x^i, s_x^i \in V$, that we may assume homogeneous without loss of generality, such that

$$u = \sum_{x \in X, i \in \mathbb{N}} r_x^i \partial^i x + \partial^i x s_x^i.$$

As $R(u) = u$, then also

$$u = \sum_{x \in X, i \in \mathbb{N}} R(r_x^i) \partial^i x + \partial^i x R(s_x^i).$$

In order to show that u can be expressed as a linear combination of products of elements from X, it is enough to notice that $R(r_x^i), R(s_x^i)$ are homogeneous elements from U of lesser degree than u, and proceed by induction on the degree.

7.3 (Not quite) proving that the VOA orbifold construction preserves finiteness

Let V be a simple finitely generated VOA, G a reductive group acting on V by automorphisms. Then both the following statements hold:

- $(V_+^G)) = (U))$ for some finite set $U \subset V_+^G$;
- $(V_+^G)) = XV$ for some finite set $X \subset (V_+^G))$.

We are however not able to show any of the following increasingly weaker statements

- $(V_+^G)) = XV$ for some finite set $X \subset V_+^G$,
- $V_+^G V = XV$ for some finite set $X \subset V_+^G$,
- $V_+^G V + VV_+^G = XV + VX$ for some finite set $X \subset V_+^G$,

which would suffice to apply Theorem 10 to ensure finiteness of V^G. Such statements depend on a stronger Noetherianity property than we are able to show.

Notice that the above proof of right Noetherianity of a simple finitely generated VOA requires considering nonzero associators, thus resulting in a strictly noncommutative statement. Noncommutative VOAs are however typically nonassociative, and this may prevent subspaces of the form XV from being right ideals.

It is not clear how one should proceed to adapt Hilbert's strategy to the VOA setting. I would like to list a few (bad and good) facts one must necessarily cope with.

- XV can fail to be an ideal of V.
- Furthermore, it is easy to construct examples of $X \subset V$ such that $\mathrm{gr} XV$ is not an ideal of V. The ideal property is likely to fail for subspaces $\mathrm{gr}(XV + VX)$ too.

However the proof of many statements does not require the full strength of ideals:

- $A \subset B, \mathrm{gr} A = \mathrm{gr} B \implies A = B$ holds for subspaces, not just ideals.
- If $\mathbf{1} \in [\![a, b]\!]$, then $aV + Vb$ contains some $\partial^N V$. However this does not seem to guarantee fullness.
- If $X \subset V$ is non-empty, then $(XV)V$ may fail to be an ideal, but is however full.
- If $A \subset V$ is a subspace such that $\mathrm{gr} A$ contains $\mathrm{gr} E_n(V)$, then A contains $E_n(V)$.

It is also possible that strong finite generation of subspaces of the form XV may fail in general, but can be proved in the special case of $X = V^G_+$.

Problem: understand what conditions ensure that a subspace $XV + VX$ contain a nonzero ideal.

References

1. Bakalov, B., D'Andrea, A., Kac, V.G.: Theory of finite pseudoalgebras. Adv. Math. **162**, 1–140 (2001)
2. Bakalov, B., Kac, V.G.: Field algebras. Int. Math. Res. Not. IMRN **3**, 123–159 (2003)
3. D'Andrea, A.: A remark on simplicity of vertex algebras and Lie conformal algebras. J. Algebra **319**(5), 2106–2112 (2008)
4. D'Andrea, A.: Commutativity and associativity of vertex algebras. In: Doebner, H.-D., Dobrev, V.K. (eds.) Lie Theory and its Applications in Physics VII. Heron Press, Sofia. Bulg. J. Phys. **35**(s1), 43–50 (2008)
5. D'Andrea, A., Kac, V.G.: Structure theory of finite conformal algebras. Selecta Math. (N.S.) **4**(3), 377–418 (1998)
6. De Sole, A., Kac, V.G.: Finite vs. affine W-algebras. Jpn. J. Math. **1**, 137–261 (2006)
7. Dong, C., Li, H.-S., Mason, G.: Vertex Lie algebras, vertex Poisson algebras and vertex algebras. Recent developments in infinite-dimensional Lie algebras and conformal field theory (Charlottesville, VA, 2000), pp. 69–96, Contemp. Math. Vol. 297. AMS, Providence, RI (2002)
8. Kac, V.G.: Vertex algebras for beginners. Univ. Lect. Series Vol. 10, AMS (1996). 2nd ed. (1998)
9. Li, H.-S.: Abelianizing vertex algebras. Comm. Math. Phys. **259**(2), 391–411 (2005)
10. Linshaw, A.R.: A Hilbert theorem for vertex algebras. Transform. Groups **15**(2), 427–448 (2010)
11. Ritt, J.F.: Differential algebras. Colloquium Publications Vol. 33. AMS (1950)

On classical finite and affine $\mathscr{W}$-algebras

Alberto De Sole

Abstract This paper is meant to be a short review and summary of recent results on the structure of finite and affine classical $\mathscr{W}$-algebras, and the application of the latter to the theory of generalized Drinfeld-Sokolov hierarchies.

1 Introduction

In Classical (Hamiltonian) Mechanics the phase space, describing the possible configurations of a physical system, is a Poisson manifold M. The physical observables are the smooth functions on M with real values, and they thus form a *Poisson algebra* (PA). The Hamiltonian equations, describing the time evolution of the system, are written in terms of the Poisson bracket: $\frac{du}{dt} = \{h,u\}$, where $h(x) \in C^\infty(M)$ is the Hamiltonian function (corresponding to the energy observable).

When we quantize a classical mechanic theory we go to Quantum Mechanics. The observables become non commutative objects, and the Poisson bracket is replaced by the commutator of these objects. Hence, the physical observables in quantum mechanics form an *associative algebra* (AA) A. The phase space is then described as a representation V of A, and the Schroedinger's equation, describing the evolution of the physical system, is written in terms of this representation: $\frac{d\psi}{dt} = H(\psi)$, where $H \in A$ is the Hamiltonian operator.

Going from a finite to an infinite number of degrees of freedom, we pass from classical and quantum mechanics to classical and quantum field theory respectively. In some sense, the algebraic structure of the space of observables in a conformal field theory is that of a *vertex algebra* (VA) [3], and its quasi-classical limit is known as *Poisson vertex algebra* (PVA) [10].

A. De Sole (✉)
Università di Roma La Sapienza, Piazzale Aldo Moro 5, 00185 Roma, Italy
e-mail: desole@mat.uniroma1.it

M. Gorelik, P. Papi (eds.): *Advances in Lie Superalgebras.* Springer INdAM Series 7,
DOI 10.1007/978-3-319-02952-8_4,

We can summarize the above observations in the following diagram of the algebraic structures of the four fundamental physical theories:

$$
\begin{array}{ccccc}
 & & \overset{quantization}{\cdots\cdots\to} & & \\
 & PVA & \xleftarrow[cl.limit]{} & VA & \\
affiniz. \uparrow & \Big\downarrow Zhu & & Zhu \Big\downarrow & \uparrow affiniz. \\
 & PA & \xleftarrow{cl.limit} & AA & \\
 & & \underset{quantization}{\cdots\cdots\to} & &
\end{array}
\tag{1}
$$

The arrows in the above diagram have the following meaning. If we have a filtered associative algebra, its associated graded is automatically a Poisson algebra called its *classical limit*. Similarly, if we have a filtered vertex algebra, its associated graded is a Poisson vertex algebra. Furthermore, starting from a positive energy vertex algebra (respectively Poisson vertex algebra) we can construct an associative algebra (resp. Poisson algebra) governing its representation theory, known as its *Zhu algebra*, [36]. On the other hand, the processes of going from a classical theory to a quantum theory ("quantization"), or from finitely many to infinitely many degrees of freedom ("affinization"), do not correspond to canonical functors, and they are represented in the diagram with dotted arrows.

$\mathscr{W}$-algebras provide a very rich family of examples, parametrized by a simple Lie algebra $\mathfrak{g}$ and a nilpotent element $f \in \mathfrak{g}$, which appear in all the 4 fundamental aspects in diagram (1):

$$
\begin{array}{ccccc}
 & & \overset{quantization}{\cdots\cdots\to} & & \\
 & \mathscr{W}_z^{cl}(\mathfrak{g},f) & \xleftarrow[cl.limit]{} & \mathscr{W}_k(\mathfrak{g},f) & \\
affiniz. \uparrow & \Big\downarrow Zhu & & Zhu \Big\downarrow & \uparrow affiniz. \\
 & \mathscr{W}^{cl,fin}(\mathfrak{g},f) & \xleftarrow{cl.limit} & \mathscr{W}^{fin}(\mathfrak{g},f) & \\
 & & \underset{quantization}{\cdots\cdots\to} & &
\end{array}
\tag{2}
$$

Each of these classes of algebras was introduced and studied separately, with different applications in mind, and the relations between them became fully clear only later.

Classical finite $\mathscr{W}$-algebras

The classical finite $\mathscr{W}$-algebra $\mathscr{W}^{cl,fin}(\mathfrak{g},f)$ is a Poisson algebra, which can be viewed as the algebra of functions on the so-called *Slodowy slice* $\mathscr{S}(\mathfrak{g},f)$. It was introduced by Slodowy while studying the singularities associated to the coadjoint nilpotent orbits of $\mathfrak{g}$, [33].

Finite $\mathscr{W}$-algebras

The first appearance of the finite $\mathscr{W}$-algebras $\mathscr{W}^{fin}(\mathfrak{g}, f)$ was in a paper of Kostant, [27]. He constructed the finite $\mathscr{W}$-algebra for a principal nilpotent $f \in \mathfrak{g}$ (in which case it is commutative), and proved that it is isomorphic to the center of the universal enveloping algebra $U(\mathfrak{g})$. The construction was then extended in [28] for an even nilpotent element $f \in \mathfrak{g}$. The general definition of finite $\mathscr{W}$-algebras $\mathscr{W}^{fin}(\mathfrak{g}, f)$, for an arbitrary nilpotent element $f \in \mathfrak{g}$, appeared much later, [31]. Starting with the work of Premet, there has been a revival of interest in finite $\mathscr{W}$-algebras in connection to geometry and representation theory of simple finite-dimensional Lie algebras, and the theory of primitive ideals (see [6, 30–32]).

Classical $\mathscr{W}$-algebras

The classical (affine) $\mathscr{W}$-algebras $\mathscr{W}_z^{cl}(\mathfrak{g}, f)$ (depending on the parameter $z \in \mathbb{F}$, where $\mathbb{F}$ is the base field) were introduced, for a principal nilpotent element f, in the seminal paper of Drinfeld and Sokolov [16]. They were introduced as Poisson algebras of function on an infinite dimensional Poisson manifold, and they were used to study KdV-type integrable bi-Hamiltonian hierarchies of PDE's, nowadays known as Drinfled Sokolov hierarchies. Subsequently, in the 90's, there was an extensive literature extending the Drinfeld-Sokolov construction of classical $\mathscr{W}$-algebras and the corresponding generalized Drinfeld-Sokolov hierarchies to other nilpotent elements, [4, 8, 9, 17, 20, 21]. Only very recently, in [13], the classical $\mathscr{W}$-algebras $\mathscr{W}_z^{cl}(\mathfrak{g}, f)$ were described as Poisson vertex algebras, and the theory of generalized Drinfeld-Sokolov hierarchies was formalized in a more rigorous and complete way.

$\mathscr{W}$-algebras

The first (quantum affine) $\mathscr{W}$-algebra which appeared in literature was the so called Zamolodchikov $\mathscr{W}_3$-algebra [35], which is the $\mathscr{W}$-algebra associated to $\mathfrak{sl}_3$ and its principal nilpotent element f. It was introduced as a "non-linear" infinite dimensional Lie algebra extending the Virasoro Lie algebra, describing the symmetries of a conformal filed theory. After the work of Zamolodchikov, a number of papers on affine $\mathscr{W}$-algebras appeared in physics literature, mainly as "extended conformal algebras", i.e. vertex algebra extensions of the Virasoro vertex algebra. A review of the subject up to the early 90's may be found in the collection of a large number of reprints on $\mathscr{W}$-algebras [5]. The most important results of this period are in the work by Feigin and Frenkel [18, 19], where the general construction of $\mathscr{W}$-algebras, via a quantization of the Drinfeld-Sokolov reduction, was introduced in the case of the principal nilpotent element f. For example, if $\mathfrak{g} = s\ell_n$, we get the Virasoro vertex algebra for $n = 2$, and Zamolodchikov's $\mathscr{W}_3$ algebra for $n = 3$. The construction was finally generalized to an arbitrary nilpotent element f in [24–26]. In these paper, $\mathscr{W}$-algebras were applied to representation theory of superconformal algebras.

A complete understanding of the links among the four different appearances of $\mathscr{W}$-algebras in diagram (2) is quite recent. In [22] Gan and Ginzburg described the finite $\mathscr{W}$-algebras as a quantization of the Poisson algebra of functions on the Slodowy slice. They thus proved that the classical finite $\mathscr{W}$-algebra $\mathscr{W}^{cl,fin}(\mathfrak{g},f)$ can be obtained as the classical limit of the finite $\mathscr{W}$-algebra $\mathscr{W}^{fin}(\mathfrak{g},f)$.

As mentioned earlier, the construction of the $\mathscr{W}$-algebra $\mathscr{W}_k(\mathfrak{g},f)$, for a principal nilpotent element f, due to Feigin and Frenkel [18], was obtained as a "quantization" of the Drinfeld-Sokolov construction of the classical $\mathscr{W}$-algebra $\mathscr{W}^{cl}_z(\mathfrak{g},f)$. But it is only in [13] that the classical $\mathscr{W}$-algebra $\mathscr{W}^{cl}_z(\mathfrak{g},f)$ is described as a Poisson vertex algebra which can be obtained as classical limit of the $\mathscr{W}$-algebra $\mathscr{W}_k(\mathfrak{g},f)$.

Furthermore, in [10] it is proved there that the (H-twisted) Zhu algebra $Zhu_H\mathscr{W}_k(\mathfrak{g},f)$ is isomorphic to the corresponding finite $\mathscr{W}$-algebra $\mathscr{W}^{fin}(\mathfrak{g},f)$. Hence, their categories of irreducible representations are equivalent. (This result was independently proved in [1] for a principal nilpotent f.) A similar result for classical $\mathscr{W}$-algebras holds as well. It is also proved in the Appendix of [10] (in collaboration with A. D'Andrea, C. De Concini and R. Heluani) that the quantum Hamiltonian reduction definition of finite $\mathscr{W}$-algebras is equivalent to the definition via the Whittaker models, which goes back to [27].

In the present paper we describe in more detail the "classical" part of diagram (2): in Sect. 2 we describe the Poisson structure of the Slodowy slice and we introduce the classical finite $\mathscr{W}$-algebra $\mathscr{W}^{cl,fin}(\mathfrak{g},f)$. In order to describe its affine analogue,B we first need to describe the classical finite $\mathscr{W}$-algebra $\mathscr{W}^{cl,fin}(\mathfrak{g},f)$ as a Hamiltonian reduction, which is done in Sect. 2.4. By taking the affine analogue of this construction, we obtain the classical $\mathscr{W}$-algebra $\mathscr{W}^{cl}_z(\mathfrak{g},f)$. Finally, in Sect. 3.5 we describe, following [13], how classical $\mathscr{W}$-algebras are used to study the generalized Drinfeld-Sokolov bi-Hamiltonian hierarchies.

2 Classical finite $\mathscr{W}$-algebras

2.1 Poisson manifolds

Recall that, by definition, a *Poisson manifold* is a manifold $M = M^n$ together with a Poisson bracket $\{\cdot,\cdot\}$ on the algebra of functions $C^\infty(M)$, making it a Poisson algebra. By the Leibniz rule, we can write the Poisson bracket as

$$\{f(x),g(x)\} = \sum_{i,j} K_{ij}(x)\frac{\partial f}{\partial x_i}\frac{\partial g}{\partial x_j}.$$

The bivector $\eta = \sum_{i,j} K_{ij}(x)\frac{\partial}{\partial x_i}\wedge\frac{\partial}{\partial x_j} \in \Gamma(\bigwedge^2 TM)$ is the Poisson structure of the manifold. With every function $h \in C^\infty(M)$ on a Poisson manifold M we associate a *Hamiltonian vector field* $X_h = \sum_{i,j=1}^n K(x)_{ij}\frac{\partial h(x)}{\partial x_i}\frac{\partial}{\partial x_j} = \{h,\cdot\}$, and the corresponding *Hamiltonian flow* (or evolution):

$$\frac{dx}{dt} = \{h,x\} = K(x)\nabla_x h. \tag{3}$$

(This is the *Hamiltonian equation* associated to the Hamiltonian function h.) If we start from a point $x \in M$ and we follow all the possible Hamiltonian flows (3) through x, we cover the *symplectic leaf* through x. The Poisson manifold M is then disjoint union of its symplectic leaves: $M = \sqcup_\alpha S_\alpha$ (which are symplectic manifolds).

It is natural to ask when a Poisson structure η on a Poisson manifold M induces a Poisson structure on a submanifold N. Some sufficient condition is given by the following

Proposition 1 (Va94) *Suppose that, for every point $x \in N$, denoting by (S,ω) the symplectic leaf of M through x, we have*

(i) *the restriction of the symplectic form $\omega(x)$ of T_xS to $T_xN \cap T_xS$ is non-degenerate;*

(ii) *N is transverse to S, i.e. $T_xN + T_xS = T_xM$.*

Then, the Poisson structure on M naturally induces a Poisson structure on N, and the symplectic leaf of N through x is $N \cap S$.

If $\mathfrak{g}$ is a Lie algebra, the dual space $\mathfrak{g}^*$ has a natural structure of a Poisson manifold. Indeed, the Lie bracket $[\cdot\,,\cdot]$ on $\mathfrak{g}$ extends uniquely to a Poisson bracket on the symmetric algebra $S(\mathfrak{g})$: if $\{x_i\}_{i=1}^n$ is a basis of $\mathfrak{g}$, we have

$$\{P,Q\} = \sum_{i,j=1}^{n} \frac{\partial P}{\partial x_i}\frac{\partial Q}{\partial x_j}[x_i,x_j]\,. \tag{4}$$

We can think of $S(\mathfrak{g})$ as the algebra of polynomial functions on $\mathfrak{g}^*$. Hence, $\mathfrak{g}^*$ has an induced structure of a Poisson manifold. In coordinates, if we think of $\{x_i\}_{i=1}^n$ as linear functions, or local coordinates, on $\mathfrak{g}^*$, and we let $\{\xi_i = \frac{\partial}{\partial x_i}\}_{i=1}^n$ be the dual basis of $\mathfrak{g}^*$, then, by (4), the Poisson structure $\eta \in \Gamma(\bigwedge^2 T\mathfrak{g}^*)$ evaluated at $\xi \in \mathfrak{g}^*$ is

$$\eta(\xi) = \sum_{i,j=1}^{n} \xi([x_i,x_j])\xi_i \wedge \xi_j = \sum_{j=1}^{n} ad^*(x_j)(\xi) \wedge \xi_j \in \textstyle\bigwedge^2(\mathfrak{g}^*)\,. \tag{5}$$

By (5), the Hamiltonian vector field associated to $a \in \mathfrak{g}$ is ad^*a, and the corresponding Hamiltonian flow through $\xi \in \mathfrak{g}^*$ is $Ad^*(e^{ta})(\xi)$. Hence, the symplectic leaves of $\mathfrak{g}^*$ are the coadjoint orbits $S = Ad^*G(\xi)$ where G is the connected Lie group with Lie algebra $\mathfrak{g}$. The Poisson structure on the coadjoint orbits $Ad^*G(\xi)$ is known as Kirillov-Kostant Poisson structure. Its inverse is a symplectic structure. At the point $\xi \in \mathfrak{g}^*$ it coincides with the following non-degenerate skewsymmetric form $\omega(\xi)$ on $ad^*\mathfrak{g}(\xi)$:

$$\omega(\xi)(ad^*(a)(\xi), ad^*(b)(\xi)) = \xi([a,b])\,. \tag{6}$$

2.2 The Poisson structure on the Slodowy slice

Let $\mathfrak{g}$ be a reductive finite dimensional Lie algebra, and let $f \in \mathfrak{g}$ be a nilpotent element. By the Jacoboson-Morozov Theorem, f can be included in an $\mathfrak{sl}_2$-triple

$\{e,h=2x,f\}$, see e.g. [7]. Let $(\cdot|\cdot)$ be a non-degenerate invariant symmetric bilinear form on $\mathfrak{g}$, and let $\Phi:\mathfrak{g}\xrightarrow{\sim}\mathfrak{g}^*$ be the isomorphism associated to this bilinear form: $\Phi(a)=(a|\cdot)$. We also let $\chi=\Phi(f)=(f|\cdot)\in\mathfrak{g}^*$.

The *Slodowy slice* [33] associated to this $\mathfrak{sl}_2$-triple element is, by definition, the following affine space

$$\mathscr{S}=\Phi(f+\mathfrak{g}^e)=\{\chi+\Phi(a)\,|\,a\in\mathfrak{g}^e\}\subset\mathfrak{g}^*. \tag{7}$$

Let $\xi=\Phi(f+r)$, $r\in\mathfrak{g}^e$, be a given point of the Slodowy slice. The tangent space to the coadjoint orbit $Ad^*G(\xi)$ at ξ is $T_\xi(Ad^*G(\xi))=ad^*(\mathfrak{g})(\xi)=\Phi([f+r,\mathfrak{g}])$, while the tangent space to the Slodowy slice at ξ is $T_\xi(\mathscr{S})\simeq\Phi(\mathfrak{g}^e)$. Recalling (6), one can check that the assumptions of Proposition 1 hold, [22]:

(i) The restriction of the symplectic form (6) to $T_\xi(Ad^*G(\xi))\cap T_\xi(\mathscr{S})$ is non-degenerate. In other words, if $a\in\mathfrak{g}$ is such that $[f+r,a]\in\mathfrak{g}^e$ and $a\perp[f+r,\mathfrak{g}]\cap\mathfrak{g}^e$, then $a=0$.

(ii) The Slodowy slice $\mathscr{S}$ intersect transversally the coadjoint orbit at ξ, i.e. $[f+r,\mathfrak{g}]+\mathfrak{g}^e=\mathfrak{g}$.

It then follows by Proposition 1 that $\mathscr{S}\subset\mathfrak{g}^*$ is a Poisson submanifold, i.e. it has a Poisson structure induced by the Kirillov-Kostant structure on $\mathfrak{g}^*$.

Definition 1 The *classical finite $\mathscr{W}$ algebra* $\mathscr{W}^{cl,fin}(\mathfrak{g},f)\simeq S(\mathfrak{g}^f)$ is the algebra of polynomial functions on the Slodowy slice $\mathscr{S}$.

Clearly, the dual space to $\Phi(\mathfrak{g}^e)$ is $\mathfrak{g}^f$. Hence, by the definition (7) of $\mathscr{S}$, we can identify $\mathscr{W}^{cl,fin}(\mathfrak{g},f)$, as a polynomial algebra, with the symmetric algebra over $\mathfrak{g}^f$. In fact, we can write down an explicit formula for the Poisson bracket of the classical finite $\mathscr{W}$-algebra. We have the direct sum decomposition: $\mathfrak{g}=[e,\mathfrak{g}]\oplus\mathfrak{g}^f$, and, for $a\in\mathfrak{g}$, we denote by $a^\sharp$ its projection on $\mathfrak{g}^f$. Let $\{q_i\}_{i=1}^k$ be a basis of $\mathfrak{g}^f$ consisting of adx-eigenvectors, and let $\{q^i\}_{i=1}^k$ be the dual basis of $\mathfrak{g}^e$. For $i\in\{1,\dots,k\}$, we let $\delta(i)\in\frac{1}{2}\mathbb{Z}$ be adx-eigenvalue of q^i. By representation theory of $\mathfrak{sl}_2$, a basis of $\mathfrak{g}$ is

$$\left\{q^i_n:=(ad\,f)^n q^i\,\middle|\,n=0,\dots,2\delta(i),\,i=1,\dots,k\right\}, \tag{8}$$

and let $\left\{q_i^n\,\middle|\,n=0,\dots,2\delta(i),\,i=1,\dots,k\right\}$, be the dual basis of $\mathfrak{g}$. (Here and further we let $q^i_0=q^i$ and $q_i^0=q^i$.)

Theorem 1 ([15]) *The Poisson bracket on the classical finite $\mathscr{W}$-algebra $\mathscr{W}^{cl,fin}(\mathfrak{g},f)$ is ($p,q\in\mathfrak{g}^f$):*

$$\{p,q\}_{\mathscr{S}}=[p,q]+\sum_{s=1}^{\infty}\sum_{i_1,\dots,i_s=1}^{k}\sum_{m_1,\dots,m_s=0}^{d}[p,q^{i_1}_{m_1}]^\sharp[q^{m_1+1}_{i_1},q^{i_2}_{m_2}]^\sharp\cdots[q^{m_s+1}_{i_s},q]^\sharp.$$

Example 1 If $q\in\mathfrak{g}^f_0=\mathfrak{g}^e_0$, then $[q_i^{m+1},q]^\sharp=0$ for all i,m. Hence, $\{p,q\}_{\mathscr{S}}=[p,q]\in\mathfrak{g}^f$.

2.3 Classical Hamiltonian reduction

In order to define, in Sect. 3, the classical $\mathcal{W}$-algebra $\mathcal{W}_z^{cl}(\mathfrak{g},f)$, i.e. affine analogue of the classical finite $\mathcal{W}$-algebra $\mathcal{W}^{cl,fin}(\mathfrak{g},f)$, it is convenient to describe the Poisson structure on the Slodowy slice $\mathcal{S}$ via a Hamiltonian reduction of $\mathfrak{g}^*$. In this section we describe, in a purely algebraic setting, the general construction of the classical Hamiltonian reduction of a Hamiltonian action of a Lie group N on a Poisson manifold P. In the next Sect. 2.4 we then describe the classical finite $\mathcal{W}$-algebra $\mathcal{W}^{cl,fin}(\mathfrak{g},f)$ as a Hamiltonian reduction.

Recall that the classical Hamiltonian reduction is associated to a Poisson manifold M, a Lie group N with a Hamiltonian action on M, and a submanifold $\mathcal{O} \subset \mathfrak{n}^*$ which is invariant by the coadjoint action of N. The corresponding *Hamiltonian reduction* is, by definition, $\mu^{-1}(\mathcal{O})/N$, where $\mu : M \to \mathfrak{n}^*$ is the moment map associated to the Hamiltonian action of N on M. One shows that, indeed, $\mu^{-1}(\mathcal{O})/N$ has a Poisson structure induced by that of M [23].

On a purely algebraic level, going to the algebras of functions, the classical Hamiltonian reduction can be defined as follows. Let $(P,\cdot,\{\cdot\,,\cdot\})$ be a unital Poisson algebra. Let $\mathfrak{n}$ be a Lie algebra. Let $\phi : \mathfrak{n} \to P$ be a Lie algebra homomorphism, and denote by $\phi : S(\mathfrak{n}) \to P$ the corresponding Poisson algebra homomorphism. Let $I \subset S(\mathfrak{n})$ be a subset which is invariant by the adjoint action of $\mathfrak{n}$, i.e. such that $ad(\mathfrak{n})(I) \subset I$. Consider the ideal $P\phi(I)$ of P generated by $\phi(I)$. Note that, in general, $P\phi(I)$ is NOT a Poisson ideal, so the quotient space $P/P\phi(I)$ has an induced structure of a commutative associative algebra, but NOT of a Poisson algebra.

Definition 2 The *Hamiltonian reduction* of the Poisson algebra P associated to the Lie algebra homomorphism $\phi : \mathfrak{n} \to P$ and to the $ad\mathfrak{n}$-invariant subset $I \subset S(\mathfrak{n})$ is, as a space,

$$\mathcal{W}(P,\mathfrak{n},I) := \big(P/P\phi(I)\big)^{\mathfrak{n}} = \Big\{f \in P \Big| \{\phi(\mathfrak{n}),f\} \subset P\phi(I)\Big\}\Big/P\phi(I). \tag{9}$$

Proposition 2 *The Hamiltonian reduction $\mathcal{W}(P,\mathfrak{n},I)$ has an induced structure of a Poisson algebra.*

Proof. First, it follows by the Leibniz rule that $\big\{f \in P \big| \{\phi(\mathfrak{n}),f\} \subset P\phi(I)\big\} \subset P$ is a subalgebra with respect to the commutative associative product of P, and, since by assumption the set I is $ad(\mathfrak{n})$-invariant, $P\phi(I) \subset \big\{f \in P \big| \{\phi(\mathfrak{n}),f\} \subset P\phi(I)\big\}$ is its ideal. Hence, the corresponding quotient $W(P,\mathfrak{n},S)$ has an induced commutative associative product. We use the same argument for the Poisson structure: we claim that

(i) $\big\{f \in P \big| \{\phi(\mathfrak{n}),f\} \subset P\phi(I)\big\} \subset P\phi(I)$ is a Lie subalgebra,

(ii) and that $P\phi(I) \subset \big\{f \in P \big| \{\phi(\mathfrak{n}),f\} \subset P\phi(I)\big\}$ is its Lie algebra ideal.

Suppose that $f,g \in P$ are such that $(ad\phi(\mathfrak{n}))(f) \subset P\phi(I)$ and $(ad\phi(\mathfrak{n}))(g) \subset P\phi(I)$. Then, by the Jacobi identity,

$$\begin{aligned}
&\{\phi(\mathfrak{n}),\{f,g\}\} \subset \{\{\phi(\mathfrak{n}),f\},g\}+\{f,\{\phi(\mathfrak{n}),g\}\} \subset \{P\phi(I),g\}+\{f,P\phi(I)\} \\
&\subset P\{\phi(I),g\}+P\{f,\phi(I)\}+P\phi(I) \subset P\{\phi(S(\mathfrak{n})),g\}+P\{\phi(S(\mathfrak{n})),f\}+P\phi(I) \\
&\subset P\{\phi(n),g\}+P\{\phi(\mathfrak{n}),f\}+P\phi(I) \subset P\phi(I)\,.
\end{aligned}$$

In the second inclusion we used the assumption on f and g, in the third inclusion we used the Leibniz rule, in the fourth inclusion we used the fact that, by construction, $I \subset S(\mathfrak{n})$, in the fifth inclusion we used the Leibniz rule, and in the last inclusion we used again the assumption on f and g. This proves claim (i). For claim (ii), let $f \in P$ be such that $\{\phi(\mathfrak{n}),f\} \subset P\phi(I)$. We have, with the same line of arguments as above,

$$\begin{aligned}
&\{P\phi(I),f\} \subset P\{\phi(I),f\}+P\phi(I) \subset P\{\phi(S(\mathfrak{n})),f\}+P\phi(I) \\
&\subset P\{\phi(\mathfrak{n}),f\}+P\phi(I) \subset P\phi(I)\,.
\end{aligned}$$

2.4 The Slodowy slice via Hamiltonian reduction

We want to describe the classical finite $\mathscr{W}$-algebra $\mathscr{W}^{cl,fin}(\mathfrak{g},f)$ introduced in Sect. 2.2 as a Hamiltonian reduction of the Poisson algebra $S(\mathfrak{g})$.

We have the adx-eigenspace decomposition $\mathfrak{g} = \bigoplus_{i\in\frac{1}{2}\mathbb{Z}} \mathfrak{g}_i$. Let ω be the following non-degenerate skewsymmetric bilinear form on $\mathfrak{g}_{\frac{1}{2}}$:

$$\omega(u,v) = (f|[u,v])\,. \tag{10}$$

Let $\ell \subset \mathfrak{g}_{\frac{1}{2}}$ be a maximal isotropic subspace. Consider the nilpotent subalgebra

$$\mathfrak{n} = \ell \oplus \mathfrak{g}_{\geq 1} \subset \mathfrak{g}\,. \tag{11}$$

Since $\ell \subset \mathfrak{g}_{\frac{1}{2}}$ is isotropic w.r.t. the bilinear form (10), we have $(f|[\mathfrak{n},\mathfrak{n}]) = 0$. Hence, the subset

$$I = \{n-(f|n)\,|\,n\in\mathfrak{n}\} \subset S(\mathfrak{n})\,, \tag{12}$$

is invariant by the adjoint action of $\mathfrak{n}$. Hence, we can consider the corresponding Hamiltonian reduction (9) applied to the data $(S(\mathfrak{g}),\mathfrak{n},I)$.

Theorem 2 ([15]) *The classical finite $\mathscr{W}$-algebra $\mathscr{W}^{cl,fin}(\mathfrak{g},f)$ is isomorphic to the Hamiltonian reduction of the Poisson algebra $S(\mathfrak{g})$, associated to the Lie algebra $\mathfrak{n} \subset S(\mathfrak{g})$ given by (11), and the $ad\mathfrak{n}$-invariant subset $I \subset S(\mathfrak{n})$ in (12):*

$$\begin{aligned}
&\mathscr{W}^{cl,fin}(\mathfrak{g},f) \simeq \mathscr{W}(S(\mathfrak{g}),\mathfrak{n},I) \\
&= \Big\{p\in S(\mathfrak{g})\,\Big|\,\{\mathfrak{n},p\} \subset \langle n-(f|n)\rangle_{n\in\mathfrak{n}}\Big\}\Big/ S(\mathfrak{g})\langle n-(f|n)\rangle_{n\in\mathfrak{n}}
\end{aligned}$$

In geometric terms, the restriction map $\mu : \mathfrak{g}^* \to \mathfrak{n}^*$ is a map of Poisson manifolds (the moment map), and the corresponding dual map $\mu^* : \mathfrak{n} \to \mathfrak{g}$, is the inclusion map. The element $\chi = (f|\cdot)|_{\mathfrak{n}} \in \mathfrak{n}^*$ is a character of $\mathfrak{n}$, in the sense that $\chi([\mathfrak{n},\mathfrak{n}]) = 0$ (by the assumption that ℓ is maximal isotropic). Hence, χ is fixed by the coadjoint action of N, the Lie group of $\mathfrak{n}$. We can then consider the Hamiltonian reduction of the Poisson manifold $\mathfrak{g}^*$, by the Hamiltonian action of N on $\mathfrak{g}^*$ given by the moment map μ, associated to the N-fixed point $\chi \in \mathfrak{g}^*$:

$$Ham.Red.(\mathfrak{g}^*,N,\chi) = \mu^{-1}(\chi)/N = \Phi(f+\mathfrak{n}^\perp)/N\,.$$

Theorem 2 can be then viewed as the algebraic analogue of the following result of Gan and Ginzburg:

Theorem 3 ([22]) *The coadjoint action $N \times \mathscr{S} \to \Phi(f+\mathfrak{n}^\perp)$ is an isomorphism of affine varieties. The corresponding bijection*

$$\mathscr{S} \simeq \mu^{-1}(\chi)/N = \text{Ham.Red.}(\mathfrak{g}^*,N,\chi)\,,$$

is an isomorphism of Poisson manifolds.

3 Classical $\mathscr{W}$-algebras

3.1 Poisson vertex algebras

In this section we introduce the notions of Lie conformal algebra and of Poisson vertex algebra. They are, in some sense, the "affine analogue" of a Lie algebra and of a Poisson algebra respectively.

Definition 3 A *Lie conformal algebra* is a $\mathbb{F}[\partial]$-module R with a bilinear λ-bracket $[\cdot\,{}_\lambda\,\cdot] : R \times R \to \mathbb{F}[\lambda] \otimes R$ satisfying the following axioms:

(i) sesquilinearity: $[\partial a_\lambda b] = -\lambda [a_\lambda b]$, $[a_\lambda \partial b] = (\partial+\lambda)[a_\lambda b]$;
(ii) skewsymmetry: $[a_\lambda b] = -[b_{-\lambda-\partial} a]$ (where ∂ is moved to the left);
(iii) Jacobi identity: $[a_\lambda [b_\mu c]] - [b_\mu [a_\lambda c]] = [[a_\lambda b]_{\lambda+\mu} c]$.

Example 2 Let $\mathfrak{g}$ be a Lie algebra with a symmetric invariant bilinear form $(\cdot|\cdot)$. The corresponding *current Lie conformal algebra* is, by definition, $R = \mathbb{F}[\partial]\mathfrak{g} \oplus \mathbb{F}$, with λ-bracket (s is a fixed element of $\mathfrak{g}$):

$$[a_\lambda b] = [a,b] + (a|b)\lambda + z(s|[a,b])\,, \tag{13}$$

for $a,b \in \mathfrak{g}$, extended to R by saying that $\mathbb{F}$ is central, and by sesquilinearity. (The term $(a|b)\lambda$ is a 2-cocycle, defining a central extension, and $z(s|[a,b])$ is a trivial 2-cocycle.)

Definition 4 A *Poisson vertex algebra* is a commutative associative differential algebra $\mathscr{V}$ (with derivation ∂) endowed with a Lie conformal algebra λ-bracket $\{\cdot\,{}_\lambda\,\cdot\}$, satisfying the Leibniz rule

$$\{a_\lambda bc\} = \{a_\lambda b\}c + \{a_\lambda c\}b\,. \tag{14}$$

Note that by the skewsymmetry axiom and the left Leibniz rule (14) we get the right Leibniz rule:

$$\{ab_\lambda c\} = \{a_{\lambda+\partial}c\}_\rightarrow b + \{b_{\lambda+\partial}c\}_\rightarrow a\,, \tag{15}$$

where the arrow means that ∂ should be moved to the right.

Example 3 If R is a Lie conformal algebra, then $S(R)$ has a natural structure of a Poisson vertex algebra, with λ-bracket extending the one in R by the left and right Leibniz rules.

Example 4 As a special case of Example 3, consider the current Lie conformal algebra $R = \mathbb{F}[\partial]\mathfrak{g} \oplus \mathbb{F}$ associated to the Lie algebra $\mathfrak{g}$ and the symmetric invariant bilinear form $(\cdot\,,\cdot)$, defined in Example 2. The *affine Poisson vertex algebra* is, by definition, $\mathscr{V}_z(\mathfrak{g}) = S(\mathbb{F}[\partial]\mathfrak{g})$, with λ-bracket (13), extended by sesquilinearity and Leibniz rule. If $\{u_i\}_{i=1}^n$ is a basis of $\mathfrak{g}$, the general formula for the λ-bracket is:

$$\begin{aligned}\{P_\lambda Q\}_z = \sum_{i,j=1}^{n} \sum_{m,n\in\mathbb{Z}_+} \frac{\partial Q}{\partial u_j^{(n)}}(\lambda+\partial)^n \\ \Big([u_i,u_j] + (u_i|u_j)(\lambda+\partial) + z(s|[u_i,u_j])\Big)(-\lambda-\partial)^m \frac{\partial Q}{\partial u_j^{(n)}}(\lambda+\partial)^n\,.\end{aligned}$$

This is the "affine analogue" of the usual Poisson algebra structure on $S(\mathfrak{g})$. It is a 1-parameter family of Poisson vertex algebras, depending on the parameter $z \in \mathbb{F}$. (Having a 1-parameter family is important for applications to the theory of integrable systems.)

3.2 Hamiltonian reduction of Poisson vertex algebras

The classical Hamiltonian reduction construction described in Sect. 2.3 has an "affine analogue" for Poisson vertex algebras.

Let $(\mathscr{V}, \partial, \cdot, \{\cdot\,_\lambda\,\cdot\})$ be a Poisson vertex algebra. Let R be a Lie conformal algebra. Let $\phi : R \to \mathscr{V}$ be a Lie conformal algebra homomorphism, which we extend to a Poisson vertex algebra homomorphism $\phi : S(R) \to \mathscr{V}$. Let $I \subset S(R)$ be a subset which is invariant by the adjoint action of R, i.e. such that $[R_\lambda I] \subset \mathbb{F}[\lambda] \otimes I$. Consider the differential algebra ideal $\langle\phi(I)\rangle_\mathscr{V} \subset \mathscr{V}$ of $\mathscr{V}$ generated by $\phi(I)$. The quotient space $\mathscr{V}/\langle\phi(I)\rangle_\mathscr{V}$ has an induced structure of a (commutative associative) differential algebra (but NOT of a Poisson vertex algebra).

Definition 5 The *Hamiltonian reduction* of the Poisson vertex algebra $\mathscr{V}$ associated to the Lie conformal algebra homomorphism $\phi : R \to \mathscr{V}$ and to the R-invariant subset $I \subset S(R)$ is, as a space,

$$\begin{aligned}\mathscr{W}(\mathscr{V},R,I) &:= \big(\mathscr{V}/\langle\phi(I)\rangle_\mathscr{V}\big)^R \\ &= \Big\{f \in \mathscr{V} \,\Big|\, \{\phi(a)_\lambda f\} \in \mathbb{F}[\lambda] \otimes \langle\phi(I)\rangle_\mathscr{V}\ \forall a \in R\Big\} \Big/ \langle\phi(I)\rangle_\mathscr{V}\,.\end{aligned} \tag{16}$$

Proposition 3 *The Hamiltonian reduction $\mathscr{W}(\mathscr{V},R,I)$ has an induced structure of a Poisson vertex algebra.*

Proof. It is analogue to the proof of Proposition 2.

3.3 Construction of the classical $\mathcal{W}$-algebras

In analogy with the construction, in Theorem 2, of the classical finite $\mathcal{W}$-algebra $\mathcal{W}^{cl,fin}(\mathfrak{g},f)$ via Hamiltonian reduction, we define its "affine analogue", the classical $\mathcal{W}$-algebra $\mathcal{W}_z^{cl}(\mathfrak{g},f)$, via a Hamiltonian reduction of the affine Poisson vertex algebra $\mathcal{V}_z(\mathfrak{g})$ as in Example 4.

Let, as in Sect. 2.4, $\mathfrak{g}$ be a reductive finite dimensional Lie algebra, with a non-degenerate symmetric bilinear form $(\cdot\,|\,\cdot)$. Let $(e,h=2x,f)$ be an $\mathfrak{sl}_2$-triple in $\mathfrak{g}$, and consider the adx-eigenspace decomposition $\mathfrak{g}=\bigoplus_{i\in\frac{1}{2}\mathbb{Z}}\mathfrak{g}_i$. Recall the definitions of the bilinear form ω on $\mathfrak{g}_{\frac{1}{2}}$ in (10), of the nilpotent subalgebra $\mathfrak{n}\subset\mathfrak{g}$ in (11), and of the $ad\mathfrak{n}$-invariant subset $I=\{n-(f|n)\}_{n\in\mathfrak{n}}\subset S(\mathfrak{n})$, as in (12). Consider the current Lie conformal algebra $\mathbb{F}[\partial]\mathfrak{n}$ with λ-bracket $\{a_\lambda b\}=[a,b]$ for $a,b\in\mathfrak{n}$, and extended to R by sesquilinearity. It is obviously a Lie conformal subalgebra of $\mathcal{V}_z(\mathfrak{g})$. Hence, the inclusion $S(\mathbb{F}[\partial]\mathfrak{n})\subset\mathcal{V}_z(\mathfrak{g})$ is a homomorphism of Poisson vertex algebras. Furthermore, the set $I=\{n-(f|n)\}_{n\in\mathfrak{n}}\subset S(\mathbb{F}[\partial]\mathfrak{n})$ is invariant by the λ-adjoint action of $\mathbb{F}[\partial]\mathfrak{n}$. Let $\langle n-(f|n)\rangle_{n\in\mathfrak{n}}\subset\mathcal{V}_z(\mathfrak{g})$ be the differential algebra ideal of $\mathcal{V}_z(\mathfrak{g})$ generated by $I=\{n-(f|n)\}_{n\in\mathfrak{n}}$. We can consider the Hamiltonian reduction associated to the data $(\mathcal{V}_z(\mathfrak{g},f),\mathbb{F}[\partial]\mathfrak{n},I)$.

Definition 6 ([13]) The *classical $\mathcal{W}$-algebra* is the following 1-parameter family (parametrized by $z\in\mathbb{F}$) of Poisson vertex algebras

$$\begin{aligned}\mathcal{W}_z^{cl}(\mathfrak{g},f)&\simeq\mathcal{W}(\mathcal{V}_z(\mathfrak{g}),\mathbb{F}[\partial]\mathfrak{n},I)\\&=\Big\{P\in S(\mathbb{F}[\partial]\mathfrak{g})\,\Big|\,\{a_\lambda P\}_z\in\mathbb{F}[\lambda]\otimes\langle n-(f|n)\rangle_{n\in\mathfrak{n}}\,\forall a\in\mathfrak{n}\Big\}\Big/\langle n-(f|n)\rangle_{n\in\mathfrak{n}}\,.\end{aligned}$$

It is convenient to give a description of the $\mathcal{W}$-algebra as a subspace, rather than as a quotient space. Let $\ell'\subset\mathfrak{g}_{\frac{1}{2}}$ be a maximal isotropic subspace (w.r.t. ω) complementary to ℓ. Hence, $\mathfrak{p}=\ell'\oplus\mathfrak{g}_{\leq 0}\subset\mathfrak{g}$ is a subspace complementary to $\mathfrak{n}$:

$$\mathfrak{g}=\mathfrak{n}\oplus\mathfrak{p}=\mathfrak{n}^\perp\oplus\mathfrak{p}^\perp\,.$$

(We can thus identify $\mathfrak{p}^*=\mathfrak{n}^\perp$ and $\mathfrak{n}^*=\mathfrak{p}^\perp$.) Consider the differential algebra homomorphism $\rho:\mathcal{V}_z(\mathfrak{g})\to S(\mathbb{F}[\partial]\mathfrak{p})$ given by

$$\rho(a)=\pi_{\mathfrak{p}}(a)+(f|a)\;\;\forall a\in\mathfrak{g}\,.$$

Clearly, $\ker(\rho)=\langle n-(f|n)\rangle_{n\in\mathfrak{n}}$. Then, the classical $\mathcal{W}$-algebra can be equivalently be defined as follows:

$$\mathcal{W}_z^{cl}(\mathfrak{g},f)\simeq\Big\{P\in S(\mathbb{F}[\partial]\mathfrak{p})\,\Big|\,\rho(\{a_\lambda P\}_z)=0\,\forall a\in\mathfrak{n}\Big\}\,,$$

with λ-bracket

$$\{P_\lambda Q\}_{z,\rho}=\rho\{P_\lambda Q\}_z\,.$$

We can find explicit formulas for the λ-brackets of generators of $\mathcal{W}^{cl}(\mathfrak{g},f)$, in analogy with the result of Theorem 1.

Theorem 4 ([15]) *As a differential algebra, the classical $\mathcal{W}$-algebra $\mathcal{W}_z^{cl}(\mathfrak{g}, f)$ is isomorphic to the algebra of differential polynomials over $\mathfrak{g}^f$, i.e. $\mathcal{W}_z^{cl}(\mathfrak{g}, f) \simeq S(\mathbb{F}[\partial]\mathfrak{g}^f)$. Moreover, the λ bracket on $\mathcal{W}_z^{fin}(\mathfrak{g}, f)$ is, using the notation in Theorem 1 ($q_{i_0}, q_{j_0} \in \mathfrak{g}^f$):*

$$
\begin{aligned}
\{p_\lambda q\}_z = & \sum_{s,t=0}^{\infty} \sum_{i_1,\dots,i_s=1}^{k} \sum_{j_1,\dots,j_t=1}^{k} \sum_{m_1,\dots,m_s=0}^{d} \sum_{n_1,\dots,n_t=0}^{d} \\
& \left([q_{j_0}, q_{n_1}^{j_1}]^\sharp - \delta_{n_1,0}\delta_{j_1,j_0}(\lambda+\partial)\right) \dots \left([q_{j_{t-1}}^{n_{t-1}+1}, q_{n_t}^{j_t}]^\sharp - \delta_{n_t,n_{t-1}+1}\delta_{j_t,j_{t-1}}(\lambda+\partial)\right) \\
& \left([q_{i_s}^{m_s+1}, q_{j_t}^{n_t+1}]^\sharp + (q_{i_s}^{m_s+1} | q_{j_t}^{n_t+1})(\lambda+\partial) + z(s|[q_{i_s}^{m_s+1}, q_{j_t}^{n_t+1}]\right) \\
& \left([q_{i_{s-1}}^{m_{s-1}+1}, q_{m_s}^{i_s}]^\sharp + \delta_{m_s,m_{s-1}+1}\delta_{i_s,i_{s-1}}(\lambda+\partial)\right) \dots \left([q_{i_0}, q_{m_1}^{i_1}]^\sharp + \delta_{m_1,0}\delta_{i_1,i_0}\lambda\right).
\end{aligned}
\tag{17}
$$

(In the RHS, when s or t is 0, we replace m_0+1 or n_0+1 by 0.)

In the special case of classical $\mathcal{W}$-algebras for principal and minimal nilpotent elements, Eq. (17) was proved in [14].

3.4 Application of Poisson vertex algebras to the theory of Hamiltonian equations

Poisson vertex algebras can be used in the study of Hamiltonian partial differential equations in classical field theory, and their integrability [2] (in the same way as Poisson algebras are used to study Hamiltonian equations in classical mechanics).

The basic observation is that, if $\mathcal{V}$ is a Poisson vertex algebra with λ-bracket $\{\cdot_\lambda\cdot\}$, then $\mathcal{V}/\partial\mathcal{V}$ is a Lie algebra, with Lie bracket

$$\{\textstyle\int f, \int g\} = \int \{f_\lambda g\}\big|_{\lambda=0},$$

and we have a representation of the Lie algebra $\mathcal{V}/\partial\mathcal{V}$ on $\mathcal{V}$, with the following action

$$\{\textstyle\int f, g\} = \{f_\lambda g\}\big|_{\lambda=0},$$

We can then introduce Hamiltonian equations and integrals of motion in the same way as in classical mechanics.

Definition 7 Given a Poisson vertex algebra $\mathcal{V}$ with λ-bracket $\{\cdot_\lambda\cdot\}$, the *Hamiltonian equation* with Hamiltonian functional $\int h \in \mathcal{V}/\partial\mathcal{V}$ is:

$$\frac{du}{dt} = \{h_\lambda u\}\big|_{\lambda=0}. \tag{18}$$

An *integral of motion* for the Hamiltonian equation (18) is an element $\int g \in \mathcal{V}$ such that

$$\{\textstyle\int h, \int g\} = \int \{h_\lambda g\}\big|_{\lambda=0} = 0.$$

The element $g \in \mathcal{V}$ is then called a *conserved density*. The usual requirement to have *integrability* is that of having an infinite sequence $\int g_0 = \int h, \int g_1, \int g_2, \dots$ of linearly independent integrals of motion in involution:

$$\textstyle\int \{g_{m\lambda} g_n\}\big|_{\lambda=0} = 0 \; forall m,n \in \mathbb{Z}_+.$$

Example 5 The famous *KdV equation*, describing the evolution of waves in shallow water is

$$\frac{\partial u}{\partial t} = 3u\frac{\partial u}{\partial x} + c\frac{\partial^3 u}{\partial x^3}.$$

It is a bi-Hamiltonian equation, since it can be written in two compatible Hamiltonian forms:

$$\frac{du}{dt} = \left\{\frac{1}{2}(u^3 + cuu'')_\lambda u\right\}_0\Big|_{\lambda=0} = \left\{\frac{1}{2}u^2{}_\lambda u\right\}_1\Big|_{\lambda=0},$$

on the differential algebra $\mathcal{V} = S(\mathbb{F}[\partial]u)$, with PVA λ-brackets

$$\{u_\lambda u\}_0 = \lambda\,,\ \{u_\lambda u\}_1 = u' + 2u\lambda + c\lambda^3\,.$$

Compatibility means that $\{\cdot_\lambda\cdot\}_z = \{\cdot_\lambda\cdot\}_0 + \{\cdot_\lambda\cdot\}_1$ is a 1-parameter family of PVA λ-brackets.

The usual "trick" to construct a sequence $\int g_n$, $n \in \mathbb{Z}_+$, of integrals of motion in involution is the so called *Lenard-Magri* scheme: assuming we have a bi-Hamiltonian equation

$$\frac{du}{dt} = \{h_{1\lambda}u\}_0 = \{h_{0\lambda}u\}_1,$$

we try to solve the following recursion equation for $\int g_n$, $n \geq 0$ (starting with $g_0 = h_0$ and $g_1 = h_1$),

$$\{g_{0\lambda}u\}_0 = 0\,,\ \{g_{n+1\lambda}u\}_0 = \{g_{n\lambda}u\}_1\,. \tag{19}$$

(There are various "cohomological" arguments indicating that, often, such recursive equations can be solved for every n, see e.g. [11, 12].) In this case, it was a simple observation of Magri [29] that the solutions $\int g_n$, $n \in \mathbb{Z}_+$, are integrals of motion in involution w.r.t. both PVA λ-brackets $\{\cdot_\lambda\cdot\}_0$ and $\{\cdot_\lambda\cdot\}_1$, and therefore we get the integrable hierarchy of bi-Hamiltonian equations

$$\frac{du}{dt_n} = \{g_{n\lambda}u\}_0\,.$$

3.5 *Generalized Drinfeld-Sokolov bi-Hamiltonian integrable hierarchies*

Following the ideas of [16], we can prove that the Lenard-Magri scheme can be applied to construct integrable hierarchies of bi-Hamiltonian equations attached to the classical $\mathcal{W}$-algebras $\mathcal{W}_z^{cl,fin}(\mathfrak{g}, f)$. For example, for $\mathfrak{g} = \mathfrak{sl}_2$, we get the KdV hierarchy (cf. Example 5).

The basic assumption is that there exists a homogeneous (w.r.t. the *adx*-eigenspace decomposition) element $s \in \ker(ad\mathfrak{n})$ such that $f + s$ is a semisimple element of $\mathfrak{g}$. Hence, $f + zs$ is a semisimple element of $\mathfrak{g}((z^{-1}))$, and we have the direct sum decomposition $\mathfrak{g}((z^{-1})) = \mathfrak{h} \oplus \mathfrak{h}^\perp$, where

$$\mathfrak{h} := Kerad(f + zs) \text{ and } \mathfrak{h}^\perp := Imad(f + zs)\,. \tag{20}$$

We define a $\frac{1}{2}\mathbb{Z}$-grading of $\mathfrak{g}((z^{-1}))$ by letting $\deg(z) = -d-1$, if $s \in \mathfrak{g}_d$. In particular, $f + zs$ is homogenous of degree -1. We have the induced decompositions of $\mathfrak{h}$ and $\mathfrak{h}^\perp$ (since $f + zs$ is homogeneous):

$$\mathfrak{h} = \widehat{\bigoplus}_{i\in\frac{1}{2}\mathbb{Z}}\mathfrak{h}_i \text{ and } \mathfrak{h}^\perp = \widehat{\bigoplus}_{i\in\frac{1}{2}\mathbb{Z}}\mathfrak{h}_i^\perp . \tag{21}$$

Consider the Lie algebra

$$\tilde{\mathfrak{g}} = \mathbb{F}\partial \ltimes \big(\mathfrak{g}((z^{-1})) \otimes \mathscr{V}(\mathfrak{p})\big),$$

where ∂ acts only on the second factor of the tensor product. Clearly, $\mathfrak{g}((z^{-1}))_{>0} \otimes \mathscr{V}(\mathfrak{p}) \subset \tilde{\mathfrak{g}}$ is a pro-nilpotent subalgebra. Hence, for $U(z) \in \mathfrak{g}((z^{-1}))_{>0} \otimes \mathscr{V}(\mathfrak{p})$, we have a well defined automorphism $e^{adU(z)}$ of $\tilde{\mathfrak{g}}$. Let $\{q_i\}_{i\in J}$ be a basis of $\mathfrak{p}$, and let $\{q^i\}_{i\in J}$ be the dual basis of $\mathfrak{n}^\perp$.

Theorem 5 ([13])

(a) *There exist unique* $U(z) \in \mathfrak{h}^\perp_{>0} \otimes \mathscr{V}(\mathfrak{p})$ *and* $h(z) \in \mathfrak{h}_{>-1} \otimes \mathscr{V}(\mathfrak{p})$ *such that*

$$e^{adU(z)}\Big(\partial + (f+zs)\otimes 1 + \sum_{i\in J} q^i \otimes q_i\Big) = \partial + (f+zs)\otimes 1 + h(z). \tag{22}$$

(b) *For* $0 \neq a(z) \in Z(\mathfrak{h})$, *the coefficients* g_n, $n \in \mathbb{Z}_+$, *of the Laurent series*

$$g(z) = (a(z)\otimes 1 | h(z)) \in \mathscr{V}(\mathfrak{p})((z^{-1})), \tag{23}$$

lie in $\mathscr{W}_z^{cl,fin}(\mathfrak{g}, f) \subset \mathscr{V}(\mathfrak{p})$ *modulo* $\partial\mathscr{V}(\mathfrak{p})$, *and they satisfy the Lenard-Magri recursion equations (19) for the PVA* λ*-brackets* $\{\cdot_\lambda\cdot\}_0 = \{\cdot_\lambda\cdot\}_{z=0}$ *and* $\{\cdot_\lambda\cdot\}_1 = \frac{d}{dz}\{\cdot_\lambda\cdot\}_z\big|_{z=0}$

Hence, we get an integrable hierarchy of bi-Hamiltonian equations, called the *generalized Drinfeld-Sokolov hierarchy* ($w \in \mathscr{W}$),

$$\frac{dw}{dt_n} = \rho\{g_{n\lambda} w\}_0\big|_{\lambda=0}, \quad n \in \mathbb{Z}_+ .$$

Acknowledgements The present paper is based on lectures given by the author for the conference *Lie superalgberas*, at INdAM, Roma, Italy, in December 2012, and for the conference *Symmetries in Mathematics and Physics*, at IMPA, Rio de Janeiro, Brazil, in June 2013. The paper was partially written while the author was visiting IHES, France, which we thank for the hospitality. The author was supported in part by the national FIRB grant RBFR12RA9W.

References

1. Arakawa, T.: Representation theory of W-algebras. Invent. Math. **169**(2), 219–320 (2007)
2. Barakat, A., De Sole, A., Kac, V.: Poisson vertex algebras in the theory of Hamiltonian equations. Jpn. J. Math. **4**(2), 141–252 (2009)

3. Borcherds, R.: Vertex algebras, Kac-Moody algebras and the Monster. Proc. Natl. Acad. Sci. USA **83**, 3068–3071 (1986)
4. Burruoughs, N., de Groot, M., Hollowood, T., Miramontes, L.: Generalized Drinfeld-Sokolov hierarchies II: the Hamiltonian structures. Comm. Math. Phys. **153**, 187–215 (1993)
5. Bouwknegt, P., Schoutens, K.: W-symmetry. Advanced Ser. in Math. Phys. Vol. 22. World Sci. (1995)
6. Brundan, J., Kleshchev, A.: Shifted Yangians and finite W-algebras. Adv. Math. **200**(1), 136–195 (2006)
7. Collingwood, D., McGovern, W.: Nilpotent orbits in semisimple Lie algebras. Van Nostrand Reinhold Mathematics Series. Van Nostrand Reinhold Co., New York (1993)
8. Delduc, F., Fehér, L.: Regular conjugacy classes in the Weyl group and integrable hierarchies, J. Phys. A **28**(20), 5843–5882 (1995)
9. de Groot, M., Hollowood, T., Miramontes, L.: Generalized Drinfeld-Sokolov hierarchies. Comm. Math. Phys. **145**, 57–84 (1992)
10. De Sole, A., Kac, V.G.: Finite vs. affine W-algebras. Jpn. J. Math. **1**(1), 137–261 (2006)
11. De Sole, A., Kac, V.G.: The variational Poisson cohomology. Jpn. J. Math. **8**(1), 1–145 (2013)
12. De Sole, A., Kac, V.G.: Essential variational Poisson cohomology. Comm. Math. Phys. **313**(3), 837–864 (2012)
13. De Sole, A., Kac, V.G., Valeri, D.: Classical $\mathcal{W}$-algebras and generalized Drinfeld-Sokolov bi-Hamiltonian systems within the theory of Poisson vertex algebras. To appear in Communications in Mathematical Physics, arXiv:1207.6286v3
14. De Sole, A., Kac, V.G., Valeri, D.: Classical W-algebras and generalized Drinfeld-Sokolov hierarchies for minimal and short nilpotents. Preprint arXiv:1306.1684
15. De Sole, A., Kac, V.G., Valeri, D.: W-algebras. in preparation.
16. Drinfeld, V.G., Sokolov, V.V.: Lie algebras and equations of Korteweg-de Vries type. Soviet J. Math. **30**, 1975–2036 (1985)
17. Fehér, L., Harnad, J., Marshall, I.: Generalized Drinfeld-Sokolov reductions and KdV type hierarchies: Comm. Math. Phys. **154**(1), 181–214 (1993)
18. Feigin, B.L., Frenkel, E.: Quantization of Drinfeld-Sokolov reduction. Phys. Lett., B **246**, 75–81 (1990)
19. Feigin, B.L., Frenkel, E.: Affine Kac-Moody algebras, bosonization and resolutions. Lett. Math. Phys. **19**, 307–317 (1990)
20. Fernández-Pousa, C., Gallas, M., Miramontes, L., Sánchez Guillén, J., $\mathcal{W}$-algebras from soliton equations and Heisenberg subalgebras. Ann. Physics **243**(2), 372–419 (1995)
21. Fernández-Pousa, C., Gallas, M., Miramontes, L., Sánchez Guillén, J., Integrable systems and $\mathcal{W}$-algebras. VIII J. A. Swieca Summer School on Particles and Fields, Rio de Janeiro (1995)
22. Gan, W.L., Ginzburg, V.: Quantization of Slodowy slices. Int. Math. Res. Not. **5**, 243–255 (2002)
23. Guillemin, V., Sternberg, S.: Symplectic techniques in physics, 2nd ed. Cambridge University Press, Cambridge (1990)
24. Kac, V.G., Roan, S.-S., Wakimoto, M.: Quantum reduction for affine superalgebras. Comm. Math. Phys. **241**, 307–342 (2003)
25. Kac, V. G., Wakimoto, M.: Quantum reduction and representation theory of superconformal algebras. Adv. Math. **185**, 400-458 (2004). Corrigendum, Adv. Math. **193**, 453–455 (2005)
26. Kac, V. G., Wakimoto, M.: Quantum reduction in the twisted case. in Progress in Math. **237**, 85–126 (2005)
27. Kostant, B., On Whittaker vectors and representation theory. Inv. Math **48**, 101–184 (1978)
28. Lynch, T.E.: Generalized Whittaker vectors and representation theory. Ph.D. Thesis, Massachusetts Institute of Technology (1979)
29. Magri, F.: A simple model of the integrable Hamiltonian equation. J. Math. Phys. **19**(5), 1156–1162 (1978)
30. Matumoto, H.: Whittaker modules associated with highest weight modules. Duke Math. J. **60**, 59-113 (1990)

31. Premet, A.: Special transverse slices and their enveloping algebras. Adv. Math. **170**, 1–55 (2002)
32. Premet, A.: Enveloping algebras of Slodowy slices and the Joseph ideal. J. Eur. Math. Soc. (JEMS) **9**(3), 487–543 (2007)
33. Slodowy, P.: Simple singularities and simple algebraic groups. Lecture Notes in Mathematics, Vol. 815. Springer-Verlag, Berlin Heidelberg New York (1980)
34. Vaisman, I.: Lectures on the geometry of Poisson manifolds. Progress in Mathematics, Vol. 118. Birkhäuser Verlag, Basel (1994)
35. Zamolodchikov, A.: Infinite extra symmetries in two-dimensional conformal quantum field theory. Teor. Mat. Fiz. **65**(3), 347–359 (1985)
36. Zhu, Y.: Modular invariance of characters of vertex operator algebras. Journal AMS **9**, 237–302 (1996)

Q-type Lie superalgebras

Maria Gorelik and Dimitar Grantcharov

Abstract The purpose of this paper is to collect some recent results on the representation theory of Lie superalgebras of type Q. Results on the centres, simple weight modules and crystal bases of these superalgebras are included.

1 Introduction

This paper is devoted to the Lie superalgebras of type Q, also known as queer or strange Lie superalgebras. These Lie superalgebras, introduced by V. Kac in [13], have attracted considerable attention of both mathematicians and physicists in the last 40 years. They are especially interesting due to their resemblance to the general linear Lie algebras $\mathfrak{gl}_n$ on the one hand, and because of the unique properties of their structure and representations on the other. By the term "Q-type superalgebras" we mean four series of Lie superalgebras: $\mathfrak{q}(n)$ $(n \geq 2)$ and its subquotients $\mathfrak{sq}(n)$, $\mathfrak{pq}(n)$, $\mathfrak{psq}(n)$ (the last one is a simple Lie superalgebra for $n \geq 3$, and in the notation of [13] it is $Q(n)$).

The Q-type Lie superalgebras are rather special in several aspects: their Cartan subalgebras $\mathfrak{h}$ are not abelian and have non-trivial odd part $\mathfrak{h}_{\bar{1}}$; they possess a non-degenerate invariant bilinear form which is *odd*; and they do not have quadratic Casimir elements. Because $\mathfrak{h}_{\bar{1}} \neq 0$, the study of highest weight modules of the Q-type Lie superalgebras requires nonstandard technique, including Clifford algebra methods. The latter are necessary due to the fact that the highest weight space of an irreducible highest weight module $L(\lambda)$ has a Clifford module structure. This

M. Gorelik
Department of Mathematics, Weizmann Institute of Science, Rehovot 76100, Israel
e-mail: maria.gorelik@weizmann.ac.il

D. Grantcharov (✉)
Department of Mathematics, University of Texas at Arlington, Arlington, TX 76021, USA
e-mail: grandim@uta.edu

M. Gorelik, P. Papi (eds.): *Advances in Lie Superalgebras.* Springer INdAM Series 7,
DOI 10.1007/978-3-319-02952-8_5,

peculiarity leads to the existence of two different candidates for a role of Verma module of the highest weight $\lambda \in \mathfrak{h}_{\bar{0}}^*$: a module $M(\lambda)$ which is induced from a simple $\mathfrak{h}_{\bar{0}}$-module $\mathbb{C}_\lambda$ and a module $N(\lambda)$ which is induced from a simple $\mathfrak{h}$-module. The character of $M(\lambda)$ nicely depends on λ, and following the kind suggestion of J. Bernstein, we call $M(\lambda)$ a *Verma module* and $N(\lambda)$ a *Weyl module*. Each Verma module $M(\lambda)$ has a finite filtration with the factors isomorphic to $N(\lambda)$ up to a parity change; each Weyl module $N(\lambda)$ has a unique simple quotient, which we denote by $L(\lambda)$. The simple highest weight $\mathfrak{gl}_n$-module of highest weight λ will be denoted by $\dot{L}(\lambda)$.

Note that from categorical point of view it is more natural to call $N(\lambda)$ Verma modules since they are proper standard modules whereas $M(\lambda)$ are standard modules, see [2].

The representation theory of finite dimensional $L(\lambda)$ is well developed. In [24] A. Sergeev established several important results, including a character formula of $L(\lambda)$ for the so called *tensor modules*, i.e. submodules of tensor powers $(\mathbb{C}^{n|n})^{\otimes r}$ of the natural $\mathfrak{q}(n)$-module $\mathbb{C}^{n|n}$. The characters of all simple finite-dimensional $\mathfrak{q}(n)$-modules have been found by I. Penkov and V. Serganova in 1996 (see [21] and [22]) via an algorithm using a supergeometric version of the Borel-Weil-Bott Theorem. This result was reproved by J. Brundan, [1] using a different approach. Very recently, using Brundan's idea and weight diagrams a character formula and a dimension formula for a finite dimensional $L(\lambda)$ were provided by Y. Su and R.B. Zhang in [28]. On the other hand the character formula problem for infinite dimensional $L(\lambda)$ remains largely open, see the conjecture in [1].

The centres of the universal enveloping algebras of the Q-type Lie supealgebras were described by Sergeev and the first author in [5, 26]. An equivalence of categories of strongly typical $\mathfrak{q}(n)$-modules and categories of $\mathfrak{gl}_n$-modules were established recently in [3].

The simple weight modules with finite weight multiplicities of all finite dimensional simple Lie superalgebras were partly classified by Dimitrov, Mathieu, and Penkov in [4]. The most interesting missing case in the classification of [4] is the case of the queer Lie superalgebras $\mathfrak{psq}(n)$. The classification in this case was completed in [6] using a new combinatorial tool - the star action. This action is a mixture of the dot action and the regular action of W depending on the atypicality of the weights.

The combinatorics of the queer Lie superalgebras is also very interesting. One important aspect of the Sergeev duality is the semisimplicity of the category of tensor modules of $\mathfrak{q}(n)$. This naturally raises the question of uniqueness and existence of a crystal bases theory for this category. The crystal bases theory and the combinatorial description of the crystals of the simple tensor modules were obtained in a series of papers of the second author and J. Jung, S.-J. Kang, M. Kashiwara, M. Kim, [7–9].

The goal of the paper is to present a survey on the recent results on the representation theory of the Q-type Lie superalgebras discussed above.

1.1 Content of the paper

The organization of the paper is as follows. In Sect. 2 we include some important definitions and preliminary results. Section 3 is devoted to the description of the centers of the Lie superalgebras of type Q. In Sect. 4 we collect the main results related to the classification of all simple weight $\mathfrak{q}(n)$-modules with finite weight multiplicities. Automorphisms and affine Lie superalgebras of type Q are discussed in Sect. 5. The last section deals with the crystal base theory of the category of tensor representations of $\mathfrak{q}(n)$.

2 Preliminaries

The symbol $\mathbb{Z}_{\geq 0}$ stands for the set of non-negative integers and $\mathbb{Z}_{>0}$ for the set of positive integers.

Let $V = V_{\bar{0}} \oplus V_{\bar{1}}$ be a $\mathbb{Z}_2$-graded vector space. We denote by $\dim V$ the total dimension of V. For a homogeneous element $u \in V$ we denote by $p(u)$ its $\mathbb{Z}_2$-degree; in all formulae where this notation is used, u is assumed to be $\mathbb{Z}_2$-homogeneous. For a subspace $N \subset V$ we set $N_i := N \cap V_i$ for $i = \bar{0}, \bar{1}$. Let Π be the functor which switches parity, i.e. $(\Pi V)_{\bar{0}} = V_{\bar{1}}, (\Pi V)_{\bar{1}} = V_{\bar{0}}$. We denote by $V^{\oplus r}$ the direct sum of r-copies of V.

For a Lie superalgebra $\mathfrak{g}$ we denote by $\mathscr{U}(\mathfrak{g})$ its universal enveloping algebra and by $\mathscr{S}(\mathfrak{g})$ its symmetric algebra.

Throughout the paper the base field is $\mathbb{C}$ and $\mathfrak{g} = \mathfrak{g}_{\bar{0}} \oplus \mathfrak{g}_{\bar{1}}$ denote one (unless otherwise specified, an arbitrary one) of Q-type Lie superalgebras $\mathfrak{q}(n), \mathfrak{sq}(n)$ for $n \geq 2$, $\mathfrak{pq}(n), \mathfrak{psq}(n)$ for $n \geq 3$.

2.1 Q-type Lie superalgebras

Recall that $\mathfrak{q}(n)$ consists of the matrices with the block form

$$X_{A,B} := \begin{pmatrix} A & B \\ B & A \end{pmatrix}$$

where A, B are arbitrary $n \times n$ matrices; $\mathfrak{q}(n)_{\bar{0}} = \{X_{A,0}\} \cong \mathfrak{gl}_n$, $\mathfrak{q}(n)_{\bar{1}} = \{X_{0,B}\}$ and

$$[X_{A,0}, X_{A',0}] = X_{[A,A'],0}, \quad [X_{A,0}, X_{0,B}] = X_{0,[A,B]}, \quad [X_{0,B}, X_{0,B'}] = X_{0,BB'+B'B}.$$

Define $\mathrm{tr}' : \mathfrak{q}(n) \to \mathbb{C}$ by $\mathrm{tr}'(X_{A,B}) = \mathrm{tr}\, B$. In this notation,

$$\begin{aligned} \mathfrak{sq}(n) &:= \{x \in \mathfrak{q}(n) \mid \mathrm{tr}' x = 0\}, \\ \mathfrak{pq}(n) &:= \mathfrak{q}(n)/(\mathrm{Id}), \\ \mathfrak{psq}(n) &:= \mathfrak{sq}(n)/(\mathrm{Id}), \end{aligned}$$

where Id is the identity matrix.

These definitions are illustrated by the following diagram:

$$\begin{array}{ccc} \mathfrak{sq}(n) & \hookrightarrow & \mathfrak{q}(n) \\ \downarrow & & \downarrow \\ \mathfrak{psq}(n) & \hookrightarrow & \mathfrak{pq}(n) \end{array}$$

Clearly, the category of $\mathfrak{pq}(n)$-modules (resp., $\mathfrak{psq}(n)$-modules) is the subcategory of $\mathfrak{q}(n)$-modules (resp., of $\mathfrak{sq}(n)$-modules) which are killed by the identity matrix Id.

The map $(x,y) \mapsto \mathrm{tr}'(xy)$ gives an odd non-degenerate invariant symmetric bilinear form on $\mathfrak{q}(n)$ and on $\mathfrak{psq}(n)$.

For the quotient algebras $\mathfrak{pq}(n), \mathfrak{psq}(n)$ we denote by $X_{A,B}$ the image of the corresponding element in the appropriate algebra.

For Q-type Lie superalgebras the set of even roots $(\Delta^+_{\bar{0}})$ coincides with the set of odd roots $(\Delta^+_{\bar{1}})$. This phenomenon has two obvious consequence. The first one is that all triangular decompositions of a Q-type Lie superalgebra are conjugate with respect to inner automorphisms (this does not hold for other simple Lie superalgebras). The second one is that the Weyl vector $\rho := \frac{1}{2}(\sum_{\alpha\in\Delta^+_{\bar{0}}} \alpha - \sum_{\alpha\in\Delta^+_{\bar{1}}} \alpha)$ is equal to zero. We set $\rho_0 := \frac{1}{2}\sum_{\alpha\in\Delta^+_{\bar{0}}} \alpha$.

We choose the natural triangular decomposition: $\mathfrak{q}(n) = \mathfrak{n}^- \oplus \mathfrak{h} \oplus \mathfrak{n}^+$ where $\mathfrak{h}_{\bar{0}}$ consists of the elements $X_{A,0}$ where A is diagonal, $\mathfrak{h}_{\bar{1}}$ consists of the elements $X_{0,B}$ where B is diagonal, and $\mathfrak{n}^+$ (resp., $\mathfrak{n}^-$) consists of the elements $X_{A,B}$ where A,B are strictly upper-triangular (resp., lower-triangular). We consider the induced triangular decompositions of $\mathfrak{sq}(n), \mathfrak{pq}(n), \mathfrak{psq}(n)$.

2.2 Notation

In the standard notation the set of roots of $\mathfrak{gl}_n = \mathfrak{q}(n)_{\bar{0}}$ can be written as

$$\Delta^+ = \{\varepsilon_i - \varepsilon_j\}_{1\le i<j\le n}$$

and the set of simple roots as $\pi := \{\varepsilon_1 - \varepsilon_2, \dots, \varepsilon_{n-1} - \varepsilon_n\}$. Each root space has dimension $(1|1)$.

For $\alpha \in \Delta^+$ let $s_\alpha : \mathfrak{h}^*_{\bar{0}} \to \mathfrak{h}^*_{\bar{0}}$ be the corresponding reflection: $s_{\varepsilon_i-\varepsilon_j}(\varepsilon_i) = \varepsilon_j$, $s_{\varepsilon_i-\varepsilon_j}(\varepsilon_k) = \varepsilon_k$ for $k \neq i,j$. Denote by W the Weyl group of $\mathfrak{g}_{\bar{0}}$ that is the group generated by $s_\alpha : \alpha \in \Delta^+$. Recall that W is generated by $s_\alpha : \alpha \in \pi$.

The space $\mathfrak{h}^*_{\bar{0}}$ has the standard non-degenerate W-invariant bilinear form: $(\varepsilon_i, \varepsilon_j) = \delta_{ij}$.

Let E_{rs} be the elementary matrix: $E_{rs} = (\delta_{ir}\delta_{sj})^n_{i,j=1}$.

The elements

$$h_i := X_{E_{ii},0}$$

form the standard basis of $\mathfrak{h}_{\bar{0}}$ for $\mathfrak{g} = \mathfrak{q}(n), \mathfrak{sq}(n)$. We use the notation h_i also for the image of h_i in the quotient algebras $\mathfrak{pq}(n), \mathfrak{psq}(n)$.

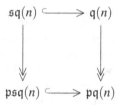

The elements $H_i := X_{0,E_{ii}}$ $(i = 1, \ldots, n)$ form a convenient basis of $\mathfrak{h}_{\bar{1}} \subset \mathfrak{q}(n)$; they satisfy the relations $[H_i, H_j] = 2\delta_{ij} h_i$.

For each positive root $\alpha = \varepsilon_i - \varepsilon_j$ we define $\overline{\alpha} = \varepsilon_i + \varepsilon_j$, and

$$\begin{array}{ll} h_\alpha := h_i - h_j,\ h_{\overline{\alpha}} := h_i + h_j, & H_\alpha := H_i - H_j, \\ e_\alpha := X_{E_{ij},0}, & E_\alpha := X_{0,E_{ij}}, \\ f_\alpha := X_{E_{ji},0}, & F_\alpha := X_{0,E_{ji}}. \end{array}$$

All above elements are non-zero in $\mathfrak{sq}(n), \mathfrak{pq}(n), \mathfrak{psq}(n)$ (since we excluded the cases $\mathfrak{pq}(2), \mathfrak{psq}(2)$).

The elements $h_\alpha, e_\alpha, f_\alpha$ $(\alpha \in \Delta^+)$ span $\mathfrak{sl}_n = [\mathfrak{gl}_n, \mathfrak{gl}_n]$; the elements E_α (resp., F_α) form the natural basis of $\mathfrak{n}^+_{\bar{1}}$ (resp., of $\mathfrak{n}^-_{\bar{1}}$) and the elements H_α span $\mathfrak{h}_{\bar{1}} \cap \mathfrak{sq}(n)$.

For each α the elements $h_\alpha, e_\alpha, f_\alpha, h_{\overline{\alpha}}, H_\alpha, E_\alpha, F_\alpha$ span $\mathfrak{sq}(2)$ and one has

$$[e_\alpha, f_\alpha] = h_\alpha,\ \ [E_\alpha, F_\alpha] = h_{\overline{\alpha}},\ \ [H_\alpha, H_\alpha] = 2h_{\overline{\alpha}}$$
$$[E_\alpha, f_\alpha] = [e_\alpha, F_\alpha] = H_\alpha.$$

Set

$$Q(\pi) := \sum_{\alpha \in \Delta^+} \mathbb{Z}\alpha, \quad Q^+(\pi) := \sum_{\alpha \in \Delta^+} \mathbb{Z}_{\geq 0}\alpha.$$

Define a partial order on $\mathfrak{h}^*_{\bar{0}}$ by $\nu \geq \mu$ iff $\nu - \mu \in Q^+(\pi)$.

2.3 The algebra $\mathscr{U}(\mathfrak{h})$

Let $\mathfrak{g}$ be a Q-type Lie superalgebra. Denote by HC the Harish-Chandra projection HC : $\mathscr{U}(\mathfrak{g}) \to \mathscr{U}(\mathfrak{h})$ along the decomposition $\mathscr{U}(\mathfrak{g}) = \mathscr{U}(\mathfrak{h}) \oplus (\mathscr{U}(\mathfrak{g})\mathfrak{n}^+ + \mathfrak{n}^-\mathscr{U}(\mathfrak{g}))$.

The algebra $\mathscr{U}(\mathfrak{h})$ is a Clifford superalgebra over the polynomial algebra $\mathscr{S}(\mathfrak{h}_0)$: $\mathscr{U}(\mathfrak{h})$ is generated by the odd space $\mathfrak{h}_{\bar{1}}$ endowed by the $\mathscr{S}(\mathfrak{h}_0)$-valued symmetric bilinear form $b(H, H') = [H, H']$. For each $\lambda \in \mathfrak{h}^*_{\bar{0}}$ the evaluation of $\mathscr{U}(\mathfrak{h})$ at λ is a complex Clifford superalgebra. Notice that a non-degenerate complex Clifford superalgebra is either the matrix algebra (if $\dim \mathfrak{h}_{\bar{1}}$ is even) or the algebra $Q(n)$ (this is an associative algebra whose Lie algebra is $\mathfrak{q}(n)$), see [5] for details. In particular, it possesses a supertrace which is even if $\dim \mathfrak{h}_{\bar{1}}$ is even and odd if $\dim \mathfrak{h}_{\bar{1}}$ is odd.

For $\lambda \in \mathfrak{h}^*_{\bar{0}}$ let $\mathbb{C}(\lambda)$ be the corresponding one-dimensional $\mathfrak{h}_{\bar{0}}$-module. Set

$$\mathscr{C}\ell(\lambda) := \mathscr{U}(\mathfrak{h}) \otimes_{\mathfrak{h}_{\bar{0}}} \mathbb{C}_\lambda.$$

Clearly, $\mathscr{C}\ell(\lambda)$ is isomorphic to a complex Clifford algebra generated by $\mathfrak{h}_{\bar{1}}$ endowed by the evaluated symmetric bilinear form $b_\lambda(H, H') := [H, H'](\lambda)$. Set

$$c(\lambda) := \dim \operatorname{Ker} b_\lambda.$$

For $\mathfrak{g} = \mathfrak{q}(n)$, $c(\lambda)$ is the number of zeros among $h_1(\lambda), \ldots, h_n(\lambda)$. The complex Clifford algebra $\mathscr{C}\ell(\lambda)$ is non-degenerate if and only if $c(\lambda) = 0$.

Denote by $E(\lambda)$ a simple $\mathscr{C}\ell(\lambda)$-module (up to a grading shift, such a module is unique). One has $\dim E(\lambda) = 2^{[\frac{\dim \mathfrak{h}_{\bar{1}} + 1 - c(\lambda)}{2}]}$.

2.4 Highest weight modules

Set $\mathfrak{b} := \mathfrak{h} + \mathfrak{n}^+, \mathfrak{b}^- := \mathfrak{h} + \mathfrak{n}^-$. Endow $\mathscr{C}\ell(\lambda)$ with the $\mathfrak{b}$-module structure via the trivial action of $\mathfrak{n}^+$. Set

$$M(\lambda) := \mathrm{Ind}_{\mathfrak{b}}^{\mathfrak{g}} \mathscr{C}\ell(\lambda), \quad N(\lambda) := \mathrm{Ind}_{\mathfrak{b}}^{\mathfrak{g}} E(\lambda).$$

Clearly, $M(\lambda)$ has a finite filtration with the factors isomorphic to $N(\lambda)$ up to parity change. We call $M(\lambda)$ a *Verma module* and $N(\lambda)$ a *Weyl module.*

For a diagonalizable $\mathfrak{h}_{\overline{0}}$-module N and a weight $\mu \in \mathfrak{h}_{\overline{0}}^*$ denote by N^μ the corresponding weight space. Say that a module N has the highest weight λ if $N = \sum_{\mu \leq \lambda} N^\mu$ and $N^\lambda \neq 0$. If all weight spaces N^μ are finite-dimensional we put $\mathrm{ch} N := \sum_\mu \dim N^\mu e^\mu$.

If N has a highest weight we denote by $\overline{N}$ the sum of all submodules which do not meet the highest weight space of N. Recall that $L(\lambda) = N(\lambda)/\overline{N}(\lambda)$.

The following conjecture is based on a discussion with V. Mazorchuk.

Conjecture *For any Q-type Lie superalgebra, and any nonzero weight* λ, $\mathrm{ch} L(\lambda)_{\overline{0}} = \mathrm{ch} L(\lambda)_{\overline{1}}$.

The above conjecture is verified for all but finitely many λ.

2.5 Example: $n = 2$

For $\mathfrak{sq}(2)$ the Cartan algebra is spanned by the even elements $h := h_\alpha, h' := h_{\overline{\alpha}}$ and the odd element $H := H_\alpha$.

The module $N(\lambda)$ is simple if $\lambda(h') \neq 0$ and $\lambda(h) \notin \mathbb{Z}_{>0}$. If $\lambda(h') = 0$, the simple $\mathfrak{sq}(2)$-module coincides with the simple $\mathfrak{gl}_2$-module $L_{\mathfrak{gl}_2}(\lambda)$; if $\lambda(h') \neq 0, \lambda(h) \in \mathbb{Z}_{>0}$, then $L(\lambda) = L_{\mathfrak{gl}(2)}(\lambda)^{\oplus 2}$ if $\lambda(h) = 1$ and $L(\lambda) = L_{\mathfrak{gl}(2)}(\lambda)^{\oplus 2} \oplus L_{\mathfrak{gl}(2)}(\lambda - \alpha)^{\oplus 2}$ if $\lambda(h) \neq 1$. This can be illustrated by the following diagrams: the module $L(\lambda)$ for $\lambda(h') \neq 0, \lambda(h) = 1$ is of the form

and the module $L(\lambda)$ for $\lambda(h') \neq 0, \lambda(h) = 4$ is of the form

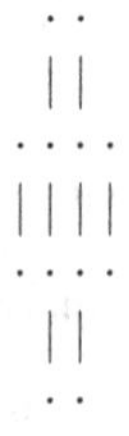

where the dots on the same level represent the vectors of the same weight and the difference between levels is equal to α; the vertical lines correspond to the action of f_α (so the dots in the same column represent a simple $\mathfrak{gl}_2$-module).

All simple module over $\mathfrak{q}(2), \mathfrak{sq}(2)$ and their quotients $\mathfrak{pq}(2), \mathfrak{psq}(2)$ are classified by V. Mazorchuk in [18].

3 Centres

3.1 Centre of enveloping algebra

A weight λ is called atypical if there exists $\alpha \in \Delta$ such that $h_{\overline{\alpha}}(\lambda) = 0$. The centres of the universal enveloping algebras of Q-type Lie algebras is given by the following theorem.

Theorem *Let $\mathfrak{g}$ be a Q-type Lie superalgebra, $\mathfrak{g} \neq \mathfrak{pq}(2), \mathfrak{psq}(2)$. The restriction of* HC *to $\mathscr{Z}(\mathfrak{g})$ is an algebra isomorphism $\mathscr{Z}(\mathfrak{g}) \xrightarrow{\sim} Z$ where Z is the set of W-invariant polynomial functions on $\mathfrak{h}_{\overline{0}}^*$ which are constant along each straight line parallel to a root α and lying in the hyperplane $h_{\overline{\alpha}}(\lambda) = 0$. In other words,*

$$Z := \mathscr{S}(\mathfrak{h}_0)^W \cap \bigcap_{\alpha \in \Delta} Z_\alpha,$$

where

$$Z_\alpha := \{f \in \mathscr{S}(\mathfrak{h}_0) \mid h_{\overline{\alpha}}(\lambda) = 0 \implies f(\lambda) = f(\lambda - c\alpha)\ \forall c \in \mathbb{C}\}.$$

The theorem is proven in [5, 26]. One has $\mathscr{Z}(U(\mathfrak{q}(n))) = \mathscr{Z}(U(\mathfrak{sq}(n)))$ and $\mathscr{Z}(U(\mathfrak{pq}(n))) = \mathscr{Z}(U(\mathfrak{psq}(n)))$.

3.2 Strongly typical weights

An element a of an associative superalgebra U is called *anticentral* if $ax - (-1)^{p(x)(p(a)+1)}xa = 0$. We denote by $\mathscr{A}(U)$ the set of anticentral elements of U.

Let $\mathfrak{g}$ be a Q-type Lie superalgebra. The anticentre of the Clifford algebra $U(\mathfrak{h})$ is equal to $\mathscr{S}(\mathfrak{h}_0)T_{\mathfrak{h}}$, where the parity of $T_{\mathfrak{h}}$ is equal to the parity of $\dim \mathfrak{h}_{\overline{1}}$ and

$$t_{\mathfrak{h}} := T_{\mathfrak{h}}^2 = \begin{cases} \pm h_1 \dots h_n & \text{for } \mathfrak{g} = \mathfrak{q}(n), \mathfrak{pq}(n) \\ \pm \sum h_1 \dots \hat{h}_i \dots h_n & \text{for } \mathfrak{g} = \mathfrak{sq}(n), \mathfrak{psq}(n). \end{cases}$$

The Harish-Chandra projection provides a linear monomorpism $\mathrm{HC} : \mathscr{A}(U(\mathfrak{g})) \xrightarrow{\sim} \mathscr{A}(U(\mathfrak{h}))$ and the image is equal to $\mathscr{S}(\mathfrak{h}_0)^W T_{\mathfrak{g}}$, where the parity of $T_{\mathfrak{g}}$ is equal to the parity of $\dim \mathfrak{h}_{\overline{1}}$ and

$$\mathrm{HC}(T_{\mathfrak{g}}) = T_{\mathfrak{h}} \prod_{\alpha \in \Delta_{\overline{0}}^+} h_{\overline{\alpha}}.$$

We say that $\lambda \in \mathfrak{h}_{\overline{0}}^*$ is *strongly typical* if $(t_{\mathfrak{h}} \prod_{\alpha \in \Delta_{\overline{0}}^+} h_{\overline{\alpha}})(\lambda) \neq 0$. Note that λ is strongly typical if and only if $T_{\mathfrak{g}} M(\lambda) \neq 0$.

3.3 Equivalence of categories

Let $\mathcal{O}^{\mathfrak{g}}$ (resp., $\mathcal{O}^{\mathfrak{g}_0}$) be the $\mathcal{O}$-category for $\mathfrak{g}$ and $\mathfrak{g}_0$-respectively. We have the natural restriction functor $Res : \mathcal{O}^{\mathfrak{g}} \to \mathcal{O}^{\mathfrak{g}_0}$ which sends a $\mathfrak{g}$-module $M = M_{\overline{0}} \oplus M_{\overline{1}}$ to the $\mathfrak{g}_0$-module M_0, and its left adjoint functor $Ind : \mathcal{O}^{\mathfrak{g}_0} \to \mathcal{O}^{\mathfrak{g}}$.

The action of the centres of the universal enveloping algebras lead to the block decomposition $\mathcal{O}^{\mathfrak{g}} = \bigoplus \mathcal{O}^{\mathfrak{g}}_{\chi}, \mathcal{O}^{\mathfrak{g}_0} = \bigoplus \mathcal{O}^{\mathfrak{g}_0}_{\hat{\chi}}$ indexed by the central characters χ and $\hat{\chi}$ respectively. This gives the projection and inclusions functors $proj_{\chi} : \mathcal{O}^{\mathfrak{g}} \to \mathcal{O}^{\mathfrak{g}}_{\chi}, incl_{\chi} : \mathcal{O}^{\mathfrak{g}}_{\chi} \to \mathcal{O}^{\mathfrak{g}}$ and $proj_{\hat{\chi}} : \mathcal{O}^{\mathfrak{g}} \to \mathcal{O}^{\mathfrak{g}}_{\hat{\chi}}, incl_{\hat{\chi}} : \mathcal{O}^{\mathfrak{g}}_{\hat{\chi}} \to \mathcal{O}^{\mathfrak{g}}$.

We say that $\lambda \in \mathfrak{h}^*_{\overline{0}}$ is regular (resp., dominant, integral) if $\lambda(h_\alpha) \neq 0$ (resp., $\lambda(h_\alpha) \notin \mathbb{Z}_{<0}, \lambda(h_\alpha) \in \mathbb{Z}$) for each $\alpha \in \Delta^+$.

If $\mathfrak{g} = \mathfrak{q}(n)$, then a weight $\lambda = (\lambda_1, \dots, \lambda_n)$ ($\lambda_i := \lambda(h_i)$) is a regular dominant strongly typical weight if and only if $\lambda_j - \lambda_i \notin \mathbb{Z}_{\geq 0}$ for $j > i$, $\lambda_i + \lambda_j \neq 0$ for $j > i$, and $\lambda_i \neq 0$ for all i.

Let χ (resp., $\hat{\chi}$) be the $\mathfrak{g}$ (resp., $\mathfrak{g}_0$) central character which corresponds to a strongly typical weight λ (so $L(\lambda) \in \mathcal{O}^{\mathfrak{g}}_{\chi}, \dot{L}(\lambda) \in \mathcal{O}^{\mathfrak{g}_0}_{\hat{\chi}}$). We set $\tilde{\mathcal{O}}^{\mathfrak{g}}_{\chi} := \mathcal{O}^{\mathfrak{g}}_{\chi}$ if $\dim \mathfrak{h}_{\overline{1}}$ is odd; if $\dim \mathfrak{h}_{\overline{1}}$ is even one has a decomposition $\mathcal{O}^{\mathfrak{g}}_{\chi} = \tilde{\mathcal{O}}^{\mathfrak{g}}_{\chi} \oplus \Pi(\tilde{\mathcal{O}}^{\mathfrak{g}}_{\chi})$, where Π is the parity change functor. Note that for an integral weight λ the blocks $\mathcal{O}^{\mathfrak{g}_0}_{\hat{\chi}}, \tilde{\mathcal{O}}^{\mathfrak{g}}_{\chi}$ are indecomposable.

The functors $F := proj_{\chi} \circ Ind \circ incl_{\hat{\chi}} : \mathcal{O}^{\mathfrak{g}_0}_{\hat{\chi}} \to \tilde{\mathcal{O}}^{\mathfrak{g}}_{\chi}$ and $G := proj_{\hat{\chi}} \circ Res \circ incl_{\chi} : \tilde{\mathcal{O}}^{\mathfrak{g}}_{\chi} \to \mathcal{O}^{\mathfrak{g}_0}_{\hat{\chi}}$ are adjoint. The main result of [3] is that for a regular dominant strongly typical weight λ both functors F and G decompose in direct sums of k copies of some functors $F_1 : \mathcal{O}^{\mathfrak{g}_0}_{\hat{\chi}} \to \tilde{\mathcal{O}}^{\mathfrak{g}}_{\chi}$ and $G_1 : \tilde{\mathcal{O}}^{\mathfrak{g}}_{\chi} \to \mathcal{O}^{\mathfrak{g}_0}_{\hat{\chi}}$ respectively and the functors F_1, G_1 are mutually inverse equivalences of categories.

4 Bounded, cuspidal and weight modules of $\mathfrak{q}(n)$

4.1 Bounded weights

We call a weight $\lambda \in \mathfrak{h}^*_{\overline{0}}$ *bounded* if the set of weight multiplicities of $L(\lambda)$ is uniformly bounded, i.e. there exists a constant C such that the $\dim L(\lambda)^{\nu} < C$ for all ν. Conditions when λ is bounded are obtained in [6]. These conditions are formulated in terms of the $*$-action, see below.

4.1.1 Definition

For $\lambda \in \mathfrak{h}^*_{\overline{0}}$ and $\alpha \in \pi$ we set $s_\alpha \cdot \lambda := s_\alpha(\lambda + \rho_0) - \rho_0$ and

$$s_\alpha * \lambda = \begin{cases} s_\alpha \lambda & \text{if } \lambda(h_{\overline{\alpha}}) \neq 0, \\ s_\alpha \cdot \lambda & \text{if } \lambda(h_{\overline{\alpha}}) = 0. \end{cases}$$

For $i = 1, \dots, n-1$ we set $s_i * \lambda := s_{\alpha_i} * \lambda$.

Note that $s_\alpha * s_\alpha * \lambda = \lambda$ and $s_\alpha * s_\beta * \lambda = s_\beta * s_\alpha * \lambda$ if $(\alpha, \beta) = 0$. Therefore the group $\widetilde{W}$ generated by the symbols $s_1, \dots, s_{n-1}$ subject to the relations $s_i^2 = 1,\ s_i s_j = s_j s_i$ for $i - j > 1$ acts on $\mathfrak{h}_{\bar{0}}^*$ via $*$-action. Note that $\widetilde{W}$ is an infinite Coxeter group.

4.1.2 Description of bounded weights

Recall that $\mathfrak{gl}_n$-module $\dot{L}(\lambda)$ is finite-dimensional if and only if $s_i \cdot \lambda < \lambda$ for each $i = 1, \dots, n-1$ (the partial order was introduced in §2.2).

The $\mathfrak{q}(n)$-module $L(\lambda)$ is finite-dimensional if and only if for each $i = 1, \dots, n-1$ one has $(\lambda, \varepsilon_i - \varepsilon_{i+1}) \in \mathbb{Z}_{>0}$ or $(\lambda, \varepsilon_i) = (\lambda, \varepsilon_{i+1}) = 0$, see [20], which can be rewritten as $s_i * \lambda < \lambda$ for each $i = 1, \dots, n-1$.

For each weight μ there exists a sequence $\mu = \mu_0 < \mu_1 < \mu_2 < \dots < \mu_s$ such that $\mu_{i+1} = s_{k_i} * \mu_i$ for some $k_i \in \{1, 2, \dots, n-1\}$ and μ_s is $W*$-maximal (i.e., $s_i * \mu \not< \mu$ for each i). We call such sequence a *$W*$-increasing string* starting at μ.

Bounded weights for $\mathfrak{gl}_n$ were described in [17]. For an integral weight μ the conditions on μ being bounded can be reformulated as follows: μ is bounded if and only if

(i) there exists a unique increasing W-string $\mu = \mu_0 < \mu_1 < \mu_2 < \dots < \mu_s$;
(ii) the set $\{i : s_i \cdot \mu_j = \mu_j\}$ is empty for $j < s$ and has cardinality at most one for $j = s$.

In [6] we proved that the same description for bounded weights is valid for $\mathfrak{q}(n)$ if we change the dot action by the $*$-action. The non-integral bounded weights can be also described in terms of the $*$-action.

4.2 Example: the case $n = 3$

4.2.1 The case $\mathfrak{gl}_3$

Consider first the case $\mathfrak{gl}_3$. There are three types of W-orbits $W \cdot \lambda$ for integral λ: the trivial orbit for $\lambda + \rho_{\bar{0}} = (a, a, a)$ (these weights are not bounded), the regular orbits which contain six elements (each element has a trivial stabilizer) and singular orbits which contain three elements (the stabilizer of each element is $\mathbb{Z}_2$). The regular orbits are of the form:

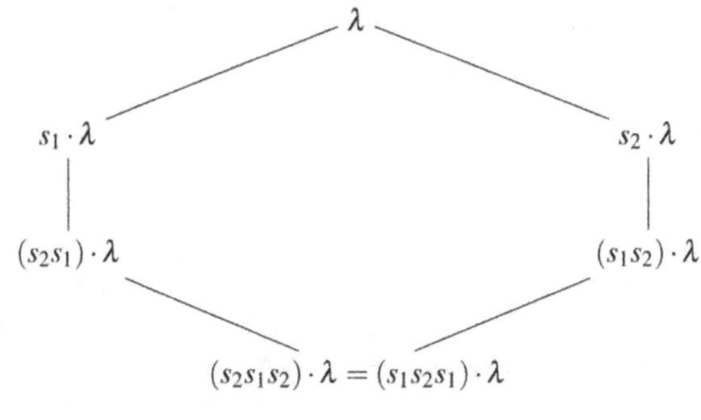

The non-trivial singular orbits are of the form:

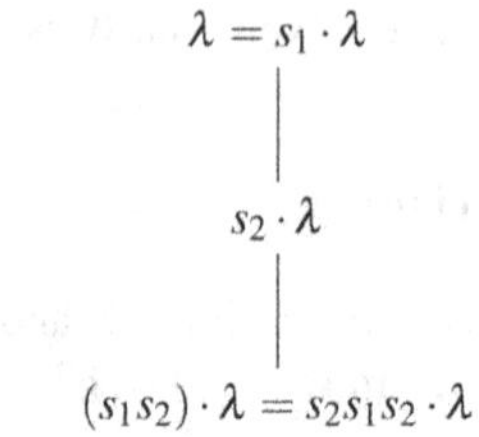

(or the same with interchanged s_1, s_2).

The edges of the diagrams correspond to simple reflections s_1, s_2 and the upper vertex in a given edge is bigger with respect to the partial order. We say that a vertex is a *top* (resp., *bottom*) vertex if there is no edge ascending (resp., descending) from this vertex.

Note that $\dot{L}(\lambda)$ is finite-dimensional if and only if λ is represented by a top vertex which belongs to $n-1$ edges. Hence $\dot{L}(\lambda)$ is finite-dimensional if and only if λ is the top vertex in the regular orbit (the diagram above).

The increasing strings are represented by the paths going in upward direction, for instance $s_1 s_2 s_1 \cdot \lambda < s_2 s_1 \cdot \lambda < s_1 \lambda < \lambda$. The condition (i) for a given vertex means that there exists a unique ascending path; the condition (ii) means that each vertex in this path, except the top one, belongs to $n-1$ edges and the top one belongs to at least $n-2$ edges.

We see that all non-bottom vertices represent bounded weights for $\mathfrak{gl}_3$.

4.2.2 The case $\mathfrak{q}(3)$

We now look at the $\widetilde{W}$-orbits in the case $\mathfrak{q}(3)$. There are 6 types of $\widetilde{W}$-orbits, which we describe below (up to the interchange s_1 and s_2).

(1) The trivial orbit corresponds to the case $\lambda = (a,a,a), a \neq 0$; these weights are not bounded.

(2) The orbits of the form

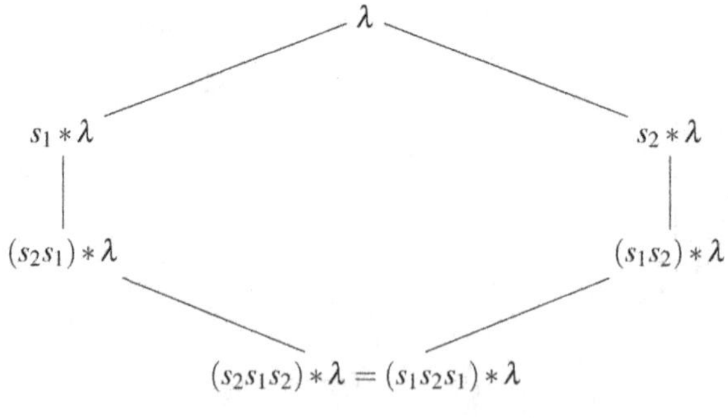

(3) The orbit

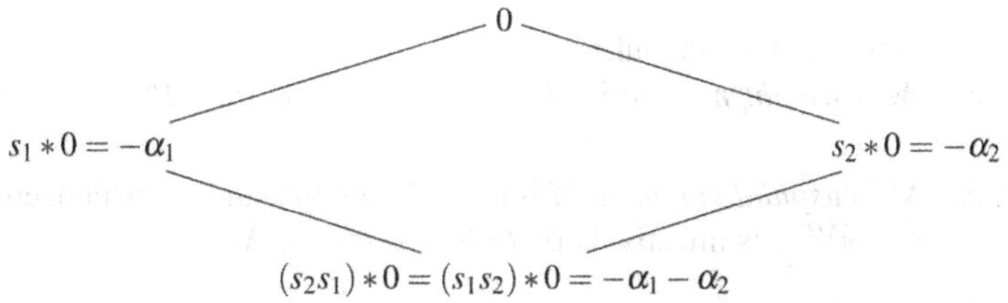

(4) The orbits of the form

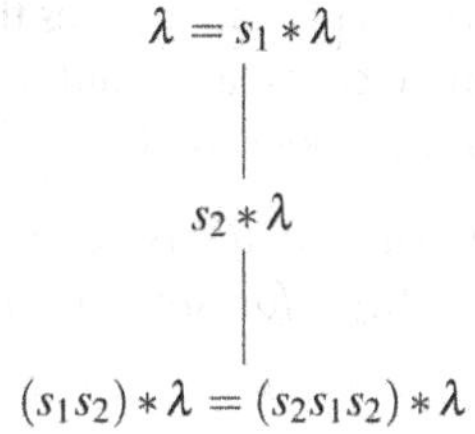

(5) The orbits of the form

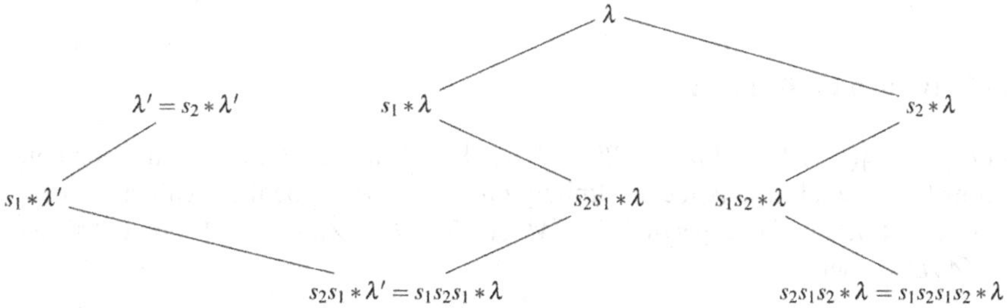

with $\lambda' = \lambda + \alpha_1$.

(6) The orbits of the form

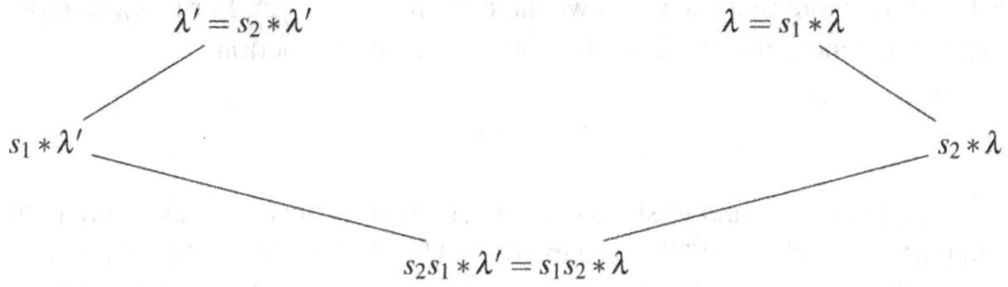

with $\lambda' = \lambda + \alpha_1$.

We see that $L(\lambda)$ is finite-dimensional if and only if λ is a top vertex in one of the orbits (2), (3) or the right top (the highest) vertex in (5); the bounded weights correspond to the non-bottom vertices in the orbits (2)–(6).

4.3 Cuspidal and weight modules

Definition Let M be a $\mathfrak{q}(n)$-module.

(i) We call M a *weight module* if $M = \bigoplus_{\lambda \in \mathfrak{h}_{\bar{0}}^*} M^\lambda$ and $\dim M^\lambda < \infty$ for every $\lambda \in \mathfrak{h}_{\bar{0}}^*$.

(ii) We call M a *cuspidal module* if M is a weight module and every nonzero even root vector $e_\alpha \in \mathfrak{q}(n)^\alpha$ acts injectively on M for every $\alpha \in \Delta$.

Remark In many cases the condition $\dim M^\lambda < \infty$ in (i) is not included in the definition of a weight module. We include this condition for convenience.

The following theorem is proved in [4] and reduces the classification of all simple weight modules of $\mathfrak{q}(n)$ to all simple cuspidal modules of $\mathfrak{q}(n)$. For the definition of "parabolically induced" we refer the reader to [4].

Theorem *Every simple weight $\mathfrak{q}(n)$-module is parabolically induced from a cuspidal module over $\mathfrak{q}(n_1) \oplus \ldots \oplus \mathfrak{q}(n_k)$, for some positive integers $n_1, \ldots, n_k$ with $n_1 + \ldots + n_k = n$.*

To classify all simple cuspidal modules we used the so called "twisted localization" technique - we present every simple cuspidal as a twisted localization of a highest weight module. Some details are listed below.

4.3.1 Twisted localization

Set $U := U(\mathfrak{q}(n))$. Then $\mathbf{F}_\alpha := \{f_\alpha^n \mid n \in \mathbb{Z}_{\geq 0}\} \subset U$ satisfies Ore's localization conditions because $\operatorname{ad} f_\alpha$ acts locally finitely on U. Let $\mathscr{D}_\alpha U$ be the localization of U relative to $\mathbf{F}_\alpha$, and for a $\mathfrak{q}(n)$-module M, set $\mathscr{D}_\alpha M = \mathscr{D}_\alpha U \otimes_U M$. For $x \in \mathbb{C}$ and $u \in \mathscr{D}_\alpha U$ we set

$$\Theta_x(u) := \sum_{i \geq 0} \binom{x}{i} (\operatorname{ad} f_\alpha)^i(u)\, f_\alpha^{-i},$$

where $\binom{x}{i} = \frac{x(x-1)\ldots(x-i+1)}{i!}$. Since $\operatorname{ad} f_\alpha$ is locally nilpotent on $\mathscr{D}_\alpha U$, the sum above is actually finite. Note that for $x \in \mathbb{Z}$ we have $\Theta_x(u) = f_\alpha^x u f_\alpha^{-x}$. For a $\mathscr{D}_\alpha U$-module M by $\Phi_\alpha^x M$ we denote the $\mathscr{D}_\alpha U$-module M twisted by the action

$$u \cdot v^x := (\Theta_x(u) \cdot v)^x,$$

where $u \in \mathscr{D}_\alpha U$, $v \in M$, and v^x stands for the element v considered as an element of $\Phi_\alpha^x M$. In particular, $v^x \in M^{\lambda + x\alpha}$ whenever $v \in M^\lambda$. Set $\mathscr{D}_\alpha^x M := \Phi_\alpha^x(\mathscr{D}_\alpha M)$.

The classification of the simple cuspidal $\mathfrak{q}(n)$-modules is obtained in the following theorem proved in [6]. This together with the description of the bounded weights completes the classification of all simple weight $\mathfrak{q}(n)$-modules. The uniqueness part of the theorem involves the definition of a bounded weight type and it is skipped here.

4.3.2 Theorem

Let M be a simple cuspidal $\mathfrak{q}(n)$–module. Then there is a bounded weight λ, and a tuple $(x_1,\ldots,x_{n-1})$ of $n-1$ complex nonintegral numbers such that $M \simeq \mathscr{D}^{x_1}_{\alpha_1}\mathscr{D}^{x_2}_{\alpha_1+\alpha_2}\cdots\mathscr{D}^{x_{n-1}}_{\alpha_1+\ldots+\alpha_{n-1}}L(\lambda)$.

5 Automorphisms of $\mathfrak{q}(n)$ and affine Lie superalgebra $\mathfrak{q}(n)^{(2)}$

5.1 Automorphisms of Q-type Lie superalgebra

Although the Q-type superalgebras are not invariant with respect to the supertransposition, they are invariant with respect to the *q-supertransposition* $\sigma_q : \begin{pmatrix} A & B \\ B & A \end{pmatrix} \mapsto \begin{pmatrix} A^t & \zeta B^t \\ \zeta B^t & A^t \end{pmatrix}$, where $\zeta = \zeta_4 \in \mathbb{C}$ is a fixed primitive 4th root of unity.

Let $\mathfrak{g}$ be a Q-type Lie superalgebra. The natural homomorphisms $\mathrm{GL}_n(\mathbb{C}) \to \mathrm{Aut}\,\mathfrak{g}$ given by $X \mapsto \mathrm{Ad}_R(X,X)$, where

$$\mathrm{Ad}_R(X,X)\begin{pmatrix} A & B \\ B & A \end{pmatrix} = \begin{pmatrix} XAX^{-1} & XBX^{-1} \\ XBX^{-1} & XAX^{-1} \end{pmatrix}$$

induces an embedding $\mathrm{PSL}_n(\mathbb{C}) \to \mathrm{Aut}\,\mathfrak{g}$. In the light of [10, 23] one has $\mathrm{Aut}\,\mathfrak{g} = \mathrm{PSL}_n(\mathbb{C}) \times \mathbb{Z}_4$, where $\mathbb{Z}_4$ is generated by $-\sigma_q$.

5.2 Affine Lie superalgebra $\mathfrak{q}(n)^{(2)}$

Recall that semisimple Lie algebras are finite-dimensional Kac-Moody algebras and affine Lie algebras are Kac-Moody algebras of finite growth. Each affine Lie algebra is a (twisted or non-twisted) affinization of a finite-dimensional Kac-Moody algebra; this means that an affine Lie algebra can be described in terms of a finite-dimensional Kac-Moody algebra and its finite order automorphism, see [14, Chap. VI–VIII]. The Cartan matrices of finite-dimensional and affine Lie algebras are symmetrizable [14, Chap. IV].

The superalgebra generalization of Kac-Moody algebras was introduced in [13]; see [12,30] for details. Call a Kac-Moody superalgebra *affine* if it has a finite growth and *symmetrizable* if it has a symmetrizable Cartan matrix. The affine symmetrizable Lie superalgebras are classified in [23, 29], and, as in Lie algebra case, they are (twisted or non-twisted) affinizations of finite-dimensional Kac-Moody superalgebras. Non-symmetrizable affine Lie superalgebras were described in [12]. The classification includes two degenerate superalgebras, one family of constant rank and one series $\mathfrak{q}(n)^{(2)}$. The superalgebras of the series are twisted affinizations of $\mathfrak{sq}(n)$ corresponding to the automorphism $\sigma_q^2 : a \mapsto (-1)^{p(a)}a$.

A symmetrizable affine Lie superalgebra has an even non-degenerate invariant bilinear form and a Casimir element. The Q-type Lie superalgebras and the affine

Kac-Moody algebra $\mathfrak{q}(n)^{(2)}$ do not have even invariant bilinear forms, but have odd ones.

An interesting feature of $\mathfrak{q}(n)^{(2)}$ is that any Verma module is reducible; more precisely if $M(\lambda)$ is a Verma module over $\mathfrak{q}(n)^{(2)}$ and $\overline{M}(\lambda)$ is its maximal submodule (i.e., $L(\lambda) = M(\lambda)/\overline{M}(\lambda)$), then $\overline{M}(\lambda)^{\lambda-2\delta} \neq 0$, where δ is the minimal imaginary root of $\mathfrak{q}(n)^{(2)}$. The proof is given in [11].

5.2.1 Description of $\mathfrak{q}(n)^{(2)}$

We introduce $\mathfrak{q}(n)^{(1)} := \mathscr{L}(\mathfrak{sq}(n)) \oplus \mathbb{C}D$, where $\mathscr{L}(\mathfrak{sq}(n)) := \mathfrak{sq}(n) \otimes \mathbb{C}[t^{\pm 1}]$ is the loop superalgebra, D acts on $\mathscr{L}(\mathfrak{sq}(n))$ by $[D, x \otimes t^k] = kx \otimes t^k$. Note that $\mathfrak{sq}(n)$, $\mathfrak{q}(n)^{(1)}$ are not Kac-Moody superalgebras since their Cartan subalgebras contain odd elements.

Let ε be an automorphism of $\mathfrak{sq}(n)$ given by $\varepsilon(x) := (-1)^{p(x)}x$, i.e. $\varepsilon = \sigma_q^2$. We extend ε to $\mathfrak{q}(n)^{(1)}$ by $\varepsilon(t) = -t$, $\varepsilon(D) = D$. Then $\mathfrak{q}(n)^{(2)}$ is the quotient of the subalgebra $(\mathfrak{q}(n)^{(1)})^{\varepsilon}$ of elements fixed by ε by the abelian ideal $\sum_{i\neq 0} \mathbb{C}X_{I,0} \otimes t^{2i}$, where I stands for the $n \times n$ identity matrix (see Sect. 2.1 for notation). We may identify $\mathfrak{q}(n)^{(2)}$ with the vector space

$$\mathfrak{sl}_n \otimes \mathbb{C}[t^{\pm 1}] \oplus \mathbb{C}K \oplus \mathbb{C}D,$$

where $K := X_{I,0}$ and

$$\begin{aligned} \mathfrak{q}(n)^{(2)}_{\bar{0}} &= \mathfrak{gl}_n \oplus \mathbb{C}D \oplus \Big(\sum_{k\in\mathbb{Z}\setminus\{0\}} \mathfrak{sl}_n \times t^{2k} \Big) \\ &= \mathfrak{sl}_n \otimes \mathbb{C}[t^{\pm 2}] \oplus \mathbb{C}K \oplus \mathbb{C}D, \quad \mathfrak{q}(n)^{(2)}_{\bar{1}} = \mathfrak{sl}_n \otimes t\mathbb{C}[t^{\pm 2}]; \end{aligned}$$

the commutator is given by

$$[x \otimes t^k, y \otimes t^m] = \begin{cases} (xy - yx) \otimes t^{k+m}, \\ \qquad \text{if } km \text{ is even,} \\ [x \otimes t^k, y \otimes t^m] = \iota(xy + yx) \otimes t^{k+m} + 2\delta_{-k,m} \operatorname{tr}(xy)K, \\ \qquad \text{if } km \text{ is odd,} \end{cases}$$

where $\iota : \mathfrak{gl}_n \to \mathfrak{sl}_n$ is the natural map $\iota(x) := x - \operatorname{tr}(x)I/n$.

6 Crystal bases of $\mathfrak{q}(n)$

6.1 The quantum queer superalgebra

Let $\mathbf{F} = \mathbb{C}((q))$ be the field of formal Laurent series in an indeterminate q and let $\mathbf{A} = \mathbb{C}[[q]]$ be the subring of $\mathbf{F}$ consisting of formal power series in q. For $k \in \mathbb{Z}_{\geq 0}$, we define

$$[k] = \frac{q^k - q^{-k}}{q - q^{-1}}, \quad [0]! = 1, \quad [k]! = [k][k-1]\cdots[2][1].$$

For an integer $n \geq 2$, let $P = \mathbb{Z}\varepsilon_1 \oplus \cdots \oplus \mathbb{Z}\varepsilon_n$ and $P^\vee = \mathbb{Z}h_1 \oplus \cdots \oplus \mathbb{Z}h_n$. Then $\mathfrak{h}_{\bar{0}} = \mathbb{C} \otimes_{\mathbb{Z}} P^\vee$.

Definition The *quantum queer superalgebra* $U_q(\mathfrak{q}(n))$ is the unital superalgebra over $\mathbf{F}$ generated by the symbols e_i, f_i, E_i, F_i $(i = 1, \ldots, n-1)$, q^h $(h \in P^\vee)$, H_j $(j = 1, \ldots, n)$ with the following defining relations.

$$
\begin{aligned}
& q^0 = 1, \ q^h q^{h'} = q^{h+h'} \ (h, h' \in P^\vee), \\
& q^h e_i q^{-h} = q^{\alpha_i(h)} e_i \ (h \in P^\vee), \\
& q^h f_i q^{-h} = q^{-\alpha_i(h)} f_i \ (h \in P^\vee), \\
& q^h H_j = H_j q^h, \\
& e_i f_j - f_j e_i = \delta_{ij} \frac{q^{h_i - h_{i+1}} - q^{-h_i + h_{i+1}}}{q - q^{-1}}, \\
& e_i e_j - e_j e_i = f_i f_j - f_j f_i = 0 \quad \text{if } |i-j| > 1, \\
& e_i^2 e_j - (q + q^{-1}) e_i e_j e_i + e_j e_i^2 = 0 \quad \text{if } |i-j| = 1, \\
& f_i^2 f_j - (q + q^{-1}) f_i f_j f_i + f_j f_i^2 = 0 \quad \text{if } |i-j| = 1, \\
& H_i^2 = \frac{q^{2h_i} - q^{-2h_i}}{q^2 - q^{-2}}, \\
& H_i H_j + H_j H_i = 0 \ (i \neq j), \\
& H_i e_i - q e_i H_i = E_i q^{-h_i}, \\
& H_i f_i - q f_i H_i = -F_i q^{h_i}, \\
& e_i F_j - F_j e_i = \delta_{ij} (H_i q^{-h_{i+1}} - H_{i+1} q^{-h_i}), \\
& E_i f_j - f_j E_i = \delta_{ij} (H_i q^{h_{i+1}} - H_{i+1} q^{h_i}), \\
& e_i E_i - E_i e_i = f_i F_i - F_i f_i = 0, \\
& e_i e_{i+1} - q e_{i+1} e_i = E_i E_{i+1} + q E_{i+1} E_i, \\
& q f_{i+1} f_i - f_i f_{i+1} = F_i F_{i+1} + q F_{i+1} F_i, \\
& e_i^2 E_j - (q + q^{-1}) e_i E_j e_i + E_j e_i^2 = 0 \quad \text{if } |i-j| = 1, \\
& f_i^2 F_j - (q + q^{-1}) f_i F_j f_i + F_j f_i^2 = 0 \quad \text{if } |i-j| = 1.
\end{aligned}
$$

The generators e_i, f_i $(i = 1, \ldots, n-1)$, q^h $(h \in P^\vee)$ are regarded as *even* and E_i, F_i $(i = 1, \ldots, n-1)$, H_j $(j = 1, \ldots, n)$ as *odd*. From the defining relations, it is easy to see that the even generators together with H_1 generate the whole algebra $U_q(\mathfrak{q}(n))$.

The superalgebra $U_q(\mathfrak{q}(n))$ is a bialgebra with the comultiplication $\Delta : U_q(\mathfrak{q}(n)) \to U_q(\mathfrak{q}(n)) \otimes U_q(\mathfrak{q}(n))$ defined by

$$
\begin{aligned}
& \Delta(q^h) = q^h \otimes q^h \quad \text{for } h \in P^\vee, \\
& \Delta(e_i) = e_i \otimes q^{-h_i + h_{i+1}} + 1 \otimes e_i,
\end{aligned}
$$

$$\Delta(f_i) = f_i \otimes 1 + q^{h_i - h_{i+1}} \otimes f_i,$$
$$\Delta(H_1) = H_1 \otimes q^{h_1} + q^{-h_1} \otimes H_1.$$

6.2 The category $\mathscr{O}^{\geq 0}_{\mathrm{int}}$.

A $U_q(\mathfrak{q}(n))$-module M is called a *weight module* if M has a weight space decomposition $M = \bigoplus_{\mu \in P} M^\mu$, where

$$M^\mu = \left\{ m \in M\,; q^h m = q^{\mu(h)} m \;\text{ for all } h \in P^\vee \right\},$$

and $\dim M^\mu < \infty$ for every μ. The *set of weights of* M is defined to be

$$\mathrm{wt}(M) = \{\mu \in P; M^\mu \neq 0\}.$$

6.2.1 Definition

A weight $U_q(\mathfrak{q}(n))$-module V is called a *highest weight module with highest weight* $\lambda \in P$ if V^λ is finite-dimensional and satisfies the following conditions:

(i) V is generated by V^λ,
(ii) $e_i v = E_i v = 0$ for all $v \in V^\lambda$. $i = 1, \ldots, n-1$.

6.2.2 Strict partitions

Set

$$P^{\geq 0} = \{\lambda = \lambda_1 \varepsilon_1 + \cdots + \lambda_n \varepsilon_n \in P; \lambda_j \in \mathbb{Z}_{\geq 0} \;\text{ for all } j = 1, \ldots, n\},$$
$$\Lambda^+ = \{\lambda = \lambda_1 \varepsilon_1 + \cdots + \lambda_n \varepsilon_n \in P^{\geq 0}; \lambda_i \geq \lambda_{i+1} \text{ and } \lambda_i = \lambda_{i+1} \text{ implies } \lambda_i = \lambda_{i+1} = 0 \text{ for all } i = 1, \ldots, n-1\}.$$

Note that each element $\lambda \in \Lambda^+$ corresponds to a *strict partition* $\lambda = (\lambda_1 > \lambda_2 > \cdots > \lambda_r > 0)$. Thus we will often call $\lambda \in \Lambda^+$ a strict partition.

6.2.3 Example

Let

$$\mathbf{V} = \bigoplus_{j=1}^{n} \mathbf{F} v_j \oplus \bigoplus_{j=1}^{n} \mathbf{F} v_{\overline{j}}$$

be the vector representation of $U_q(\mathfrak{q}(n))$. The action of $U_q(\mathfrak{q}(n))$ on $\mathbf{V}$ is given as follows:

$$e_i v_j = \delta_{j,i+1} v_i,\;\; e_i v_{\overline{j}} = \delta_{j,i+1} v_{\overline{i}},\;\; f_i v_j = \delta_{j,i} v_{i+1},\;\; f_i v_{\overline{j}} = \delta_{j,i} v_{\overline{i+1}},$$
$$E_i v_j = \delta_{j,i+1} v_{\overline{i}},\; E_i v_{\overline{j}} = \delta_{j,i+1} v_i,\; F_i v_j = \delta_{j,i} v_{\overline{i+1}},\; F_i v_{\overline{j}} = \delta_{j,i} v_{i+1},$$
$$q^h v_j = q^{\varepsilon_j(h)} v_j,\; q^h v_{\overline{j}} = q^{\varepsilon_j(h)} v_{\overline{j}},\; H_i v_j = \delta_{j,i} v_{\overline{j}},\;\; H_i v_{\overline{j}} = \delta_{j,i} v_j.$$

Note that $\mathbf{V}$ is an irreducible highest weight module with highest weight ε_1 and $\mathrm{wt}(\mathbf{V}) = \{\varepsilon_1, \ldots, \varepsilon_n\}$.

6.2.4 Definition

We define $\mathscr{O}_{\mathrm{int}}^{\geq 0}$ to be the category of finite-dimensional weight modules M satisfying the following conditions:

(i) $\mathrm{wt}(M) \subset P^{\geq 0}$;
(ii) for any $\mu \in P^{\geq 0}$ and $i \in \{1, \ldots, n\}$ such that $\mu(h_i) = 0$, we have $H_i|_{M^\mu} = 0$.

The following proposition is proved in [7].

6.2.5 Proposition

Any irreducible $U_q(\mathfrak{q}(n))$-module in $\mathscr{O}_{\mathrm{int}}^{\geq 0}$ appears as a direct summand of a tensor power of $\mathbf{V}$.

6.3 Crystal bases in $\mathscr{O}_{\mathrm{int}}^{\geq 0}$

Let M be a $U_q(\mathfrak{q}(n))$-module in $\mathscr{O}_{\mathrm{int}}^{\geq 0}$. For $i = 1, 2, \ldots, n-1$, we define the *even Kashiwara operators* on M in the usual way. That is, for a weight vector $u \in M_\lambda$, consider the *i-string decomposition* of u:

$$u = \sum_{k \geq 0} f_i^{(k)} u_k,$$

where $e_i u_k = 0$ for all $k \geq 0$, $f_i^{(k)} = f_i^k / [k]!$, and define the even Kashiwara operators $\tilde{e}_i$, $\tilde{f}_i$ $(i = 1, \ldots, n-1)$ by

$$\tilde{e}_i u = \sum_{k \geq 1} f_i^{(k-1)} u_k,$$
$$\tilde{f}_i u = \sum_{k \geq 0} f_i^{(k+1)} u_k.$$

On the other hand, we define the *odd Kashiwara operators* $\widetilde{H}_1$, $\widetilde{E}_1$, $\widetilde{F}_1$ by

$$\begin{aligned}
\widetilde{H}_1 &= q^{h_1 - 1} H_1, \\
\widetilde{E}_1 &= -(e_1 H_1 - q H_1 e_1) q^{h_1 - 1}, \\
\widetilde{F}_1 &= -(H_1 f_1 - q f_1 H_1) q^{h_2 - 1}.
\end{aligned}$$

For convenience we will use the following notation $e_{\bar{1}} = E_1, f_{\bar{1}} = F_1, \tilde{e}_{\bar{1}} = \widetilde{E}_1, \tilde{f}_{\bar{1}} = \widetilde{F}_1$.

Recall that an abstract $\mathfrak{gl}(n)$-crystal is a set B together with the maps $\tilde{e}_i, \tilde{f}_i \colon B \to B \sqcup \{0\}$, $\varphi_i, \varepsilon_i \colon B \to \mathbb{Z} \sqcup \{-\infty\}$ $(i = 1, \ldots, n-1)$, and $\mathrm{wt} \colon B \to P$ satisfying the conditions given in [16]. For an abstract $\mathfrak{gl}(n)$-crystal B and $\lambda \in P$, we set $B^\lambda = \{b \in$

$B \mid \mathrm{wt}(b) = \lambda\}$. We say that an abstract $\mathfrak{gl}(n)$-crystal is a $\mathfrak{gl}(n)$-*crystal* if it is realized as a crystal basis of a finite-dimensional integrable $U_q(\mathfrak{gl}(n))$-module. In particular, we have $\varepsilon_i(b) = \max\{n \in \mathbb{Z}_{\geq 0}\,;\, \tilde{e}_i^n b \neq 0\}$ and $\varphi_i(b) = \max\{n \in \mathbb{Z}_{\geq 0}\,;\, \tilde{f}_i^n b \neq 0\}$ for any b in a $\mathfrak{gl}(n)$-crystal B.

6.3.1 Crystal basis

Definition Let $M = \bigoplus_{\mu \in P^{\geq 0}} M^\mu$ be a $U_q(\mathfrak{q}(n))$-module in the category $\mathcal{O}_{\text{int}}^{\geq 0}$. A *crystal basis* of M is a triple $(L, B, l_B = (l_b)_{b \in B})$, where

(i) L is a free $\mathbf{A}$-submodule of M such that

(1) $\mathbf{F} \otimes_{\mathbf{A}} L \xrightarrow{\sim} M$,
(2) $L = \bigoplus_{\mu \in P^{\geq 0}} L^\mu$, where $L^\mu = L \cap M^\mu$,
(3) L is stable under the Kashiwara operators $\tilde{e}_i$, $\tilde{f}_i$ $(i = 1, \ldots, n-1)$, $\widetilde{H}_1$, $\widetilde{E}_1$, $\widetilde{F}_1$.

(ii) B is a finite $\mathfrak{gl}(n)$-crystal together with the maps $\widetilde{E}_1, \widetilde{F}_1 : B \to B \sqcup \{0\}$ such that

(1) $\mathrm{wt}(\widetilde{E}_1 b) = \mathrm{wt}(b) + \alpha_1$, $\mathrm{wt}(\widetilde{F}_1 b) = \mathrm{wt}(b) - \alpha_1$,
(2) for all $b, b' \in B$, $\widetilde{F}_1 b = b'$ if and only if $b = \widetilde{E}_1 b'$.

(iii) $l_B = (l_b)_{b \in B}$ is a family of $\mathbb{C}$-vector spaces such that

(1) $l_b \subset (L/qL)^\mu$ for $b \in B^\mu$,
(2) $L/qL = \bigoplus_{b \in B} l_b$,
(3) $\widetilde{H}_1 l_b \subset l_b$,
(4) for $i = 1, \ldots, n-1, \bar{1}$, we have
 (1) if $\tilde{e}_i b = 0$ then $\tilde{e}_i l_b = 0$, and otherwise $\tilde{e}_i$ induces an isomorphism $l_b \xrightarrow{\sim} l_{\tilde{e}_i b}$.
 (2) if $\tilde{f}_i b = 0$ then $\tilde{f}_i l_b = 0$, and otherwise $\tilde{f}_i$ induces an isomorphism $l_b \xrightarrow{\sim} l_{\tilde{f}_i b}$.

As proved in [8], for every crystal basis (L, B, l_B) of a $U_q(\mathfrak{q}(n))$-module M we have $\widetilde{E}_1^2 = \widetilde{F}_1^2 = 0$ as endomorphisms on L/qL.

6.3.2 Example

Let

$$\mathbf{V} = \bigoplus_{j=1}^{n} \mathbf{F} v_j \oplus \bigoplus_{j=1}^{n} \mathbf{F} v_{\bar{j}}$$

be the vector representation of $U_q(\mathfrak{q}(n))$. The action of $U_q(\mathfrak{q}(n))$ on $\mathbf{V}$ is given as follows:

$e_i v_j = \delta_{j,i+1} v_i$, $e_i v_{\bar{j}} = \delta_{j,i+1} v_{\bar{i}}$, $f_i v_j = \delta_{j,i} v_{i+1}$, $f_i v_{\bar{j}} = \delta_{j,i} v_{\overline{i+1}}$, $E_i v_j = \delta_{j,i+1} v_{\bar{i}}$, $E_i v_{\bar{j}} = \delta_{j,i+1} v_i$, $F_i v_j = \delta_{j,i} v_{\overline{i+1}}$, $F_i v_{\bar{j}} = \delta_{j,i} v_{i+1}$, $q^h v_j = q^{\varepsilon_j(h)} v_j$, $q^h v_{\bar{j}} = q^{\varepsilon_j(h)} v_{\bar{j}}$, $H_i v_j = \delta_{j,i} v_{\bar{j}}$, $H_i v_{\bar{j}} = \delta_{j,i} v_j$.

Set

$$\mathbf{L} = \bigoplus_{j=1}^{n} \mathbf{A} v_j \oplus \bigoplus_{j=1}^{n} \mathbf{A} v_{\bar{j}},$$

$l_j = \mathbb{C} v_j \oplus \mathbb{C} v_{\bar{j}}$, and let $\mathbf{B}$ be the crystal graph given below:

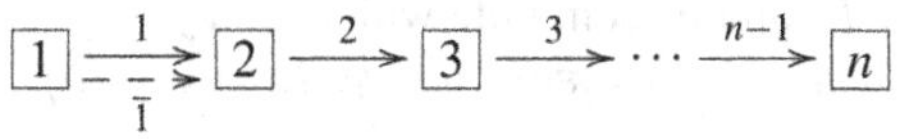

Here, the actions of $\tilde{f}_i$ $(i = 1, \ldots, n-1)$ are expressed by i-arrows and of $\widetilde{F}_1$ by an $\bar{1}$-arrow. Then $(\mathbf{L}, \mathbf{B}, l_{\mathbf{B}} = (l_j)_{j=1}^{n})$ is a crystal basis of $\mathbf{V}$.

6.3.3 Tensor product rule

The *tensor product rule* for the crystal bases in the category $\mathcal{O}_{\text{int}}^{\geq 0}$ is given by the following theorem (Theorem 2.7 in [8]).

Theorem *Let M_j be a $U_q(\mathfrak{g})$-module in $\mathcal{O}_{\text{int}}^{\geq 0}$ with crystal basis (L_j, B_j, l_{B_j}) $(j = 1, 2)$. Set $B_1 \otimes B_2 = B_1 \times B_2$ and*

$$l_{B_1 \otimes B_2} = (l_{b_1} \otimes l_{b_2})_{b_1 \in B_1, b_2 \in B_2}.$$

Then

$$(L_1 \otimes_{\mathbf{A}} L_2, B_1 \otimes B_2, l_{B_1 \otimes B_2})$$

is a crystal basis of $M_1 \otimes_{\mathbf{F}} M_2$, where the action of the Kashiwara operators on $B_1 \otimes B_2$ are given as follows.

$$\begin{aligned}
\tilde{e}_i(b_1 \otimes b_2) &= \begin{cases} \tilde{e}_i b_1 \otimes b_2 & \text{if } \varphi_i(b_1) \geq \varepsilon_i(b_2), \\ b_1 \otimes \tilde{e}_i b_2 & \text{if } \varphi_i(b_1) < \varepsilon_i(b_2), \end{cases} \\
\tilde{f}_i(b_1 \otimes b_2) &= \begin{cases} \tilde{f}_i b_1 \otimes b_2 & \text{if } \varphi_i(b_1) > \varepsilon_i(b_2), \\ b_1 \otimes \tilde{f}_i b_2 & \text{if } \varphi_i(b_1) \leq \varepsilon_i(b_2), \end{cases}
\end{aligned} \tag{1}$$

$$\begin{aligned}
\widetilde{E}_1(b_1 \otimes b_2) &= \begin{cases} \widetilde{E}_1 b_1 \otimes b_2 & \text{if } \langle h_1, \mathrm{wt}\, b_2 \rangle = 0, \\ & \quad \langle h_2, \mathrm{wt}\, b_2 \rangle = 0, \\ b_1 \otimes \widetilde{E}_1 b_2 & \text{otherwise}, \end{cases} \\
\widetilde{F}_1(b_1 \otimes b_2) &= \begin{cases} \widetilde{F}_1 b_1 \otimes b_2 & \text{if } \langle h_1, \mathrm{wt}\, b_2 \rangle = 0, \\ & \quad \langle h_2, \mathrm{wt}\, b_2 \rangle = 0, \\ b_1 \otimes \widetilde{F}_1 b_2 & \text{otherwise}. \end{cases}
\end{aligned} \tag{2}$$

6.3.4 Abstract crystal

Definition An *abstract* $\mathfrak{q}(n)$*-crystal* is a $\mathfrak{gl}(n)$-crystal together with the maps $\widetilde{E}_1, \widetilde{F}_1 : B \to B \sqcup \{0\}$ satisfying the following conditions:

(a) $\mathrm{wt}(B) \subset P^{\geq 0}$,
(b) $\mathrm{wt}(\widetilde{E}_1 b) = \mathrm{wt}(b) + \alpha_1$, $\mathrm{wt}(\widetilde{F}_1 b) = \mathrm{wt}(b) - \alpha_1$,
(c) for all $b, b' \in B$, $\widetilde{F}_1 b = b'$ if and only if $b = \widetilde{E}_1 b'$,
(d) if $3 \leq i \leq n-1$, we have

  (i) the operators $\widetilde{E}_1$ and $\widetilde{F}_1$ commute with $\tilde{e}_i$ and $\tilde{f}_i$;
  (ii) if $\widetilde{E}_1 b \in B$, then $\varepsilon_i(\widetilde{E}_1 b) = \varepsilon_i(b)$ and $\varphi_i(\widetilde{E}_1 b) = \varphi_i(b)$.

Recall that W is the Weyl group of $\mathfrak{gl}_n$. The *Weyl group action* on an abstract $\mathfrak{q}(n)$-crystal B is the action of W on $\mathfrak{gl}_n$-crystal B, which is given in [15].

Let B_1 and B_2 be abstract $\mathfrak{q}(n)$-crystals. The *tensor product* $B_1 \otimes B_2$ of B_1 and B_2 is defined to be the $\mathfrak{gl}(n)$-crystal $B_1 \otimes B_2$ together with the maps $\widetilde{E}_1$, $\widetilde{F}_1$ defined by (2). Then it is an abstract $\mathfrak{q}(n)$-crystal. Note that $\otimes$ satisfies the associative axiom on the set of abstract $\mathfrak{q}(n)$-crystals.

6.3.5 Example

(a) If (L, B, l_B) is a crystal basis of a $U_q(\mathfrak{q}(n))$-module M in the category $\mathscr{O}_{\mathrm{int}}^{\geq 0}$, then B is an abstract $\mathfrak{q}(n)$-crystal.
(b) The crystal graph $\mathbf{B}$ is an abstract $\mathfrak{q}(n)$-crystal.
(c) By the tensor product rule, $\mathbf{B}^{\otimes N}$ is an abstract $\mathfrak{q}(n)$-crystal. When $n = 3$, the $\mathfrak{q}(n)$-crystal structure of $\mathbf{B} \otimes \mathbf{B}$ is given below:

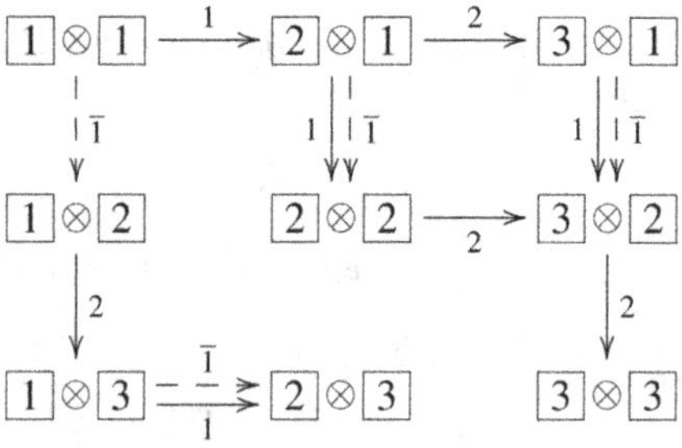

Let B be an abstract $\mathfrak{q}(n)$-crystal. For $i = 1, \ldots, n-1$, we set

$$w_i = s_2 \cdots s_i s_1 \cdots s_{i-1}.$$

Then w_i is the shortest element in W such that $w_i(\alpha_i) = \alpha_1$. We define the *odd Kashiwara operators* $\widetilde{E}_i$, $\widetilde{F}_i$ $(i = 2, \ldots, n-1)$ by

$$\widetilde{E}_i = S_{w_i^{-1}} \widetilde{E}_1 S_{w_i}, \quad \widetilde{F}_i = S_{w_i^{-1}} \widetilde{F}_1 S_{w_i}.$$

6.3.6 Definition

Let B be an abstract $\mathfrak{q}(n)$-crystal.

(a) An element $b \in B$ is called a $\mathfrak{gl}(n)$-*highest weight vector* if $\tilde{e}_i b = 0$ for $1 \leq i \leq n$.
(b) An element $b \in B$ is called a *highest weight vector* if $\tilde{e}_i b = \widetilde{E}_i b = 0$ for $1 \leq i \leq n$.
(c) An element $b \in B$ is called a *lowest weight vector* if $w_0 b$ is a $\mathfrak{q}(n)$-highest weight vector, where w_0 is the element of W of longest length.

The description of the set of the highest (and hence of the lowest) weight vectors in $\mathbf{B}^{\otimes N}$ for $N > 0$ is given by the following proposition (see Theorem 4.6 (c) in [8]).

6.3.7 Proposition

An element b_0 in $\mathbf{B}^{\otimes N}$ is a highest weight vector if and only if $b_0 = 1 \otimes \tilde{f}_1 \cdots \tilde{f}_{j-1} b$ for some j, and some highest weight vector b in $\mathbf{B}^{\otimes (N-1)}$ such that $\mathrm{wt}(b_0) = \mathrm{wt}(b) + \varepsilon_j$ is a strict partition.

The following uniqueness and existence theorem is one of the main results in [8].

6.3.8 Theorem

(a) Let $\lambda \in \Lambda^+$ be a strict partition and let M be a highest weight $U_q(\mathfrak{q}(n))$-module with highest weight λ in the category $\mathscr{O}_{\mathrm{int}}^{\geq 0}$. If (L, B, l_B) is a crystal basis of M, then L^λ is invariant under $\widetilde{K}_i := q^{h_i - 1} H_i$ for all $i = 1, \ldots, n$. Conversely, if M^λ is generated by a free $\mathbf{A}$-submodule L_λ^0 invariant under $\widetilde{K}_i$ $(i = 1, \ldots, n)$, then there exists a unique crystal basis (L, B, l_B) of M such that

(i) $L^\lambda = L_\lambda^0$,
(ii) $B^\lambda = \{b_\lambda\}$,
(iii) $L_\lambda^0 / q L_\lambda^0 = l_{b_\lambda}$,
(iv) B is connected.

Moreover, B, as an abstract $\mathfrak{q}(n)$-crystal depends only on λ. Hence we may write $B = B(\lambda)$.

(b) The $\mathfrak{q}(n)$-crystal $B(\lambda)$ has a unique highest weight vector b_λ and unique lowest weight vector l_λ.

A combinatorial description of the crystal bases in terms of semistandard decomposition tableaux has been obtained in [9].

Acknowledgements We would like to thank V. Mazorchuk and V. Serganova for the fruitful discussions. Also, we are grateful to INdAM for the excellent workshop "Lie Superalgebras" where some of the results in the paper were reported. D.G. was partially supported by NSA grant H98230-13-1-0245.

References

1. Brundan, J.: Kazhdan-Lusztig polynomials and character formulae for the Lie superalgebra $\mathfrak{q}(n)$. Adv. Math. **182**, 28–77 (2004)
2. Frisk, A.: Typical blocks of the category $\mathcal{O}$ for the queer Lie superalgebra. J. Algebra Appl. **6**(5), 731–778 (2007)
3. Frisk, A., Mazorchuk, V.: Regular strongly typical blocks of $\mathcal{O}^{\mathfrak{q}}$. Commun. Math. Phys. **291**, 533–542 (2009)
4. Dimitrov, I., Mathieu, O., Penkov, I.: On the structure of weight modules. Trans. Amer. Math. Soc. **352**, 2857–2869 (2000)
5. Gorelik, M.: Shapovalov determinants of Q-type Lie superalgebras. Int. Math. Res. Pap., Article ID 96895, 1–71 (2006)
6. Gorelik, M., Grantcharov, D.: Bounded highest weight modules of $\mathfrak{q}(n)$. To appear in IMRN.
7. Grantcharov, D., Jung, J.H., Kang, S.-J., Kim, M.: Highest weight modules over quantum queer superalgebra $U_q(\mathfrak{q}(n))$. Commun. Math. Phys. **296**, 827–860 (2010)
8. Grantcharov, D., Jung, J.H., Kang, S.-J., Kashiwara, M., Kim, M.: Crystal bases for the quantum queer superalgebra. To appear in J. Eur. Math Soc.
9. Grantcharov,D., Jung, J.H., Kang, S.-J., Kashiwara, M., Kim, M.: Crystal bases for the quantum queer superalgebra and semistandard decomposition tableaux. To appear in Trans. Amer. Math. Soc.
10. Grantcharov, D., Pianzola, A.: Automorphisms and twisted loop algebras of finite dimensional simple Lie superalgebras. Int. Math. Res. Not. **73**, 3937–3962 (2004)
11. Gorelik, M., Serganova, V.: On representations of the affine superalgebra $\mathfrak{q}(n)^{(2)}$. Moscow Math. J. **8**(1), 91–109 (2008)
12. Hoyt, C., Serganova, V.: Kac-Moody Lie superalgebras of finite growth. Comm. in Algebra **35**(3), 851–874 (2007)
13. Kac, V.G.: Lie superalgebras. Adv. in Math. **26**, 8–96 (1977)
14. Kac, V.G.: Infinite dimensional Lie algebras, 3rd ed. Cambridge University Press, Cambridge (1990)
15. Kashiwara, M.: On crystal bases of the q-analogue of universal enveloping algebras. Duke Math. J. **63**, 465–516 (1991)
16. Kashiwara, M.: Crystal base and Littelmann's refined Demazure character formula. Duke Math. J. **71**, 839–858 (1993)
17. Mathieu, O.: Classification of irreducible weight modules. Ann. Inst. Fourier **50**, 537–592 (2000)
18. Mazorchuk, V.: Classification of simple $\mathfrak{q}(2)$-supermodules. Tohoku Math. J. (2) **62**(3), 401–426 (2010)
19. Nazarov, M., Sergeev, A.: Centralizer construction of the Yangian of the queer Lie superalgebra. In: Studies in Lie Theory, A. Joseph Festschrift. Progress in Math. **243** (2006)
20. Penkov, I.: Characters of typical irreducible finite-dimensional $\mathfrak{q}(n)$-modules. Functional Anal. Appl. **20**(1), 30–37 (1986)
21. Penkov, I., Serganova, V.: Characters of irreducible G-modules and cohomology of G/P for the Lie supergroup $G = Q(N)$. J. Math. Sci. **84**, 1382–1412, New York (1997)
22. Penkov, I., Serganova, V.: Characters of finite-dimensional irreducible $\mathfrak{q}(n)$-modules. Lett. Math. Phys. **40**, 147–158 (1997)
23. Serganova, V.: Automorphisms of simple Lie superalgebras. Math. USSR Izv. **24**, 539–551 (1985)
24. Sergeev, A.: Tensor algebra of the identity representation as a module over the Lie superalgebras Gl(n,m) and $Q(n)$ (in Russian). Mat. Sb. (N.S.) **123(165)**, 422–430 (1984)
25. Sergeev, A.: Invariant polynomial functions on Lie superalgebras. C.R. Acad. Bulgare Sci. **35**(5), 573–576 (1982)
26. Sergeev, A.: The centre of enveloping algebra for Lie superalgebra $Q(n,\mathbb{C})$. Lett. Math. Phys. **7**(3), 177–179 (1983)

27. Sergeev, A.: Invariant polynomials on simple Lie superalgebras. Representation Theory **3**, 250–280 (1999)
28. Su, Y., Zhang, R.B.: Character and dimension formulae for queer Lie superalgebra. arXiv:1305.2906
29. van de Leur, J.W.: A classification of contragredient Lie superalgebras of finite growth. Comm. in Algebra **17**, 1815–1841 (1989)
30. Wakimoto, M.: Infinite Dimensional Lie Algebras. Translation of Mathematical Monographs Vol. 195 (2001)

Weight modules of $D(2,1,\alpha)$

Crystal Hoyt

Abstract Let $\mathfrak{g}$ be a basic Lie superalgebra. A weight module M over $\mathfrak{g}$ is called finite if all of its weight spaces are finite dimensional, and it is called bounded if there is a uniform bound on the dimension of a weight space. The minimum bound is called the degree of M. For $\mathfrak{g} = D(2,1,\alpha)$, we prove that every simple weight module M is bounded and has degree less than or equal to 8. This bound is attained by a cuspidal module M if and only if M belongs to a $(\mathfrak{g},\mathfrak{g}_{\bar{0}})$-coherent family $L(\lambda)^{\mu}_{\Gamma}$ for some typical module $L(\lambda)$. Cuspidal modules which correspond to atypical modules have degree less than or equal to 6 and greater than or equal to 2.

1 Introduction

Basic Lie superalgebras are a natural generalization of simple finite dimensional Lie algebras, and their finite dimensional modules have been studied extensively [9,14]. In this case, every simple $\mathfrak{g}$-module is a highest weight module with respect to each choice of simple roots of $\mathfrak{g}$, and moreover, there exist various character formulas which help one "count" the dimension of each weight space of the module. A natural generalization of this setting is to the study of (possibly) infinite dimensional modules that have finite dimensional weight spaces, namely, finite weight modules. However, since these modules are not necessarily highest weight with respect to any choice of the set of simple roots, the question arises how to characterize and classify all such simple modules.

In [13], Mathieu gave an answer to this problem for simple Lie algebras by relating to each simple finite weight module M a corresponding simple highest weight module $L(\lambda)$ such that M is the *twisted localization* of $L(\lambda)$, that is, $M \cong L(\lambda)^{\mu}_{\Gamma}$. Grantcharov extended this result to cover classical Lie superalgebras in [8]. Using

C. Hoyt (✉)
Department of Mathematics, Technion - Israel Institute of Technology, Haifa 32000, Israel
e-mail: hoyt@tx.technion.ac.il

M. Gorelik, P. Papi (eds.): *Advances in Lie Superalgebras*. Springer INdAM Series 7,
DOI 10.1007/978-3-319-02952-8_6,

this characterization one can gain information about a simple finite weight module M from the corresponding simple highest weight module $L(\lambda)$, including the calculation of its degree (see (1)). Moreover, this is a major step towards the classification of simple finite weight modules.

In addition, one must determine which simple highest weight modules $L(\lambda)$ can appear in this correspondence. These are the modules which have uniformly bounded weight multiplicities, the so called "bounded modules". Then one should determine the simplicity conditions for the modules $L(\lambda)_{\Gamma}^{\mu}$ and the isomorphisms between them. For simple Lie algebras this problem was completely solved by Mathieu in [13], but the general problem remains open for basic Lie superalgebras. For modules with a strongly typical central character, this description can be derived from results of Gorelik, Penkov and Serganova [3, 15, 16], however the situation is not surprisingly more difficult when the central character is atypical.

One can further reduce the classification problem to that of classifying "cuspidal modules" using Theorem 1 (Fernando [2]; Dimitrov, Mathieu, Penkov [1]). A cuspidal module is a simple finite weight $\mathfrak{g}$-module that is not parabolically induced from any proper parabolic subalgebra $\mathfrak{p} \subset \mathfrak{g}$. These modules can be characterized in terms of their support and in terms of the action of $\mathfrak{g}$.

In this paper, we focus on the family of Lie superalgebras $D(2,1,\alpha)$ which are defined by one complex parameter $\alpha \in \mathbb{C}\setminus\{0,-1\}$ and study their finite weight modules. For $\mathfrak{g} = D(2,1,\alpha)$, every simple weight module is a finite weight module, and moreover it is bounded! Indeed, here $\mathfrak{g}_{\bar{0}} = \mathfrak{sl}_2 \times \mathfrak{sl}_2 \times \mathfrak{sl}_2$, so a simple weight module V for $\mathfrak{g}_{\bar{0}}$ is just the tensor product $V = V_1 \otimes V_2 \otimes V_3$ of simple $\mathfrak{sl}_2$ weight modules. Now each simple $\mathfrak{sl}_2$ weight module V_i has one dimensional weight spaces, since the Casimir element $h^2 + 2h + fe$ acts by a scalar. Hence, V also has one dimensional weight spaces. Now any simple weight module of $\mathfrak{g}$ can be realized as the quotient of an induced module $\mathrm{Ind}_{\mathfrak{g}_{\bar{0}}}^{\mathfrak{g}} V = U(\mathfrak{g}) \otimes_{\mathfrak{g}_{\bar{0}}} V$, where V is a simple $\mathfrak{g}_{\bar{0}}$ weight module. The claim then follows from the fact that $U(\mathfrak{g}_{\bar{1}})$ is finite dimensional (here $\dim U(\mathfrak{g}_{\bar{1}}) = 256$).

For $\mathfrak{g} = D(2,1,\alpha)$, we prove that every Verma module is bounded and has degree less than or equal to 8. It then follows from a result of Grantcharov that simple (finite) weight modules are also bounded of degree less than or equal to 8. We show that the dimensions of the weight spaces of a cuspidal $\mathfrak{g}$-module are constant on $Q_{\bar{0}}$-cosets and we calculate this degree. We prove that a cuspidal $\mathfrak{g}$-module has degree 8 if and only if it is "typical". We prove that if M is an "atypical" cuspidal $\mathfrak{g}$-module then $2 \leq \deg M \leq 6$. We determine the conditions on λ and Γ that are necessary for $L(\lambda)_{\Gamma}^{\mu}$ to be a cuspidal module. It remains to determine the restrictions on $\mu \in \mathbb{C}^3$ and isomorphisms between these modules.

2 Basic Lie superalgebras

A simple finite dimensional Lie superalgebra $\mathfrak{g} = \mathfrak{g}_{\bar{0}} \oplus \mathfrak{g}_{\bar{1}}$ is called *basic* if $\mathfrak{g}_{\bar{0}}$ is a reductive Lie algebra, and there exists an even non-degenerate (symmetric) invariant bilinear form on $\mathfrak{g}$. These are the Lie superalgebras: $\mathfrak{sl}(m|n)$ for $m \neq n$, $\mathfrak{psl}(n|n)$,

$\mathfrak{osp}(m|2n)$, $F(4)$, $G(3)$ and $D(2,1,\alpha)$, $a \in \mathbb{C} \setminus \{0,-1\}$, and finite-dimensional simple Lie algebras. A basic Lie superalgebra can be represented by a Dynkin diagram, though not uniquely.

Let $\mathfrak{g}$ be a basic Lie superalgebra, and fix a Cartan subalgebra $\mathfrak{h} \subset \mathfrak{g}_{\bar{0}} \subset \mathfrak{g}$. We have a root space decomposition $\mathfrak{g} = \mathfrak{h} \oplus \bigoplus_{\alpha \in \Delta} \mathfrak{g}_\alpha$. For each set of simple roots $\Pi \subset \Delta$, we have a corresponding set of positive roots $\Delta^+ = \Delta_{\bar{0}}^+ \cup \Delta_{\bar{1}}^+$ and a triangular decomposition $\mathfrak{g} = \mathfrak{n}^- \oplus \mathfrak{h} \oplus \mathfrak{n}^+$. This induces a triangular decomposition of $\mathfrak{g}_{\bar{0}}$, namely, $\mathfrak{g}_{\bar{0}} = \mathfrak{n}_{\bar{0}}^- \oplus \mathfrak{h} \oplus \mathfrak{n}_{\bar{0}}^+$. Let $\Pi_{\bar{0}}$ denote the corresponding set of simple roots for $\mathfrak{g}_{\bar{0}}$. The root lattice of $\mathfrak{g}$ (resp. $\mathfrak{g}_{\bar{0}}$) is defined to be $Q = \sum_{\alpha \in \Pi} \mathbb{Z}\alpha$ (resp. $Q_{\bar{0}} = \sum_{\alpha \in \Pi_{\bar{0}}} \mathbb{Z}\alpha$). Let $\rho_0 = \frac{1}{2}\sum_{\alpha \in \Delta_{\bar{0}}^+} \alpha$, $\rho_1 = \frac{1}{2}\sum_{\alpha \in \Delta_{\bar{1}}^+} \alpha$ and $\rho = \rho_0 - \rho_1$. Denote by $U(\mathfrak{g})$ (resp. $U(\mathfrak{g}_{\bar{0}})$) the universal enveloping algebra of $\mathfrak{g}$ (resp. $\mathfrak{g}_{\bar{0}}$). See [9, 14] for definitions and more details.

2.1 Finite weight modules

A $\mathfrak{g}$-module M is called a *weight module* if it decomposes into a direct sum of weight spaces $M = \oplus_{\mu \in \mathfrak{h}^*} M_\mu$, where $M_\mu = \{m \in M \mid h.m = \mu(h)m \text{ for all } h \in \mathfrak{h}\}$. A weight module M is called *finite* if dim $M_\mu < \infty$ for all $\mu \in \mathfrak{h}^*$. Define the *support of M* to be the set

$$\operatorname{Supp} M = \{\mu \in \mathfrak{h}^* \mid \dim M_\mu \neq 0\}.$$

The module M is called *bounded* if the exists a constant c such that dim $M_\mu < c$ for all $\mu \in \mathfrak{h}^*$. Recall that a $\mathfrak{g}$-module $M = M_{\bar{0}} \oplus M_{\bar{1}}$ is also $\mathbb{Z}/2\mathbb{Z}$-graded. The *degree of M* is defined to be

$$\deg M = max_{\mu \in \mathfrak{h}^*} \dim(M)_\mu, \tag{1}$$

and we define the *graded degree of M* to be (d_0, d_1), where

$$d_i = max_{\mu \in \mathfrak{h}^*} \dim(M_{\bar{i}})_\mu \text{ for } i \in \{0,1\}.$$

Remark 1 Clearly, $\max\{d_0, d_1\} \leq \deg M \leq d_0 + d_1$. However, if M is a weight module that can be generated by a single weight vector (i.e. simple or highest weight module), then each weight space of M is either purely even or purely odd, and so in this case $\deg M = \max\{d_0, d_1\}$.

Let $M(\lambda)$ denote the Verma module of highest weight $\lambda \in \mathfrak{h}^*$ with respect to a set of simple roots Π, and let $L(\lambda)$ denote its unique simple quotient [14]. It is clear that $M(\lambda)$ and $L(\lambda)$ are finite weight modules, but they are not always bounded.

For each $\beta \in \Pi$ an odd isotropic root (i.e. $(\beta,\beta) = 0$), we have an odd reflection of the set of simple roots $r_\beta : \Pi \to \Pi'$ satisfying $\Pi' = (\Pi \setminus \{\beta\}) \cup \{-\beta\}$ [12]. Moreover, for a simple highest weight module $L_\Pi(\lambda)$ there exists $\lambda' \in \mathfrak{h}^*$ such that $L_{\Pi'}(\lambda') = L_\Pi(\lambda)$. In particular, $\lambda' = \lambda - \beta$ if $(\lambda,\beta) \neq 0$, while $\lambda' = \lambda$ otherwise [11]. Using even and odd reflections one can move between all the different choices of simple roots for a basic Lie superalgebra $\mathfrak{g}$ [17]. Moreover, one can move between two different Dynkin diagrams of $\mathfrak{g}$ using only odd reflections.

A simple highest weight module $L(\lambda)$ is called *typical* if $(\lambda+\rho,\alpha)\neq 0$ for all $\alpha\in\Delta_{\bar{1}}$, and *atypical* otherwise. The notion of typicality is preserved by an odd reflection of the set of simple roots, that is, given an odd reflection $r_\beta : \Pi \to \Pi'$ and $L_{\Pi'}(\lambda') = L_\Pi(\lambda)$, then $L_{\Pi'}(\lambda')$ is typical iff $L_\Pi(\lambda)$ is typical [11, 17].

It was shown by Penkov and Serganova that if $\mathfrak{g}$ is a basic Lie superalgebra, then the category of representations of $\mathfrak{g}$ with a fixed generic typical central character is equivalent to the category of representations of $\mathfrak{g}_{\bar{0}}$ with a certain corresponding central character [15, 16]. This equivalence of categories was extended to representations of $\mathfrak{g}$ with a fixed strongly typical central character by Gorelik in [3]. In the case that the root system of $\mathfrak{g}$ is reduced ($\alpha, k\alpha\in\Delta$ implies $k=\pm 1$) all typical central characters are strongly typical.

2.2 Cuspidal modules

A $\mathbb{Z}$-*grading* of $\mathfrak{g}$ is a decomposition $\mathfrak{g}=\oplus_{j\in\mathbb{Z}}\mathfrak{g}(j)$ satisfying $[\mathfrak{g}(i),\mathfrak{g}(j)]\subset\mathfrak{g}(i+j)$ and $\mathfrak{h}\subset\mathfrak{g}(0)$. A subalgebra $\mathfrak{p}\subset\mathfrak{g}$ is called a *parabolic subalgebra* if there exists a $\mathbb{Z}$-grading of $\mathfrak{g}$ such that $\mathfrak{p}=\oplus_{j\geq 0}\mathfrak{g}(j)$. In this case, $\mathfrak{l}=\mathfrak{g}(0)$ is a *Levi subalgebra* and $\mathfrak{n}=\oplus_{j\geq 1}\mathfrak{g}(j)$ is the nilradical of $\mathfrak{p}$.

Let $\mathfrak{p}=\mathfrak{l}\oplus\mathfrak{n}$ be a parabolic subalgebra of $\mathfrak{g}$, and let S be a simple $\mathfrak{p}$-module. Then $M_\mathfrak{p}(S) := \mathrm{Ind}_\mathfrak{p}^\mathfrak{g} S$ has a unique simple quotient $L_\mathfrak{p}(S)$. The module $L_\mathfrak{p}(S)$ is said to be *parabolically induced*. A simple $\mathfrak{g}$-module is called *cuspidal* if it is not parabolically induced from any proper parabolic subalgebra $\mathfrak{p}\subset\mathfrak{g}$.

Theorem 1 (Fernando [2]; Dimitrov, Mathieu, Penkov [1]) *Let $\mathfrak{g}$ be a basic Lie superalgebra. Any simple finite weight $\mathfrak{g}$-module is obtained by parabolic induction from a cuspidal module of a Levi subalgebra.*

This theorem is an important step towards the classification of all simple finite weight modules. It reduces the general classification problem to that of classifying cuspidal modules.

Theorem 2 (Fernando [2]) *If $\mathfrak{g}$ is a finite dimensional simple Lie algebra that admits a cuspidal module, then $\mathfrak{g}$ is of type A or C.*

Theorem 3 (Dimitrov, Mathieu, Penkov [1]) *Only the following basic Lie superalgebras admit a cuspidal module: $\mathfrak{psl}(n|n)$, $\mathfrak{osp}(m|2n)$ with $m\leq 6$, $D(2,1,\alpha)$ with $\alpha\in\mathbb{C}\setminus\{0,-1\}$, $\mathfrak{sl}(n)$, $\mathfrak{sp}(2n)$.*

The following theorem gives a characterization of cuspidal $\mathfrak{g}_{\bar{0}}$-modules.

Theorem 4 (Fernando [2]) *Let $\mathfrak{g}_{\bar{0}}$ be a reductive Lie algebra, and let M be a simple finite weight $\mathfrak{g}_{\bar{0}}$-module. Then M is cuspidal iff Supp M is exactly one Q coset iff ad x_α is injective for all $\alpha\in\Delta$, $x_\alpha\in\mathfrak{g}_\alpha$.*

Corollary 1 *Let $\mathfrak{g}_{\bar{0}}$ be a reductive Lie algebra. If M is a cuspidal $\mathfrak{g}_{\bar{0}}$-module, then M is bounded and dim M_μ = deg M for all $\mu\in$ Supp M.*

These conditions are too strict when $\mathfrak{g}$ is a basic Lie superalgebra, and so the following definitions were introduced in [1]. A finite weight module M is called *torsion-free* if the monoid generated by

$$\text{inj } M = \{\alpha \in \Delta_{\bar{0}} \mid x_\alpha \in \mathfrak{g}_\alpha \text{ acts injectively on } M\}$$

is a subgroup of finite index in Q. A finite weight module M is called *dense* if Supp M is a finite union of Q'-cosets, for some subgroup Q' of finite index in Q.

Theorem 5 (Dimitrov, Mathieu, Penkov [1]) *Let $\mathfrak{g}$ be a basic Lie superalgebra, and let M be a simple finite weight $\mathfrak{g}$-module. Then M is cuspidal iff M is dense iff M is torsion free.*

The following lemmas are proven in [13].

Lemma 1 *Let $\mathfrak{g}_{\bar{0}}$ be a reductive Lie algebra. Any bounded $\mathfrak{g}_{\bar{0}}$-module has finite length.*

Lemma 2 *Let $\mathfrak{g}$ be a basic Lie superalgebra. If M is a simple finite weight $\mathfrak{g}$-module, then for each $\alpha \in \Delta_{\bar{0}}$ the action of $x \in \mathfrak{g}_\alpha$ on M is either injective or locally nilpotent.*

For each $\mathfrak{g}_{\bar{0}}$-module N, let

$$\widetilde{N} := \text{Ind}_{\mathfrak{g}_{\bar{0}}}^{\mathfrak{g}}(N).$$

Theorem 6 (Dimitrov, Mathieu, Penkov [1]) *(i) For any finite cuspidal $\mathfrak{g}_{\bar{0}}$-module N, the module $\widetilde{N}$ contains at least one and only finitely many non-isomorphic cuspidal submodules.*
(ii) For any finite cuspidal $\mathfrak{g}$-module M, there is at least one and only finitely many non-isomorphic cuspidal $\mathfrak{g}_{\bar{0}}$-modules such that $M \subset \widetilde{N}$.

2.3 Coherent families

Let $C(\mathfrak{h})$ denote the centralizer of $\mathfrak{h}$ in $U(\mathfrak{g}_{\bar{0}})$. A $(\mathfrak{g}, \mathfrak{g}_{\bar{0}})$*-coherent family of degree d* is a finite weight $\mathfrak{g}$-module M such that dim $M_\mu = d$ for all $\lambda \in \mathfrak{h}^*$ and the function $\mu \mapsto \text{Tr } u|_{M_\mu}$ is polynomial in μ, for all $u \in C(\mathfrak{h})$ [6,7,13].

Example 1 (Mathieu [13]) Let $\mathfrak{g} = \mathfrak{sl}_2$ and fix $a \in \mathbb{C}$. Define a module $V(a) = \oplus_{s \in \mathbb{C}} \mathbb{C}x^s$ with the following $\mathfrak{sl}_2$ action.

$$e \mapsto x^2\, d/dx + ax \qquad e.x^s = (a+s)x^{s+1}$$
$$f \mapsto -d/dx + a(1/x) \qquad f.x^s = (a-s)x^{s-1}$$
$$h \mapsto 2x\, d/dx \qquad h.x^s = (2s)x^s$$

For each $a \in \mathbb{C}$, $V(a)$ is a $\mathfrak{sl}_2$-coherent family. For each $[\mu] \in \mathbb{C}/\mathbb{Z}$ with representative $\mu \in \mathfrak{h}^*$,

$$V(a)^{[\mu]} := \oplus_{n \in \mathbb{Z}} \mathbb{C}x^{\mu+n}$$

is a submodule, which is simple and cuspidal if and only if $\mu \pm a \notin \mathbb{Z}$.

Let $\Gamma = \{\gamma_1, \ldots, \gamma_k\} \subset \Delta_{\bar{0}}^+$ be a set of commuting roots, and for each $\gamma_i \in \Gamma$ choose $f_i \in \mathfrak{g}_{-\gamma_i}$. Let U_Γ be the localization of $U(\mathfrak{g})$ at the set $\{f_i^n \mid n \in \mathbb{N}, \gamma_i \in \Gamma\}$. If $\Gamma \subset \operatorname{inj} L(\lambda)$, define the *localization of* $L(\lambda)$ *at* Γ to be the module $L(\lambda)_\Gamma := U_\Gamma \otimes_{U(\mathfrak{g})} L(\lambda)$. Then $L(\lambda)$ is a submodule of $L(\lambda)_\Gamma$ and $\deg L(\lambda)_\Gamma = \deg L(\lambda)$.

Now for each $\mu \in \mathbb{C}^k$, we define a new module $L(\lambda)_\Gamma^\mu$ whose underlying vector space is $L(\lambda)_\Gamma$, but with a new action of $\mathfrak{g}$ defined as follows. For $u \in U_\Gamma$ and $x \in L(\lambda)_\Gamma^\mu$,

$$u \cdot x := \Phi_\Gamma^\mu(u)v,$$

where

$$\Phi_\Gamma^\mu(u) = \sum_{0 \leq i_1, \ldots, i_k} \binom{\mu_1}{i_1} \cdots \binom{\mu_k}{i_k} \operatorname{ad}(f_1)^{i_1} \ldots \operatorname{ad}(f_k)^{i_k}(u) f_1^{-i_1} \ldots f_k^{-i_k}.$$

Note that this sum is finite for each choice of u. The module $L(\lambda)_\Gamma^\mu$ is called the *twisted localization of* $L(\lambda)$ *with respect to* Γ *and* μ, and it is a $(\mathfrak{g}, \mathfrak{g}_{\bar{0}})$-coherent family of degree $d = \deg L(\lambda)$.

Theorem 7 (Mathieu [13]; Grantcharov [8]) *Let* $\mathfrak{g}$ *be a basic Lie superalgebra. Each simple finite weight* $\mathfrak{g}$*-module* M *is a twisted localization of a simple highest weight module* $L_{\mathfrak{b}}(\lambda)$ *for some Borel subalgebra* $\mathfrak{b} \subset \mathfrak{g}$ *and* $\lambda \in \mathfrak{h}^*$. *In particular,* $M \cong L_{\mathfrak{b}}(\lambda)_\Gamma^\mu$ *for some* $\mu \in \mathbb{C}$ *and set of commuting even roots* Γ.

Remark 2 If M is a cuspidal or bounded module, then $L_{\mathfrak{b}}(\lambda)$ is necessarily bounded.

3 The Lie superalgebra $D(2,1,\alpha)$

For each $\alpha \in \mathbb{C} \setminus \{0, -1\}$, the Lie superalgebra $\mathfrak{g} = D(2,1,\alpha)$ can be realized as a contragredient Lie superalgebra $\mathfrak{g}(A)$ with Cartan matrix

$$A = \begin{pmatrix} 0 & 1 & \alpha \\ 1 & 0 & -\alpha-1 \\ \alpha & -\alpha-1 & 0 \end{pmatrix}, \tag{2}$$

set of simple roots $\Pi = \{\beta_1, \beta_2, \beta_3\}$ with parity $(1,1,1)$, generating set

$$\{e_i \in \mathfrak{g}_{\beta_i}, f_i \in \mathfrak{g}_{-\beta_i}, h_i \in \mathfrak{h} \mid i = 1,2,3\}$$

and defining relations [9]. Then $\mathfrak{g}_{\bar{0}} = \mathfrak{sl}_2 \times \mathfrak{sl}_2 \times \mathfrak{sl}_2$ is 9-dimensional with

$$\Pi_{\bar{0}} = \{\beta_1 + \beta_2, \beta_1 + \beta_3, \beta_2 + \beta_3\},$$

and $\mathfrak{g}_{\bar{1}}$ is the 8-dimensional $\mathfrak{g}_{\bar{0}}$-module given by tensoring three copies of the standard representation of $\mathfrak{sl}_2$. Our choice of Π induces a triangular decomposition $\Delta = \Delta^+ \cup \Delta^-$ where

$$\Delta_{\bar{0}}^+ = \{\beta_1 + \beta_2, \beta_1 + \beta_3, \beta_2 + \beta_3\} \text{ and } \Delta_{\bar{1}}^+ = \{\beta_1, \beta_2, \beta_3, \beta_1 + \beta_2 + \beta_3\}. \tag{3}$$

Remark 3 For $\mathfrak{g} = D(2,1,\alpha)$, the even root lattice $Q_{\bar{0}}$ is a sublattice of index 2 in Q.

$D(2,1,\alpha)$ has four different Dynkin diagrams.

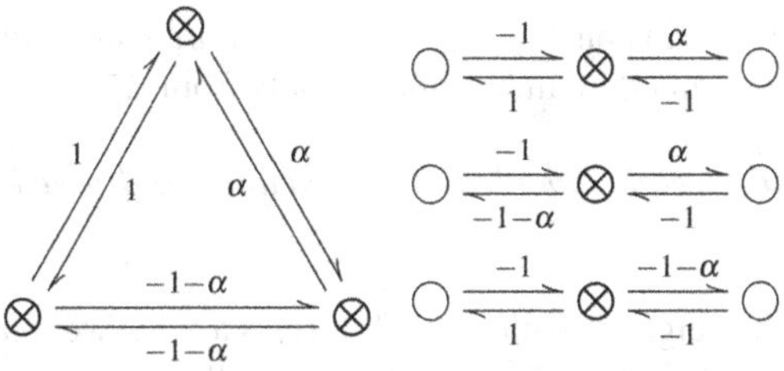

Remark 4 The Cartan matrix A in (2) is equivalent to the diagram on the left, and corresponds to Δ^+ given in (3). Most of our computations will be with respect to this choice of the set of simple roots, since here $\rho = 0$ and since no set of simple roots of $\mathfrak{g} = D(2,1,\alpha)$ contains a set of simple roots for $\mathfrak{g}_{\bar{0}}$.

3.1 Finite weight modules for $D(2,1,\alpha)$

In this section we study finite weight modules for the basic Lie superalgebra $D(2,1,\alpha)$.

It was shown in [1] that every simple finite weight module is obtained by parabolic induction from a cuspidal module $L_{\mathfrak{p}}(S)$ with $\mathfrak{p} = \mathfrak{l} \oplus \mathfrak{n}$, such that either $\mathfrak{l}$ is a proper reductive subalgebra of $\mathfrak{g}_{\bar{0}} = \mathfrak{sl}_2 \times \mathfrak{sl}_2 \times \mathfrak{sl}_2$ or $\mathfrak{l} = \mathfrak{g}$. Cuspidal modules for $\mathfrak{sl}_2$ are classified in Example 1, and all cuspidal modules for $\mathfrak{sl}_2 \times \mathfrak{sl}_2$ can be obtained by tensoring two cuspidal $\mathfrak{sl}_2$-modules together, namely,

$$V(a_1)^{[\mu_1]} \otimes V(a_2)^{[\mu_2]} \text{ with } \mu_i \pm a_i, \notin \mathbb{Z},\ i = 1,2.$$

So it remains to describe the cuspidal modules for $\mathfrak{g} = D(2,1,\alpha)$. The following theorems will help us realize these modules.

Theorem 8 *Let $\mathfrak{g} = D(2,1,\alpha)$. For each set of positive roots Δ^+ and each $\lambda \in \mathfrak{h}^*$, the Verma module $M(\lambda)$ is bounded. Moreover, $M(\lambda)$ has degree* 8 *and graded degree* $(8,8)$.

Proof. The dimensions of the weight spaces of $M(\lambda)$ are given by the coefficients of the character formula $\operatorname{ch} M(\lambda) = \frac{e^{\lambda+\rho}}{e^{\rho}R} = e^{\lambda}\frac{R_1}{R_0}$, where $R_0 = \Pi_{\alpha\in\Delta_{\bar{0}}^+}(1-e^{-\alpha})$ and $R_1 = \Pi_{\alpha\in\Delta_{\bar{1}}^+}(1+e^{-\alpha})$. Since $e^{\rho}R$ is invariant under odd reflections, it is sufficient to prove the theorem with respect to Δ^+ from (3).

$$\begin{aligned}\operatorname{ch} M(\lambda) &= e^{\lambda}\frac{(1+e^{-\beta_1})(1+e^{-\beta_2})(1+e^{-\beta_3})(1+e^{-\beta_1-\beta_2-\beta_3})}{(1-e^{-\beta_1-\beta_2})(1-e^{-\beta_1-\beta_3})(1-e^{-\beta_2-\beta_3})}\\ &= e^{\lambda}(1+e^{-\beta_1})(1+e^{-\beta_2})(1+e^{-\beta_3})(1+e^{-\beta_1-\beta_2-\beta_3})\\ &\quad\cdot\left(\sum_{\substack{k_1,k_2,k_3\in\mathbb{N}:\\ k_1+k_2+k_3 \text{ is even}}} e^{-k_1\beta_1-k_2\beta_2-k_3\beta_3}\right)\end{aligned}$$

So for $m_1, m_2, m_3 \in \mathbb{N}$ sufficiently large, the $(\lambda - m_1\beta_1 - m_2\beta_2 - m_3\beta_3)$ weight space of $M(\lambda)$ is purely even with dimension $\binom{4}{0} + \binom{4}{2} + \binom{4}{4} = 8$ when $m_1 + m_2 + m_3$ is even, (which corresponds to a choice of an even number of odd roots from $\Delta_{\bar{1}}^-$), and it is purely odd with dimension $\binom{4}{1} + \binom{4}{3} = 8$ when $m_1 + m_2 + m_3$ is odd, (corresponding to a choice of an odd number of odd roots from $\Delta_{\bar{1}}^-$).

Corollary 2 *A highest weight module is bounded and has degree less than or equal to* 8.

Since every simple weight module of $D(2,1,\alpha)$ is a finite weight module, we can combine Theorem 7 with Corollary 2 to obtain the following.

Theorem 9 *For* $\mathfrak{g} = D(2,1,\alpha)$, *any simple weight module M is bounded and has degree less than or equal to* 8.

Remark 5 In Theorem 9, we do not assume that M is a highest weight module.

Let Δ^+ be as in (3), and let $\alpha_1 = \beta_2 + \beta_3$, $\alpha_2 = \beta_1 + \beta_3$, $\alpha_3 = \beta_1 + \beta_2$, so that $\Pi_{\bar{0}} = \{\alpha_1, \alpha_2, \alpha_3\}$. Then for each $\alpha_i \in \Pi_{\bar{0}}$, $i = 1,2,3$, we have an $\mathfrak{sl}_2$-triple $\{E_i, F_i, H_i\}$ with $E_i \in \mathfrak{g}_{\alpha_i}$, $F_i \in \mathfrak{g}_{-\alpha_i}$ and $H_i = [E_i, F_i]$ satisfying $\alpha_i(H_i) = 2$. For each $\lambda \in \mathfrak{h}^*$, $i = 1,2,3$, let $\lambda_i = \lambda(h_i)$ and $c_i = \lambda(H_i)$. Then

$$c_1 = \frac{\lambda_2 + \lambda_3}{-\alpha - 1},\ c_2 = \frac{\lambda_1 + \lambda_3}{\alpha},\ c_3 = \lambda_1 + \lambda_2. \tag{4}$$

Theorem 10 *Let* $\mathfrak{g} = D(2,1,\alpha)$ *and let* Δ^+ *be as in* (3). *Then for each* $\lambda \in \mathfrak{h}^*$, *we have that inj* $L(\lambda) = \Delta_{\bar{0}}^-$ *if and only if* $c_1, c_2, c_3 \notin \mathbb{Z}_{\geq 1}$ *and at most one of* $\lambda_1, \lambda_2, \lambda_3$ *equals zero.*

Proof. Let Π denote the set of simple roots of Δ^+. Then F_i acts injectively on $L(\lambda)$ if and only if given a set of simple roots Π' containing α_i which can be obtained by a sequence of odd reflections from Π, we have that $\lambda'(H_i) \notin \mathbb{Z}_{\geq 0}$, where $\lambda' \in \mathfrak{h}^*$ satisfies $L_{\Pi'}(\lambda') = L_{\Pi}(\lambda)$.

Now if $\lambda_i, \lambda_j = 0$, then $f_i v = 0$ and $f_j v = 0$ imply $F_k v = [f_i, f_j]v = 0$. Hence, if F_1, F_2, F_3 act injectively on $L(\lambda)$, then at most one of $\lambda_1, \lambda_2, \lambda_3$ equals zero. By reflecting at β_j we get $\Pi' = \{-\beta_j, \alpha_i, \alpha_k\}$ where $i \neq j \neq k$. If β_j is a typical root ($\lambda_j \neq 0$) then F_i acts injectively iff $(\lambda - \beta_j)(H_i) = c_i - 1 \notin \mathbb{Z}_{\geq 0}$, and F_k acts injectively iff $(\lambda - \beta_j)(H_k) = c_k - 1 \notin \mathbb{Z}_{\geq 0}$. Hence, if at least two of $\lambda_1, \lambda_2, \lambda_3$ are non-zero, it follows that F_1, F_2, F_3 act injectively iff $c_1, c_2, c_3 \notin \mathbb{Z}_{\geq 1}$.

Remark 6 Let Δ^+ be as in (3), then $L(\lambda)$ is typical if and only if $\lambda_1, \lambda_2, \lambda_3 \neq 0$ and $\lambda_1 + \lambda_2 + \lambda_3 \neq 0$.

Theorem 11 *Let* $\mathfrak{g} = D(2,1,\alpha)$, *and suppose that* $L(\lambda)$ *is a simple highest weight module with respect to some decomposition* $\Delta = \Delta^+ \cup \Delta^-$, *which satisfies inj* $L(\lambda) = \Delta_{\bar{0}}^-$. *Then* $L(\lambda) = M(\lambda)$ *iff* $L(\lambda)$ *is typical.*

Proof. Since each of these properties is a module property that is preserved under odd reflections, it suffices to prove the theorem with respect to Δ^+ in (3). By Theorem 10, we have inj $L(\lambda) = \Delta_{\bar{0}}^-$ implies $\lambda(H_i) \notin \mathbb{Z}_{\geq 1}$ for each $\alpha_i \in \Delta_{\bar{0}}^+$. Hence,

$$\frac{2(\lambda+\rho,\alpha_i)}{(\alpha_i,\alpha_i)} = \frac{2(\lambda,\alpha_i)}{(\alpha_i,\alpha_i)} = \lambda(H_i) \quad \notin \mathbb{Z}_{\geq 1},$$

where the first equality is due to the fact that $\rho = 0$ for our choice of Δ^+, and the second equality is given by the identification of $\mathfrak{h}$ with $\mathfrak{h}^*$. The claim now follows from computing the Shapovalov determinant using the formula given in [5, Sect. 1.2.8]. Since $(\beta,\beta) = 0$ for $\beta \in \Delta_{\bar{1}}$, we conclude that $M(\lambda)$ is simple if and only if $(\lambda+\rho,\beta) \neq 0$ for all $\beta \in \Delta_{\bar{1}}$.

3.2 Cuspidal modules for $D(2,1,\alpha)$

The following theorem gives a characterization of cuspidal modules for $D(2,1,\alpha)$.

Theorem 12 *Let $\mathfrak{g} = D(2,1,\alpha)$, and let M be a simple weight $\mathfrak{g}$-module. The following are equivalent:*

1. *M is cuspidal;*
2. *Supp M is exactly one Q coset;*
3. *x_α acts injectively for all $\alpha \in \Delta_{\bar{0}}$, $x_\alpha \in \mathfrak{g}_\alpha$.*

Corollary 3 *Let $\mathfrak{g} = D(2,1,\alpha)$. If M is a cuspidal $\mathfrak{g}$-module, then dim $M_\lambda =$ dim M_μ for all $\lambda - \mu \in Q_{\bar{0}}$.*

For $\mathfrak{g} = D(2,1,\alpha)$ it follows from [3], that for typical central characters we have a $1-1$ correspondence between cuspidal $\mathfrak{g}$-modules and cuspidal $\mathfrak{g}_{\bar{0}}$-modules. Here we describe cuspidal $\mathfrak{g}_{\bar{0}}$-modules.

Theorem 13 *Let $\mathfrak{g}_{\bar{0}} = \mathfrak{sl}_2 \times \mathfrak{sl}_2 \times \mathfrak{sl}_2$. Then the cuspidal $\mathfrak{g}_{\bar{0}}$-modules are as follows:*

$$V_a^\mu := V(a_1)^{[\mu_1]} \otimes V(a_2)^{[\mu_2]} \otimes V(a_3)^{[\mu_3]} \quad a,\mu \in \mathbb{C}^3,\ \mu_i \pm a_i, \notin \mathbb{Z},\ i = 1,2,3 \quad (5)$$

Moreover, Supp $V_a^\mu = Q + \mu$ and deg $V_a^\mu = 1$.

We have two ways to realize cuspidal modules. The first method is by decomposing the modules $\widetilde{N}$ appearing in Theorem 6, since each simple subquotient of $\widetilde{N}$ is cuspidal. It follows from the PBW theorem that deg $\widetilde{V_a^\mu} = 2^7$, so we see from Theorem 9 that $\widetilde{V_a^\mu}$ is far from simple. Alternatively, one could determine simplicity conditions for the modules $L(\lambda)_\Gamma^\mu$ appearing in Theorem 7.

Here we calculate the degree of a cuspidal $\mathfrak{g}$-module $L(\lambda)_\Gamma^\mu$ using the results from Sect. 3.1 and Shapovalov determinants for the module $M(\lambda)$ [4,5,10].

Theorem 14 *Let $\mathfrak{g} = D(2,1,\alpha)$, and suppose that $L(\lambda)_\Gamma^\mu$ is a (simple) cuspidal $\mathfrak{g}$-module for some simple highest weight module $L(\lambda)$ for $\Delta = \Delta^+ \cup \Delta^-$, where $\Gamma \subset \Delta_{\bar{0}}^-$ and $\mu \in \mathbb{C}^3$. Then*

1. $\Gamma = inj\ L(\lambda) = \Delta_{\bar{0}}^{-}$, *and if* Δ^{+} *is as in (3) then Theorem 10 applies;*
2. $deg\ L(\lambda)_{\Gamma}^{\mu} = 8$ *iff* $L(\lambda)$ *is typical iff* $L(\lambda)$ *has graded degree* $(8,8)$*;*
3. *if* $L(\lambda)$ *is atypical, then* $2 \leq deg\ L(\lambda)_{\Gamma}^{\mu} \leq 6$*;*
4. *if* $(\lambda, \beta) = 0$ *for some simple odd root* β*, then* $deg\ L(\lambda)_{\Gamma}^{\mu} \leq 4$.

Acknowledgements I would like to thank Maria Gorelik, Ivan Penkov and Vera Serganova for helpful discussions. Supported in part at the Technion by an Aly Kaufman Fellowship.

References

1. Dimitrov, I., Mathieu, O., Penkov, I.: On the structure of weight modules. Trans. Amer. Math. Soc. **352**, 2857–2869 (2000)
2. Fernando, S.: Lie algebra modules with finite-dimensional weight spaces, I. Trans. Amer. Math. Soc. **322**, 757–781 (1990)
3. Gorelik, M.: Strongly typical representations of the basic classical Lie superalgebras. J. Amer. Math. Soc. **15**, 167–184 (2001)
4. Gorelik, M.: The Kac construction of the centre of $U(\mathfrak{g})$ for Lie superalgebras JNMP **11**, 325–349 (2004)
5. Gorelik, M., Kac, V.: On Simplicity of Vacuum modules. Advances in Math. **211**, 621-677 (2007)
6. Grantcharov, D.: Coherent families of weight modules of Lie superalgebras and an explicit description of the simple admissible $\mathfrak{sl}(n+1|1)$-modules. J. Algebra **265**, 711–733 (2003)
7. Grantcharov, D.: On the structure and character of weight modules. Forum Math. **18**, 933–950 (2006)
8. Grantcharov, D.: Explicit realizations of simple weight modules of classical Lie superalgebras. Contemp. Math. **499**, 141–148 (2009)
9. Kac, V.G.: Lie superalgebras. Advances in Math. **26**, 8–96 (1977)
10. Kac, V.G.: Representations of classical Lie superalgebras. Lect. Notes Math. Vol. 676, pp. 597–626. Springer-Verlag, Berlin Heidelberg New York (1978)
11. Kac, V.G., Wakimoto, M.: Integrable highest weight modules over affine superalgebras and number theory. Lie Theory and Geometry, Progress in Math. **123**, 415–456 (1994)
12. Leites, D., Savel'ev, M., Serganova, V., Embeddings of Lie superalgebra $\mathfrak{osp}(1|2)$ and non-linear supersymmetric equations. Group Theoretical Methods in Physics **1**, 377–394 (1986)
13. Mathieu, O.: Classification of irreducible weight modules. Ann. Inst. Fourier **50**, 537–592 (2000)
14. Musson, I.: Lie superalgebras and enveloping algebras. Graduate Studies in Mathematics, Vol. 131 (2012)
15. Penkov, I.: Generic representations of classical Lie superalgebras and their localization Monatshefte f. Math. **118**, 267–313 (1994)
16. Penkov, I., Serganova, V.: Representation of classical Lie superalgebras of type I, Indag. Mathem. N.S. **3**, 419–466 (1992)
17. Serganova, V.: Kac-Moody superalgebras and integrability, in Developments and Trends in Infinite-Dimensional Lie Theory. Progress in Mathematics, Vol. 288, pp. 169–218. Birkhäuser, Boston (2011)

On SUSY curves

Rita Fioresi and Stephen Diwen Kwok

Abstract We give a summary of some elementary results in the theory of super Riemann surfaces (SUSY curves), which are mostly known, but are not readily available in the literature. In particular, we give the classification of all genus 0 SUSY-1 curves and touch on the case of genus 1. We also briefly discuss the related topic of Π-projective spaces.

1 Introduction

In this note we give a summary of some elementary results in the theory of super Riemann surfaces (SUSY curves), which are mostly known, but are not readily available in the literature. Our main source is Manin, who has provided a terse introduction to this subject in [12]. More recently Freund and Rabin have given important results on the uniformization (see [14]) and Witten has written an account of the state of the art of this subject, from the physical point of view, in [19].

The paper is organized as follows.

In Sects. 2 and 3, we are g oing to recall briefly the main definitions of supergeometry and study in detail the examples of super projective space and Π-projective line, which are very important in the theory of SUSY curves.

In Sect. 4 we discuss some general facts on SUSY curves, including the *theta characteristic*, while in Sect. 5 we prove some characterization results concerning genus zero and genus one SUSY curves.

R. Fioresi (✉)
Dipartimento di Matematica, Università di Bologna, Piazza di Porta S. Donato 5, 40126 Bologna, Italy
e-mail: rita.fioresi@unibo.it

S.D. Kwok
Dipartimento di Matematica, Università di Bologna, Piazza di Porta S. Donato 5, 40126 Bologna, Italy
e-mail: stephendiwen.kwok@unibo.it

M. Gorelik, P. Papi (eds.): *Advances in Lie Superalgebras*. Springer INdAM Series 7,
DOI 10.1007/978-3-319-02952-8_7,

2 Preliminaries

We are going to briefly recall some basic definitions of analytic supergeometry. For more details see [6, 9, 12, 13, 18] and the classical references [3, 5].

Let our ground field be $\mathbb{C}$.

A *superspace* $S = (|S|, \mathcal{O}_S)$ is a topological space $|S|$ endowed with a sheaf of superalgebras $\mathcal{O}_S$ such that the stalk at a point $x \in |S|$, denoted by $\mathcal{O}_{S,x}$, is a local superalgebra.

A *morphism* $\phi : S \longrightarrow T$ of superspaces is given by $\phi = (|\phi|, \phi^*)$, where $|\phi| : |S| \longrightarrow |T|$ is a map of topological spaces and $\phi^* : \mathcal{O}_T \longrightarrow \phi_* \mathcal{O}_S$ is such that $\phi_x^*(\mathbf{m}_{|\phi|(x)}) \subseteq \mathbf{m}_x$ where $\mathbf{m}_{|\phi|(x)}$ and $\mathbf{m}_x$ are the maximal ideals in the stalks $\mathcal{O}_{T,|\phi|(x)}$ and $\mathcal{O}_{S,x}$ respectively.

The superspace $\mathbb{C}^{p|q}$ is the topological space $\mathbb{C}^p$ endowed with the following sheaf of superalgebras. For any open subset $U \subset \mathbb{C}^p$

$$\mathcal{O}_{\mathbb{C}^{p|q}}(U) = \mathrm{Hol}_{\mathbb{C}^p}(U) \otimes \wedge(\xi_1, \dots, \xi_q),$$

where $\mathrm{Hol}_{\mathbb{C}^p}$ denotes the complex analytic sheaf on $\mathbb{C}^p$ and $\wedge(\xi_1, \dots, \xi_q)$ is the exterior algebra in the variables $\xi_1, \dots, \xi_q$.

A *supermanifold* of dimension $p|q$ is a superspace $M = (|M|, \mathcal{O}_M)$ which is locally isomorphic to $\mathbb{C}^{p|q}$, as superspaces. A *morphism* of supermanifolds is simply a morphism of superspaces.

All of our supermanifolds and supermanifold morphisms are assumed to be smooth.

We now look at an important example of supermanifold, namely the *projective superspace*.

Let $\mathbb{P}^m = \mathbb{C}^{m+1} \setminus \{0\} / \sim$ be the ordinary complex projective space of dimension m with homogeneous coordinates $z_0, \dots, z_m$; $[z_0, \dots, z_m]$ denotes as usual an equivalence class in $\mathbb{P}^m$. Let $\{U_i\}_{i=1,\dots,m}$ be the affine cover $U_i = \{[z_0, \dots, z_m] \,|\, z_i \neq 0\}$, $U_i \cong \mathbb{C}^m$. On each U_i we take the global ordinary coordinates $u_0^i, \dots, \hat{u}_i^i, \dots u_m^i$, $u_k := z_k / z_i$ ($\hat{u}_i^i$ means we are omitting the variable u_i^i from the list). We now want to define the sheaf of superalgebras $\mathcal{O}_{U_i}$ on the topological space U_i:

$$\mathcal{O}_{U_i}(V) = \mathrm{Hol}_{U_i}(V) \otimes \wedge(\xi_1^i, \dots \xi_n^i), \qquad V \text{ open in } U_i,$$

where Hol_{U_i} is the sheaf of holomorphic functions on U_i and $\xi_1^i, \dots \xi_n^i$ are odd variables.

As one can readily check $\mathcal{U}_i = (U_i, \mathcal{O}_{U_i})$ is a supermanifold, isomorphic to $\mathbb{C}^{m|n}$. We now define the morphisms $\phi_{ij} : \mathcal{U}_i \cap \mathcal{U}_j \mapsto \mathcal{U}_i \cap \mathcal{U}_j$, where the domain is thought as an open submanifold of $\mathcal{U}_i$, while the codomain as an open submanifold of $\mathcal{U}_j$. The ϕ_{ij}'s are completely determined by the ordinary morphisms, together with the choice of m even and n odd sections in $\mathcal{O}_{U_i}(U_i \cap U_j)$. We write:

$$\phi_{ij} : (u_0^i, \dots, \hat{u}_i^i, \dots u_m^i, \xi_1^i, \dots, \xi_n^i) \mapsto \left(\frac{u_1^i}{u_j^i}, \dots, \frac{1}{u_j^i}, \dots, \frac{u_m^i}{u_j^i}, \frac{\xi_1^i}{u_j^i}, \dots, \frac{\xi_n^i}{u_j^i} \right) \qquad (1)$$

where on the right hand side the $1/u^i_j$ appears in the ith position and the jth position is omitted. As customary in the literature, the formula (1) is just a synthetic way to express the pullbacks:

$$\phi_{ij}^*(u_k^j) = \frac{u_k^i}{u_j^i}, \quad 0 \le k \ne j \le m, \qquad \phi_{ij}^*(\xi_l^j) = \frac{\xi_l^i}{u_j^i}, \quad 0 \le l \le n.$$

One can easily check that the ϕ_{ij}'s satisfy the compatibility conditions:

$$\phi_{ij}\phi_{ki} = \phi_{jk}, \qquad \text{on} \quad \mathscr{U}_i \cap \mathscr{U}_j \cap \mathscr{U}_k$$

hence they allow us to define uniquely a sheaf, denoted with $\mathscr{O}_{\mathbb{P}^{m|n}}$, hence a supermanifold structure on the topological space $\mathbb{P}^m$. The supermanifold $(\mathbb{P}^m, \mathscr{O}_{\mathbb{P}^{m|n}})$ is called the *projective space of dimension* $m|n$.

One can replicate the same construction and obtain more generally a supermanifold structure for the topological space: $\mathbb{P}(V) := V \setminus \{0\}$, where V in any complex super vector space.

We now introduce the functor of points approach to supergeometry.

The *functor of points* of a supermanifold X is the functor (denoted with the same letter) $X : (\text{smflds})^o \longrightarrow (\text{sets})$, $X(T) = \text{Hom}(T,X)$, $X(f)\phi = f \circ \phi$. The functor of points characterizes completely the supermanifold X: in fact, two supermanifolds are isomorphic if and only if their functor of points are isomorphic. This is one of the statements of Yoneda's Lemma, for more details see [6, Chap. 3].

The functor of points approach allows us to retrieve some of the geometric intuition. For example, let us consider the functor $P : (\text{smflds})^o \longrightarrow (\text{sets})$ associating to each supermanifold T the locally free subsheaves of $\mathscr{O}_T^{m+1|n}$ of rank $1|0$, where $\mathscr{O}_T^{m+1|n} := (\mathbb{C}^{\oplus m} \oplus (\Pi\mathbb{C})^{\oplus n}) \otimes \mathscr{O}_T$. [1] P is defined in an obvious way on the morphism: any morphism of supermanifolds $\phi : T \longrightarrow S$ defines a corresponding morphism of the structural sheaves $\phi^* : \mathscr{O}_S \longrightarrow \phi_* \mathscr{O}_T$, so that also $P(\phi)$ is defined.

The next proposition allows us to identify the functor P with the functor of points of the super projective space $\mathbb{P}^{m|n}$.

Proposition 1 *There is a one-to-one correspondence between the two sets:*

$$P(T) \longleftrightarrow \mathbb{P}^{m|n}(T), \qquad T \in (\text{smflds})$$

which is functorial in T. In other words P is the functor of points of $\mathbb{P}^{m|n}(T)$.

Proof. We briefly sketch the proof, leaving to the reader the routine checks. Let us start with an element in $P(T)$, that is a locally free sheaf $\mathscr{F}_T \subset \mathscr{O}_T^{m+1|n}$ of rank $1|0$. We want to associate to $\mathscr{F}_T$ a T-point of $\mathbb{P}^{m|n}$ that is a morphism $T \longrightarrow \mathbb{P}^{m|n}$. First cover T with V_i so that $\mathscr{F}_T|_{V_i}$ is free. Hence:

$$\mathscr{F}_T(V_i) = \text{span}\,\{(t_0, \ldots, t_m, \theta_1, \ldots, \theta_n)\} \subset \mathscr{O}_T^{m+1|n}(V_i)$$

[1] We may denote the linear superspace $\mathbb{C}^{\oplus m} \oplus (\Pi\mathbb{C})^{\oplus n}$ simply with $\mathbb{C}^{m|n}$ whenever it is clear it is not the complex supermanifold introduced at the beginning of this section.

where we assume that the section $t_i \in \mathcal{O}_T(V_i)$ is invertible without loss of generality, since the rank of $\mathcal{F}_T$ is $1|0$ (this assumption may require to change the cover). Hence:

$$\mathcal{F}_T(V_i) = \mathrm{span}\{(t_0/t_i, \dots, 1, \dots, t_m/t_i, \theta_1/t_i, \dots, \theta_n/t_i)\}.$$

Any other basis $(t'_1, \dots, t'_{m+1}, \theta'_1, \dots, \theta'_n)$ of $\mathcal{F}_T(V_i)$ is a multiple of $(t_0, \dots, t_m, \theta_1, \dots, \theta_n)$ by an invertible section on V_i, hence we have:

$$t'_j/t'_i = t_j/t_i \qquad \theta'_k/t'_i = \theta_k/t_i.$$

Thus the functions $t_j/t_i, \theta_k/t_i$, which a priori are only defined on open subsets where $\mathcal{F}_T$ is a free module, are actually defined on the whole of the open set where t_i is invertible, being independent of the choice of basis for $\mathcal{F}_T(V_i)$.
We have then immediately a morphism of supermanifolds $f_i : \mathcal{V}_i \longrightarrow \mathcal{U}_i \subset \mathbb{P}^{m|n}$:

$$f_i^*(u_1^i) = t_0/t_i, \dots, f_i^*(u_m^i) = t_m/t_i, \quad f_i^*(\xi_1^i) = \theta_1/t_i, \dots, f_i^*(\xi_n^i) = \theta_n/t_i \qquad (2)$$

where $\mathcal{V}_i = (V_i, \mathcal{O}_T|_{V_i})$ and $\mathcal{U}_i = (U_i, \mathcal{O}_{\mathbb{P}^{m|n}}|_{U_i})$. It is immediate to check that the f_i's agree on $\mathcal{V}_i \cap \mathcal{V}_j$, so they glue to give a morphism $f : T \longrightarrow \mathbb{P}^{m|n}$.

As for the vice versa, consider $f : T \longrightarrow \mathbb{P}^{m|n}$ and define $V_i = |f|^{-1}(U_i)$. The morphism $f|_{V_i}$ corresponds to the choice of m even and n odd sections in $\mathcal{O}_T(V_i)$: $v_1, \dots, v_m, \eta_1, \dots, \eta_n$. We can then define immediately the free sheaves $\mathcal{F}_{V_i} \subset \mathcal{O}_T|_{V_i}^{m+1|n}$ of rank $1|0$ on each of the V_i as

$$\mathcal{F}_{V_i}(V) := \mathrm{span}\{(v_1|_V, \dots, 1, \dots v_m|_V, \eta_1|_V, \dots, \eta_n|_V)\},$$

(1 in the ith position). As one can readily check the $\mathcal{F}_{V_i}$ glue to give a locally free subsheaf of $\mathcal{O}_T^{m+1|n}$.

We now want to define the *Π-projective line* which represents in some sense a generalization of the super projective space of dimension $1|1$ that we defined previously.

Let $\mathbb{P}^1 = \mathbb{C}^2 \setminus \{0\}/\sim$ be the ordinary complex projective line with homogeneous coordinates z_0, z_1. Define, as we did before, the following supermanifold structure on each U_i belonging to the open cover $\{U_0, U_1\}$ of $\mathbb{P}^1$: $\mathcal{O}_{U_i}(V) = \mathrm{Hol}_{U_i}(V) \otimes \wedge(\xi)$, V open in U_i, $i = 1, 2$, so that $\mathcal{U}_i = (U_i, \mathcal{O}_{U_i})$ is a supermanifold isomorphic to $\mathbb{C}^{1|1}$. At this point, instead of the change of chart ϕ_{12}, we define the following transition map (there is only one such):

$$\begin{aligned} \psi_{12} : \mathcal{U}_0 \cap \mathcal{U}_1 &\longrightarrow \mathcal{U}_0 \cap \mathcal{U}_1 \\ (u, \xi) &\mapsto \left(\frac{1}{u}, -\frac{\xi}{u^2}\right). \end{aligned}$$

As one can readily check, this defines a supermanifold structure on the topological space $\mathbb{P}^1$ and we call this supermanifold the *Π-projective line* $\mathbb{P}_\Pi^{1|1} = (\mathbb{P}^1, \mathcal{O}_{\mathbb{P}_\Pi^1})$.

Alternatively, this supermanifold can be constructed by taking $\mathcal{O}_{\mathbb{P}^1} \oplus \Pi\mathcal{O}(1)$ to be the structure sheaf.

In the next section we will characterize its functor of points.

3 The Π-projective line

In this section we want to take advantage of the functor of points approach in order to give a more geometric point of view on the Π-projective line and to understand in which sense it is a generalization of the super projective line, whose functor of points was described in the previous section. (See [12, Chap. 2, Sect. 8.5] for a further discussion on the geometry of the Π-projective spaces more generally). Let us start with an overview of the ordinary geometric construction of the projective line.

The topological space $\mathbb{P}^1$ consists of the 1-dimensional subspaces of $\mathbb{C}^2$, that is $\mathbb{P}^1 = \mathbb{C}^2 \setminus \{0\}/\sim$, where $(z_0, z_1) \sim (z_0', z_1')$ if and only if $(z_0, z_1) = \lambda(z_0', z_1')$, $\lambda \in \mathbb{C}^\times$. In other words, the equivalence class $[z_0, z_1] \in \mathbb{P}^1$ consists of all the points in $\mathbb{C}^2$ which are in the orbit of (z_0, z_1) under the action of $\mathbb{C}^\times$ by left (or right) multiplication.

Now we go to the functor of points of $\mathbb{P}^{1|1}$. A T-point of $\mathbb{P}^{1|1}$ locally is a $1|0$-submodule of $\mathscr{O}_T^{1|1}(V)$ (V is a suitably chosen open in T). So it is locally an equivalence class $[z_0, z_1, \eta_0, \eta_1]$ where we identify two quadruples $(z_0, z_1, \eta_0, \eta_1) \sim (z_0', z_1', \eta_0', \eta_1')$ if and only if $z_i = \lambda z_i'$ and $\eta_i = \lambda \eta_i'$, $i = 0, 1$, $\lambda \in \mathscr{O}_T(V)^\times$. In other words, exactly as we did before for the case of $\mathbb{P}^1$, we identify those elements in $\mathbb{C}^{2|1}(T)$ that belong to the same orbit of the multiplicative group of the complex field $\mathbb{G}_m^{1|0}(T) \cong \mathbb{C}^\times(T)$.[2] It makes then perfect sense to generalize this construction and look at the equivalence classes with respect to the action of the multiplicative supergroup $\mathbb{G}_m^{1|1}$, which is the supergroup with underlying topological space $\mathbb{C}^\times$, with one global odd coordinate and with group law (in the functor of points notation):

$$(a, \alpha.) \cdot (a', \alpha.') = (aa' + \alpha.\alpha.', a\alpha.' + \alpha.a').$$

$\mathbb{G}_m^{1|1}$ is naturally embedded into $\mathrm{GL}(1|1)$, the complex general linear supergroup via the morphism (in the functor of points notation):

$$\begin{aligned} \mathbb{G}_m^{1|1}(T) &\longrightarrow \mathrm{GL}(1|1)(T) \\ (a, \alpha.) &\mapsto \begin{pmatrix} a & \alpha. \\ \alpha. & a \end{pmatrix}. \end{aligned}$$

This is precisely the point of view we are taking in constructing the Π-projective line: we identify T-points in $\mathbb{C}^{2|2}$ which lie in the same $\mathbb{G}_m^{1|1}$ orbit, but instead of looking simply at rank $1|1$ submodules of $\mathbb{C}^{2|2}(T)$ we look at a more elaborate structure, which is matching very naturally the $\mathbb{G}_m^{1|1}$ action on $\mathbb{C}^{2|2}$. This structure is embodied by the condition of ϕ-invariance for a suitable odd endomorphism ϕ of $\mathbb{C}^{2|2}$, that we shall presently see. For more details see Appendix 6.

Consider now the supermanifold $\mathbb{C}^{2|2}$, and the odd endomorphism ϕ on $\mathbb{C}^{2|2}$ given in terms of the standard homogeneous basis $\{e_0, e_1 | \mathscr{E}_0, \mathscr{E}_1\}$ of the super vector space

---

[2] All of our arguments here take place for an open cover of T in which a T point corresponds to a free sheaf and not just a locally free one. For simplicity of exposition we omit to mention the cover and the necessary gluing to make all of our argument stand.

underlying $\mathbb{C}^{2|2}$ by:

$$\left(\begin{array}{cc|cc} 0 & 0 & 1 & 0 \\ 0 & 0 & 0 & 1 \\ \hline 1 & 0 & 0 & 0 \\ 0 & 1 & 0 & 0 \end{array}\right).$$

We note that $\phi^2 = 1$.

In analogy with the projective superspace, we now consider the functor P_Π : $(\mathrm{smflds})^o \longrightarrow (\mathrm{sets})$, where

$$P_\Pi(T) := \{\text{rank } 1|1 \text{ locally free, } \phi\text{-invariant subsheaves of } \mathscr{O}_T^{2|2}\}.$$

Here the action of ϕ is extended to $\mathscr{O}_T^{2|2} = \mathbb{C}^{2|2} \otimes_{\mathbb{C}} \mathscr{O}_T$, by acting on the first factor, and $\mathbb{C}^{2|2}$ is a linear superspace.

Lemma 1 *Let $\mathscr{F}_T \in P_\Pi(T)$. Then there exist an open cover $\{V_i\}$ of T, where $\mathscr{F}_T(V_i)$ is free and a basis e, $\mathscr{E}$ of $\mathscr{F}_T(V_i)$ such that $\phi(e) = \mathscr{E}, \phi(\mathscr{E}) = e$.*

Proof. Since $\mathscr{F}_T$ is locally free, there exist an open cover $\{V_i\}$ of T, where $\mathscr{F}_T(V_i)$ is free with basis, say, e', $\mathscr{E}'$. Let Ψ be the matrix of $\phi|_{V_i}$ in this basis. Since $\phi^2 = 1$, we have $\Psi^2 = 1$, which implies that Ψ has the form:

$$\Psi := \left(\begin{array}{c|c} \alpha & a \\ \hline a^{-1} & -\alpha \end{array}\right)$$

with $a \in \mathscr{O}_T^*(V_i)_0$, $\alpha \in \mathscr{O}_T(V_i)_1$. Let $P \in \mathrm{GL}(1|1)(\mathscr{O}_T(V_i))$ be the matrix:

$$P := \left(\begin{array}{c|c} a^{-1} & 0 \\ \hline -a^{-1}\alpha & 1 \end{array}\right)$$

P is invertible because a is, and one calculates that $P\Psi P^{-1} = \Phi$, so P gives the desired change of basis.

Proposition 2 *There is a one-to-one correspondence between the two sets:*

$$P_\Pi(T) \longrightarrow \mathbb{P}_\Pi^{1|1}(T), \qquad T \in (\mathrm{smflds})$$

which is functorial in T. In other words P_Π is the functor of points of $\mathbb{P}_\Pi^{1|1}$.

Proof. We briefly sketch the proof, leaving to the reader the routine checks. Let us consider a locally free sheaf $\mathscr{F}_T \subset \mathscr{O}_T^{2|2}$ of rank $1|1$ in $P_\Pi(T)$, invariant under ϕ. We want to associate to each such $\mathscr{F}_T$ a T-point of $\mathbb{P}_\Pi^{1|1}(T)$, that is, a morphism $f : T \longrightarrow \mathbb{P}_\Pi^{1|1}$.

First we cover T with V_i, so that $\mathscr{F}_T|_{V_i}$ is free. By Lemma 1 there exists a basis e, $\mathscr{E}$ of $\mathscr{F}_T(V_i)$ such that $\phi(e) = \mathscr{E}$, $\phi(\mathscr{E}) = e$.

Representing e, $\mathscr{E}$ using the basis $\{e_0, \mathscr{E}_0, e_1, \mathscr{E}_1\}$ of $\mathbb{C}^{2|2}$, we have:

$$\mathscr{F}_T(V_i) = \text{span}\{e = (s_0, \sigma_0, s_1, \sigma_1), \mathscr{E} = (\sigma_0, s_0, \sigma_1, s_1)\}$$

for some sections s_j, σ_j in $\mathscr{O}_T(V_i)$. Since the rank of $\mathscr{F}_T$ is $1|1$, either s_0 or s_1 must be invertible. Let us call V_0 the union of the V_i for which s_0 is invertible and V_1 the union of the V_i for which s_1 is invertible.

Hence, we can make a change of basis of $\mathscr{F}_T(V_i)$ by right multiplying the column vectors representing e and $\mathscr{E}$ in the given basis, by a suitable element $g_i \in \mathrm{GL}(1|1)(\mathscr{O}_T(V_i))$ obtaining:

$$\mathscr{F}_T(V_0) = \text{span } \{(1, 0, s_1 s_0^{-1} - \sigma_1 \sigma_0 s_0^{-2}, \sigma_1 s_0^{-1} - s_1 \sigma_0 s_0^{-2}),$$

$$(0, 1, \sigma_1 s_0^{-1} - s_1 \sigma_0 s_0^{-2}, s_1 s_0^{-1} - \sigma_1 \sigma_0 s_0^{-2})\},$$

$$\mathscr{F}_T(V_1) = \text{span } \{(s_0 s_1^{-1} - \sigma_0 \sigma_1 s_1^{-2}, \sigma_0 s_1^{-1} - s_0 \sigma_1 s_1^{-2}, 1, 0),$$

$$(\sigma_0 s_1^{-1} - s_0 \sigma_1 s_1^{-2}, s_0 s_1^{-1} - \sigma_0 \sigma_1 s_1^{-2}, 0, 1)\},$$

$$g_0 = \begin{pmatrix} s_0^{-1} & -\sigma_0 s_0^{-2} \\ -\sigma_0 s_0^{-2} & s_0^{-1} \end{pmatrix}, \qquad g_1 = \begin{pmatrix} s_1^{-1} & -\sigma_1 s_1^{-2} \\ -\sigma_1 s_1^{-2} & s_1^{-1} \end{pmatrix}.$$

Suppose now $\{e', \mathscr{E}'\} := \{(s_0', \sigma_0', s_1', \sigma_1'), (\sigma_0', s_0', \sigma_1', s_1')\}$ is another basis of $\mathscr{F}_T(V_i)$ such that $\phi(e') = \mathscr{E}', \phi(\mathscr{E}') = e'$. The sections in $\mathscr{O}_T(V_i)$ we have obtained, namely:

$$v_0 = s_1 s_0^{-1} - \sigma_1 \sigma_0 s_0^{-2}, \; \nu_0 = \sigma_1 s_0^{-1} - s_1 \sigma_0 s_0^{-2}$$
$$v_1 = s_0 s_1^{-1} - \sigma_0 \sigma_1 s_1^{-2}, \; \nu_1 = \sigma_0 s_1^{-1} - s_0 \sigma_1 s_1^{-2}$$

are independent of the choice of such a basis. This can be easily seen with an argument very similar to the one in Proposition 1.

Hence we have well-defined morphisms of supermanifolds $f_i : \mathscr{V}_i \longrightarrow \mathscr{U}_i \subset \mathbb{P}_\Pi^{1|1}$:

$$f_0^*(u_0) = v_0, \; f_0^*(\xi_0) = \nu_0$$
$$f_0^*(u_1) = v_1, \; f_0^*(\xi_1) = \nu_1$$

where $\mathscr{V}_i = (V_i, \mathscr{O}_T|_{V_i})$ and $\mathscr{U}_i = (U_i, \mathscr{O}_{\mathbb{P}_\Pi^{1|1}}|_{U_i})$, while (u_i, ξ_i) are global coordinates on $\mathscr{U}_i \cong \mathbb{C}^{1|1}$. A small calculation shows that the f_i's agree on $\mathscr{V}_0 \cap \mathscr{V}_1$, in fact as one can readily check:

$$(1, 0, v_0, \nu_0) \sim \left(\frac{1}{v_0}, \frac{-\nu_0}{v_0^2}, 1, 0\right)$$

and similarly for $(v_1, \nu_1, 1, 0)$, which corresponds to the transition map for $\mathbb{P}_\Pi^{1|1}$ we defined in Sec. 2. So the f_i's glue to give a morphism $f : T \longrightarrow \mathbb{P}_\Pi^{1|1}$.

For the converse, consider $f : T \longrightarrow \mathbb{P}_\Pi^{1|1}$ and define $V_i = |f|^{-1}(U_i)$. We can define immediately the sheaves $\mathscr{F}_{V_i} \subset \mathscr{O}_{V_i}^{2|2}$ on each of the V_i as we did in the proof of

Proposition 1:

$$\mathscr{F}_{V_0} = \text{span}\ \{(1,0,t_0,\tau_0),(0,1,\tau_0,t_0)\}$$
$$\mathscr{F}_{V_1} = \text{span}\ \{(t_1,\tau_1,1,0),(\tau_1,t_1,0,1)\}$$

where $t_i = f^*(u_i), \tau_i = f^*(\xi_i)$. The $\mathscr{F}_{V_i}$ so defined are free of rank $1|1$ (by inspection there are no nontrivial relations between the generators), and ϕ-invariant by construction. Finally, one checks that the relations $t_1 = t_0^{-1}, \tau_1 = -t_0^{-2}\tau_0$ imply that the $\mathscr{F}_{V_i}$ glue on $V_0 \cap V_1$ to give a locally free rank $1|1$ subsheaf of $\mathscr{O}_T^{2|2}$.

4 Super Riemann surfaces

In this section we give the definition of super Riemann surface and we examine some elementary, yet important properties.

Much of the material we discuss in this section is contained, though not so explicitly, in Chap. 2 of [12]; see also the article [2]. Further key results on the geometry of super Riemann surfaces may be found in the work of Rabin and his collaborators (see e.g. [4, 14, 15]). The moduli of SUSY-1 curves is a well-studied subject; we refer the reader to [16] where the deformation theory of complex analytic superspaces and coherent sheaves on superspaces is developed and used to prove the existence of the universal deformation of a given SUSY-1 curve $(X,\mathscr{D})$.

Definition 1 A $1|1$-*super Riemann surface* is a pair $(X,\mathscr{D})$, where X is a $1|1$-dimensional complex supermanifold, and $\mathscr{D}$ is a locally direct (and consequently locally free, by the super Nakayama's lemma) rank $0|1$ subsheaf of the tangent sheaf TX such that:

$$\mathscr{D}\otimes\mathscr{D} \longrightarrow TX/\mathscr{D}$$
$$Y\otimes Z \mapsto [Y,Z]\ \ (mod\,\mathscr{D})$$

is an isomorphism of sheaves. Here $[\,,\,]$ denotes the super Lie bracket of vector fields. The distinguished subsheaf $\mathscr{D}$ is called a *SUSY-1 structure* on X, and $1|1$-super Riemann surfaces are thus alternatively referred to as *SUSY-1 curves*. We shall refer to SUSY 1-structures simply as SUSY structures.

We say that X has genus g if the underlying topological space $|X|$ has genus g.

Definition 2 Let $(X,\mathscr{D})$, $(X',\mathscr{D}')$ be SUSY-1 curves, and $F: X \to X'$ a biholomorphic map of supermanifolds. F is a *isomorphism of SUSY curves*, or simply a *SUSY isomorphism*, if $(dF)_p(\mathscr{D}_p) = \mathscr{D}'_{|F|(p)}$ for all reduced points $p \in |X|$. Here $(dF)_p$ denotes the differential of F at p, $\mathscr{D}_p \subset T_pX$ (resp. $\mathscr{D}'_q$) the stalk of the subsheaf $\mathscr{D}$ (resp. $\mathscr{D}'$) at p (resp. q).

Example 1 Let us consider the supermanifold $\mathbb{C}^{1|1}$, with global coordinates z, ζ together with the odd vector field:

$$V = \partial_\zeta + \zeta\partial_z.$$

If $\mathscr{D} = \text{span}\{V\}$, $\mathscr{D}$ is a SUSY structure on $\mathbb{C}^{1|1}$ since V, V^2 span $T\mathbb{C}^{1|1}$. As we will see, this is the *unique* (up to SUSY isomorphism) SUSY structure on $\mathbb{C}^{1|1}$.

We now want to relate the SUSY structures on a supermanifold and the canonical bundle of the reduced underlying manifold. It is important to remember that for a supermanifold X of dimension $1|1$, $\mathscr{O}_{X,0} = \mathscr{O}_{X,\text{red}}$; that is, the even part of its structural sheaf coincides with its reduced part. This is of course not true for a generic supermanifold.

We start by showing that any SUSY structure can be locally put into a *canonical form*.

Lemma 2 *Let $(X, \mathscr{D})$ be a SUSY-1 curve, p a topological point in X_{red}. Then there exists an open set U containing p and a coordinate system $W = (w, \eta)$ for U such that $\mathscr{D}|_U = \text{span}\ \{\partial_\eta + \eta\partial_w\}$.*

Proof. Since $\mathscr{D}$ is locally free, there exists a neighborhood U of p on which $\mathscr{D} =$ span $\{D\}$, where D is some odd vector field; by shrinking U, we may assume it is also a coordinate domain with coordinates (z, ζ). Since X has only one odd coordinate, we have $D = f(z)\partial_\zeta + g(z)\zeta\partial_z$ for some holomorphic even functions f, g. So:

$$D^2 = [D,D]/2 = g(z)\partial_z + gf'\zeta\partial_\zeta.$$

Since D, D^2 form a local basis for the $\mathscr{O}_X$-module TX, we have

$$a_{11}D + a_{12}D^2 = \partial_z, \qquad a_{21}D + a_{22}D^2 = \partial_\zeta.$$

If we substitute the expression for D and D^2 we obtain:

$$g(a_{11}\zeta + a_{12}) = 1, \qquad a_{21}f = 1 - a_{22}gf'\zeta$$

from which we conclude that both f and g must be units.

We now show that we can find a new coordinate system (possibly shrinking U) so that D can be put in the desired form. We will assume such a coordinate system exists, then determine a formula for it and this formula will give us the existence. Let $w = w(z)$, $\eta = h(z)\zeta$ be the new coordinate system, where w and h are holomorphic functions. By the chain rule, we have:

$$\partial_z = w'(z)\partial_w + h'(z)\zeta\partial_\eta, \qquad \partial_\zeta = h(z)\partial_\eta.$$

We now set $D = \partial_\eta + \eta\partial_w$ and substituting we have:

$$D = \partial_\eta + \eta\partial_w = fh\partial_\eta + gh^{-1}\eta w'(z)\partial_w$$

which holds if and only if the system of equations:

$$fh = 1, \qquad gw' = h$$

has a solution for w, h. By shrinking our original coordinate domain, we may assume it is simply connected. Then since f and g are units, the system has a solution, by standard facts from complex analysis. We leave to the reader the easy check that (w, η) is indeed a coordinate system.

Definition 3 Let $(X, \mathscr{D})$ be a SUSY-1 curve, U an open set. Any coordinate system (w, η) on U having the property of Lemma 2 is said to be *compatible* with the SUSY structure $\mathscr{D}$. Any open set U that admits a $\mathscr{D}$-compatible coordinate system is said to be *compatible* with $\mathscr{D}$.

Definition 4 Let X_{red} be an ordinary Riemann surface, $K_{X_{\text{red}}}$ its canonical bundle. A *theta characteristic* is a pair $(\mathscr{L}, \alpha)$, where $\mathscr{L}$ is a holomorphic line bundle on X_{red}, and α a holomorphic isomorphism of line bundles $\alpha. : \mathscr{L} \otimes \mathscr{L} \longrightarrow K_{X_{\text{red}}}$. An *isomorphism* of theta characteristics $(\mathscr{L}, \alpha)$, $(\mathscr{L}', \alpha')$ is an isomorphism $\phi : \mathscr{L} \to \mathscr{L}'$ of line bundles such that $\alpha' \circ \phi^{\otimes 2} = \alpha$. Some authors also call a theta characteristic of X_{red} a *square root* of the canonical bundle $K_{X_{\text{red}}}$.

Definition 5 A *super Riemann pair*, or *SUSY pair* for short, is a pair $(X_r, (\mathscr{L}, \alpha))$ where X_r is an ordinary Riemann surface, and $(\mathscr{L}, \alpha)$ is a theta characteristic on X_r. An *isomorphism* of SUSY pairs $F : (X_r, (\mathscr{L}, \alpha)) \to (X_r', (\mathscr{L}', \alpha'))$ is a pair (f, ϕ) where $f : X_r \to X_r'$ is a biholomorphism of ordinary Riemann surfaces, and $\phi : \mathscr{L}' \to f_*(\mathscr{L})$ is an isomorphism of theta characteristics on X_r'.

For the sake of brevity, we will occasionally omit writing the isomorphism α in describing a super Riemann pair. The following theorem, a version of which is stated in Chap. 2, Prop. 2.3 (cf. [12, Chap. 2, Ex. 2.4]), shows that the data of super Riemann surface and of super Riemann pair are completely equivalent.

Theorem 1 *Let $(X, \mathscr{D})$ be a SUSY-1 curve. Then $(X_{\text{red}}, \mathscr{O}_{X,1})$ is a SUSY pair, where $\mathscr{O}_{X,1}$ is regarded as an $\mathscr{O}_{X,0}$-line bundle. Furthermore, if $F := (f, f^\#) : (X, \mathscr{D}) \to (X', \mathscr{D}')$ is a SUSY-isomorphism, then $(f, f^\#|_{\mathscr{O}_{X,1}}) : (X_{\text{red}}, \mathscr{O}_{X,1}) \to (X'_{\text{red}}, \mathscr{O}_{X',1})$ is an isomorphism of SUSY pairs.*

Conversely, suppose $(X_r, (\mathscr{L}, \alpha))$ is a SUSY pair. Then there exists a structure of SUSY-1 curve $(X_{\mathscr{L}}, \mathscr{D}_{\mathscr{L}})$ on X_r, such that the SUSY pair associated to $(X_{\mathscr{L}}, \mathscr{D})$ equals $(X_r, (\mathscr{L}, \alpha))$. Any isomorphism of SUSY pairs $(X_r, (\mathscr{L}, \alpha)) \to (X_r', (\mathscr{L}', \alpha'))$ induces a SUSY-isomorphism $X_{\mathscr{L}} \to X_{\mathscr{L}'}$.

Proof. First we show that if X is a $1|1$-complex supermanifold with a SUSY structure, then the $\mathscr{O}_{X,0}$-line-bundle $\mathscr{O}_{X,1}$ is a square root of the canonical bundle $K_{X_{\text{red}}}$. By Lemma 2, X has an open cover by compatible coordinate charts. If (z, ζ) and (w, η) are two such coordinate charts, then

$$D_z = \partial_\zeta + \zeta \partial_z, \qquad D_w = \partial_\eta + \eta \partial_w$$

with $D_z = h(z) D_w$, $h(z) \neq 0$. If $w = f(z)$ and $\eta = g(z)\zeta$ a small calculation implies that $f'(z) = g^2$, that is, $\mathscr{O}_{X,1}^{\otimes 2}$ and $K_{X_{\text{red}}}$ have the same transition functions for this covering, hence there is an isomorphism $\mathscr{O}_{X,1}^{\otimes 2} \to K_{X_{\text{red}}}$. If $F := (f, f^\#) : (X, \mathscr{D}) \to (X', \mathscr{D}')$ is a SUSY-isomorphism of SUSY-1 curves with underlying Riemann surface X_{red}, then one checks that $f^\#|_{\mathscr{O}_{X',1}} : \mathscr{O}_{X',1} \to f_*(\mathscr{O}_{X,1})$ is an isomorphism of line bundles. Covering X with an atlas of compatible coordinate charts, transferring this atlas to a compatible atlas on X' by F, and comparing the transition functions for $f_*(\mathscr{O}_{X,1})$ and $\mathscr{O}_{X',1}$ in this atlas as above, we obtain the desired isomorphism of theta characteristics.

Conversely, if we have a theta characteristic $\alpha : \mathscr{L}^{\otimes 2} \longrightarrow K_{X_{\mathrm{red}}}$, we define a sheaf of supercommutative rings $\mathscr{O}_{X_{\mathscr{L}}}$ on $|X|$, the topological space underlying X_{red}, by setting:

$$\mathscr{O}_{X_{\mathscr{L}}} = \mathscr{O}_{X_{\mathrm{red}}} \oplus \mathscr{L}$$

with multiplication $(f,s)\cdot(g,t) = (fg, ft+gs)$. One checks that $\mathscr{O}_{X_{\mathscr{L}}}$ so defined is a sheaf of local supercommutative rings, using the standard fact that a supercommutative ring A is local if and only if its even part A_0 is local. By taking a local basis χ of $\mathscr{L}$ in a trivialization, and sending $(f, g\chi)$ to $f+g\eta$, we see that $\mathscr{O}_{X_{\mathscr{L}}}$ so defined is locally isomorphic to $\mathscr{O}_{X_{\mathrm{red}}} \otimes \Lambda[\eta]$ and hence $(X_{\mathrm{red}}, \mathscr{O}_{X_{\mathscr{L}}})$ is a supermanifold. The SUSY structure is defined as follows. Let z be a coordinate for X_{red} on an open set U. By shrinking U we may assume $\mathscr{O}_{\mathscr{L}}(U)$ is free. Then there is some basis ζ of $\mathscr{O}_{\mathscr{L}}(U)$ such that $\alpha(\zeta \otimes \zeta) = dz$; then z, ζ so defined are coordinates for $X_{\mathscr{L}}$ on U. We set the SUSY structure on U to be that spanned by $D_Z := \partial_\zeta + \zeta\partial_z$.

We will show the local SUSY structure thus defined is independent of our choices and hence is global on X. Suppose w is another coordinate on U, and η a basis of $\mathscr{O}_{\mathscr{L}}(U)$ such that $\alpha(\eta\otimes\eta) = dw$. Then $w = f(z), \eta = g(z)\zeta$, with f' a unit in U. Since $dw = f'(z)\,dz$, we have:

$$\begin{aligned}\alpha(\eta\otimes\eta) &= g^2\alpha(\zeta\otimes\zeta)\\ &= f'(z)dz\end{aligned}$$

from which it follows that $g^2 = f'$; in particular, g is also a unit. Then by the chain rule, $\partial_\zeta + \zeta\partial_z = g(\partial_\eta + \eta\partial_w)$, hence D_Z and D_W span the same SUSY structure on U.

Now suppose $(X'_r, \mathscr{L}')$ is another SUSY pair, isomorphic to $(X,\mathscr{L})$ by (f,ϕ). Then ϕ will induce an isomorphism of analytic supermanifolds $\psi : X_{\mathscr{L}} \longrightarrow X_{\mathscr{L}'}$, since $f : X_{\mathscr{L},\mathrm{red}} \to X_{\mathscr{L}',\mathrm{red}}$ is an isomorphism, and $f_*(\mathscr{O}_{X_{\mathscr{L}},1}) \cong \mathscr{O}_{X_{\mathscr{L}'},1}$ via the isomorphism ϕ of theta characteristics. Now we check we have a SUSY isomorphism. This may be done locally: given a point $p \in X_{\mathrm{red}}$ one chooses coordinates (z,ζ) and (z',ζ') around p that are compatible with the SUSY-1 structures on $X_{\mathscr{L}}$ and $X_{\mathscr{L}'}$, so that $D_Z := \partial_\zeta + \zeta\partial_z$ (resp. $D_{Z'} := \partial_{\zeta'} + \zeta'\partial_{z'}$) locally generate the SUSY structures. In these coordinates the reader may check readily that $d\psi(D_Z|_p) = D_{Z'}|_p$.

Theorem 1 has the following important immediate consequence.

Corollary 1 *A* 1*-dimensional complex manifold* X_{red} *carries a SUSY structure if and only if* X *admits a theta characteristic.*

Remark 1 One can prove, via a direct argument using cocycles, that any compact Riemann surface S admits a theta characteristic, using the fact that the Chern class is $c_1(K_S) = 2-2g$ (i.e. it is divisible by 2 hence K_S admits a square root). Hence SUSY-1 curves exist in abundance: any compact Riemann surface admits at least one structure of SUSY-1 curve.

Theorem 1 has also the following important consequences:

Proposition 3 *Up to SUSY-isomorphism, there is a unique SUSY-1 structure on* $\mathbb{C}^{1|1}$*, namely, that defined by the odd vector field:*

$$V = \partial_\zeta + \zeta \partial_z,$$

where (z, ζ) *are the standard linear coordinates on* $\mathbb{C}^{1|1}$*.*

Proof. The reduced manifold of $\mathbb{C}^{1|1}$ is $\mathbb{C}$. It is well known that all holomorphic line bundles on $\mathbb{C}$ are trivial. This implies there exists only one theta characteristic for $\mathbb{C}$ up to isomorphism, namely $(\mathcal{O}_{\mathbb{C}}, \mathbb{1}^{\otimes 2})$. The essential point in verifying this uniqueness is that any automorphism of trivial line bundles on $\mathbb{C}$ is completely determined by an invertible entire function on $\mathbb{C}$, and such a function always has an invertible entire square root. Hence by Theorem 1, there is only one SUSY-1 structure on $\mathbb{C}$ up to isomorphism. For the last statement of the theorem, see Example 1.

The next example shows that Lemma 2 is a purely local result.

Example 2 Consider the vector field:

$$Z = \partial_\zeta + e^z \zeta \partial_z.$$

Z is an odd vector field on $\mathbb{C}^{1|1}$ defining a SUSY structure on $\mathbb{C}^{1|1}$. The previous proposition implies that the SUSY structure defined by Z is isomorphic to that defined by $V = \partial_\zeta + \zeta \partial_z$. However, it does *not* imply that there exists a global coordinate system (w, η) for $\mathbb{C}^{1|1}$ in which Z takes the form $\partial_\eta + \eta \partial_w$. In fact suppose such a global coordinate system $w = f(z), \eta = g(z)\zeta$ existed. Then:

$$\left(\begin{array}{c|c} f' & 0 \\ \hline g'\zeta & g \end{array}\right) \left(\begin{array}{c} e^z\zeta \\ \hline 1 \end{array}\right) = \left(\begin{array}{c} \eta \\ \hline 1 \end{array}\right)$$

from which we conclude that $g = 1, f' = e^{-z}$. Hence $f = -e^{-z} + c$, but since f is not one-to-one, this contradicts the assumption that (w, η) is a coordinate system on all of $\mathbb{C}^{1|1}$. This shows that Lemma 2 cannot be globalized even in the simple case of $\mathbb{C}^{1|1}$, even though $\mathbb{C}^{1|1}$ has a unique SUSY structure up to SUSY isomorphism.

5 Super Riemann surfaces of genus zero and one

In this section we want to provide some classification results on SUSY curves of genus zero and one. The next proposition provides a complete classification of compact super Riemann surfaces of genus zero and shows the existence of a genus zero $1|1$ compact complex supermanifold, namely the Π-projective line, that does not admit a SUSY structure.

Proposition 4 1. $\mathbb{P}^{1|1}$ *admits a unique SUSY structure, up to SUSY-isomorphism. More generally, if* X *is a supermanifold of dimension* $1|1$ *of genus zero, then* X *admits a SUSY structure if and only if* X *is isomorphic to* $\mathbb{P}^{1|1}$.
2. $\mathbb{P}^1_\Pi$ *admits no SUSY structure.*

Proof. To prove (1), recall the well-known classification of line bundles on $\mathbb{P}^1$: $Pic(\mathbb{P}^1)$ is a free abelian group of rank one, generated by the isomorphism class of the hyperplane bundle $\mathscr{O}(1)$, and $K_{\mathbb{P}^1} \cong \mathscr{O}(-2)$.

Hence, up to isomorphism, there is a unique theta characteristic on $\mathbb{P}^1$, namely $(\mathscr{O}(-1), \psi)$ where ψ is any fixed isomorphism of line bundles $\mathscr{O}(-1)^{\otimes 2} \to \mathscr{O}(-2)$. Similar to Proposition 3, the proof of this uniqueness reduces to the problem of lifting a given global automorphism of $\mathscr{O}(-1)^{\otimes 2} \cong \mathscr{O}(-2)$ to an automorphism of $\mathscr{O}(-1)$. This requires the fact that $End(L) = L^* \otimes L = \mathscr{O}$ for any line bundle L, and that $H^0(\mathbb{P}^1, \mathscr{O}) = \mathbb{C}$. In particular, any global automorphism of $\mathscr{O}(-1)^{\otimes 2}$ is given by multiplication by an invertible scalar, which has an invertible square root in $\mathbb{C}$; this is the desired automorphism of $\mathscr{O}(-1)$.

Considering now the statement (2), by Theorem 1, if $X = \mathbb{P}^{1|1}_\Pi$ admitted a SUSY-1 structure, we would have $\mathscr{O}_{X,1} \cong \mathscr{O}(-1)$. Using the coordinates from Sect. 2, we see that $\mathscr{O}_{X,1} \cong \mathscr{O}(-2)$. This is a contradiction.

We now turn to the study of genus one SUSY curves.

In ordinary geometry a compact Riemann surface X of genus one, that is an elliptic curve, is obtained by quotienting $\mathbb{C}$ by a lattice $L \cong \mathbb{Z}^2$. It is easily seen that any such lattice L is equivalent, under scalar multiplication, to a lattice of the form $L_0 := \text{span}\{1, \tau\}$, where τ lies in the upper half plane. Two lattices $L_0 = \text{span}\{1, \tau\}$ and $L_0' = \text{span}\{1, \tau'\}$, are equivalent, i.e., yield isomorphic elliptic curves, if and only if τ and τ' lie in the same orbit of the group $\Gamma = \text{PSL}_2(\mathbb{Z})$, where the action is via linear fractional transformations:

$$\tau \mapsto \frac{a\tau + b}{c\tau + d}.$$

A fundamental domain for this action is:

$$D = \{\tau \in \mathbb{C} \,|\, Im(\tau) > 0, |Re(\tau)| \leq 1/2, |\tau| \geq 1\}.$$

We now want to generalize this picture to the super setting. Our main reference will be [14]. The reader may wish to consult Chap. 2., Sect. 7 of [12], where the universal families of genus 1 SUSY-1 curves are constructed and embedded into Π-projective spaces by means of super theta functions.

We start by observing the ordinary action of $\mathbb{Z}^2 \cong L_0 = \langle A_0, B_0 \rangle$ on $\mathbb{C}$ is given explicitly by:

$$A_0 : z \mapsto z + 1, \qquad B_0 : z \mapsto z + \tau.$$

In [14] Freund and Rabin take a similar point of view in constructing a super Riemann surface: they define even super elliptic curves as quotients of $\mathbb{C}^{1|1}$ by $\mathbb{Z}^2 =$

$\langle A, B\rangle$, acting by:

$$A : (z,\zeta) \mapsto (z+1, \pm\zeta)$$
$$B : (z,\zeta) \mapsto (z+\tau, \pm\zeta).$$

In this section, we will justify their choice of these particular actions by showing they are the only reasonable generalizations of the classical actions of $\mathbb{Z}^2$ on $\mathbb{C}$.

In Sect. 4 we proved that on $\mathbb{C}^{1|1}$ there exists, up to isomorphism, only one SUSY structure, corresponding to the vector field $V = \partial_\zeta + \zeta\partial_z$. We now want to characterize all possible SUSY automorphisms preserving this SUSY structure.

We start with some lemmas.

Lemma 3 *Let X be a $1|1$ complex supermanifold and ω, ω' be holomorphic $1|0$ differential forms on X such that $ker(\omega), ker(\omega')$ are $0|1$ distributions. Then $ker(\omega) = ker(\omega')$ if and only if $\omega' = t\omega$ for some invertible even function $t(z)$.*

Proof. The $\Leftarrow$ implication is clear. To prove the $\Rightarrow$ implication, we can reduce to a local calculation. Suppose now that $ker(\omega) = ker(\omega')$. Given any point p in X, fix an open neighborhood $U \ni p$ where $TX|_U$ is free. As $\mathscr{D}$ is locally a direct summand, it is locally free of rank $0|1$, by the super Nakayama's lemma (see [18]) and $\mathscr{D}|_U$ has a local complement $\mathscr{E} \subset TX|_U$ (shrinking U, if needed).

Let us use the notation $\mathscr{O}_U = \mathscr{O}_X|_U$ and $\mathscr{O}(U) = \mathscr{O}_U(U)$. As $\mathscr{E}$ is also a direct summand of $TX|_U$, it is also a free $\mathscr{O}_U$-module (again possibly shrinking U) hence must be of rank $1|0$. Hence we have a local splitting $TX|_U = \mathscr{D} \oplus \mathscr{E}$ of free $\mathscr{O}_U$-modules. Let Z be a basis for $\mathscr{D}|_U$, W a basis for $\mathscr{E}$; then W, Z form a basis for $TX|_U$. $\omega|_U : \mathscr{O}_{TX}(U) \to \mathscr{O}_U(U)$ induces an even linear functional $\omega_p : T_pX \to \mathbb{C}$ on the tangent space at p, and the splitting $TX|_U = \mathscr{D} \oplus \mathscr{E}$ induces a corresponding splitting $T_pX = D_p \oplus E_p$ of super vector spaces, with $dim(D_p) = 0|1$, $dim(E_p) = 1|0$, such that $ker(\omega_p) = D_p$ and $\text{span}(W_p) = E_p$. By linear algebra, $\omega_p|_{E_p}$ is an isomorphism; in particular, $\omega_p(W_p)$ is a basis for $\mathbb{C} = \mathscr{O}_p/\mathfrak{M}_p$. The super Nakayama's lemma then implies that $\omega(W)$ generates $\mathscr{O}_U$ as $\mathscr{O}_U$-module (again shrinking U if necessary), which is true if and only if $\omega(W)$ is a unit; the same is true for $\omega(W')$.

We now show that the ratio $\omega'(W)/\omega(W) \in \mathscr{O}^*_{U,0}$ is independent of the local complement $\mathscr{E}$ and the choice of W, so that it defines an invertible even function t on all of X. Suppose $\mathscr{E}'$ is another local complement to $\mathscr{D}$ on U, and W' a local basis for $\mathscr{E}'$. We have as before that Z, W' form a basis of $TX|_U$. Then $\omega(W'), \omega'(W')$ are invertible in $\mathscr{O}_U$ by the above argument, and $W' = bW + \beta Z$ with $b \in \mathscr{O}^*_{U,0}$, $\beta \in \mathscr{O}_{U,1}$.

$$\begin{aligned}\frac{\omega(W')}{\omega'(W')} &= \frac{b\omega(W) + \beta\omega(Z)}{b\omega'(W) + \beta\omega'(Z)} \\ &= \frac{\omega(W)}{\omega'(W)}.\end{aligned}$$

Note here that we have used the hypothesis $ker(\omega) = ker(\omega')$ to conclude $\omega(Z) = \omega'(Z) = 0$.

Finally, we verify that $\omega' = t\omega$; this can again be done locally since t is now known to be globally defined. The argument is left to the reader.

The odd vector field $V = \partial_\zeta + \zeta \partial_z$ defining our SUSY structure $\mathscr{D}$ is dual to the differential form $\omega := dz - \zeta\, d\zeta$. As one can readily check $\mathscr{D} = \mathrm{span}\{V\} = ker(\omega)$.

Lemma 4 *An automorphism $F : \mathbb{C}^{1|1} \to \mathbb{C}^{1|1}$ is a SUSY automorphism if and only if $F^*(\omega) = t(z)\omega$ for some invertible even function $t(z)$.*

Proof. Unraveling the definitions, one sees that F preserves the SUSY structure if and only if

$$ker(F^*(\omega))_p = ker(\omega)_p$$

for each $p \in X$. We claim that the latter is true if and only if $ker(F^*(\omega)) = ker(\omega)$. One implication is clear. Conversely, suppose that $ker(F^*(\omega))_p = ker(\omega)_p$ for each $p \in X$. By a standard argument using the super Nakayama's Lemma, $ker(F^*(\omega)) = ker(\omega)$ in a neighborhood of p for any point p, hence $ker(\omega) = ker(F^*(\omega))$. The result then follows by Lemma 3.

We are now ready for the result characterizing all of the SUSY automorphisms of $\mathbb{C}^{1|1}$.

Proposition 5 *Let (z, ζ) be the standard linear coordinates on $\mathbb{C}^{1|1}$, and let $\mathbb{C}^{1|1}$ have the natural SUSY-1 structure defined by the vector field $V = \partial_\zeta + \zeta \partial_z$. The SUSY automorphisms of $\mathbb{C}^{1|1}$ are precisely the endomorphisms F of $\mathbb{C}^{1|1}$ such that:*

$$F(z,\zeta) = (az+b, \pm\sqrt{a}\zeta)$$

where $a \in \mathbb{C}^*, b \in \mathbb{C}$, and $\sqrt{a}$ denotes either of the two square roots of a.

Proof. Let F be such an automorphism, and z, ζ the standard coordinates on $\mathbb{C}^{1|1}$. Then $F(z,\zeta) = (f(z), g(z)\zeta)$ for some entire functions f, g of z. Similarly, $F^{-1}(z,\zeta) = (h(z), k(z)\zeta))$ for some entire functions h, k. Since F and F^{-1} are inverses, f is a biholomorphic automorphism of $\mathbb{C}^{1|0}$, hence linear by standard facts from complex analysis: $f(z) = az+b$ for some $a, b \in \mathbb{C}$, $a \neq 0$; the same is true for h.

So by the Lemma 4, F preserves the SUSY-1 structure on $\mathbb{C}^{1|1}$ if and only if $F^*(\omega) = t(z)\omega$. We calculate:

$$\begin{aligned} F^*(dz - \zeta\, d\zeta) &= df - F^*(\zeta)\, d(g\zeta) \\ &= f'\, dz - g^2 \zeta\, d\zeta. \end{aligned}$$

Equating this with $t(z)\omega$, we see $t = f' = g^2$. Thus $g^2 = a$, so in particular g is constant. Hence:

$$F(z,\zeta) = (az+b, c\zeta)$$

where $c^2 = a$, $a \in \mathbb{C}\backslash\{0\}$, $b \in \mathbb{C}$. Conversely, one checks that any morphism $\mathbb{C}^{1|1} \to \mathbb{C}^{1|1}$ of the above form is an automorphism, and that it preserves the SUSY structure.

From our previous proposition, we conclude immediately that the only actions of $\mathbb{Z}^2$ on $\mathbb{C}^{1|1}$ that restrict to the usual action on the reduced space $\mathbb{C}$ are of the form:

$$A:(z,\zeta)\mapsto(z+1,\pm\zeta)$$
$$B:(z,\zeta)\mapsto(z+\tau,\pm\zeta),$$

since the actions of A and B must be by automorphisms of the form:

$$(z,\zeta)\mapsto(az+b,\pm\sqrt{a}\zeta)$$

and in this case, a must be taken to be 1. This justifies the choice made in [14].

Remark 2 One can show that the functor of SUSY-preserving automorphisms of the functor of points of $\mathbb{C}^{1|1}$ is represented by a complex supergroup $\mathrm{Aut}(\mathbb{C}^{1|1})$. However, in this work, we have only described the $\mathbb{C}$-points of this supergroup.

Remark 3 Using Theorem 1, we see that the SUSY structures on X_{red} correspond one-to-one to isomorphism classes of theta characteristics on X_{red}. It is well-known from the theory of elliptic curves over $\mathbb{C}$ that an elliptic curve X_{red} has four distinct theta characteristics, up to isomorphism. Regarding X_{red} as an algebraic group, these theta characteristics correspond to the elements of the subgroup of order 2 in X_{red}.

As noted in [11], one can define the *parity* of a theta characteristic $\mathscr{L}$ as $dim H^0(X_{\mathrm{red}},\mathscr{L})\,(mod\ 2)$. This is a fundamental invariant of the theta characteristic (cf. [1] where the parity is shown to be stable under holomorphic deformation). The isomorphism class of the trivial theta characteristic $\mathscr{O}_{X_{\mathrm{red}}}$ is distinguished from the other three by its parity: it has odd parity, the others have even parity. The odd case is therefore fundamentally different from the perspective of supergeometry, and is best studied in the context of *families* of super Riemann surfaces; families of odd super elliptic curves are considered in, for instance, [11, 14, 15, 19].

In [14], Rabin and Freund also describe a projective embedding of the SUSY curve defined by $\mathbb{C}^{1|1}/\langle A,B\rangle$ using the classical Weierstrass function $\wp$ and the function $\wp_1$ defined as $\wp_1^2=\wp-e_1$ (as usual $e_1=\wp(\omega_i)$ with $\omega_1=1/2$, $\omega_2=\tau/2$ and $\omega_3=(1+\tau)/2$). If U_0,U_1,U_2 is the open cover of $\mathbb{P}^{2|3}(\mathbb{C})$ described in Sect. 2, on U_2 the embedding is defined as:

$$\begin{array}{ccc}\mathbb{C}^{1|1}/\langle A,B\rangle & \longrightarrow & U_2\subset\mathbb{P}^{2|3}(\mathbb{C})\\ (z,\zeta) & \mapsto & [\wp(z),\wp'(z),1,\wp_1(z)\zeta,\wp_1'(z)\zeta,\wp_1(z)\wp(z)\zeta].\end{array}$$

In [14], they describe also the equations of the ideal in U_2 corresponding to the SUSY curve in this embedding:

$$y^2=4x^3-a_1x^2-a_2,\qquad 2(x-e_1)\eta_2=y\eta_1$$
$$y\eta_2=2(x-e_2)(x-e_3)\eta_1\qquad \eta_3=x\eta_1$$

where $(x,y,\eta_1,\eta_2,\eta_3)$ are the global coordinates on $U_2\cong\mathbb{C}^{2|3}$. One can readily compute the homogeneous ideal in the ring $\mathbb{C}[x_0,x_1,x_2,\xi_1,\xi_2,\xi_3]$ associated with the given projective embedding. It is generated by the equations:

$$x_1^2x_2=4x_0^3-a_1x_0^2x_2-a_2x_2^3,\qquad 2(x_0x_2-e_1x_2^2)\xi_2=x_1x_2\xi_1$$
$$x_1x_2\xi_2=2(x_0-e_2x_2)(x_0-e_3x_2)\xi_1\qquad \xi_3x_2=x_0\xi_1.$$

6 Π-Projective geometry revisited

We devote this appendix to reinterpret the Π-projective line, discussed in Sect. 3, through the superalgebra $\mathbb{D}$.

Let $\mathbb{D}$ denote the *super skew field*, $\mathbb{D} = \mathbb{C}[\theta]$, θ odd and $\theta^2 = -1$. As a complex super vector space of dimension $1|1$, $\mathbb{D} = \{a + b\theta \mid a, b \in \mathbb{C}\}$, thus it has a canonical structure of analytic supermanifold, and its functor of points is:

$$T \mapsto \mathbb{D}(T) := (\mathbb{D} \otimes \mathscr{O}(T))_0 = \mathbb{D}_0 \otimes \mathscr{O}(T)_0 \oplus \mathbb{D}_1 \otimes \mathscr{O}(T)_1.$$

Let $\mathbb{D}^\times$ be the analytic supermanifold obtained by restricting the structure sheaf of the supermanifold $\mathbb{D}$ to the open subset $\mathbb{D} \setminus \{0\}$.

$\mathbb{D}^\times$ is an analytic supergroup and its functor of points is:

$$T \mapsto \mathbb{D}^\times(T) := (\mathbb{D} \otimes \mathscr{O}(T))_0^*$$

where $(\mathbb{D} \otimes \mathscr{O}(T))_0^*$ denotes the invertible elements in $(\mathbb{D} \otimes \mathscr{O}(T))_0$;

As a supergroup $\mathbb{D}^\times$ is isomorphic to $\mathbb{G}_m^{1|1}$, which is the supergroup with underlying topological space $\mathbb{C}^\times$, described in Sect. 3. The isomorphism between $\mathbb{G}_m^{1|1}$ and $\mathbb{D}^\times$ simply reads as:

$$(a, \alpha.) \mapsto a + \theta\alpha.$$

Notice that $\mathbb{G}_m^{1|1}$ (hence $\mathbb{D}^\times$) is naturally embedded into $\mathrm{GL}(1|1)$, the complex general linear supergroup via the morphism (in the functor of points notation):

$$\begin{aligned} \mathbb{D}^\times(T) &\cong \mathbb{G}_m^{1|1}(T) \longrightarrow \mathrm{GL}(1|1)(T) \\ a + \theta\alpha. &\cong (a, \alpha.) \quad \mapsto \quad \begin{pmatrix} a & \alpha. \\ \alpha. & a \end{pmatrix}. \end{aligned}$$

Before we continue this important characterization of Π-projective geometry due to Deligne, let us point out that while $\mathbb{G}_m^{1|1} \cong \mathbb{D}^\times$ are commutative supergroups, the commutative algebra $\mathbb{D} = \mathbb{C}[\theta]$ is not a commutative superalgebra, because if it were, then $\theta^2 = 0$ and not $\theta^2 = -1$ as we have instead. This is an important fact, which makes Π-projective supergeometry more similar to non commutative geometry than to regular supergeometry.

We now want to relate more closely the Π-projective supergeometry with $\mathbb{D}$.

Lemma 5 *A right action of $\mathbb{D}$ on a complex super vector space V is equivalent to the choice of an odd endomorphism ϕ of V such that $\phi^2 = 1$.*

Proof. Let V be a right $\mathbb{D}$-module. A right action of $\mathbb{D} = \mathbb{C}[\theta]$ is an antihomomorphism $f : \mathbb{D} \to \underline{\mathrm{End}}(V)$, which corresponds to a left action of the opposite algebra $\mathbb{D}^o = \mathbb{C}[\theta^o]$, $(\theta^o)^2 = 1$ ($\underline{\mathrm{End}}(V)$ denotes all of the endomorphisms of V, not just the parity preserving ones). Such actions are specified once we know the odd endomorphisms ψ and ϕ corresponding respectively to θ and θ^o. Hence explicitly right multiplication by θ gives rise to an odd endomorphism ϕ such that $\phi^2 = 1$, by:

$$\phi(v) := (-1)^{|v|} v \cdot \theta.$$

Conversely, given a super vector space V and an odd endomorphism ϕ of square 1, we can define a right $\mathbb{D}$-module structure on V by:

$$v \cdot (a + b\theta) := v \cdot a + (-1)^{|v|}\phi(v) \cdot b.$$

Given any complex supermanifold X, there is a sheaf $\underline{\mathbb{D}}$ of superalgebras, defined by $\underline{\mathbb{D}}(U) := \mathscr{O}_X(U) \otimes_{\mathbb{C}} \mathbb{D}$, for any open set $U \subseteq |X|$. Then a sheaf of right (resp. left) $\mathbb{D}$-modules on X is a sheaf of right (resp. left) modules for the sheaf $\underline{\mathbb{D}}$; a *morphism* $\mathscr{F} \to \mathscr{F}'$ of sheaves of $\mathbb{D}$-modules is simply a sheaf morphism that intertwines the $\mathbb{D}$-actions on $\mathscr{F}, \mathscr{F}'$. A sheaf of $\mathbb{D}$-modules $\mathscr{F}$ is *locally free of* $\mathbb{D}$*-rank* n if $\mathscr{F}$ is locally isomorphic to $\underline{\mathbb{D}}^n$.

Lemma 6 *Let X be a complex supermanifold and let U be an open set. If V is a free $\underline{\mathbb{D}}(U)$-module of $\mathbb{D}$-rank* 1, $\mathrm{Aut}_{\mathbb{D}}(V) \cong \mathbb{D}^{\times}(U)$.

Proof. Since V is free of $\mathbb{D}$-rank 1, we may reduce to the case $V = \underline{\mathbb{D}}(U)$ as right $\mathbb{D}$-modules, where this identification is obvious: $f \mapsto f(1) \in \underline{\mathbb{D}}^{\times}(U)$ for $f \in \mathrm{Aut}_{\mathbb{D}}(V)$.

We are now ready to reinterpret the functor of points of the Π-projective line.

Proposition 6 *Let the notation be as above.*

$$P_{\Pi}(T) = \{\textit{locally free, } \mathbb{D}\textit{-rank 1 right } \mathbb{D}\textit{-subsheaves } \mathscr{F}_{\mathscr{T}} \subseteq \mathscr{O}_T^{2|2}\}$$

In other words, the functor of points of the Π-projective line associates to each supermanifold T the set of locally free right $\mathbb{D}$-subsheaves of rank $1|0$ *of* $\mathscr{O}_T^{2|2}$.

Proof. (Sketch). The right action on $\mathscr{F}_T$ of $\mathbb{D}$ corresponds to the right action of $\mathbb{D}$ on $\mathbb{C}^{2|2} \otimes \mathscr{O}_T$ occurring on the first term through the left multiplication by the odd endomorphism ϕ, (see Lemma 5). Hence the ϕ-invariant subsheaves are in one to one correspondence with the right $\mathbb{D}$-subsheaves of $\mathscr{O}_T^{2|2}$. Notice furthermore that by Lemma 6, the change of basis of the free module $\mathscr{F}_T(V_i)$ we used in 2, corresponds to right multiplication by an element of $\mathbb{G}_m^{1|1} \cong \mathbb{D}^{\times}$, that is the natural left action of $\mathbb{D}^{\times}$ on the locally free, $\mathbb{D}$-rank 1 sheaf $\mathscr{F}_T(V_i)$ by automorphisms.

Remark 4 The generalization from $\mathbb{C}^{\times}$ action on $\mathbb{C}^n$ (see the construction of ordinary projective space Sect. 2) to $\mathbb{G}_m^{1|1} \cong \mathbb{D}^{\times}$ action on $\mathbb{C}^{n|n}$ gives us naturally the odd endomorphism ϕ, which is used to construct the Π-projective space and ultimately it is the base on which Π-projective geometry is built. The introduction of the skew-field $\mathbb{D}$, $\mathbb{D}^{\times} \cong \mathbb{G}_m^{1|1}$ is not merely a computational device, but suggest a more fundamental way to think about Π-projective geometry. We are unable to provide a complete treatment here, but we shall do so in a forthcoming paper.

References

1. Atiyah, M.F.: Spin structures on Riemann surfaces. Ann. Sci. École Norm. Sup. **4**(4), 47–62 (1971)
2. Beilinson, A.A., Manin, Y.I., Schechtman, V.V.: Sheaves of the Virasoro and Neveu-Schwarz algebras. In: K-theory, arithmetic and geometry (Moscow, 1984–1986), pp. 52–66, Lecture Notes in Math., Vol. 1289. Springer-Verlag, Berlin Heidelberg New York (1987)
3. Berezin, F.A.: Introduction to Superanalysis. D. Reidel Publishing Company, Holland (1987)
4. Bergvelt, M.J., Rabin, J.M.: Super curves, their Jacobians, and super KP equations. Duke Math. J. **98**, 1–57 (1999)
5. Berezin, F.A., Supermanifolds, D.A. Leites.: Dokl. Akad. Nauk SSSR, **224**(3), 505–508 (1975)
6. Carmeli, C., Caston, L., Fioresi, R.: Mathematical Foundation of Supersymmetry, with an appendix with I. Dimitrov. EMS Ser. Lect. Math.. European Math. Soc., Zurich (2011)
7. Crane, L., Rabin, J.M.: Super Riemann surfaces: uniformization and Teichmuller theory. Comm. Math. Phys. **113**(4), 601–623 (1988)
8. Deligne, P.: personal communication to Y.I. Manin (1987)
9. Deligne, P., Morgan, J.: Notes on supersymmetry (following J. Bernstein). In: Quantum Fields and Strings. A Course for Mathematicians, Vol. 1. AMS (1999)
10. Leites, D.A.: Introduction to the theory of supermanifolds, Russian Math. Surveys **35**(1), 1–64 (1980)
11. Supersymmetric elliptic curves, Funct, A.M. Levin.: Analysis and Appl. **21**(3), 243–244 (1987)
12. Manin, Y.I.: Topics in Noncommutative Geometry. Princeton University Press, New Jersey (1991)
13. Manin, Y.I.: Gauge Field Theory and Complex Geometry; translated by N. Koblitz and J.R. King. Springer-Verlag, Berlin Heidelberg New York (1988)
14. Freund, P.G.O., Rabin, J.M.: Supertori are elliptic curves. Comm. Math. Phys. **114**(1), 131–145 (1988)
15. Rabin, J.M.: Super elliptic curves. J. Geom. Phys. **15**, 252–80 (1995)
16. Vaintrob, A.Y.: Deformation of complex superspaces and coherent sheaves on them. J. Mat. Sci. **51**(1), 2140–2188 (1990)
17. Varadarajan, V.S.: Lie Groups, Lie Algebras, and Their Representations. Graduate Text in Mathematics. Springer-Verlag, New York (1984)
18. Varadarajan, V.S.: Supersymmetry for Mathematicians: An Introduction Courant Lecture Notes, Vol. 1. AMS (2004)
19. Witten, E.: Notes on super Riemann surfaces and their moduli. arXiv:1209.2459

Dirac operators and the very strange formula for Lie superalgebras

Victor G. Kac, Pierluigi Möseneder Frajria and Paolo Papi

Abstract Using a super-affine version of Kostant's cubic Dirac operator, we prove a very strange formula for quadratic finite-dimensional Lie superalgebras with a reductive even subalgebra.

1 Introduction

The goal of this paper is to provide an approach to the strange and very strange formulas for a wide class of finite-dimensional Lie superalgebras. Let us recall what these formulas are in the even case. Let $\mathfrak{g}$ be a finite-dimensional complex simple Lie algebra. Fix a Cartan subalgebra $\mathfrak{h} \subset \mathfrak{g}$ and let Δ^+ be a set of positive roots for the set Δ of $\mathfrak{h}$-roots in $\mathfrak{g}$. Let $\rho = \frac{1}{2}\sum_{\alpha\in\Delta^+}\alpha$ be the corresponding Weyl vector. Freudenthal and de Vries discovered in [6] the following remarkable relation between the square length of ρ in the Killing form κ and the dimension of $\mathfrak{g}$:

$$\kappa(\rho,\rho) = \frac{\dim\mathfrak{g}}{24}.$$

They called this the *strange formula*. It can be proved in several very different ways (see e.g. [2,5]), and it plays an important role in the proof of the Macdonald identities for the powers of the η-function. Indeed, the strange formula enters as a transition factor between the Euler product $\varphi(x) = \prod_{i=1}^{\infty}(1-x^i)$ and Dedekind's η-function $\eta(x) = x^{\frac{1}{24}}\varphi(x)$.

V.G. Kac
Department of Mathematics, MIT, 77 Mass. Ave, Cambridge, MA 02139, USA
e-mail: kac@math.mit.edu

P. Möseneder Frajria (✉)
Politecnico di Milano, Polo regionale di Como, via Valleggio 11, 22100 Como, Italy
e-mail: pierluigi.moseneder@polimi.it

P. Papi
Dipartimento di Matematica, Sapienza Università di Roma, P.le A. Moro 2, 00185 Roma, Italy
e-mail: papi@at.uniroma1.it

M. Gorelik, P. Papi (eds.): *Advances in Lie Superalgebras.* Springer INdAM Series 7,
DOI 10.1007/978-3-319-02952-8_8,

In [8] Kac gave a representation theoretic interpretation of the Macdonald identities as denominator identities for affine Lie algebras. Moreover, using modular forms, he provided in [10] general specializations of them and a corresponding transition identity, named the *very strange formula*. Here the representation theoretic interpretation of the formulas involves an affine Lie algebra, which is built up from a simple Lie algebra endowed with a finite order automorphism. To get the very strange formula from a "master formula" it was also required that the characteristic polynomial of the automorphism has rational coefficients. A more general form, with no rationality hypothesis, is proved in [14], where it is also used to estimate the asymptotic behavior at cusps of the modular forms involved in the character of a highest weight module. Let us state this version of the very strange formula, for simplicity of exposition, in the case of inner automorphisms. Let σ be an automorphism of order m of type $(s_0, s_1, \dots, s_n; 1)$ (see [11, Chap. 8]). Let $\mathfrak{g} = \oplus_{\bar{j}\in\mathbb{Z}/m\mathbb{Z}}\mathfrak{g}^{\bar{j}}$ be the eigenspace decomposition with respect to σ. Define $\lambda_s \in \mathfrak{h}^*$ by $\kappa(\lambda_s, \alpha_i) = \frac{s_i}{2m}$, $1 \le i \le n$, where $\{\alpha_1, \dots, \alpha_n\}$ is the set of simple roots of $\mathfrak{g}$. Then

$$\kappa(\rho - \lambda_s, \rho - \lambda_s) = \frac{\dim\mathfrak{g}}{24} - \frac{1}{4m^2}\sum_{j=1}^{m-1} j(m-j)\dim\mathfrak{g}^{\bar{j}}. \tag{1}$$

Much more recently, we provided a vertex algebra approach to this formula (in a slightly generalized version where an elliptic automorphism is considered, cf. [12]) as a byproduct of our attempt to reproduce Kostant's theory of the cubic Dirac operator in affine setting. Our proof is based on two main ingredients:

(a) An explicit vertex algebra isomorphism $V^k(\mathfrak{g}) \otimes F(\overline{\mathfrak{g}}) \cong V^{k+g,1}(\mathfrak{g})$, where $V^k(\mathfrak{g})$ is the universal affine vertex algebra of noncritical level k, g is the dual Coxeter number, $V^{k,1}(\mathfrak{g})$ is the universal super-affine vertex algebra and $F(\overline{\mathfrak{g}})$ is the fermionic vertex algebra on $\mathfrak{g}$ viewed as a purely odd space.
(b) A nice formula for the λ-bracket of the Kac-Todorov Dirac field $G \in V^{k+g,1}(\mathfrak{g})$ with itself.

Indeed, using (a), we can let the zero mode G_0 of G act on representations of $V^k(\mathfrak{g}) \otimes F(\overline{\mathfrak{g}})$. Since we are able to compute $G_0 v \otimes 1$, where v is a highest weight vector of a highest weight module for the affinization of $\mathfrak{g}$, the expression for $[G_\lambda G]$ obtained in step (b) yields a formula which can be recast in the form (1) (cf. [12, Sect. 6]).

Now we discuss our work in the super case. A finite-dimensional Lie superalgebra $\mathfrak{g} = \mathfrak{g}_0 \oplus \mathfrak{g}_1$ is called quadratic if it carries a supersymmetric bilinear form (i.e. symmetric on $\mathfrak{g}_0$, skewsymmetric on $\mathfrak{g}_1$, and $\mathfrak{g}_0$ is orthogonal to $\mathfrak{g}_1$), which is non-degenerate and invariant. We say that a complex quadratic Lie superalgebra $\mathfrak{g}$ is of *basic type* if $\mathfrak{g}_0$ is a reductive subalgebra of $\mathfrak{g}$. In Theorem 1 we prove a *very strange formula* (cf. (55)) for basic type Lie superalgebras endowed with an indecomposable elliptic automorphism (see Definition 1) which preserves the invariant form. When the automorphism is the identity, this formula specializes to the strange formula (54), which has been proved for Lie superalgebras of defect zero in [10] and for general basic classical Lie superalgebras in [13], using case by case combinatorial calculations. The proof using the Weyl character formula as in [6] or the proof using modular forms as in [14] are not applicable in this setting.

Although the proof proceeds along the lines of what we did in [12] for Lie algebras, we have to face several technical difficulties. We single out two of them. First, we have to build up a twisted Clifford-Weil module for $F(\overline{\mathfrak{g}})$; this requires a careful choice of a maximal isotropic subspace in $\mathfrak{g}^{\overline{0}}$. In Sect. 2 we prove that the class of Lie superalgebras of basic type is closed under taking fixed points of automorphisms and in Sect. 3 we show that Lie superalgebras of basic type admit a triangular decomposition. This implies the existence of a "good" maximal isotropic subspace.

Secondly, the isomorphism in (a) is given by formulas which are different from the even case, and this makes the computation of the square of the Dirac field under λ-bracket subtler. We have also obtained several simplifications of the exposition given in [12].

Some of our results have been (very sketchily) announced in [15].

2 Setup

Throughout the paper, $\mathfrak{g} = \mathfrak{g}_0 \oplus \mathfrak{g}_1$ is a finite dimensional Lie superalgebra of basic type. This means that

1. $\mathfrak{g}_0$ is a reductive subalgebra of $\mathfrak{g}$, i.e., the adjoint representation of $\mathfrak{g}_0$ on $\mathfrak{g}$ is completely reducible;
2. $\mathfrak{g}$ is *quadratic*, i.e., $\mathfrak{g}$ admits a nondegenerate invariant supersymmetric bilinear form $(\cdot,\cdot)$.

Note that condition (1) implies that $\mathfrak{g}_0$ is a reductive Lie algebra and that $\mathfrak{g}_1$ is completely reducible as a $\mathfrak{g}_0$-module. Examples are given by the simple basic classical Lie superalgebras and the contragredient finite dimensional Lie superalgebras with a symmetrizable Cartan matrix (in particular, $gl(m,n)$). There are of course examples of different kind, like a symplectic vector space regarded as a purely odd abelian Lie superalgebra. An inductive classification is provided in [1].

We say that $\mathfrak{g}$ is $(\cdot,\cdot)$-irreducible if the form restricted to any proper ideal is degenerate.

Definition 1 An automorphism σ of $\mathfrak{g}$ is said indecomposable if $\mathfrak{g}$ cannot be decomposed as an orthogonal direct sum of two nonzero σ-stable ideals.

We say that σ is elliptic if it is diagonalizable with modulus 1 eigenvalues.

Let σ be an indecomposable elliptic automorphism of $\mathfrak{g}$ which is parity preserving and leaves the form invariant. If $j \in \mathbb{R}$, set $\overline{j} = j + \mathbb{Z} \in \mathbb{R}/\mathbb{Z}$. Set $\mathfrak{g}^{\overline{j}} = \{x \in \mathfrak{g} \mid \sigma(x) = e^{2\pi\sqrt{-1}j}x\}$. Let $\mathfrak{h}^0$ be a Cartan subalgebra of $\mathfrak{g}^{\overline{0}}$.

Proposition 1 *If $\mathfrak{g}$ is of basic type, then $\mathfrak{g}^{\overline{0}}$ is of basic type.*

Proof. Since σ is parity preserving, it induces an automorphism of $\mathfrak{g}_0$. Since $\mathfrak{g}_0$ is reductive, we have that $\mathfrak{g}_0^{\overline{0}}$ is also reductive. Since σ preserves the invariant bilinear form, we have that $(\mathfrak{g}^{\bar{i}}, \mathfrak{g}^{\bar{j}}) \neq 0$ if and only if $\bar{i} = -\bar{j}$. Thus $(\cdot,\cdot)_{|\mathfrak{g}^{\bar{0}} \times \mathfrak{g}^{\bar{0}}}$ is nondegenerate. Since $\mathfrak{g}_0^{\bar{0}}$ is reductive, $\mathfrak{h}^0$ is abelian, thus it is contained in a Cartan subalgebra $\mathfrak{h}$ of $\mathfrak{g}$. Since $\mathfrak{g}_1$ is completely reducible as a $\mathfrak{g}_0$-module, $\mathfrak{h}$ acts semisimply on $\mathfrak{g}_1$, hence $\mathfrak{h}^0$ acts semisimply on $\mathfrak{g}_1^{\bar{0}}$. Thus $\mathfrak{g}_1^{\bar{0}}$ is a semisimple $\mathfrak{g}_0^{\bar{0}}$-module. □

It is well-known (see e.g. [11]) that we can choose as Cartan subalgebra for $\mathfrak{g}$ the centralizer $\mathfrak{h}$ of $\mathfrak{h}^0$ in $\mathfrak{g}_0$. In particular we have that $\sigma(\mathfrak{h}) = \mathfrak{h}$. If $\mathfrak{a}$ is any Lie superalgebra, we let $\mathfrak{z}(\mathfrak{a})$ denote its center.

3 The structure of basic type Lie superalgebras

The goal of this section is to prove that a basic type Lie superalgebra $\mathfrak{g}$ admits a triangular decomposition. We will apply this result in the next sections to $\mathfrak{g}^{\bar{0}}$, which, by Proposition 1, is of basic type.

Since $\mathfrak{g}_0$ is reductive, we can fix a Cartan subalgebra $\mathfrak{h} \subset \mathfrak{g}_0$ and a set of positive roots for $\mathfrak{g}_0$. If $\lambda \in \mathfrak{h}^*$, let h_λ be, as usual, the unique element of $\mathfrak{h}$ such that $(h_\lambda, h) = \lambda(h)$ for all $h \in \mathfrak{h}$. Let $V(\lambda)$ denote the irreducible representation of $\mathfrak{g}_0$ with highest weight $\lambda \in \mathfrak{h}^*$.

Then we can write

$$\mathfrak{g} = (\mathfrak{h} + [\mathfrak{g}_0, \mathfrak{g}_0]) \oplus \sum_{\lambda \in \mathfrak{h}^*} V(\lambda). \tag{2}$$

Decompose now the Cartan subalgebra $\mathfrak{h}$ of $\mathfrak{g}$ as

$$\mathfrak{h} = \mathfrak{h}' \oplus \mathfrak{h}'', \qquad \mathfrak{h}' = \mathfrak{h} \cap [\mathfrak{g}, \mathfrak{g}].$$

Let M_{triv} be the isotypic component of the trivial $[\mathfrak{g}_0, \mathfrak{g}_0]$-module in $\mathfrak{g}_1$. Decompose it into isotypic components for $\mathfrak{g}_0$ as

$$M_{triv} = \oplus_{\lambda \in \Lambda} M(\lambda) = M(0) \oplus M'_{triv}, \tag{3}$$

where $M'_{triv} = \oplus_{0 \neq \lambda \in \Lambda} M(\lambda)$. Then

$$\begin{aligned} \mathfrak{g}^{(1)} &:= [\mathfrak{g}, \mathfrak{g}] \\ &= (\mathfrak{h}' + [\mathfrak{g}_0, \mathfrak{g}_0]) \oplus \sum_{0 \neq \lambda \in \mathfrak{h}^*} V(\lambda) \\ &= (\mathfrak{h}' + [\mathfrak{g}_0, \mathfrak{g}_0]) \oplus M'_{triv} \oplus \sum_{\lambda \in \mathfrak{h}^*, \dim V(\lambda) > 1} V(\lambda). \end{aligned} \tag{4}$$

Lemma 1 *If in decomposition* (2) *we have that* $\dim V(\lambda) = 1$ *for some* $\lambda \in \mathfrak{h}^*$, *then* $\lambda(\mathfrak{h}') = 0$.

Proof. Let $h \in \mathfrak{h}'$. If $h \in [\mathfrak{g}_0, \mathfrak{g}_0]$ the claim is obvious; if $h \in [\mathfrak{g}_1, \mathfrak{g}_1]$ then we may assume that $h = [x_\mu, x_{-\mu}] = h_\mu$, μ being a $\mathfrak{h}$-weight of $\mathfrak{g}_1$. Assume first $\mu \pm \lambda \neq 0$. Then, for $v_\lambda \in V(\lambda)$, we have

$$0 = [v_{-\lambda}, [v_\lambda, x_\mu]] = \mu(h_\lambda) x_\mu - [v_\lambda, [v_{-\lambda}, x_\mu]] = \mu(h_\lambda) x_\mu \tag{5}$$

so that $\mu(h_\lambda) = 0$ or $\lambda(h_\mu) = 0$. It remains to deal with the case $\mu = \pm\lambda$. We have

$$[[v_\lambda, v_{-\lambda}], v_{\pm\lambda}] = \pm||\lambda||^2 v_{\pm\lambda}.$$

This finishes the proof, since $||\lambda||^2 = 0$. Indeed, $[v_\lambda, v_\lambda]$ is a weight vector of weight 2λ. This implies either $\lambda = 0$, and we are done, or $[v_\lambda, v_\lambda] = 0$. In the latter case $||\lambda||^2 v_\lambda = [[v_\lambda, v_{-\lambda}], v_\lambda] = 0$ by the Jacobi identity. □

In turn, by Lemma 1,

$$\begin{aligned}\mathfrak{g}^{(2)} &:= [\mathfrak{g}^{(1)}, \mathfrak{g}^{(1)}] \\ &= (\mathfrak{h}' + [\mathfrak{g}_0, \mathfrak{g}_0]) \oplus (\sum_{\lambda\in\mathfrak{h}^*, \dim V(\lambda)>1} V(\lambda)).\end{aligned}$$

Finally define

$$\underline{\mathfrak{g}} = \mathfrak{g}^{(2)}/(\mathfrak{z}(\mathfrak{g}) \cap \mathfrak{g}^{(2)}).$$

Lemma 2

1. *The radical of the restriction of the invariant form to* $\mathfrak{g}^{(2)}$ *equals* $\mathfrak{z}(\mathfrak{g}) \cap \mathfrak{g}^{(2)}$.
2. $\underline{\mathfrak{g}}$ *is an orthogonal direct sum of quadratic simple Lie superalgebras.*

Proof. It suffices to show that if $x \in \mathfrak{g}^{(2)}$ belongs to the radical of the (restricted) form, then it belongs to the center of $\mathfrak{g}$. We know that $(x, [y,z]) = 0$ for all $y, z \in \mathfrak{g}^{(1)}$; invariance of the form implies that $[x,y]$ belongs to the radical of the form restricted to $\mathfrak{g}^{(1)}$. This in turn means that $([x,y],[w,t]) = 0$ for all $w,t \in \mathfrak{g}$. Therefore, $0 = ([x,y],[w,t]) = ([[x,y],w],t) \forall t \in \mathfrak{g}$. Since the form on $\mathfrak{g}$ is nondegenerate, we have that $[x,y] \in \mathfrak{z}(\mathfrak{g})$ for any $y \in \mathfrak{g}^{(1)}$. If $x \in \mathfrak{g}_1$ and $y \in \mathfrak{g}_0^{(1)}$, then $[x,y] \in \mathfrak{z}(\mathfrak{g}) \cap (\sum_{\lambda\in\mathfrak{h}^*, \lambda_{|\mathfrak{h}'}\neq 0} V(\lambda)) = \{0\}$. This imples that x commutes with $\mathfrak{g}_0^{(1)}$. Since $x \in \sum_{\lambda\in\mathfrak{h}^*, \lambda_{|\mathfrak{h}'}\neq 0} V(\lambda)$, we have $x = 0$. If $x \in \mathfrak{g}_0$, then, if $y \in \sum_{\lambda\in\mathfrak{h}^*, \lambda_{|\mathfrak{h}'}\neq 0} V(\lambda)$, we have $[x,y] \in \mathfrak{z}(\mathfrak{g}) \cap (\sum_{\lambda\in\mathfrak{h}^*, \lambda_{|\mathfrak{h}'}\neq 0} V(\lambda)) = \{0\}$. If $y \in \mathfrak{g}_0^{(2)}$, then $[x,y] \in [\mathfrak{g}_0, \mathfrak{g}_0] \cap \mathfrak{z}(\mathfrak{g}) = \{0\}$. So $x \in \mathfrak{z}(\mathfrak{g}^{(2)})$. This implies that $x \in \mathfrak{h}'$, so it commutes also with M_{triv} and $\mathfrak{h}$, hence $x \in \mathfrak{z}(\mathfrak{g})$, as required.

To prove the second statement, it suffices to show that there does not exist an isotropic ideal in $\underline{\mathfrak{g}}$. Indeed, if this is the case and $\mathfrak{i}$ is a minimal ideal, then by minimality either $\mathfrak{i} \subseteq \mathfrak{i}^\perp$ or $\mathfrak{i} \cap \mathfrak{i}^\perp = \{0\}$. Since we have excluded the former case, we have $\underline{\mathfrak{g}} = \mathfrak{i} \oplus \mathfrak{i}^\perp$ with $\mathfrak{i}$ a simple Lie superalgebra endowed with a non degenerate form and we can conclude by induction.

Suppose that $\mathfrak{i}$ is an isotropic ideal. If $x \in \mathfrak{g}^{(2)}$, we let $\pi(x)$ be its image in $\underline{\mathfrak{g}}$. If $\mathfrak{i}_1 \neq \{0\}$, we have that there is $\pi(V(\lambda)) \subset \mathfrak{i}_1$. We can choose an highest weight vector v_λ in $V(\lambda)$ and a vector $v_{-\lambda} \in \mathfrak{g}_1$ of weight $-\lambda$ such that $(v_\lambda, v_{-\lambda}) = 1$. Then $\pi(h_\lambda) = [\pi(v_\lambda), \pi(v_{-\lambda})] \in \mathfrak{i}$. Note that $h_\lambda \notin \mathfrak{z}(\mathfrak{g})$. In fact, since $\dim V(\lambda) > 1$, $\lambda_{|\mathfrak{h}\cap[\mathfrak{g}_0,\mathfrak{g}_0]} \neq 0$. On the other hand, if $[h_\lambda, \mathfrak{g}^{(2)})] \neq 0$, then there is $0 \neq v_\mu \in (\mathfrak{g}^{(2)})_\mu$ such that $[h_\lambda, v_\mu] = \lambda(h_\mu)v_\mu \neq 0$. In particular $\pi(v_\mu) \in \mathfrak{i}$. Choose $v_{-\mu} \in (\mathfrak{g}^{(2)})_{-\mu}$ such $(v_\mu, v_{-\mu}) \neq 0$. Then $\pi(v_{-\mu}) = -\frac{1}{\lambda(h_\mu)}[\pi(h_\lambda), \pi(v_{-\mu})] \in \mathfrak{i}$. But then $\mathfrak{i}$ is not isotropic. It follows that $h_\lambda \in \mathfrak{z}(\mathfrak{g}^{(2)}) = \mathfrak{z}(\mathfrak{g}) \cap \mathfrak{g}^{(2)}$, which is absurd. Then $\mathfrak{i} = \mathfrak{i}_0$, hence $\mathfrak{i} \subset \mathfrak{z}(\underline{\mathfrak{g}}_0)$. Since $[\mathfrak{i}, \underline{\mathfrak{g}}_1] = 0$, we have that $\pi^{-1}(\mathfrak{i}) \subset \mathfrak{z}(\mathfrak{g}^{(2)})$. Since $\mathfrak{z}(\mathfrak{g}^{(2)}) = \mathfrak{z}(\mathfrak{g}) \cap \mathfrak{g}^{(2)}$, we have $\mathfrak{i} = \{0\}$. □

At this point we have the following decomposition:

$$\mathfrak{g} = (\mathfrak{h} + [\mathfrak{g}_0, \mathfrak{g}_0]) \oplus \sum_{\lambda \in \mathfrak{h}^*, \dim V(\lambda) > 1} V(\lambda) \oplus M_{triv}.$$

Let, as in the proof of Lemma 2, $\pi : \mathfrak{g}^{(2)} \to \underline{\mathfrak{g}}$ be the projection. Since $\mathfrak{z}(\mathfrak{g}) \cap [\mathfrak{g}_0, \mathfrak{g}_0] = 0$, we see that $[\mathfrak{g}_0, \mathfrak{g}_0] = \pi([\mathfrak{g}_0, \mathfrak{g}_0])$, hence we can see the set of positive roots for $\mathfrak{g}_0$ as a set of positive roots for $\underline{\mathfrak{g}}_0$. By Lemma 2, we have

$$\underline{\mathfrak{g}} = \bigoplus_{i=1}^{k} \underline{\mathfrak{g}}(i), \tag{6}$$

with $\underline{\mathfrak{g}}(i)$ simple ideals. It is clear that $\mathfrak{g}$ acts on $\underline{\mathfrak{g}}$ and that the projection π intertwines the action of $\mathfrak{g}_0$ on $\mathfrak{g}_1^{(2)}$ with that on $\underline{\mathfrak{g}}_1$. Since $[\mathfrak{g}_0, \mathfrak{g}_0] = \pi([\mathfrak{g}_0, \mathfrak{g}_0])$, we see that $\underline{\mathfrak{g}}(i)_1$ is a $[\mathfrak{g}_0, \mathfrak{g}_0]$-module. Since the decomposition (6) is orthogonal, we see that the $[\mathfrak{g}_0, \mathfrak{g}_0]$-modules $\underline{\mathfrak{g}}(i)_1$ are inequivalent. It follows that that $\mathfrak{z}(\mathfrak{g}_0)$ stabilizes $\underline{\mathfrak{g}}(i)_1$, thus $\underline{\mathfrak{g}}(i)_1$ is a $\mathfrak{g}_0$-module.

We now discuss the $\mathfrak{g}_0$-module structure of $\underline{\mathfrak{g}}(i)_1$. By the classification of simple Lie superalgebras, either $\underline{\mathfrak{g}}(i)_1 = V(i)$ with $V(i)$ self dual irreducible $\underline{\mathfrak{g}}_0$-module or there is a polarization (with respect to $(\cdot,\cdot)$) $\underline{\mathfrak{g}}(i)_1 = V \oplus V^*$ with V an irreducible $\underline{\mathfrak{g}}_0$-module. In the first case $\underline{\mathfrak{g}}(i)_1$ is an irreducible $\mathfrak{g}_0$-module. In the second case, since the action of $\mathfrak{h}$ is semisimple, $\underline{\mathfrak{g}}(i)$ decomposes as $V_1(i) \oplus V_2(i)$, with $V_j(i)$ $(j = 1, 2)$ irreducible $\mathfrak{g}_0$-modules. If $V_1(i)$ is not self dual, then the decomposition $\underline{\mathfrak{g}}(i) = V_1(i) \oplus V_2(i)$ is a polarization. If $V_1(i) = V_1(i)^*$ and $V_2(i) = V_2(i)^*$ then the center of $\mathfrak{g}_0$ acts trivially on $\underline{\mathfrak{g}}(i)_1$, thus V and V^* are actually $\mathfrak{g}_0$-modules. We can therefore choose $V_1(i) = V$ and $V_2(i) = V^*$.

The simple ideals $\underline{\mathfrak{g}}(i)$ are basic classical Lie superalgebras. By the classification of such algebras (see [9]) there is a contragredient Lie superalgebra $\tilde{\mathfrak{g}}(i)$ such that $\underline{\mathfrak{g}}(i) = [\tilde{\mathfrak{g}}(i), \tilde{\mathfrak{g}}(i)] / \mathfrak{z}([\tilde{\mathfrak{g}}(i), \tilde{\mathfrak{g}}(i)])$. Choose Chevalley generators $\{\tilde{e}_j, \tilde{f}_j\}_{j \in J(i)}$ for $\tilde{\mathfrak{g}}(i)$. Let $\underline{e}_j, \underline{f}_j$ be their image in $\underline{\mathfrak{g}}(i)$.

We claim that $\underline{e}_j, \underline{f}_j$ are $\mathfrak{h}$-weight vectors. The vectors $\underline{e}_j, \underline{f}_j$ are root vectors for $\underline{\mathfrak{g}}(i)$. If the roots of $\underline{e}_j, \underline{f}_j$ have multiplicity one, then, $\underline{e}_j, \underline{f}_j$ must be $\mathfrak{h}$-stable, hence they are $\mathfrak{h}$-weight vectors. If there are roots of higher multiplicity then $\underline{\mathfrak{g}}(i)$ is of type $A(1,1)$. Let $\overline{d}$ be the derivation on $\underline{\mathfrak{g}}(i)$ defined by setting $\overline{d}(\underline{\mathfrak{g}}_0) = 0$, $\overline{d}(v) = v$ for $v \in V_1(i)$, and $\overline{d}(v) = -v$ for $v \in V_2(i)$. Then it is not hard to check that $\tilde{\mathfrak{g}}(i) = \mathbb{C}d \oplus \mathbb{C}c \oplus \underline{\mathfrak{g}}(i)$ with bracket defined as in Exercise 2.10 of [11]. If the roots of $\underline{e}_j, \underline{f}_j$ have multiplicity two, then $\tilde{e}_j, \tilde{f}_j$ are odd root vectors of $\tilde{\mathfrak{g}}(i)$. In particular $\tilde{e}_j$ is in $V_1(i)$ and $\tilde{f}_j$ is in $V_2(i)$. This implies that $\underline{e}_j$ is in $V_1(i)$ and $\underline{f}_j$ is in $V_2(i)$. Since $\mathfrak{z}(\mathfrak{g}_0)$

acts as multiple of the identity on $V_1(i)$ and $V_2(i)$, we see that $\underline{e}_j, \underline{f}_j$ are $\mathfrak{h}$-weight vectors also in this case.

Since π restricted to $[\mathfrak{g}_0, \mathfrak{g}_0] + \mathfrak{g}_1^{(2)}$ is an isomorphism, we can define e_j, f_j to be the unique elements of $[\mathfrak{g}_0, \mathfrak{g}_0] + \mathfrak{g}_1^{(2)}$ such that $\pi(e_j) = \underline{e}_j$ and $\pi(f_j) = \underline{f}_j$. Since $\underline{e}_j, \underline{f}_j$ are $\mathfrak{h}$-weight vectors, we have that e_j, f_j are root vectors for $\mathfrak{g}$.

Set $J = \cup_i J(i)$. We can always assume that the positive root vectors of $\underline{\mathfrak{g}}_0$ are in the algebra spanned by $\{\underline{e}_j \mid j \in J\}$.

Let $\alpha_j \in \mathfrak{h}^*$ be the weight of e_j. We note that the weight of f_j is $-\alpha_j$ for any $j \in J$. One way to check this is the following: if $j \in J(i)$, there is an invariant form $< \cdot, \cdot >$ on $\tilde{\mathfrak{g}}(i)$ such that $< \tilde{e}_j, \tilde{f}_j > \neq 0$. Since $\underline{\mathfrak{g}}(i)$ is simple, the form $(\cdot,\cdot)$ is a (nonzero) multiple of the form induced by $< \cdot, \cdot >$. In particular $(\underline{e}_j, \underline{f}_j) \neq 0$. Since $(e_j, f_j) = (\underline{e}_j, \underline{f}_j)$, we see that the root of f_j is $-\alpha_j$.

Subdivide Λ in (3) as $\Lambda = \{0\} \cup \Lambda^+ \cup \Lambda^-$ with $\Lambda^+ \cap \Lambda^- = \emptyset$ and $\Lambda^- = -\Lambda^+$ (which is possible since the form $(\cdot,\cdot)$ is nondegenerate on M'_{triv}). Choose a basis $\{e_i^\lambda \mid i = 1, \dots, \dim M(\lambda)\}$ in $M(\lambda)$ for $\lambda \in \Lambda^+$ and let $\{f_i^\lambda\} \subset M(-\lambda)$ be the dual basis. Also $(\cdot,\cdot)_{|M(0)\times M(0)}$ is nondegenerate, hence we can find a polarization $M(0) = M^+ \oplus M^-$. Let $\{e_i^0\}$ and $\{f_i^0\}$ be a basis of M^+ and its dual basis in M^-, respectively.

We now check that relations

$$[e_i, f_j] = \delta_{ij} h_i,\ i, j \in J \quad [e_i, f_j^\lambda] = [e_i^\lambda, f_j] = 0,\ j \in J \quad [e_i^\lambda, f_j^\mu] = \delta_{\lambda,\mu}\delta_{i,j} h_i^\lambda,$$

hold for $\{e_j, f_j\}_{j\in J} \cup \{e_i^\lambda, f_i^\lambda\}_{\lambda \in \Lambda^+\cup\{0\}}$. Assume now $i \neq j$, $i, j \in J$. Then, since $[\underline{e}_i, \underline{f}_j] = 0$, $[e_i, f_j] \in \mathfrak{z}(\mathfrak{g}) \cap \mathfrak{g}^{(2)} \subset \mathfrak{h}'$. This implies that $\alpha_i = \alpha_j$ so $[e_i, f_j] \in \mathbb{C}h_{\alpha_i}$. If e_i is even then α_i is a root of $\mathfrak{g}_0$ so $\pi(h_{\alpha_i}) \neq 0$, hence $[e_i, f_j] = 0$. If e_i is odd and $\pi(h_{\alpha_i}) = 0$, since $[\underline{e}_i, \underline{f}_i] = (\underline{e}_i, \underline{f}_i)\pi(h_{\alpha_i}) = 0$, we have that $[\underline{e}_i, \underline{f}_j] = 0$ for any $j \in J$. In particular, $\underline{e}_i$ is a lowest weight vector for $\underline{\mathfrak{g}}_0$. On the other hand, since $h_{\alpha_i} \in \mathfrak{z}(\mathfrak{g})$, we have in particular that $\alpha(h_{\alpha_i}) = 0$ for any root of $\mathfrak{g}_0$. This implies that $\mathbb{C}\underline{e}_i$ is stable under the adjoint action of $\underline{\mathfrak{g}}_0$. This is absurd since $\underline{\mathfrak{g}}_1$ does not have one-dimensional $\underline{\mathfrak{g}}_0$-submodules.

If $\lambda, \mu \in \Lambda^+ \cup \{0\}$, then $[e_i^\lambda, f_j^\mu]$ is in the center of $\mathfrak{g}_0$, hence $[e_i^\lambda, f_j^\mu] = \delta_{i,j}\delta_{\lambda,\mu} h_\lambda$. Moreover it is obvious that $[e_h^\lambda, f_j] = [e_j, f_h^\lambda] = 0$ if e_j, f_j are even. It remains to check that $[e_h^\lambda, f_j] = [e_j, f_h^\lambda] = 0$ when e_j, f_j are odd.

This follows from the more general

Lemma 3

$$[M_{triv}, \mathfrak{g}_1^{(2)}] = 0.$$

Proof. Choose $x \in M(\lambda)$ and $y \in V(\mu)$ with $\dim V(\mu) > 1$. It is enough to show that $([x,y], z) = 0$ for any $z \in \mathfrak{g}_0$. Observe that, since $\mathbb{C}x$ and $V(\mu)^*$ are inequivalent as $\mathfrak{g}_0$-modules, we have that $(x, V(\mu)) = 0$. Since $([x,y], z) = (x, [y,z])$ and $[y,z]$ is in $V(\mu)$, we have the claim. □

The outcome of the above construction is that we have a triangular decomposition

$$\mathfrak{g} = \mathfrak{n} + \mathfrak{h} + \mathfrak{n}_-, \tag{7}$$

where $\mathfrak{n}$ (resp. $\mathfrak{n}_-$) is the algebra generated by $\{e_j, | j \in J\} \cup \{e_i^\lambda \mid \lambda \in \Lambda^+ \cup \{0\}\}$ (resp. $\{f_j, | j \in J\} \cup \{f_i^\lambda \mid \lambda \in \Lambda^+ \cup \{0\}\}$). By Lemma 3, we see that $[e_h^\lambda, e_j] = 0$. It follows that

$$\mathfrak{n} = \mathfrak{n}_{triv} \oplus \mathfrak{e},$$

where $\mathfrak{n}_{triv}$ is the algebra generated by $\{e_i^\lambda \mid \lambda \in \Lambda^+ \cup \{0\}\}$ and $\mathfrak{e}$ is the algebra generated by $\{e_j, | j \in J\}$. Notice that

$$\mathfrak{n}_{triv} = M^+ \oplus \sum_{\lambda \in \Lambda^+} M(\lambda).$$

This follows from the fact that the right hand side is an abelian subalgebra. Since $\mathfrak{e} \subset \mathfrak{g}^{(2)}$, we have the orthogonal decomposition

$$\mathfrak{n} = M^+ \oplus (\sum_{\lambda \in \Lambda^+} M(\lambda)) \oplus (\mathfrak{n} \cap \mathfrak{g}^{(2)}). \tag{8}$$

Choose any maximal isotropic subspace $\mathfrak{h}^+$ in $\mathfrak{h}$. The previous constructions imply the following fact.

Lemma 4 $\mathfrak{h}^+ + \mathfrak{n}$ *is a maximal isotropic subspace in* $\mathfrak{g}$.

Proof. We first prove that $\mathfrak{n}$ is isotropic. By (8), it is enough to check that $M^+ \oplus (\sum_{\lambda \in \Lambda^+} M(\lambda))$ and $(\mathfrak{n} \cap \mathfrak{g}^{(2)})$ are isotropic. By construction M^+ is isotropic. Moreover, if $\lambda \neq -\mu$, then $M(\lambda)^*$ and $M(\mu)$ are inequivalent, thus $(M(\lambda), M(\mu)) = 0$. This implies that $M(\lambda)$ is isotropic if $\lambda \neq 0$ and $(M(\lambda), M(\mu)) = 0$ if $\lambda \neq \mu$, $\lambda, \mu \in \Lambda^+ \cup \{0\}$.

If $x, y \in \mathfrak{n} \cap \mathfrak{g}^{(2)}$ and $\pi(x) \in \underline{\mathfrak{g}}(i)$, $\pi(y) \in \underline{\mathfrak{g}}(j)$ with $i \neq j$, then $(x, y) = (\pi(x), \pi(y)) = 0$. If $i = j$, let $p : [\tilde{\mathfrak{g}}(i), \tilde{\mathfrak{g}}(i)] \to \underline{\mathfrak{g}}(i)$ be the projection. Let $\tilde{\mathfrak{n}}(i)$ be the algebra spanned by the $\{\tilde{e}_j\}_{j \in J(i)}$. Then $\pi(x), \pi(y) \in p(\tilde{\mathfrak{n}}(i))$. Recall that the weights of $\tilde{\mathfrak{n}}(i)$ are a set of positive roots for $\tilde{\mathfrak{g}}(i)$, and, since $\tilde{\mathfrak{g}}(i)$ is contragredient α and $-\alpha$ cannot be both positive roots for $\tilde{\mathfrak{g}}(i)$. This implies that $\tilde{\mathfrak{n}}(i)$ is an isotropic subspace of $\tilde{\mathfrak{g}}(i)$ for any invariant form of $\tilde{\mathfrak{g}}(i)$. Since $\underline{\mathfrak{g}}(i)$ is simple, $(\cdot, \cdot)$ is induced by an invariant form on $\tilde{\mathfrak{g}}(i)$ so $p(\tilde{n}(i))$ is isotropic.

Clearly, $(\mathfrak{h}, \mathfrak{n}) = (\mathfrak{h}, \mathfrak{n} \cap \mathfrak{g}_0) = 0$, since $\mathfrak{n} \cap \mathfrak{g}_0$ is the nilradical of a Borel subalgebra. Note that $(\mathfrak{n}, \mathfrak{g}) = (\mathfrak{n}, \mathfrak{n}_-)$ so $\mathfrak{n}$ and $\mathfrak{n}_-$ are non degenerately paired. Thus $\mathfrak{n}$ is a maximal isotropic subspace of $\mathfrak{n} + \mathfrak{n}_-$. Since $\mathfrak{h}$ and $\mathfrak{n} + \mathfrak{n}_-$ are orthogonal, the result follows. □

Proposition 2

$$\mathfrak{g} = \mathfrak{n} \oplus \mathfrak{h} \oplus \mathfrak{n}_-. \tag{9}$$

Proof. Having (7) at hand, it remains to prove that the sum is direct. This follows from Lemma 4: indeed, if $x \in \mathfrak{n} \cap \mathfrak{n}_-$, then x would be in the radical of the form. □

4 The universal super affine vertex algebra

Set $\bar{\mathfrak{g}} = P\mathfrak{g}$, where P is the parity reversing functor. In the following, we refer the reader to [3] for basic definitions and notation regarding Lie conformal and vertex algebras. In particular, for the reader's convenience we recall Wick's formula, which will be used several times in the following. Let V be a vertex algebra and $a,b,c \in V$; then

$$[a_\lambda : bc :] =: [a_\lambda b]c : + p(a,b) : b[a_\lambda c] : + \int_0^\lambda [[a_\lambda b]_\mu c]d\mu. \tag{10}$$

Consider the Lie conformal superalgebra $R = (\mathbb{C}[T] \otimes \mathfrak{g}) \oplus (\mathbb{C}[T] \otimes \bar{\mathfrak{g}}) \oplus \mathbb{C}K \oplus \mathbb{C}\bar{K}$ with λ-brackets ($a,b \in \mathfrak{g}$, $\bar{a},\bar{b}$ are the corresponding elements of $\bar{\mathfrak{g}}$):

$$[a_\lambda b] = [a,b] + \lambda(a,b)K, \tag{11}$$

$$[a_\lambda \bar{b}] = \overline{[a,b]}, \quad [\bar{a}_\lambda b] = p(b)\overline{[a,b]}, \tag{12}$$

$$[\bar{a}_\lambda \bar{b}] = (b,a)\bar{K}, \tag{13}$$

K, $\bar{K}$ being even central elements. Let $V(R)$ be the corresponding universal enveloping vertex algebra, and denote by $V^{k,1}(\mathfrak{g})$ its quotient by the ideal generated by $K - k|0\rangle$ and $\overline{K} - |0\rangle$. The vertex algebra $V^{k,1}(\mathfrak{g})$ is called the universal super affine vertex algebra of level k. The relations are the same used in [12] for even variables. We remark that the order of a,b in the r.h.s. of (13) is relevant.

Recall that one defines the current Lie conformal superalgebra $Cur(\mathfrak{g})$ as

$$Cur(\mathfrak{g}) = (\mathbb{C}[T] \otimes \mathfrak{g}) + \mathbb{C}\mathscr{K}$$

with $T(\mathscr{K}) = 0$ and the λ-bracket defined for $a,b \in 1 \otimes \mathfrak{g}$ by

$$[a_\lambda b] = [a,b] + \lambda(a,b)\mathscr{K}, \quad [a_\lambda \mathscr{K}] = [\mathscr{K}_\lambda \mathscr{K}] = 0.$$

Let $V(\mathfrak{g})$ be its universal enveloping vertex algebra. The quotient $V^k(\mathfrak{g})$ of $V(\mathfrak{g})$ by the ideal generated by $\mathscr{K} - k|0\rangle$ is called the level k affine vertex algebra.

If A is a superspace equipped with a skewsupersymmetric bilinear form $< \cdot, \cdot >$ one also has the Clifford Lie conformal superalgebra

$$C(A) = (\mathbb{C}[T] \otimes A) + \mathbb{C}\overline{\mathscr{K}}$$

with $T(\overline{\mathscr{K}}) = 0$ and the λ-bracket defined for $a,b \in 1 \otimes A$ by

$$[a_\lambda b] = < a,b > \overline{\mathscr{K}}, \quad [a_\lambda \overline{\mathscr{K}}] = [\overline{\mathscr{K}}_\lambda \overline{\mathscr{K}}] = 0.$$

Let V be the universal enveloping vertex algebra of $C(A)$. The quotient of V by the ideal generated by $\overline{\mathscr{K}} - |0\rangle$ is denoted by $F(A)$. Applying this construction to $\bar{\mathfrak{g}}$ with the form $< \cdot, \cdot >$ defined by $< a,b > = (b,a)$ one obtains the vertex algebra $F(\bar{\mathfrak{g}})$.

We define the Casimir operator of $\mathfrak{g}$ as $\Omega_\mathfrak{g} = \sum_i x^i x_i$ if $\{x_i\}$ is a basis of $\mathfrak{g}$ and $\{x^i\}$ its dual basis w.r.t. $(\cdot,\cdot)$ (see [9, pag. 85]). Since $\Omega_\mathfrak{g}$ commutes with any element of $U(\mathfrak{g})$, the generalized eigenspaces of its action on $\mathfrak{g}$ are ideals in $\mathfrak{g}$. Observe that $\Omega_\mathfrak{g}$ is a symmetric operator: indeed

$$\begin{aligned}(\Omega_\mathfrak{g}(a),b) &= \sum_i([x^i,[x_i,a]],b) = \sum_i -p(x^i,[x_i,a])([[x_i,a],x^i],b) \\ &= \sum_i p(x^i)([a,x_i],[x^i,b]) = \sum_i p(x^i)(a,[x_i,[x^i,b]]) = (a,\Omega_\mathfrak{g}(b)).\end{aligned}$$

Since $\Omega_\mathfrak{g}$ is symmetric, the generalized eigenspaces provide an orthogonal decomposition of $\mathfrak{g}$. Moreover, since σ preserves the form, we have that $\sigma \circ \Omega_\mathfrak{g} = \Omega_\mathfrak{g} \circ \sigma$, hence σ stabilizes the generalized eigenspaces. Since σ is assumed to be indecomposable, it follows that that $\Omega_\mathfrak{g}$ has a unique eigenvalue. Let $2g$ be such eigenvalue. If the form $(\cdot,\cdot)$ is normalized as in [13, (1.3)], then the number g is called the dual Coxeter number of $\mathfrak{g}$.

Lemma 5 *If $\Omega_\mathfrak{g} - 2gI \neq 0$ then $g = 0$. Moreover, in such a case, $\Omega_\mathfrak{g}(\mathfrak{g})$ is a central ideal.*

Proof. Let $\mathfrak{g} = \sum_{S=1}^k \mathfrak{g}(S)$ be a orthogonal decomposition in $(\cdot,\cdot)$-irreducible ideals. Clearly $\Omega_\mathfrak{g}(\mathfrak{g}(i)) \subset \mathfrak{g}(i)$, hence we can assume without loss of generality that $\mathfrak{g}$ is $(\cdot,\cdot)$-irreducible.

If x is in the center of $\mathfrak{g}$, then x is orthogonal to $[\mathfrak{g},\mathfrak{g}]$, so, if $\mathfrak{g} = [\mathfrak{g},\mathfrak{g}]$, then $\mathfrak{g}$ must be centerless. In particular, in this case, $\mathfrak{g} = \underline{\mathfrak{g}}$ is a sum of simple ideals, but, being $(\cdot,\cdot)$-irreducible, it is simple. Since $\Omega_\mathfrak{g} - 2gI$ is nilpotent, we have that $(\Omega_\mathfrak{g} - 2gI)(\mathfrak{g})$ is a proper ideal of $\mathfrak{g}$, hence $\Omega_\mathfrak{g} = 2gI$.

Thus, if $\Omega_\mathfrak{g} - 2gI \neq 0$, we must have $\mathfrak{g} \neq [\mathfrak{g},\mathfrak{g}]$. Since $\mathfrak{g} \neq [\mathfrak{g},\mathfrak{g}]$, the form becomes degenerate when restricted to $[\mathfrak{g},\mathfrak{g}]$ and its radical is contained in the center of $\mathfrak{g}$. It follows that the center of $\mathfrak{g}$ is nonzero. Clearly $\Omega_\mathfrak{g}$ acts trivially on the center, hence $g = 0$.

Since $\mathfrak{g}^{(2)}$ is an ideal of $\mathfrak{g}$, clearly $\Omega_\mathfrak{g}$ acts on it. Since $\Omega_\mathfrak{g}(\mathfrak{z}(\mathfrak{g})) = 0$, this action descends to $\underline{\mathfrak{g}}$. Recall that we have $\underline{\mathfrak{g}} = \bigoplus_{i=1}^k \underline{\mathfrak{g}}(i)$, with $\underline{\mathfrak{g}}(i)$ simple ideals. We already observed that these ideals are inequivalent as $\mathfrak{g}_0$-modules, thus $\Omega_\mathfrak{g}(\underline{\mathfrak{g}}(i)) \subset \underline{\mathfrak{g}}(i)$. Since $\Omega_\mathfrak{g}(\underline{\mathfrak{g}}(i))$ is a proper ideal, we see that $\Omega_\mathfrak{g}(\underline{\mathfrak{g}}) = 0$. Thus $\Omega_\mathfrak{g}(\mathfrak{g}^{(2)}) \subset \mathfrak{z}(\mathfrak{g}) \cap \mathfrak{g}^{(2)}$. We now check that $\Omega_\mathfrak{g}(M_{triv}) = 0$. Let $x \in M(\lambda)$. If $x_i \in \mathfrak{g}_1^{(2)}$, by Lemma 3, $[x_i, x] = 0$. If $x_i \in M(\mu)$ with $\mu \neq -\lambda$, then $[x_i, x] = 0$. If $x_i \in M(-\lambda)$, then $[x^i, [x_i, x]] = (x_i, x)[x^i, h_\lambda] = (x_i, x)\|\lambda\|^2 x_i = 0$. It follows that $\Omega_\mathfrak{g}(x) = \Omega_{\mathfrak{g}_0}(x) = \|\lambda\|^2 x = 0$. The final outcome is that $\Omega_\mathfrak{g}(\mathfrak{g}_1) \subset \mathfrak{z}(\mathfrak{g}) \cap \mathfrak{g}^{(2)} \subset \mathfrak{h}'$. Thus, since $\Omega_\mathfrak{g}$ preserves parity,

$$\Omega_\mathfrak{g}(\mathfrak{g}_1) = \{0\}. \tag{14}$$

It follows that $\Omega_\mathfrak{g}(\mathfrak{g}_0) = \Omega_\mathfrak{g}(\mathfrak{g})$ is an ideal of $\mathfrak{g}$ contained in $\mathfrak{g}_0$. Since $\Omega_\mathfrak{g}$ is nilpotent, $\Omega_\mathfrak{g}(\mathfrak{g})$ is a nilpotent ideal, hence it intersects trivially $[\mathfrak{g}_0, \mathfrak{g}_0]$. It follows that $[\Omega_\mathfrak{g}(\mathfrak{g}), \mathfrak{g}_0] = 0$. Since $\Omega_\mathfrak{g}(\mathfrak{g})$ is an ideal contained in $\mathfrak{g}_0$, $[\Omega_\mathfrak{g}(\mathfrak{g}), \mathfrak{g}_1] \subset \mathfrak{g}_1 \cap \mathfrak{g}_0 = \{0\}$ as well. The result follows. □

Remark 1 Note that we have proved the following fact: if a basic type Lie super algebra $\mathfrak{g}$ is centerless and $(\cdot,\cdot)$-irreducible then $\mathfrak{g}$ is simple (cf. [1, Theorem 2.1]).

Set $C_\mathfrak{g} = \Omega_\mathfrak{g} - 2gI_\mathfrak{g}$.

Proposition 3 *Assume $k + g \neq 0$. Let $\{x_i\}$ be a basis of $\mathfrak{g}$ and let $\{x^i\}$ be its dual basis w.r.t. $(\cdot,\cdot)$. For $x \in \mathfrak{g}$ set*

$$\widetilde{x} = x - \frac{1}{2}\sum_i : \overline{[x, x_i]}\overline{x}^i : + \frac{1}{4(k+g)} C_\mathfrak{g}(x). \tag{15}$$

The map $x \mapsto \widetilde{x}$, $\overline{y} \mapsto \overline{y}$, *induces an isomorphism of vertex algebras* $V^k(\mathfrak{g}) \otimes F(\overline{\mathfrak{g}}) \cong V^{k+g,1}(\mathfrak{g})$.

Proof. Set $\alpha = \frac{1}{4(k+g)}$. Fix $a,b \in \mathfrak{g}$. Since, by Lemma 5, $C_{\mathfrak{g}}(b)$ is central, we get, from Wick formula (10), that

$$\begin{aligned}
[a_\lambda \widetilde{b}] &= [a_\lambda b] - \tfrac{1}{2}\sum_i [a_\lambda :\overline{[b,x_i]}\overline{x}^i:] + \alpha\lambda(a, C_{\mathfrak{g}}(b))K \\
&= [a_\lambda b] - \tfrac{1}{2}\sum_i (:[a_\lambda \overline{[b,x_i]}]\overline{x}^i: + p(a,\overline{[b,x_i]}) :\overline{[b,x_i]}[a_\lambda \overline{x}^i]:) \\
&\quad - \frac{1}{2}\sum_i \int_0^\lambda [[a_\lambda \overline{[b,x_i]}]_\mu \overline{x}^i] d\mu + \alpha\lambda(a, C_{\mathfrak{g}}(b))K \\
&= [a_\lambda b] - \tfrac{1}{2}\sum_i (:\overline{[a,[b,x_i]]}\overline{x}^i: + p(a,\overline{[b,x_i]}) :\overline{[b,x_i]}\,\overline{[a,x^i]}:) \\
&\quad - \frac{1}{2}\sum_i \lambda(x^i,[a,[b,x_i]])\overline{K} + \alpha\lambda(a, C_{\mathfrak{g}}(b))K.
\end{aligned}$$

Using the invariance of the form and Jacobi identity, we have

$$\begin{aligned}
[a_\lambda \widetilde{b}] &= [a_\lambda b] - \tfrac{1}{2}\sum_i (:\overline{[a,[b,x_i]]}\overline{x}^i: - p(a,b) :\overline{[b,[a,x_i]]}\overline{x}^i:) \\
&\quad - \frac{1}{2}\sum_i \lambda(x^i,[a,[b,x_i]])\overline{K} + \alpha\lambda(a, C_{\mathfrak{g}}(b))K \\
&= [a_\lambda b] - \tfrac{1}{2}\sum_i :\overline{[[a,b],x_i]}\overline{x}^i: \\
&\quad - \frac{1}{2}\sum_i \lambda(a,[x^i,[x_i,b]])\overline{K} + \alpha\lambda(a, C_{\mathfrak{g}}(b))K \\
&= [a_\lambda b] - \tfrac{1}{2}\sum_i :\overline{[[a,b],x_i]}\overline{x}^i: - \frac{1}{2}\lambda(a,\Omega_{\mathfrak{g}}(b))\overline{K} + \alpha\lambda(a, C_{\mathfrak{g}}(b))K.
\end{aligned} \tag{16}$$

Next we prove that

$$[\overline{a}_\lambda \widetilde{b}] = 0. \tag{17}$$

By Lemma 5 and (10),

$$\begin{aligned}
[\overline{a}_\lambda \widetilde{b}] &= [\overline{a}_\lambda b] - \tfrac{1}{2}\sum_i [\overline{a}_\lambda :\overline{[b,x_i]}\overline{x}^i:] \\
&= p(b)\overline{[a,b]} - \tfrac{1}{2}\sum_i (:[\overline{a}_\lambda \overline{[b,x_i]}]\overline{x}^i: + p(\overline{a},\overline{[b,x_i]}) :\overline{[b,x_i]}[\overline{a}_\lambda \overline{x}^i]:) \\
&= p(b)\overline{[a,b]} - \tfrac{1}{2}\sum_i (([b,x_i],a)\overline{x}^i + p(\overline{a},\overline{[b,x_i]})(x^i,a)\overline{[b,x_i]}) \\
&= p(b)\overline{[a,b]} - \tfrac{1}{2}(p(b)\overline{[a,b]} + p(b)\overline{[a,b]}) = 0.
\end{aligned}$$

We now compute $[\widetilde{a}_\lambda \widetilde{b}]$. Using (17), we find $[\widetilde{a}_\lambda \widetilde{b}] = [a_\lambda \widetilde{b}] + \alpha[C_{\mathfrak{g}}(a)_\lambda \widetilde{b}]$, hence, by (16)

$$\begin{aligned}
[\widetilde{a}_\lambda \widetilde{b}] &= [a_\lambda b] - \tfrac{1}{2}\sum_i : \overline{[[a,b],x_i]}\overline{x}^i : -\tfrac{1}{2}\lambda(a,\Omega_{\mathfrak{g}}(b))\overline{K} + \alpha\lambda(a,C_{\mathfrak{g}}(b))K + \\
&\quad + \alpha[C_{\mathfrak{g}}(a)_\lambda \widetilde{b}] \\
&= [a_\lambda b] - \tfrac{1}{2}\sum_i : \overline{[[a,b],x_i]}\overline{x}^i : -\tfrac{1}{2}\lambda(a,\Omega_{\mathfrak{g}}(b))\overline{K} + \alpha\lambda(a,C_{\mathfrak{g}}(b))K + \\
&\quad + \alpha([C_{\mathfrak{g}}(a)_\lambda b] - \tfrac{1}{2}\sum_i : \overline{[[C_{\mathfrak{g}}(a),b],x_i]}\overline{x}^i :) \\
&\quad - \alpha(\frac{1}{2}\lambda(C_{\mathfrak{g}}(a),\Omega_{\mathfrak{g}}(b))\overline{K} + \alpha\lambda(C_{\mathfrak{g}}(a),C_{\mathfrak{g}}(b))K).
\end{aligned}$$

By Lemma (5), $[C_{\mathfrak{g}}(a),b] = 0$. Since $\Omega_{\mathfrak{g}}(b) \in [\mathfrak{g},\mathfrak{g}]$ and $C_{\mathfrak{g}}(a)$ is central, we have $(C_{\mathfrak{g}}(a),\Omega_{\mathfrak{g}}(b)) = 0$.

The term $(C_{\mathfrak{g}}(a),C_{\mathfrak{g}}(b))$ is zero as well: if $g \neq 0$, then, by Lemma 5, $C_{\mathfrak{g}}(b) = 0$, and, if $g = 0$, as above, $(C_{\mathfrak{g}}(a),C_{\mathfrak{g}}(b)) = (C_{\mathfrak{g}}(a),\Omega_{\mathfrak{g}}(b)) = 0$. Thus, we can write

$$\begin{aligned}
[\widetilde{a}_\lambda \widetilde{b}] &= [a_\lambda b] - \tfrac{1}{2}\sum_i : \overline{[[a,b],x_i]}\overline{x}^i : -\tfrac{1}{2}\lambda(a,\Omega_{\mathfrak{g}}(b))\overline{K} + \alpha\lambda(a,C_{\mathfrak{g}}(b))K + \\
&\quad + \alpha([C_{\mathfrak{g}}(a),b] + \lambda(C_{\mathfrak{g}}(a),b)K) \\
&= [a_\lambda b] - \tfrac{1}{2}\sum_i : \overline{[[a,b],x_i]}\overline{x}^i : -\tfrac{1}{2}\lambda(a,\Omega_{\mathfrak{g}}(b))\overline{K} + \alpha\lambda(a,C_{\mathfrak{g}}(b))K + \\
&\quad + \lambda(C_{\mathfrak{g}}(a),b)K).
\end{aligned}$$

In the last equality we used the fact that $C_{\mathfrak{g}}(a)$ is central.

Since $\Omega_{\mathfrak{g}}$ (hence $C_{\mathfrak{g}}$) is symmetric, we have

$$[\widetilde{a}_\lambda \widetilde{b}] = [a_\lambda b] - \tfrac{1}{2}\sum_i : \overline{[[a,b],x_i]}\overline{x}^i : -\tfrac{1}{2}\lambda(a,\Omega_{\mathfrak{g}}(b))\overline{K} + 2\alpha\lambda(a,C_{\mathfrak{g}}(b))K.$$

Note that $C_{\mathfrak{g}}([a,b]) = 0$. In fact, for any $z \in \mathfrak{g}$,

$$(C_{\mathfrak{g}}([a,b]),z) = ([a,b],C_{\mathfrak{g}}(z)) = (a,[b,C_{\mathfrak{g}}(z)]) = 0.$$

It follows that $[a_\lambda b] - \frac{1}{2}\sum_i : \overline{[[a,b],x_i]}\overline{x}^i := [a,b] + \lambda(a,b)K - \frac{1}{2}\sum_i : \overline{[[a,b],x_i]}\overline{x}^i := \widetilde{[a,b]} + \lambda(a,b)K$. Hence

$$\begin{aligned}
[\widetilde{a}_\lambda \widetilde{b}] &= \widetilde{[a,b]} + \lambda(a,b)K - \tfrac{1}{2}\lambda(a,\Omega_{\mathfrak{g}}(b))\overline{K} + 2\alpha\lambda(a,C_{\mathfrak{g}}(b))K \\
&= \widetilde{[a,b]} + \lambda(a,b)K - \lambda g(a,b)\overline{K} - \tfrac{1}{2}\lambda(a,C_{\mathfrak{g}}(b)\overline{K} \\
&\quad + 2\alpha\lambda(a,C_{\mathfrak{g}}(b)K).
\end{aligned}$$

Thus, in $V^{k+g,1}(\mathfrak{g})$, we have

$$\begin{aligned}
[\widetilde{a}_\lambda \widetilde{b}] &= \widetilde{[a,b]} + \lambda(a,b)(k+g) - \lambda g(a,b) - \tfrac{1}{2}\lambda(a,C_{\mathfrak{g}}(b))|0\rangle \\
&\quad + 2\alpha\lambda(a,C_{\mathfrak{g}}(b))(k+g)|0\rangle
\end{aligned}$$

so, recalling that $\alpha = \frac{1}{4(k+g)}$, we get

$$[\widetilde{a}_\lambda \widetilde{b}] = \widetilde{[a,b]} + \lambda k(a,b).$$

We can now finish the proof as in Proposition 2.1 of [12]. □

Remark 2 If $g \neq 0$, as in Proposition 2.1 of [12], the map $x \mapsto \widetilde{x}$, $\overline{y} \mapsto \overline{y}$, $\mathscr{K} \mapsto K - g\overline{K}$, $\overline{\mathscr{K}} \mapsto \overline{K}$ defines a homomorphism of Lie conformal algebras $Cur(\mathfrak{g}) \otimes F(\overline{\mathfrak{g}}) \to V(R)$. In particular, if $g \neq 0$, Propositon 3 holds true for any $k \in \mathbb{C}$.

Let A be a vector superspace with a non-degenerate bilinear skewsupersymmetric form $(\cdot,\cdot)$ and σ an elliptic operator preserving the parity and leaving the form invariant. If $r \in \mathbb{R}$ let $\bar{r} = r + \mathbb{Z} \in \mathbb{R}/\mathbb{Z}$. Let $A^{\bar{r}}$ be the $e^{2\pi i r}$ eigenspace of A. We set $L(\sigma,A) = \oplus_{\mu \in \frac{1}{2}+\bar{r}}(t^\mu \otimes A^{\bar{r}})$ and define the bilinear form $< \cdot,\cdot >$ on $L(\sigma,A)$ by setting $< t^\mu \otimes a, t^\nu \otimes b >= \delta_{\mu+\nu,-1}(a,b)$.

If B is any superspace endowed with a non-degenerate bilinear skewsupersymmetric form $< \cdot,\cdot >$, we denote by $\mathscr{W}(B)$ be the quotient of the tensor algebra of B modulo the ideal generated by

$$a \otimes b - p(a,b) b \otimes a - < a,b >, \quad a,b \in B.$$

We now apply this construction to $B = L(\sigma,A)$ to obtain $\mathscr{W}(L(\sigma,A))$. We choose a maximal isotropic subspace L^+ of $L(\sigma,A)$ as follows: fix a maximal isotropic subspace A^+ of $A^{\bar{0}}$, and let

$$L^+ = \bigoplus_{\mu > -\frac{1}{2}} (t^\mu \otimes A^{\bar{\mu}}) \oplus (t^{-\frac{1}{2}} \otimes A^+).$$

We obtain a $\mathscr{W}(L(\sigma,A))$-module $CW(A) = \mathscr{W}(L(\sigma,A))/\mathscr{W}(L(\sigma,A))L^+$ (here CW stands for "Clifford-Weil").

Note that $-I_A$ induces an involutive automorphism of $C(A)$ that we denote by ω. Set $\tau = \omega \circ \sigma$. Then we can define fields

$$Y(a,z) = \sum_{n \in \frac{1}{2}+\bar{r}} (t^n \otimes a) z^{-n-1}, \quad a \in A^{\bar{r}},$$

where we let $t^n \otimes a$ act on $CW(A)$ by left multiplication. Setting furthermore $Y(\overline{\mathscr{K}},z) = I_A$, we get a τ-twisted representation of $C(A)$ on $CW(A)$ (that descends to a representation of $F(A)$).

Take now $A = \mathfrak{g}$, and let σ be an automorphism of $\mathfrak{g}$ as in Sect. 2. Let $\mathfrak{g}^{\bar{0}} = \mathfrak{n}^0 \oplus \mathfrak{h}^0 \oplus \mathfrak{n}^0_-$ be the triangular decomposition provided by Proposition 2 applied to $\mathfrak{g}^{\bar{0}}$. Choosing an isotropic subspace $\mathfrak{h}^+$ of $\mathfrak{h}^0$ we can choose the maximal isotropic subspace of $\mathfrak{g}^{\bar{0}}$ provided by Lemma 4 and construct the corresponding Clifford-Weil module $CW(\overline{\mathfrak{g}})$, which we regard as a τ-twisted representation of $F(\overline{\mathfrak{g}})$.

In light of Proposition 3, given a σ-twisted representation M of $V^k(\mathfrak{g})$, we can form $\sigma \otimes \tau$-twisted representation $X(M) = M \otimes CW(\overline{\mathfrak{g}})$ of $V^{k+g,1}(\mathfrak{g})$.

In the above setting, we choose M in a particular class of representations arising from the theory of twisted affine Lie superalgebras. We recall briefly their construction and refer the reader to [12] for more details.

Let $L'(\mathfrak{g},\sigma)=\sum_{j\in\mathbb{R}}(t^j\otimes\mathfrak{g}^{\bar{j}})\oplus\mathbb{C}K$. This is a Lie superalgebra with bracket defined by

$$[t^m\otimes a, t^n\otimes b]=t^{m+n}\otimes[a,b]+\delta_{m,-n}m(a,b)K,\quad m,n\in\mathbb{R},$$

K being a central element.

Let $(\mathfrak{h}^0)'=\mathfrak{h}^0+\mathbb{C}K$. If $\mu\in((\mathfrak{h}^0)')^*$, we set $\overline{\mu}=\mu_{|\mathfrak{h}^0}$. Set $\mathfrak{n}'=\mathfrak{n}^0+\sum_{j>0}t^j\otimes\mathfrak{g}^{\bar{j}}$. Fix $\Lambda\in((\mathfrak{h}^0)')^*$. A $L'(\mathfrak{g},\sigma)$-module M is called a highest weight module with highest weight Λ if there is a nonzero vector $v_\Lambda\in M$ such that

$$\mathfrak{n}'(v_\Lambda)=0,\ hv_\Lambda=\Lambda(h)v_\Lambda \text{ for } h\in(\mathfrak{h}^0)',\ U(L'(\mathfrak{g},\sigma))v_\Lambda=M. \tag{18}$$

Let $\Delta^{\bar{j}}$ be the set of $\mathfrak{h}^0$-weights of $\mathfrak{g}^{\bar{j}}$. If $\mu\in(\mathfrak{h}^0)^*$ and $\mathfrak{m}$ is any $\mathfrak{h}^0$-stable subspace of $\mathfrak{g}$, then we let $\mathfrak{m}_\mu$ be the corresponding weight space. Denote by Δ^0 the set of roots (i. e. the nonzero $\mathfrak{h}^0$-weights) of $\mathfrak{g}^{\bar{0}}$. Set $\Delta_+^0=\{\alpha\in\Delta^0\mid\mathfrak{n}_\alpha^0\neq\{0\}\}$.

Since $\mathfrak{n}^0$ and $\mathfrak{n}_-^0$ are non degenerately paired, we have that $-\Delta_+^0=\{\alpha\in\Delta^0\mid(\mathfrak{n}_-)_\alpha\neq\{0\}\}$. By the decomposition (9) we have $\Delta^0=\Delta_+^0\cup-\Delta_+^0$.

Set

$$\rho^{\bar{0}}=\frac{1}{2}\sum_{\alpha\in\Delta_+^0}(\mathrm{sdim}\,\mathfrak{n}_\alpha^0)\alpha,\quad \rho^{\bar{j}}=\frac{1}{2}\sum_{\alpha\in\Delta^{\bar{j}}}(\mathrm{sdim}\,\mathfrak{g}_\alpha^{\bar{j}})\alpha\quad \text{if } \bar{j}\neq\bar{0}, \tag{19}$$

$$\rho_\sigma=\sum_{0\leq j\leq\frac{1}{2}}(1-2j)\rho^{\bar{j}}. \tag{20}$$

Finally set

$$z(\mathfrak{g},\sigma)=\frac{1}{2}\sum_{0\leq j<1}\frac{j(1-j)}{2}\mathrm{sdim}\,\mathfrak{g}^{\bar{j}}. \tag{21}$$

Here and in the following we denote by $\mathrm{sdim}V$ the superdimension $\dim V_0-\dim V_1$ of a superspace $V=V_0\oplus V_1$.

If X is a twisted representation of a vertex algebra V (see [12, § 3]) and $a\in V^{\bar{j}}$, we let

$$Y^X(a,z)=\sum_{n\in\bar{j}}a_{(n)}^X z^{-n-1}$$

be the corresponding field. As explained in [12], a highest weight module M for $L'(\mathfrak{g},\sigma)$ of highest weight Λ becomes automatically a σ-twisted representation of $V^k(\mathfrak{g})$ where $k=\Lambda(K)$.

Set

$$L^{\mathfrak{g}}=\frac{1}{2}\sum_i :x^ix_i:\in V^k(\mathfrak{g}), \tag{22}$$

$$L^{\overline{\mathfrak{g}}}=\frac{1}{2}\sum_i :T(\overline{x}_i)\overline{x}^i:\in F(\overline{\mathfrak{g}}). \tag{23}$$

We can now prove (cf. [12, Lemma 3.2]):

Lemma 6 *If M is a highest weight module for $L'(\mathfrak{g},\sigma)$ with highest weight Λ and $\Lambda(K)=k$ then*

$$\sum_i :C_{\mathfrak{g}}(x^i)x_i:^M_{(1)}(v_\Lambda)=\overline{\Lambda}(C_{\mathfrak{g}}(h_{\overline{\Lambda}}))v_\Lambda \tag{24}$$

$$(L^{\mathfrak{g}})^M_{(1)}(v_\Lambda)=\frac{1}{2}(\overline{\Lambda}+2\rho_\sigma,\overline{\Lambda})v_\Lambda+kz(\mathfrak{g},\sigma)v_\Lambda. \tag{25}$$

Proof. We can and do choose $\{x_i\}$ so that $x_i\in\mathfrak{g}^{\bar{s}_i}$, for some $\bar{s}_i\in\mathbb{R}/\mathbb{Z}$. Let A be any parity preserving operator on $\mathfrak{g}$ which commutes with σ. In particular A preserves $\mathfrak{g}^{\bar{j}}$ for any $j\in\mathbb{R}$. By (3.4) of [12], we have

$$\sum_i :A(x^i)x_i:^M_{(1)}=\sum_i\left(\sum_{n<-s_i}A(x^i)^M_{(n)}(x_i)^M_{(-n)}+\sum_{n\geq -s_i}p(x_i)(x_i)^M_{(-n)}A(x^i)^M_{(n)}\right)$$
$$-\sum_{r\in\mathbb{Z}_+}\binom{-s_i}{r+1}(A(x^i)_{(r)}(x_i))^M_{(-r)}. \tag{26}$$

We choose $s_i\in[0,1)$, thus

$$\sum_i :A(x^i)x_i:^M_{(1)}(v_\Lambda)=$$
$$\sum_i(p(x_i)(x_i)^M_{(s_i)}A(x^i)^M_{(-s_i)}+s_i[A(x^i),x_i]^M_{(0)}-k\binom{-s_i}{2}(A(x^i),x_i))(v_\Lambda),$$

which we can rewrite as

$$\sum_i(A(x^i)^M_{(-s_i)}(x_i)^M_{(s_i)}+(s_i-1)[A(x^i),x_i]^M_{(0)}-k(\binom{-s_i}{2}-s_i)(A(x^i),x_i))(v_\Lambda)=$$
$$\sum_{i:s_i=0}A(x^i)^M_{(0)}(x_i)^M_{(0)}(v_\Lambda)+\sum_i((s_i-1)[A(x^i),x_i]^M_{(0)}-k\binom{s_i}{2}(A(x^i),x_i))(v_\Lambda).$$

Assume now that $A=C_{\mathfrak{g}}$. Then, since $C_{\mathfrak{g}}(x^i)$ is central, $[C_{\mathfrak{g}}(x^i),x_i]=0$. Note also that $\sum_{i:s_i=j}(C_{\mathfrak{g}}(x^i),x_i)$ is the supertrace of $(C_{\mathfrak{g}})_{|\mathfrak{g}^{\bar{j}}}$. Since $C_{\mathfrak{g}}$ is nilpotent, we obtain that $\sum_{i:s_i=j}(C_{\mathfrak{g}}(x^i),x_i)=0$. Thus

$$\sum_i :C_{\mathfrak{g}}(x^i)x_i:^M_{(1)}(v_\Lambda)=\sum_{i:s_i=0}C_{\mathfrak{g}}(x^i)^M_{(0)}(x_i)^M_{(0)}(v_\Lambda).$$

We choose the basis $\{x_i\}$ by choosing, for each $\alpha\in\Delta^0\cup\{0\}$, a basis $\{(x_\alpha)_i\}$ of $\mathfrak{g}^{\bar{0}}_\alpha$. Set $\{x^i_\alpha\}$ to be its dual basis in $(\mathfrak{g})_{-\alpha}$. If $x\in\mathfrak{g}_\alpha$, then $0=C_{\mathfrak{g}}([h,x_\alpha])=\alpha(h)C_{\mathfrak{g}}(x_\alpha)$. If $\alpha\neq 0$ then this implies $C_{\mathfrak{g}}(x_\alpha)=0$. If $\alpha=0$ and $(x_\alpha)_i\in\mathfrak{g}_1$, then, by (14), we have that $C_{\mathfrak{g}}((x_\alpha)_i)=0$ as well. This implies that, if $\{h_i\}$ is an orthonormal basis of $\mathfrak{h}^0$,

$$\sum_i :C_{\mathfrak{g}}(x^i)x_i:^M_{(1)}(v_\Lambda)=\sum_i C_{\mathfrak{g}}(h_i)^M_{(0)}(h_i)^M_{(0)}(v_\Lambda)$$
$$=\sum_i\Lambda(C_{\mathfrak{g}}(h_i))\Lambda(h_i)v_\Lambda=\overline{\Lambda}(C_{\mathfrak{g}}(h_{\overline{\Lambda}}))v_\Lambda.$$

Let now $A = Id$. Clearly we can assume that the basis $\{(x_\alpha)_i\}$ of $\mathfrak{g}_\alpha^{\bar{0}}$ is the union of a basis of $\mathfrak{n}_\alpha^0$ and a basis of $(\mathfrak{n}_-^0)_\alpha$ if $\alpha \in \Delta_+^0$, while, if $\alpha = 0$, we can choose the basis $\{(x_\alpha)_i\}$ to be the union of a basis of $\mathfrak{n}_\alpha^0$, a basis of $(\mathfrak{n}_-^0)_\alpha$ and an orthonormal basis $\{h_i\}$ of $\mathfrak{h}^0$. We can therefore write

$$\sum_{i:s_i=0} x^i_{(0)}(x_i)_{(0)} = 2(h_{\rho_0})_{(0)} + \sum_i (h_i)^2_{(0)} + 2\sum_{(x_\alpha)_i \in \mathfrak{n}_\alpha^0} (x^i_\alpha)_{(0)}((x_\alpha)_i)_{(0)}.$$

We find that

$$\sum_i :x^i x_i:^M_{(1)}(v_\Lambda) = (\overline{\Lambda} + 2\rho_0, \overline{\Lambda})v_\Lambda + k(\sum_{0<j<1} \frac{j(1-j)}{2} \operatorname{sdim} \mathfrak{g}^{\bar{j}})v_\Lambda$$
$$+ \sum_{i:s_i>0} s_i[x^i, x_i]^M_{(0)}(v_\Lambda).$$

In order to evaluate $\sum_{i:s_i>0} s_i[x^i, x_i]^M_{(0)}(v_\Lambda)$, we observe that

$$\sum_{i:s_i=s} [x^i, x_i] = \sum_{i:s_i=1-s} p(x_i)[x_i, x^i] = -\sum_{i:s_i=1-s} [x^i, x_i].$$

This relation is easily derived by exchanging the roles of x_i and x^i. Hence

$$\begin{aligned}\sum_{i:s_i>0} s_i[x^i, x_i] &= \sum_{i:\frac{1}{2}>s_i>0} s_i[x^i, x_i] + \sum_{i:1>s_i>\frac{1}{2}} s_i[x^i, x_i] \\ &= \sum_{i:\frac{1}{2}>s_i>0} s_i[x^i, x_i] + \sum_{i:0<s_i<\frac{1}{2}} (s_i - 1)[x^i, x_i] \\ &= -\sum_{i:\frac{1}{2}>s_i>0} (1-2s_i)[x^i, x_i].\end{aligned}$$

We can choose x_i in $\mathfrak{g}_\alpha^{\bar{s}_i}$ so that $[x^i, x_i] = -p(x_i)h_\alpha$, hence

$$\sum_i s_i[x^i, x_i] = \sum_{i:0<s_i<\frac{1}{2}} 2(1-2s_i)h_{\rho^{\bar{s}_i}}, \tag{27}$$

hence

$$\sum_i s_i[x^i, x_i]^M_{(0)}(v_\Lambda) = \sum_{i:0<s_i<\frac{1}{2}} (1-2s_i)(2\rho^{\bar{s}_i}, \overline{\Lambda})v_\Lambda.$$

This completes the proof of (24).

5 Dirac operators

As in the previous section $\{x_i\}$ is a homogeneous basis of $\mathfrak{g}$ and $\{x^i\}$ is its dual basis. The following element of $V^{k,1}(\mathfrak{g})$:

$$G_{\mathfrak{g}} = \sum_i :x^i \overline{x}_i: - \frac{1}{3}\sum_{i,j} :\overline{[x^i, x_j]}\overline{x}^j \overline{x}_i: \tag{28}$$

is called the affine Dirac operator. It has the following properties:

$$[a_\lambda G_{\mathfrak{g}}] = \lambda k\bar{a}, \quad [\bar{a}_\lambda G_{\mathfrak{g}}] = a. \tag{29}$$

By the sesquilinearity of the λ-bracket, we also have

$$[(G_{\mathfrak{g}})_\lambda a] = p(a)k(\lambda\bar{a} + T(\bar{a})), \quad [(G_{\mathfrak{g}})_\lambda \bar{a}] = p(a)a. \tag{30}$$

If $\mathfrak{g}$ is purely even, $G_{\mathfrak{g}}$ is the Kac-Todorov-Dirac field considered in [12] and in [3] as an analogue of Kostant's cubic Dirac operator. It can be proved, by using a suitable Zhu functor $\pi : V^{k,1}(\mathfrak{g}) \to U(\mathfrak{g}) \otimes \mathscr{W}(\bar{\mathfrak{g}})$, that $\pi(G_{\mathfrak{g}})$ is the Dirac operator considered by Huang and Pandzic in [7].

Write for shortness G instead of $G_{\mathfrak{g}}$. We want to calculate $[G_\lambda G]$. We proceed in steps. Set

$$\theta(x) = \frac{1}{2}\sum_i : \overline{[x,x_i]}\bar{x}^i :, \tag{31}$$

and note that

$$G = \sum_i : x^i\bar{x}_i : -\tfrac{2}{3}\sum_i : \theta(x^i)\bar{x}_i : . \tag{32}$$

We start by collecting some formulas.

Lemma 7

$$[a_\lambda \theta(b)] = \theta([a,b]) + \frac{1}{2}\lambda(a, \Omega_{\mathfrak{g}}(b)), \tag{33}$$

$$[\bar{a}_\lambda \theta(b)] = p(b)\overline{[a,b]}, \tag{34}$$

$$[\theta(a)_\lambda b] = \theta([a,b]) + \frac{1}{2}\lambda(a, \Omega_{\mathfrak{g}}(b)), \tag{35}$$

$$[\theta(a)_\lambda \bar{b}] = \overline{[a,b]}, \tag{36}$$

$$[\theta(a)_\lambda \theta(b))] = \theta([a,b]) + \frac{1}{2}\lambda(a, \Omega_{\mathfrak{g}}(b)), \tag{37}$$

$$[\theta(a)_\lambda \sum_i : x^i\bar{x}_i :] = -\sum_i : (x^i - \theta(x^i))\overline{[x_i,a]} : + \frac{3}{2}\lambda\overline{\Omega_{\mathfrak{g}}(a)}, \tag{38}$$

$$[\theta(a)_\lambda \sum_i : \theta(x^i)\bar{x}_i :] = \frac{3}{2}\lambda\overline{\Omega_{\mathfrak{g}}(a)}, \tag{39}$$

$$\sum_i : x^i\theta(x_i) := \sum_i : \theta(x^i)x_i :, \tag{40}$$

$$\sum_i : \theta(x^i)\theta(x_i) := 0. \tag{41}$$

Proof. Formulas (33) and (34) have been proven in the proof of Proposition 3. Formulas (35) and (36) are obtained by applying sesquilinearity of the λ-bracket to (33) and (34). From (36) and (12) one derives that

$$[\bar{a}_\lambda (b - \theta(b))] = [(b - \theta(b))_\lambda \bar{a}] = 0, \tag{42}$$

hence $[\theta(a)_\lambda(b-\theta(b))]=0$. This implies (37). Using Wick's formula (10), (35), and (36) we get

$$[\theta(a)_\lambda \sum_i :x^i\bar{x}_i:] = \sum_i (:\theta([a,x^i])\bar{x}_i: + p(a,x_i):x^i\overline{[a,x_i]}:) + \frac{3}{2}\lambda\overline{\Omega_{\mathfrak{g}}(a)}.$$

Note that

$$\sum_i :\theta(x^i)\overline{[x_i,a]} := \sum_i :\theta([a,x^i])\bar{x}_i: \tag{43}$$

so (38) follows. Likewise

$$[\theta(a)_\lambda \sum_i :\theta(x^i)\bar{x}_i:] = \sum_i (:\theta([a,x^i])\bar{x}_i: + p(a,x_i):\theta(x^i)\overline{[a,x_i]}:) + \frac{3}{2}\lambda\overline{\Omega_{\mathfrak{g}}(a)}.$$

so, by (43), (39) follows as well. For (40), it is enough to apply formula (1.39) of [3] and (33). Finally,

$$\begin{aligned}\sum_i :\theta(x^i)\theta(x_i): &= \tfrac{1}{2}\sum_{i,r} :\theta(x^i):\overline{[x_i,x_r]}\bar{x}^r ::= \tfrac{1}{2}\sum_{i,r,s} :\theta(x^i)([x_i,x_r],x^s):\bar{x}_s\bar{x}^r :: \\ &= \tfrac{1}{2}\sum_{i,r,s} :\theta(x^i)(x_i,[x_r,x^s]):\bar{x}_s\bar{x}^r ::= \tfrac{1}{2}\sum_{r,s} :\theta([x_r,x^s]):\bar{x}_s\bar{x}^r ::\end{aligned}$$

and

$$\begin{aligned}\sum_{r,s} :\theta([x_r,x^s]):\bar{x}_s\bar{x}^r :: &= \sum_{r,s} -p(x_r,x^s)p(\bar{x}_s,\bar{x}_r):\theta([x^s,x_r]):\bar{x}^r\bar{x}_s :: \\ &= -\sum_{r,s} p(x^s)p(x_r):\theta([x^s,x_r]):\bar{x}^r\bar{x}_s :: \\ &= -\sum_{r,s} :\theta([x_s,x^r]):\bar{x}_r\bar{x}^s :: .\end{aligned}$$

so (41) holds. □

We start our computation of $[G_\lambda G]$. First observe that, by (30),

$$\sum_i [G_\lambda :x^i\bar{x}_i:] = \sum_i :x^i x_i: + k\sum_i :T(\bar{x}_i)\bar{x}^i: + k\frac{\lambda^2}{2}\mathrm{sdim}\mathfrak{g}. \tag{44}$$

Next we compute $[G_\lambda \sum_i :\theta(x^i)\bar{x}_i:]$. By (32), (38), and (39),

$$[\theta(a)_\lambda G] = -\sum_i :(x^i-\theta(x^i))\overline{[x_i,a]}: + \frac{1}{2}\lambda\overline{\Omega_{\mathfrak{g}}(a)}$$

and, by sesquilinearity,

$$[G_\lambda \theta(a)] = p(a)\sum_i :(x^i-\theta(x^i))\overline{[x_i,a]}: + p(a)\frac{1}{2}(\lambda+T)(\overline{\Omega_{\mathfrak{g}}(a)}) \tag{45}$$

By Wick's formula, (45), and (30),

$$\begin{aligned}[G_\lambda \sum_i : \theta(x^i)\bar{x}_i :] &= \sum_{i,j} :: (x^j - \theta(x^j))\overline{[x_j, x_i]} : \bar{x}^i :) \\ &+ \frac{1}{2}\sum_i p(x^i)(\lambda + T) : \overline{\Omega_\mathfrak{g}(x^i)}\bar{x}_i : + \sum_i : \theta(x^i)x_i : \\ &+ \sum_{i,j}\int_0^\lambda [: (x^j - \theta(x^j))\overline{[x_j, x_i]} :_\mu \bar{x}^i] d\mu \\ &+ \frac{1}{2}\sum_i p(x^i)\int_0^\lambda [(\lambda + T)(\overline{\Omega_\mathfrak{g}(x^i)})_\mu \bar{x}_i] d\mu).\end{aligned}$$

Let us compute the terms of the above sum one by one. By formula (1.40) of [3], and (42) above, we have

$$\begin{aligned}&\sum_{i,j} :: (x^j - \theta(x^j))\overline{[x_j, x_i]} : \bar{x}^i := \sum_{i,j} : (x^j - \theta(x^j)) : \overline{[x_j, x_i]}\bar{x}^i :: \\ &\quad + \sum_{i,j}\int_0^T d\lambda : (x^j - \theta(x^j))[\overline{[x_j, x_i]}_\lambda \bar{x}^i] : \\ &= 2\sum_j : (x^j - \theta(x^j))\theta(x_j) : - \sum_{i,j}\int_0^T d\lambda : (x^j - \theta(x^j))([x^i, x_i], x_j) : \\ &= 2\sum_j : (x^j - \theta(x^j))\theta(x_j) :\end{aligned}$$

Note that $\sum_i p(x^i) : \overline{\Omega_\mathfrak{g}(x^i)}\bar{x}_i := 0$. Indeed

$$\begin{aligned}\sum_i p(x_i) : \overline{\Omega_\mathfrak{g}(x^i)}\bar{x}_i : &= \sum_i : \overline{\Omega_\mathfrak{g}(x_i)}\bar{x}^i := \sum_{i,r}(\Omega_\mathfrak{g}(x_i), x^r) : \bar{x}_r \bar{x}^i : \\ &= \sum_{i,r}(x_i, \Omega_\mathfrak{g}(x^r)) : \bar{x}_r \bar{x}^i := \sum_r : \bar{x}_r \overline{\Omega_\mathfrak{g}(x^r)} : \\ &= \sum_r p(\bar{x}_r) : \overline{\Omega_\mathfrak{g}(x^r)}\bar{x}_r := -\sum_r p(x_r) : \overline{\Omega_\mathfrak{g}(x^r)}\bar{x}_r.\end{aligned}$$

Using formula (1.38) of [3] and (42) above we see that

$$\sum_{i,j}\int_0^\lambda [: (x^j - \theta(x^j))\overline{[x_j, x_i]} :_\mu \bar{x}^i] d\mu = \sum_{i,j}\int_0^\lambda (x^j - \theta(x^j))(x_j, [x^i, x_i]) d\mu = 0.$$

Finally, since $\Omega_\mathfrak{g} - 2gId$ is nilpotent, it has zero supertrace, hence

$$\sum_i p(x^i)\int_0^\lambda [(\lambda + T)(\overline{\Omega_\mathfrak{g}(x^i)})_\mu \bar{x}_i] d\mu = \sum_i p(x^i)\int_0^\lambda \lambda(x_i, \Omega_\mathfrak{g}(x^i)) d\mu = g\lambda^2 \mathrm{sdim}\mathfrak{g}.$$

It follows that

$$[G_\lambda \sum_i : \theta(x^i)\bar{x}_i :] = 2\sum_j : (x^j - \theta(x^j))\theta(x_j) : + \sum_i : \theta(x^i)x_i : + \frac{g}{2}\lambda^2 \mathrm{sdim}\mathfrak{g}.$$

Using (40) and (41), we can conclude that

$$[G_\lambda \sum_i :\theta(x^i)\overline{x}_i:] = 3\sum_i :x^i\theta(x_i): -\frac{3}{2}\sum_i :\theta(x^i)\theta(x_i): +\frac{g}{2}\lambda^2 \mathrm{sdim}\mathfrak{g},$$

Combining this with (44), the final outcome is that

$$[G_\lambda G] = \sum_i :(x^i - \theta(x^i))(x_i - \theta(x_i)): +k\sum_i :T(\overline{x}_i)\overline{x}^i: +\frac{\lambda^2}{2}(k-\frac{2g}{3})\mathrm{sdim}\mathfrak{g}. \quad (46)$$

Since $x-\theta(x) = \tilde{x} - \frac{1}{4k}C_\mathfrak{g}(x) = \tilde{x} - \frac{1}{4k}\widetilde{C_\mathfrak{g}(x)}$ we have that

$$\begin{aligned}\sum_i :(x^i - \theta(x^i))(x_i - \theta(x_i)): &= \sum_i :(\tilde{x}^i - \tfrac{1}{4k}C_\mathfrak{g}(x^i))(\tilde{x}_i - \tfrac{1}{4k}C_\mathfrak{g}(x_i)): \\ &= \sum_i :\tilde{x}^i\tilde{x}_i: -\tfrac{1}{2k}\sum_i :\widetilde{C_\mathfrak{g}(x^i)}\tilde{x}_i: .\end{aligned}$$

We used the fact, that, since $C_\mathfrak{g}$ is symmetric, $\sum_i :\widetilde{C_\mathfrak{g}(x^i)}\tilde{x}_i: = \sum_i :\tilde{x}^i\widetilde{C_\mathfrak{g}(x_i)}:$ and, since $C_\mathfrak{g}^2 = 0$, $\sum_i :\widetilde{C_\mathfrak{g}(x^i)}\widetilde{C_\mathfrak{g}(x_i)}: = 0$. Thus (46) can be rewritten as

$$[G_{\mathfrak{g}\lambda} G_\mathfrak{g}] = \sum_i (:\tilde{x}^i\tilde{x}_i: -\tfrac{1}{2k}:\widetilde{C_\mathfrak{g}(x^i)}\tilde{x}_i: +k:T(\overline{x}_i)\overline{x}^i:) + \frac{\lambda^2}{2}(k-\frac{2g}{3})\mathrm{sdim}\mathfrak{g}. \quad (47)$$

Identifying $V^{k,1}(\mathfrak{g})$ with $V^{k-g}(\mathfrak{g})\otimes F(\overline{\mathfrak{g}})$ we have that (47) can be rewritten as

$$[G_{\mathfrak{g}\lambda} G_\mathfrak{g}] = 2L^\mathfrak{g}\otimes|0\rangle - \tfrac{1}{2k}\sum_i :C_\mathfrak{g}(x^i)x_i:\otimes|0\rangle + 2k|0\rangle\otimes L^{\overline{\mathfrak{g}}} + \tfrac{\lambda^2}{2}(k-\tfrac{2}{3}g)\mathrm{sdim}\mathfrak{g}. \quad (48)$$

Recall that, given a highest weight representation M of $L'(\mathfrak{g},\sigma)$, we constructed a $\sigma\otimes\tau$-twisted representation $X = X(M)$ of $V^{k,1}(\mathfrak{g})$. Setting $(G_\mathfrak{g})_n^X = (G_\mathfrak{g})_{(n+1/2)}^X$, we can write the field $Y^X(G_\mathfrak{g}, z)$ as

$$Y^X(G_\mathfrak{g}, z) = \sum_{n\in\mathbb{Z}} G_n^X z^{-n-\frac{3}{2}}.$$

Using the fact that $(G_0^X)^2 = \frac{1}{2}[G_0^X, G_0^X]$ and (48), we have

$$(G_0^X)^2 = (L^\mathfrak{g} - \tfrac{1}{4k}\sum_i :C_\mathfrak{g}(x^i)x_i:)_{(1)}^M \otimes I_{CW(\overline{\mathfrak{g}})} - kI_M\otimes(L^{\overline{\mathfrak{g}}})_{(1)}^{CW(\overline{\mathfrak{g}})} - \tfrac{1}{16}(k-\tfrac{2}{3}g)(\mathrm{sdim}\mathfrak{g})I_X. \quad (49)$$

From now on we will write $a_{(n)}$ instead of $a_{(n)}^V$ when there is no risk of confusion for the twisted representation V.

Lemma 8 *In $CW(\overline{\mathfrak{g}})$, if $x\in\mathfrak{g}^{\overline{s}}$ and $n>0$, we have $\theta(x)_{(n)}\cdot 1 = 0$.*

Proof. Choose the basis $\{x_i\}$ of $\mathfrak{g}$, so that $x_i\in\mathfrak{g}^{\overline{s}_i}$. We can clearly assume $s, s_i\in[0,1)$. We apply formula (3.4) of [12] to get

$$\begin{aligned}\theta(x)_{(n)} = &\sum_{i, m<s+s_i-\frac{1}{2}} \overline{[x,x_i]}_{(m)}\overline{x}^i_{(n-m-1)} - p(x,x_i)p(x)p(x_i)\sum_{i, m\geq s+s_i-\frac{1}{2}} \overline{x}^i_{(n-m-1)}\overline{[x,x_i]}_{(m)} \\ &-\sum_i \binom{s+s_i-\frac{1}{2}}{1}[\overline{[x,x_i]}_{(0)}\overline{x}^i]_{(n-1)}.\end{aligned}$$

If $m < s+s_i-\frac{1}{2}$, then $n-m-1 > n-s-s_i-\frac{1}{2}$. Since $n \in \bar{s}$, $n > 0$ and $s \in [0,1)$, we have $n-s \geq 0$. Thus $n-m-1 > -s_i-\frac{1}{2}$. Since $s_i \in [0,1)$ and $n-m-1 \in -\bar{s}_i+\frac{1}{2}$, we see that $n-m-1 \geq -s_i+\frac{1}{2} > -\frac{1}{2}$. It follows that $\overline{x}^i_{(n-m-1)} \cdot 1 = 0$. If $m > s+s_i-\frac{1}{2}$, then $m > -\frac{1}{2}$ so $\overline{[x,x_i]}_{(m)} \cdot 1 = 0$. Since $[\overline{[x,x_i]}_{(0)}\overline{x}^i] = (x^i,[x,x_i])|0\rangle$, we see that, since $n > 0$, $[\overline{[x,x_i]}_{(0)}\overline{x}^i]_{(n-1)} = 0$. We therefore obtain that

$$\theta(x)_{(n)} = -p(x,x_i)p(x)p(x_i)\sum_i \overline{x}^i_{(n-s-s_i-\frac{1}{2})}\overline{[x,x_i]}_{(s+s_i-\frac{1}{2})}.$$

If $s > 0$ or $s_i > 0$ then $\overline{[x,x_i]}_{(s+s_i-\frac{1}{2})} \cdot 1 = 0$, so we can assume $s = 0$ and get that

$$\theta(x)_{(n)} = -p(x,x_i)p(x)p(x_i)\sum_{i:s_i=0} \overline{x}^i_{(n-\frac{1}{2})}\overline{[x,x_i]}_{(-\frac{1}{2})} = -\sum_{i:s_i=0} \overline{[x,x_i]}_{(-\frac{1}{2})}\overline{x}^i_{(n-\frac{1}{2})}.$$

Observing that, since $n > 0$, $\overline{x}^i_{(n-\frac{1}{2})} \cdot 1 = 0$ we get the claim. □

Lemma 9 *In $CW(\overline{\mathfrak{g}})$ we have that*

$$\sum_{i,j:s_i=s_j=0} p(\overline{[x^i,x_j]},\overline{x}^j)(\overline{x}^j)_{(-\frac{1}{2})}(\overline{[x^i,x_j]})_{(-\frac{1}{2})}(\overline{x}_i)_{(-\frac{1}{2})} \cdot 1 = 6(\overline{h}_{\rho_0})_{(-\frac{1}{2})} \cdot 1.$$

Proof. Clearly we can choose the basis $\{x_i\}$ of $\mathfrak{g}^{\bar{0}}$ to be homogeneous with respect to the triangular decomposition $\mathfrak{g}^{\bar{0}} = \mathfrak{n}^0 \oplus \mathfrak{h}^0 \oplus \mathfrak{n}^0_-$. We can also assume that the x_i are $\mathfrak{h}^0$-weight vectors. Let μ_i be the weight of x_i. Set $\mathfrak{b}^0 = \mathfrak{h}^0 \oplus \mathfrak{n}^0$ and $\mathfrak{b}^0_- = \mathfrak{h}^0 \oplus \mathfrak{n}^0_-$. Then

$$\begin{aligned}
&\sum_{i,j:s_i=s_j=0} p(\overline{[x^i,x_j]},\overline{x}^j)(\overline{x}^j)_{(-\frac{1}{2})}(\overline{[x^i,x_j]})_{(-\frac{1}{2})}(\overline{x}_i)_{(-\frac{1}{2})} \cdot 1 \\
&\quad = \sum_{i,j:x_i\in\mathfrak{b}^0_-,s_j=0} p(\overline{[x^i,x_j]},\overline{x}^j)(\overline{x}^j)_{(-\frac{1}{2})}(\overline{[x^i,x_j]})_{(-\frac{1}{2})}(\overline{x}_i)_{(-\frac{1}{2})} \cdot 1 \\
&\quad = \sum_{i,j:x_i\in\mathfrak{b}^0_-,s_j=0} (\overline{[x^i,x_j]})_{(-\frac{1}{2})}(\overline{x}^j)_{(-\frac{1}{2})}(\overline{x}_i)_{(-\frac{1}{2})} \cdot 1 \\
&\quad\quad + \sum_{i,j:x_i\in\mathfrak{b}^0_-,s_j=0} p(\overline{[x^i,x_j]},\overline{x}^j)([x^i,x_j],x^j)(\overline{x}_i)_{(-\frac{1}{2})} \cdot 1.
\end{aligned}$$

Since $[x_j,x^j] \in \mathfrak{h}^0$, we have that

$$\begin{aligned}
&\sum_{i,j:x_i\in\mathfrak{b}^0_-,s_j=0} p(\overline{[x^i,x_j]},\overline{x}^j)([x^i,x_j],x^j)(\overline{x}_i)_{(-\frac{1}{2})} \cdot 1 = \\
&\quad -\sum_{i,j:s_i=s_j=0} (x^i,[x^j,x_j])(\overline{x}_i)_{(-\frac{1}{2})} \cdot 1 = 0.
\end{aligned}$$

It follows that

$$\sum_{i,j:s_i=s_j=0} p(\overline{[x^i,x_j]},\overline{x}^j)(\overline{x}^j)_{(-\frac{1}{2})}(\overline{[x^i,x_j]})_{(-\frac{1}{2})}(\overline{x}_i)_{(-\frac{1}{2})}\cdot 1$$

$$= \sum_{i,j:x_i\in\mathfrak{b}_-^0,s_j=0} (\overline{[x^i,x_j]})_{(-\frac{1}{2})}(\overline{x}^j)_{(-\frac{1}{2})}(\overline{x}_i)_{(-\frac{1}{2})}\cdot 1$$

$$= \sum_{i,j:x_i,\in\mathfrak{b}_-^0,s_j=0} p(\overline{x}_i,\overline{x}_j)(\overline{[x^i,x_j]})_{(-\frac{1}{2})}(\overline{x}_i)_{(-\frac{1}{2})}(\overline{x}^j)_{(-\frac{1}{2})}\cdot 1$$

$$+ \sum_{i:x_i\in\mathfrak{b}_-^0} (\overline{[x^i,x_i]})_{(-\frac{1}{2})}\cdot 1.$$

Since $\sum_{i:x_i\in\mathfrak{b}_-^0}(\overline{[x^i,x_i]})_{(-\frac{1}{2})}\cdot 1=\sum_{i:x_i\in\mathfrak{n}_-}p(x_i)(\overline{h}_{-\mu_i})_{(-\frac{1}{2})}\cdot 1=2(\overline{h}_{\rho_0})_{(-\frac{1}{2})}\cdot 1$, we need only to check that

$$\sum_{i,j:x_i,\in\mathfrak{b}_-^0,s_j=0} p(\overline{x}_i,\overline{x}_j)(\overline{[x^i,x_j]})_{(-\frac{1}{2})}(\overline{x}_i)_{(-\frac{1}{2})}(\overline{x}^j)_{(-\frac{1}{2})}\cdot 1=4(\overline{h}_{\rho_0})_{(-\frac{1}{2})}\cdot 1. \tag{50}$$

Now

$$\sum_{i,j:x_i,\in\mathfrak{b}_-^0,s_j=0} p(\overline{x}_i,\overline{x}_j)(\overline{[x^i,x_j]})_{(-\frac{1}{2})}(\overline{x}_i)_{(-\frac{1}{2})}(\overline{x}^j)_{(-\frac{1}{2})}\cdot 1=$$

$$\sum_{i,j:x_i,\in\mathfrak{b}_-^0,s_j=0} p(x_i)(\overline{[x^j,x^i]})_{(-\frac{1}{2})}(\overline{x}_i)_{(-\frac{1}{2})}(\overline{x}_j)_{(-\frac{1}{2})}\cdot 1=$$

$$\sum_{i,j:x_i,x_j\in\mathfrak{b}_-^0} p(x_i)(\overline{[x^j,x^i]})_{(-\frac{1}{2})}(\overline{x}_i)_{(-\frac{1}{2})}(\overline{x}_j)_{(-\frac{1}{2})}\cdot 1=$$

$$\sum_{i,j:x_i,x_j\in\mathfrak{b}_-^0} p(x_i)(\overline{x}_i)_{(-\frac{1}{2})}(\overline{x}_j)_{(-\frac{1}{2})}(\overline{[x^j,x^i]})_{(-\frac{1}{2})}\cdot 1$$

$$+ \sum_{i,j:x_i,x_j\in\mathfrak{b}_-^0} p(x_i)(x_i,[x^j,x^i])(\overline{x}_j)_{(-\frac{1}{2})}\cdot 1$$

$$+ \sum_{i,j:x_i,x_j\in\mathfrak{b}_-^0} p(x_i)p(\overline{[x^j,x^i]},\overline{x}_i)(\overline{x}_i)_{(-\frac{1}{2})}(x_j,[x^j,x^i])\cdot 1.$$

Since $[x^j,x^i]\in\mathfrak{b}_-^0$ only when $x^j,x^i\in\mathfrak{h}^0$, we see that

$$\sum_{i,j:x_i,x_j\in\mathfrak{b}_-^0} p(x_i)(\overline{x}_i)_{(-\frac{1}{2})}(\overline{x}_j)_{(-\frac{1}{2})}(\overline{[x^j,x^i]})_{(-\frac{1}{2})}\cdot 1=0.$$

Moreover both

$$\sum_{i,j:x_i,x_j\in\mathfrak{b}_-^0} p(x_i)(x_i,[x^j,x^i])(\overline{x}_j)_{(-\frac{1}{2})}\cdot 1$$

and

$$\sum_{i,j:x_i,x_j\in\mathfrak{b}_-^0} p(x_i)p(\overline{[x^j,x^i]},\overline{x}_i)(\overline{x}_i)_{(-\frac{1}{2})}(x_j,[x^j,x^i])\cdot 1$$

are equal to $\sum_{x_i\in\mathfrak{b}_-^0}(\overline{[x^i,x_i]})_{(-\frac{1}{2})}\cdot 1=2(\overline{h}_{\rho_0})_{(-\frac{1}{2})}\cdot 1$. This proves (50), hence the statement. □

6 The very strange formula

We are interested in calculating $G_0^X(v_\Lambda \otimes 1)$, v_Λ being a highest weight vector of a $L'(\mathfrak{g},\sigma)$-module M with highest weight Λ such that $\Lambda(K) = k - g$.

Since σ preserves the form $(\cdot,\cdot)$, we have that $\sigma\Omega_\mathfrak{g} = \Omega_\mathfrak{g}\sigma$. It follows that $\Omega_\mathfrak{g}$ stabilizes $\mathfrak{g}^{\bar{j}}$ for any j. Recall, furthermore, that $\Omega_\mathfrak{g}(\mathfrak{g})$ is contained in the radical of the form restricted to $[\mathfrak{g},\mathfrak{g}]$. In particular $\Omega_\mathfrak{g}(\mathfrak{g}) \subset \mathfrak{h}$. We can therefore choose the maximal isotropic subspace $\mathfrak{h}^+$ of $\mathfrak{h}^0$ so that $\Omega_\mathfrak{g}(\mathfrak{g}^{\bar{0}}) \subset \mathfrak{h}^+$. With this choice we are now ready to prove the following result.

Proposition 4

$$G_0^X(v_\Lambda \otimes 1) = v_\Lambda \otimes (\overline{h}_{\overline{\Lambda}+\rho_\sigma})_{(-\frac{1}{2})} \cdot 1. \tag{51}$$

Proof. Since $C_\mathfrak{g}$ is symmetric, we can rewrite G_0^X as

$$G_0^X = \sum_i :\tilde{x}^i\overline{x}_i:_{(\frac{1}{2})} + \tfrac{1}{3}\sum_i :\theta(x^i)\overline{x}_i:_{(\frac{1}{2})} - \tfrac{1}{4k}\sum_i :\tilde{x}^i\overline{C_\mathfrak{g}(x_i)}:_{(\frac{1}{2})}.$$

With easy calculations one proves that

$$\sum_i :\tilde{x}^i\overline{x}_i:_{(\frac{1}{2})}(v_\Lambda \otimes 1) = v_\Lambda \otimes (\overline{h}_{\overline{\Lambda}})_{(-\frac{1}{2})} \cdot 1. \tag{52}$$

Next we observe that, since $C_\mathfrak{g}(x) \in \mathfrak{h}^+$ when $x \in \mathfrak{g}^{\bar{0}}$, we have that

$$\sum_i :\tilde{x}^i\overline{C_\mathfrak{g}(x_i)}:_{(\frac{1}{2})}(v_\Lambda \otimes 1) = 0. \tag{53}$$

It remains to check the action of $\sum_i :\theta(x^i)\overline{x}_i:_{(\frac{1}{2})}$ on 1. Choose the basis $\{x_i\}$ of $\mathfrak{g}$, so that $x_i \in \mathfrak{g}^{\bar{s}_i}$. We can clearly assume $s_i \in [0,1)$. We apply formula (3.4) of [12] to get

$$\begin{aligned}\sum_i :\theta(x^i)\overline{x}_i:_{(\frac{1}{2})} = &\sum_{i,m<-s_i} \theta(x^i)_{(m)}(\overline{x}_i)_{(-m-\frac{1}{2})} + \sum_{i,m\geq -s_i} (\overline{x}_i)_{(-m-\frac{1}{2})}\theta(x^i)_{(m)} \\ &-\sum_i \binom{-s_i}{1}\overline{[x^i,x_i]}_{(-\frac{1}{2})}.\end{aligned}$$

If $m < -s_i$ then $-m > 0$ so $(\overline{x}_i)_{(-m-\frac{1}{2})} \cdot 1 = 0$. Since $s_i \in [0,1)$, if $m > -s_i$ then $m > 0$. By Lemma 8, $\theta(x^i)_{(m)} \cdot 1 = 0$. Thus

$$\begin{aligned}\sum_i :\theta(x^i)\overline{x}_i:_{(\frac{1}{2})} &= \sum_i (\overline{x}_i)_{(s_i-\frac{1}{2})}\theta(x^i)_{(-s_i)} + \sum_i s_i\overline{[x^i,x_i]}_{(-\frac{1}{2})} \\ &= \sum_i \theta(x^i)_{(-s_i)}(\overline{x}_i)_{(s_i-\frac{1}{2})} + \sum_i p(x_i)\overline{[x_i,x^i]}_{(-\frac{1}{2})} + \sum_i s_i\overline{[x^i,x_i]}_{(-\frac{1}{2})}.\end{aligned}$$

If $s_i > 0$ then $(\bar{x}_i)_{(s_i-\frac{1}{2})} \cdot 1 = 0$. Observe also that $\sum_i p(x_i)\overline{[x_i,x^i]}_{(-\frac{1}{2})} = 0$. Thus

$$\sum_i : \theta(x^i)\bar{x}_i :_{(\frac{1}{2})} = \sum_{i:s_i=0} \theta(x^i)_{(0)}(\bar{x}_i)_{(-\frac{1}{2})} + \sum_i s_i\overline{[x^i,x_i]}_{(-\frac{1}{2})}.$$

Now, applying formula (3.4) of [12], we get

$$\theta(x^i)_{(0)} = \frac{1}{2} \sum_{j,m<s_j-\frac{1}{2}} (\overline{[x^i,x_j]})_{(m)}(\bar{x}^j)_{(-m-1)} + \frac{1}{2}p(\overline{[x^i,x_j]},\bar{x}^j) \sum_{j,m\geq s_j-\frac{1}{2}} (\bar{x}^j)_{(-m-1)}(\overline{[x^i,x_j]})_{(m)} - \frac{1}{2}\sum_j \binom{s_j-\frac{1}{2}}{1}(x^j,[x^i,x_j])I_{CW(\mathfrak{g})}.$$

hence

$$\begin{aligned}\theta(x^i)_{(0)}(\bar{x}_i)_{(-\frac{1}{2})} \cdot 1 = \frac{1}{2} &\sum_{j,m<s_j-\frac{1}{2}} (\overline{[x^i,x_j]})_{(m)}(\bar{x}^j)_{(-m-1)}(\bar{x}_i)_{(-\frac{1}{2})} \cdot 1 \\ &+ \frac{1}{2}p(\overline{[x^i,x_j]},\bar{x}^j) \sum_{j,m\geq s_j-\frac{1}{2}} (\bar{x}^j)_{(-m-1)}(\overline{[x^i,x_j]})_{(m)}(\bar{x}_i)_{(-\frac{1}{2})} \cdot 1 \\ &- \frac{1}{2}\sum_j \binom{s_j-\frac{1}{2}}{1}(x^j,[x^i,x_j])(\bar{x}_i)_{(-\frac{1}{2})} \cdot 1.\end{aligned}$$

If $m < s_j - \frac{1}{2}$, then $-m-1 \geq -s_j + \frac{1}{2} > -\frac{1}{2}$. It follows that $(\bar{x}^j)_{(-m-1)}(\bar{x}_i)_{(-\frac{1}{2})} \cdot 1 = p(\bar{x}_j,\bar{x}_i)(\bar{x}_i)_{(-\frac{1}{2})}(\bar{x}^j)_{(-m-1)} \cdot 1 = 0$. If $s_j > 0$ and $m \geq s_j - \frac{1}{2}$ or $s_j = 0$ and $m > s_j - \frac{1}{2}$, then $(\overline{[x^i,x_j]})_{(m)}(\bar{x}_i)_{(-\frac{1}{2})} \cdot 1 = p(\overline{[x^i,x_j]},\bar{x}_i)(\bar{x}_i)_{(-\frac{1}{2})}(\overline{[x^i,x_j]})_{(m)} \cdot 1 = 0$. Thus

$$\begin{aligned}\sum_{i:s_i=0} \theta(x^i)_{(0)}(\bar{x}_i)_{(-\frac{1}{2})} \cdot 1 = \frac{1}{2} &\sum_{i,j:s_i=s_j=0} p(\overline{[x^i,x_j]},\bar{x}^j)(\bar{x}^j)_{(-\frac{1}{2})}(\overline{[x^i,x_j]})_{(-\frac{1}{2})}(\bar{x}_i)_{(-\frac{1}{2})} \cdot 1 \\ &- \frac{1}{2} \sum_{i:s_i=0,j} \binom{s_j-\frac{1}{2}}{1}(x^j,[x^i,x_j])(\bar{x}_i)_{(-\frac{1}{2})} \cdot 1.\end{aligned}$$

Next we compute

$$\begin{aligned}&\sum_{i:s_i=0,j} \binom{s_j-\frac{1}{2}}{1}(x^j,[x^i,x_j])(\bar{x}_i)_{(-\frac{1}{2})} \cdot 1 = \\ &-\sum_{i:s_i=0,j} \binom{s_j-\frac{1}{2}}{1}p(x^j,x_i)([x^j,x_j],x^i)(\bar{x}_i)_{(-\frac{1}{2})} \cdot 1 = \\ &-\sum_{i,j} \binom{s_j-\frac{1}{2}}{1}([x^j,x_j],x^i)(\bar{x}_i)_{(-\frac{1}{2})} \cdot 1 =\end{aligned}$$

$$
\begin{aligned}
&-\sum_j \binom{s_j-\frac{1}{2}}{1}\overline{[x^j,x_j]}_{(-\frac{1}{2})}\cdot 1\\
&=-\sum_j s_j\overline{[x^j,x_j]}_{(-\frac{1}{2})}.
\end{aligned}
$$

The final outcome is that

$$
\begin{aligned}
\sum_i :\theta(x^i)\overline{x}_i:_{(\frac{1}{2})}\cdot 1 = \tfrac{1}{2}\sum_{i,j:s_i=s_j=0} & p(\overline{[x^i,x_j]},\overline{x}^j)(\overline{x}^j)_{(-\frac{1}{2})}(\overline{[x^i,x_j]})_{(-\frac{1}{2})}(\overline{x}_i)_{(-\frac{1}{2})}\cdot 1\\
&+\tfrac{3}{2}\sum_i s_i\overline{[x^i,x_i]}_{(-\frac{1}{2})}\cdot 1.
\end{aligned}
$$

By (27), we see that

$$\sum_{i:0\le s_i<1} s_i[x^i,x_i]=2\sum_{0<\bar{j}\le\frac{1}{2}} h_{\rho_{\bar{j}}}.$$

Combining this observation with Lemma 9, we see that

$$\tfrac{1}{3}\sum_i :\theta(x^i)\overline{x}_i:_{(\frac{1}{2})}\cdot 1=(\overline{h}_{\rho_\sigma})_{(-\frac{1}{2})}\cdot 1.$$

This, together with (52) and (53), gives the statement. □

Theorem 1 *Let $\mathfrak{g}$ be a basic type Lie superalgebra and let σ be an indecomposable elliptic automorphism preserving the bilinear form. Let $2g$ be the eigenvalue of the Casimir operator in the adjoint representation. Let ρ_σ be defined by* (20) *and $z(\mathfrak{g},\sigma)$ by* (21)*. Set $\rho=\rho_{Id}$. Then we have:*

(Strange formula)

$$||\rho||^2=\frac{g}{12}\text{sdim}\mathfrak{g}. \tag{54}$$

(Very strange formula)

$$||\rho_\sigma||^2=g(\frac{\text{sdim}\mathfrak{g}}{12}-2z(\mathfrak{g},\sigma)). \tag{55}$$

Remark 3 If $\mathfrak{z}(\mathfrak{g})$ is non-zero, then it contains an eigenvector of the Casimir operator with zero eigenvalue, hence $g=0$, and the very strange formula amounts to saying that ρ_σ is isotropic.

Proof. Let $\{v_i\}_{i\in\mathbb{Z}_+}$ be a basis of $CW(\overline{\mathfrak{g}})$ with $v_0=1$. Write $(L^{\overline{\mathfrak{g}}})^{CW(\overline{\mathfrak{g}})}_{(1)}\cdot 1=\sum_i c_iv_i$. If M_0 is a highest weight module with highest weight $\Lambda=-\rho_\sigma+k\Lambda_0$ then $L^{\overline{\mathfrak{g}}}_{(1)}(v_\Lambda\otimes 1)=\sum c_i(v_\Lambda\otimes v_i)$ with the coefficents c_i that do not depend on k. By Proposition 4, $G_0(v_\Lambda\otimes 1)=0$. Applying (49) and Lemma 6, we find that

$$
\begin{aligned}
0=(-\frac{1}{2}\|\rho_\sigma\|^2+(k-g)z(\mathfrak{g},\sigma)-\tfrac{1}{4k}\rho_\sigma(C_{\mathfrak{g}}(h_{\rho_\sigma}))-\tfrac{1}{16}(k-\frac{2g}{3})\text{sdim}\mathfrak{g})(v_\Lambda\otimes 1)\\
-\sum_i c_ik(v_\Lambda\otimes v_i).
\end{aligned}
$$

Since this equality holds for any k, we see that $c_i = 0$ if $i > 0$. Moreover the coefficient of $v_\Lambda \otimes 1$ must vanish. This coefficient is

$$\tfrac{1}{k}\left(-\tfrac{1}{4}\rho_\sigma(C_{\mathfrak{g}}(h_{\rho_\sigma}))+k(-\frac{1}{2}\|\rho_\sigma\|^2-gz(\mathfrak{g},\sigma)+\tfrac{g}{24}\mathrm{sdim}\mathfrak{g})+k^2(z(\mathfrak{g},\sigma)-\tfrac{1}{16}\mathrm{sdim}\mathfrak{g}-c_0)\right),$$

so, again by the genericity of k, we obtain

$$\rho_\sigma(C_{\mathfrak{g}}(h_{\rho_\sigma}))=0, \tag{56}$$

$$-\frac{1}{2}\|\rho_\sigma\|^2-gz(\mathfrak{g},\sigma)+\tfrac{g}{24}\mathrm{sdim}\mathfrak{g}=0, \tag{57}$$

$$z(\mathfrak{g},\sigma)-\tfrac{1}{16}\mathrm{sdim}\mathfrak{g}=c_0. \tag{58}$$

Formula (57) is (55) which specializes clearly to (54) when $\sigma = I_{\mathfrak{g}}$. □

As byproduct of the proof of Theorem 1 we also obtain

Proposition 5

1. $(L^{\overline{\mathfrak{g}}})_0^{CW(\overline{\mathfrak{g}})}\cdot 1 = z(\mathfrak{g},\sigma)-\frac{1}{16}\mathrm{sdim}\mathfrak{g}$.
2. *If M is a highest weight $L(\mathfrak{g})'$-module with highest weight Λ, then*

$$\begin{aligned}&(G_0^X)^2(v_\Lambda\otimes 1)=\\&\tfrac{1}{2}\left((\overline{\Lambda}+2\rho_\sigma,\overline{\Lambda})-\tfrac{1}{2k}\overline{\Lambda}(C_{\mathfrak{g}}(h_{\overline{\Lambda}}))+\frac{g}{12}\mathrm{sdim}\mathfrak{g}-2gz(\mathfrak{g},\sigma)\right)(v_\Lambda\otimes 1).\end{aligned} \tag{59}$$

Proof. We saw in the proof of Proposition 1 that $(L^{\overline{\mathfrak{g}}})_0^{CW(\overline{\mathfrak{g}})}\cdot 1 = c_0$ and (58) gives our formula for c_0.

Again by (49) and Lemma 6,

$$\begin{aligned}G_0^2(v_\Lambda\otimes 1)=(&\frac{1}{2}(\overline{\Lambda}+2\rho_\sigma,\overline{\Lambda})-\tfrac{1}{4k}\overline{\Lambda}(C_{\mathfrak{g}}(\overline{\Lambda}))+(k-g)z(\mathfrak{g},\sigma))v_\Lambda\otimes 1\\&-\tfrac{1}{16}(k-\frac{2g}{3})\mathrm{sdim}\mathfrak{g}(v_\Lambda\otimes 1)v_\Lambda\otimes 1-kv_\Lambda\otimes(L^{\overline{\mathfrak{g}}})_0^{CW(\overline{\mathfrak{g}})}\cdot 1).\end{aligned}$$

Using the first equality we get the second claim. □

References

1. Benayadi, S.: Quadratic Lie superalgebras with the completely reducible action of the even part on the odd part. J. Algebra **223**, 344–366 (2000)
2. Burns, J.: An elementary proof of the "strange formula" of Freudenthal and de Vries. Q. J. Math. **51**(3), 295–297 (2000)
3. De Sole, A., Kac, V.G.: Finite vs affine W–algebras. Jpn. J. Math. **1**, 137–261 (2006)
4. Chuah, M.: Finite order automorphisms on contragredient Lie superalgebras. J. Algebra, **351**, 138–159 (2012)
5. Fegan, H.D., Steer, B.: On the "strange formula" of Freudenthal and de Vries. Math. Proc. Cambridge Philos. Soc. **105**(2), 249–252 (1989)
6. Freudenthal, H., de Vries, H.: Linear Lie Groups. Academic Press, New York (1969)

7. Huang, J.-S., Pandžić, P.: Dirac cohomology for Lie superalgebras. Transform. Groups **10**, 201–209 (2005)
8. Kac, V.G.: Infinite-dimensional Lie algebras, and the Dedekind η-function. (Russian) Funkcional. Anal. i Prilozen. **8**(1), 77–78 (1974)
9. Kac, V.G.: Lie superalgebras, Advances in Math. **26**(1), 8–96 (1977)
10. Kac, V.G.: Infinite-dimensional algebras, Dedekind's η-function, classical Möbius function and the very strange formula. Adv. in Math. **30**(2), 85–136 (1978)
11. Kac, V.G.: Infinite dimensional Lie algebras, 3rd ed. Cambridge University Press, Cambridge (1990)
12. Kac, V.G., Möseneder Frajria, P., Papi, P.: Multiplets of representations, twisted Dirac operators and Vogan's conjecture in affine setting. Advances in Math **217**(6), 2485–2562 (2008)
13. Kac, V.G., Wakimoto, M.: Integrable highest weight modules over affine superalgebras and number theory. Lie theory and geometry, pp. 415–456. Progr. Math., Vol. 123, Birkhäuser, Boston
14. Kac, V.G., Peterson, D.H.: Infinite-dimensional Lie algebras, theta functions and modular forms. Adv. in Math. **53**(2), 125–264 (1984)
15. Papi, P.: Dirac operators in the affine setting. Oberwolfach Reports **6**(1), 835–837 (2009)

Parabolic category $\mathscr{O}$ for classical Lie superalgebras

Volodymyr Mazorchuk

Abstract We compare properties of (the parabolic version of) the BGG category $\mathscr{O}$ for semi-simple Lie algebras with those for classical (not necessarily simple) Lie superalgebras.

1 Introduction

Category $\mathscr{O}$ for semi-simple complex finite dimensional Lie algebras, introduced in [2], is a central object of study in the modern representation theory (see [21]) with many interesting connections to, in particular, combinatorics, algebraic geometry and topology. This category has a natural counterpart in the super-world and this super version $\tilde{\mathscr{O}}$ of category $\mathscr{O}$ was intensively studied (mostly for some particular simple classical Lie superalgebra) in the last decade, see e.g. [3–6, 15, 19] or the recent books [7, 33] for details.

The aim of the present paper is to compare some basic but general properties of the category $\mathscr{O}$ in the non-super and super cases for a rather general classical super-setup. We mainly restrict to the properties for which the non-super and super cases can be connected using the usual restriction and induction functors and the biadjunction (up to parity change) between these two functors is our main tool. The paper also complements, extends and gives a more detailed exposition for some results which appeared in [1, Sect. 7].

The original category $\mathscr{O}$ has a natural parabolic version which first appeared in [35]. We start in Sect. 2 by setting up an elementary approach (using root system geometry) to the definition of a parabolic category $\mathscr{O}$ for classical finite dimensional Lie superalgebras. In Sect. 3 we define the parabolic category $\widetilde{\mathscr{O}}^{\omega}$ and describe its basic categorical properties, including simple objects and blocks. In Sect. 4 we ad-

V. Mazorchuk (✉)
Department of Mathematics, Uppsala University, Box. 480, SE-75106 Uppsala, Sweden
e-mail: mazor@math.uu.se

M. Gorelik, P. Papi (eds.): *Advances in Lie Superalgebras.* Springer INdAM Series 7,
DOI 10.1007/978-3-319-02952-8_9, © Springer International Publishing Switzerland 2014

dress the natural stratification on $\widetilde{\mathcal{O}}^{\omega}$ and its consequences, in particular, existence of tilting modules and estimates for the finitistic dimension, proving the following:

Theorem 1 *Category $\widetilde{\mathcal{O}}^{\omega}$ has finite finitistic dimension.*

Finally, in Sect. 5 we address properties of $\widetilde{\mathcal{O}}^{\omega}$ which are based upon projective-injective modules in this category. This includes an Irving-type theorem describing socular constituents of Verma modules and an analogue of Soergel's Struktursatz. As the last application we prove that category $\widetilde{\mathcal{O}}^{\omega}$ is Ringel self-dual.

2 Preliminaries

2.1 Classical Lie superalgebras

We work over $\mathbb{C}$ and set $\mathbb{N} = \{1,2,3,\dots\}$, $\mathbb{Z}_+ = \{0,1,2,3\dots\}$. For a Lie (super)algebra $\mathfrak{a}$ we denote by $U(\mathfrak{a})$ the corresponding enveloping (super)algebra.

Let $\mathfrak{g} = \mathfrak{g}_{\overline{0}} \oplus \mathfrak{g}_{\overline{1}}$ be a Lie superalgebra over $\mathbb{C}$. From now on we assume that $\mathfrak{g}$ is *classical* in the sense that $\mathfrak{g}_{\overline{0}}$ is a finite dimensional reductive Lie algebra and $\mathfrak{g}_{\overline{1}}$ is a semi-simple finite dimensional $\mathfrak{g}_{\overline{0}}$-module. We do **not** require $\mathfrak{g}$ to be simple. We denote by $\mathfrak{g}$-smod the abelian category of $\mathfrak{g}$-supermodules. Morphisms in $\mathfrak{g}$-smod are homogeneous $\mathfrak{g}$-homomorphisms of degree 0.

Example 1 The general linear superalgebra $\mathfrak{gl}(\mathbb{C}^{m|n})$ of the super vector space $\mathbb{C}^{m|n} = \mathbb{C}^m_{\overline{0}} \oplus \mathbb{C}^n_{\overline{1}}$ with respect to the usual super-commutator of linear operators. Fix the standard bases in $\mathbb{C}^m_{\overline{0}}$ and $\mathbb{C}^n_{\overline{1}}$ and $\mathfrak{gl}(\mathbb{C}^{m|n})$ becomes isomorphic to the superalgebra $\mathfrak{gl}(m|n)$ of $(n+m)\times(n+m)$ matrices naturally divided into $n\times n$, $n\times m$, $m\times n$ and $m\times m$ blocks, under the usual super-commutator of matrices.

Example 2 The subsuperalgebra $\mathfrak{q}_n$ of $\mathfrak{gl}(n|n)$ consisting of all matrices of the form

$$\left(\begin{array}{c|c} A & B \\ \hline B & A \end{array}\right). \tag{1}$$

The even part corresponds to $B = 0$ while the odd part corresponds to $A = 0$.

Example 3 Let $\mathfrak{a}$ be any finite dimensional reductive Lie algebra and V any semi-simple finite dimensional $\mathfrak{a}$-module. Set $\mathfrak{g}_{\overline{0}} := \mathfrak{a}$, $\mathfrak{g}_{\overline{1}} := V$ and $\mathfrak{g}(\mathfrak{a},V) := \mathfrak{g}_{\overline{0}} \oplus \mathfrak{g}_{\overline{1}}$. Setting $[\mathfrak{g}_{\overline{1}},\mathfrak{g}_{\overline{1}}] = 0$ and considering the natural action of the Lie algebra $\mathfrak{g}_{\overline{0}}$ on the $\mathfrak{g}_{\overline{0}}$-module $\mathfrak{g}_{\overline{1}}$ defines on $\mathfrak{g}(\mathfrak{a},V)$ the structure of a Lie superalgebra, which is called the *generalized Takiff superalgebra* associated with $\mathfrak{a}$ and V. These superalgebras appear in [20].

2.2 Natural categories of supermodules via restriction

Consider $\mathfrak{g}_{\bar{0}}$ as a purely even Lie superalgebra. Then $\mathfrak{g}_{\bar{0}}$-smod is equivalent to the direct sum of an even and an odd copy of $\mathfrak{g}_{\bar{0}}$-mod in the obvious way:

$$\mathfrak{g}_{\bar{0}}\text{-smod} \cong (\mathfrak{g}_{\bar{0}}\text{-mod})_{\bar{0}} \oplus (\mathfrak{g}_{\bar{0}}\text{-mod})_{\bar{1}}.$$

Further, we have the usual restriction functor

$$\mathrm{Res}^{\mathfrak{g}}_{\mathfrak{g}_{\bar{0}}} : \mathfrak{g}\text{-smod} \to \mathfrak{g}_{\bar{0}}\text{-smod}.$$

For any subcategory C in $\mathfrak{g}_{\bar{0}}$-mod we now can define the category $\tilde{\mathrm{C}}$ (the notation follows [33]) as the subcategory in $\mathfrak{g}$-smod consisting of all objects and morphisms which are sent to $(\mathrm{C})_{\bar{0}} \oplus (\mathrm{C})_{\bar{1}}$ by $\mathrm{Res}^{\mathfrak{g}}_{\mathfrak{g}_{\bar{0}}}$.

The functor $\mathrm{Res}^{\mathfrak{g}}_{\mathfrak{g}_{\bar{0}}}$ is exact and has both the left adjoint

$$\mathrm{Ind}^{\mathfrak{g}}_{\mathfrak{g}_{\bar{0}}} := U(\mathfrak{g}) \otimes_{U(\mathfrak{g}_{\bar{0}})} - : \mathfrak{g}_{\bar{0}}\text{-smod} \to \mathfrak{g}\text{-smod}$$

and the right adjoint

$$\mathrm{Coind}^{\mathfrak{g}}_{\mathfrak{g}_{\bar{0}}} := \mathrm{Hom}_{U(\mathfrak{g}_{\bar{0}})}(U(\mathfrak{g}), \_) : \mathfrak{g}_{\bar{0}}\text{-smod} \to \mathfrak{g}\text{-smod}.$$

Furthermore, by [18, Theorem 3.2.3] we have

$$\mathrm{Ind}^{\mathfrak{g}}_{\mathfrak{g}_{\bar{0}}} \cong \Pi^{\dim \mathfrak{g}_{\bar{1}}} \circ \mathrm{Coind}^{\mathfrak{g}}_{\mathfrak{g}_{\bar{0}}}, \tag{2}$$

where Π is the functor which changes the parity (see e.g. [18]). If the subcategory C above is isomorphism-closed and stable under tensoring with the $\mathfrak{g}_{\bar{0}}$-module $\bigwedge \mathfrak{g}_{\bar{1}}$, then $\mathrm{Ind}^{\mathfrak{g}}_{\mathfrak{g}_{\bar{0}}}$ maps C to $\tilde{\mathrm{C}}$.

2.3 Weight (super)modules

Fix some Cartan subalgebra $\mathfrak{h}_{\bar{0}}$ in $\mathfrak{g}_{\bar{0}}$. Since the Lie algebra $\mathfrak{g}_{\bar{0}}$ is reductive, the algebra $\mathfrak{h}_{\bar{0}}$ is commutative and contains the (possibly zero) center of $\mathfrak{g}_{\bar{0}}$. A $\mathfrak{g}_{\bar{0}}$-module V is called a *weight* module (with respect to $\mathfrak{h}_{\bar{0}}$) provided that the action of $\mathfrak{h}_{\bar{0}}$ on V is diagonalizable. Put differently, the module V is weight if we have a decomposition

$$V \cong \bigoplus_{\lambda \in \mathfrak{h}_{\bar{0}}^*} V_\lambda, \quad \text{where} \quad V_\lambda := \{v \in V | h \cdot v = \lambda(h)v \text{ for all } h \in \mathfrak{h}_{\bar{0}}\}.$$

The space V_λ is called the *weight space* of V corresponding to a *weight* λ. For a weight module V the *support* of V is the set

$$\mathrm{supp}(V) = \mathrm{supp}_{\mathfrak{h}_{\bar{0}}}(V) := \{\lambda \in \mathfrak{h}_{\bar{0}}^* | V_\lambda \neq 0\}.$$

Denote by $\mathfrak{W}$ the full subcategory in $\mathfrak{g}_{\bar{0}}$-mod consisting of all weight modules. Note that $\mathfrak{W}$ is both isomorphism-closed and closed under the usual tensor product of $\mathfrak{g}_{\overline{0}}$-modules. Furthermore, since $\mathfrak{g}_{\overline{1}}$ is a semi-simple finite dimensional $\mathfrak{g}_{\overline{0}}$-module, we have $\mathfrak{g}_{\overline{1}} \in \mathfrak{W}$ and thus $\bigwedge \mathfrak{g}_{\overline{1}} \in \mathfrak{W}$. This implies that $\mathfrak{W}$ is stable under tensoring with $\bigwedge \mathfrak{g}_{\overline{1}}$.

Now we can consider the corresponding category $\widetilde{\mathfrak{W}}$ of *weight* $\mathfrak{g}$-supermodules and from Sect. 2.2 we obtain that $(\mathrm{Ind}_{\mathfrak{g}_{\overline{0}}}^{\mathfrak{g}}, \mathrm{Res}_{\mathfrak{g}_{\overline{0}}}^{\mathfrak{g}}, \mathrm{Coind}_{\mathfrak{g}_{\overline{0}}}^{\mathfrak{g}})$ restricts to an adjoint triple of functors between $\mathfrak{W}$ and $\widetilde{\mathfrak{W}}$.

Example 4 We have $\mathfrak{g}_{\overline{0}} \in \mathfrak{W}$ and $\mathfrak{g} \in \widetilde{\mathfrak{W}}$, where $\mathfrak{g}_{\overline{0}}$ and $\mathfrak{g}$ are the adjoint module and supermodule, respectively.

2.4 Parabolic and triangular decompositions

This is inspired by [9]. Consider the real vector space $\mathscr{H} := \mathbb{R}\mathrm{supp}(\mathfrak{g}_{\overline{0}})$. The set $R_{\overline{0}} := \mathrm{supp}(\mathfrak{g}_{\overline{0}}) \setminus \{0\}$ is a root system in $\mathscr{H}$ and we let W be the corresponding Weyl group and $(\cdot,\cdot)$ the usual W-invariant inner product on $\mathscr{H}$. For a fixed $\omega \in \mathscr{H}$ we have a *parabolic decomposition*

$$\mathfrak{g}_{\overline{0}} = \mathfrak{n}_{\overline{0}}^{\omega,-} \oplus \mathfrak{l}_{\overline{0}}^{\omega} \oplus \mathfrak{n}_{\overline{0}}^{\omega,+} \tag{3}$$

of $\mathfrak{g}_{\overline{0}}$, where

$$\mathfrak{n}_{\overline{0}}^{\omega,-} = \bigoplus_{\substack{\alpha\in R_{\overline{0}}\\ (\alpha,\omega)<0}} (\mathfrak{g}_{\overline{0}})_{\lambda}, \qquad \mathfrak{l}_{\overline{0}}^{\omega} = \bigoplus_{\substack{\alpha\in R_{\overline{0}}\cup\{0\}\\ (\alpha,\omega)=0}} (\mathfrak{g}_{\overline{0}})_{\lambda}, \qquad \mathfrak{n}_{\overline{0}}^{\omega,+} = \bigoplus_{\substack{\alpha\in R_{\overline{0}}\\ (\alpha,\omega)>0}} (\mathfrak{g}_{\overline{0}})_{\lambda}.$$

The subalgebra $\mathfrak{p}_{\overline{0}}^{\omega} := \mathfrak{l}_{\overline{0}}^{\omega} \oplus \mathfrak{n}_{\overline{0}}^{\omega,+}$ is a parabolic subalgebra of $\mathfrak{g}_{\overline{0}}$, $\mathfrak{n}_{\overline{0}}^{\omega,+}$ is the nilpotent radical of $\mathfrak{p}_{\overline{0}}^{\alpha}$ and $\mathfrak{l}_{\overline{0}}^{\omega}$ is the corresponding Levi subalgebra. In the case $(\alpha,\omega) \neq 0$ for all $\alpha \in R_{\overline{0}}$, we have $\mathfrak{l}_{\overline{0}}^{\omega} = \mathfrak{h}_{\overline{0}}$, moreover, $\mathfrak{b}_{\overline{0}}^{\omega} := \mathfrak{h}_{\overline{0}} \oplus \mathfrak{n}_{\overline{0}}^{\omega,+}$ is a Borel subalgebra of $\mathfrak{g}_{\overline{0}}$ and the decomposition (3) is a *triangular decomposition* of $\mathfrak{g}_{\overline{0}}$ in the sense of [32]. For nonzero $\omega_1, \omega_2 \in \mathscr{H}$ say that ω_1 and ω_2 are equivalent if the parabolic decompositions of $\mathfrak{g}_{\overline{0}}$ corresponding to ω_1 and ω_2 coincide. Then the equivalence classes are exactly the (nonzero) facets of the simplicial cone decomposition of $\mathscr{H}$ as described e.g. in [36, § 1.2].

Consider the derived Lie algebra $\mathfrak{g}_{\overline{0}}' := [\mathfrak{g}_{\overline{0}}, \mathfrak{g}_{\overline{0}}]$ of $\mathfrak{g}_{\overline{0}}$ and let $\mathfrak{h}_{\overline{0}}' := \mathfrak{g}_{\overline{0}}' \cap \mathfrak{h}_{\overline{0}}$. Set $R_{\overline{0}}' := \mathrm{supp}_{\mathfrak{h}_{\overline{0}}'}(\mathfrak{g}_{\overline{0}}) \setminus \{0\}$ and $\mathscr{H}' := \mathbb{R}R_{\overline{0}}'$. Then the set $R_{\overline{0}}'$ is again a root system in $\mathscr{H}'$ (of the same type as $R_{\overline{0}}$). Moreover, there is the obvious canonical isomorphism $R_{\overline{0}}' \cong R_{\overline{0}}$ of root systems which induces an isomorphism $\mathscr{H}' \cong \mathscr{H}$ of vector spaces. Using this isomorphism we identify $\mathscr{H}'$ and $\mathscr{H}$.

Let $R_{\overline{1}}' := \mathrm{supp}_{\mathfrak{h}_{\overline{0}}'}(\mathfrak{g}_{\overline{1}}) \subset \mathscr{H}'$. Then the same $\omega \in \mathscr{H} = \mathscr{H}'$ leads to the decomposition

$$\mathfrak{g}_{\overline{1}} = \mathfrak{n}_{\overline{1}}^{\omega,-} \oplus \mathfrak{l}_{\overline{1}}^{\omega} \oplus \mathfrak{n}_{\overline{1}}^{\omega,+}, \tag{4}$$

where

$$\mathfrak{n}_{\overline{1}}^{\omega,-} = \bigoplus_{\substack{\alpha\in R_{\overline{1}}\\ (\alpha,\omega)<0}} (\mathfrak{g}_{\overline{1}})_{\lambda}, \qquad \mathfrak{l}_{\overline{1}}^{\omega} = \bigoplus_{\substack{\alpha\in R_{\overline{1}}\\ (\alpha,\omega)=0}} (\mathfrak{g}_{\overline{1}})_{\lambda}, \qquad \mathfrak{n}_{\overline{1}}^{\omega,+} = \bigoplus_{\substack{\alpha\in R_{\overline{1}}\\ (\alpha,\omega)>0}} (\mathfrak{g}_{\overline{1}})_{\lambda}.$$

Setting
$$\mathfrak{n}^{\omega,\pm} := \mathfrak{n}_{\overline{0}}^{\omega,\pm} \oplus \mathfrak{n}_{\overline{1}}^{\omega,\pm} \quad \text{and} \quad \mathfrak{l}^{\omega} := \mathfrak{l}_{\overline{0}}^{\omega} \oplus \mathfrak{l}_{\overline{1}}^{\omega},$$

and combining (3) and (4) we obtain the following *parabolic decomposition* of $\mathfrak{g}$ corresponding to ω:
$$\mathfrak{g} := \mathfrak{n}^{\omega,-} \oplus \mathfrak{l}^{\omega} \oplus \mathfrak{n}^{\omega,+}. \tag{5}$$

Here $\mathfrak{p}^{\omega} := \mathfrak{l}^{\omega} \oplus \mathfrak{n}^{\omega,+}$ is a parabolic subalgebra with the "nilpotent radical" $\mathfrak{n}^{\omega,+}$ and the "Levi subalgebra" $\mathfrak{l}^{\omega}$.

Set $R' := R'_{\overline{0}} \cup R'_{\overline{1}}$. If the only $\alpha \in R'$ satisfying $(\alpha, \omega) = 0$ is $\alpha = 0$, then decomposition (5) is called a *triangular decomposition*. An important difference with the Lie algebra case is that even in the case of a triangular decomposition we might have $\mathfrak{l}^{\omega} \neq \mathfrak{h}_{\overline{0}}$.

Example 5 Let $\mathfrak{g} = \mathfrak{q}_n$ for $n > 1$, and $\mathfrak{h}_{\overline{0}}$ be the subalgebra of all matrices of the form (1) for which $B = 0$ and A is diagonal. Choose any ω such that $\mathfrak{n}_{\overline{0}}^{\omega,+}$ consists of all matrices of the form (1) for which $B = 0$ and A is upper triangular. Then $\mathfrak{n}_{\overline{0}}^{\omega,-}$ consists of all matrices of the form (1) for which $B = 0$ and A is lower triangular; $\mathfrak{h}_{\overline{1}}$ consists of all matrices of the form (1) for which $A = 0$ and B is diagonal; $\mathfrak{n}_{\overline{1}}^{\omega,+}$ consists of all matrices of the form (1) for which $A = 0$ and B is upper triangular; $\mathfrak{n}_{\overline{1}}^{\omega,-}$ consists of all matrices of the form (1) for which $A = 0$ and B is lower triangular. In this case the "Cartan subalgebra" $\mathfrak{h} := \mathfrak{l}^{\omega}$ is not commutative.

Equivalence classes of elements from $\mathscr{H}'$ which give rise to the same parabolic decomposition of $\mathfrak{g}$ define a simplicial cone decomposition of $\mathscr{H}'$ which refines the one defined for $\mathfrak{g}_{\overline{0}}$ above.

3 Parabolic category $\widetilde{\mathscr{O}}$ and its elementary properties

3.1 Parabolic categories $\mathscr{O}^{\omega}$ and $\widetilde{\mathscr{O}}^{\omega}$

Fix an ω as above and consider the corresponding parabolic decompositions of $\mathfrak{g}_{\overline{0}}$ and $\mathfrak{g}$, given by (3) and (5), respectively. Denote by $\mathscr{O}^{\omega} = {}^{\mathfrak{g}_{\overline{0}}}\mathscr{O}^{\omega}$ the full subcategory of $\mathfrak{g}_{\overline{0}}$-mod consisting of all modules M which are

- finitely generated,
- decompose into a direct sum of simple finite dimensional $\mathfrak{l}_{\overline{0}}^{\omega}$-modules,
- are locally $\mathfrak{n}_{\overline{0}}^{\omega,+}$-finite in the sense that $\dim(U(\mathfrak{n}_{\overline{0}}^{\omega,+})v) < \infty$ for all $v \in M$.

The category $\mathscr{O}^{\omega}$ is the $\mathfrak{p}_{\overline{0}}^{\omega}$-parabolic version of the BGG category $\mathscr{O}$. The original category $\mathscr{O}$ was defined in [2] (it corresponds to the situation when the decomposition (3) is a triangular decomposition), and the parabolic version was defined in [35]. We also refer to [21] for more details. We will drop the superscript $\mathfrak{g}_{\overline{0}}$ if it is clear from the context.

The category $\mathcal{O}^{\omega}$ is isomorphism-closed and stable under tensoring with simple finite dimensional $\mathfrak{g}_{\overline{0}}$-modules. Hence the corresponding category $\widetilde{\mathcal{O}}^{\omega} = {}^{\mathfrak{g}}\widetilde{\mathcal{O}}^{\omega}$ of $\mathfrak{g}$-modules leads us to the nice situation described at the end of Subsection 2.2 (we will drop the superscript $\mathfrak{g}$ if it is clear from the context). Alternatively, the category $\widetilde{\mathcal{O}}^{\omega}$ can be described as the full subcategory of $\mathfrak{g}$-smod consisting of all supermodules M which are

- finitely generated,
- decompose into a direct sum of simple finite dimensional $\mathfrak{l}^{\omega}_{\overline{0}}$-modules,
- are locally $\mathfrak{n}^{\omega,+}$-finite in the sense that $\dim(U(\mathfrak{n}^{\omega,+})v) < \infty$ for all $v \in M$.

Note that it is really $\mathfrak{l}^{\omega}_{\overline{0}}$ and not $\mathfrak{l}^{\omega}$ in the second condition.

3.2 Elementary categorical properties of $\widetilde{\mathcal{O}}^{\omega}$

Proposition 1

(a) $\widetilde{\mathcal{O}}^{\omega}$ *is a Serre subcategory of* $\widetilde{\mathfrak{W}}$*, in particular,* $\widetilde{\mathcal{O}}^{\omega}$ *is abelian.*
(b) *Every object of* $\widetilde{\mathcal{O}}^{\omega}$ *has finite length as a* $\mathfrak{g}$*-module.*
(c) $\widetilde{\mathcal{O}}^{\omega}$ *has enough projective modules.*
(d) $\widetilde{\mathcal{O}}^{\omega}$ *has enough injective modules.*
(e) *All morphism spaces in* $\widetilde{\mathcal{O}}^{\omega}$ *are finite dimensional.*
(f) *For every i and any* $M,N \in \widetilde{\mathcal{O}}^{\omega}$ *we have* $\dim \mathrm{Ext}^i_{\widetilde{\mathcal{O}}^{\omega}}(M,N) < \infty$.

Proof. Claim (a) follows directly from the definitions. To prove claim (b) we just observe that each $M \in \widetilde{\mathcal{O}}^{\omega}$ is in $\mathcal{O}^{\omega}$, when considered as a $\mathfrak{g}_{\overline{0}}$-module. In particular, it has finite length already as a $\mathfrak{g}_{\overline{0}}$-module (see e.g. [35, Proposition 3.3] or [2, 21]).

Because of claim (b), to prove claims (c) and (d) it is enough to prove that each simple object in $\widetilde{\mathcal{O}}^{\omega}$ has both a projective cover and an injective envelope. We prove the first claim and the second one is proved similarly. Let $L \in \widetilde{\mathcal{O}}^{\omega}$ be simple and let $P \in \mathcal{O}^{\omega}$ be a projective cover of $(\mathrm{Res}^{\mathfrak{g}}_{\mathfrak{g}_{\overline{0}}} L)_{\overline{0}}$ (see e.g. [35, Corollary 4.2] or [2, 21] for existence of projective covers in $\mathcal{O}^{\omega}$), which we may assume to be nonzero up to parity change. Then, by adjunction, we have

$$0 \neq \mathrm{Hom}_{\mathfrak{g}_{\overline{0}}}(P, (\mathrm{Res}^{\mathfrak{g}}_{\mathfrak{g}_{\overline{0}}} L)_{\overline{0}}) = \mathrm{Hom}_{\mathfrak{g}}\big(\mathrm{Ind}^{\mathfrak{g}}_{\mathfrak{g}_{\overline{0}}} P, L\big).$$

As $\mathrm{Ind}^{\mathfrak{g}}_{\mathfrak{g}_{\overline{0}}}$ is left adjoint to the exact functor $\mathrm{Res}^{\mathfrak{g}}_{\mathfrak{g}_{\overline{0}}}$, the former functor maps projective objects to projective objects. Hence $\mathrm{Ind}^{\mathfrak{g}}_{\mathfrak{g}_{\overline{0}}} P$ is a projective object in $\widetilde{\mathcal{O}}^{\omega}$ which surjects onto L.

Claim (e) follows directly from the definition and the fact that all morphism spaces in $\mathcal{O}^{\omega}$ are finite dimensional (again, see e.g. [35, Sects. 3 and 4] or [2, 21]). Claim (f) follows from claims (c) and (e) considering projective resolutions. □

3.3 Simple objects in $\widetilde{\mathcal{O}}^{\omega}$

For each simple finite-dimensional $\mathfrak{l}_{\overline{0}}^{\omega}$-module V we have the corresponding *generalized Verma module*

$$M(\omega,V) := U(\mathfrak{g}_{\overline{0}}) \bigotimes_{U(\mathfrak{l}_{\overline{0}}^{\omega}\oplus\mathfrak{n}_{\overline{0}}^{\omega,+})} V,$$

where $\mathfrak{n}_{\overline{0}}^{\omega,+}V = 0$. This module lies in $\mathcal{O}^{\omega}$ and has a unique simple quotient denoted $L(\omega,V)$. Let $\mathscr{I}_{\overline{0}}^{\omega,0}$ denote the set of isomorphism classes of simple finite-dimensional $\mathfrak{l}_{\overline{0}}^{\omega}$-modules. As $\mathfrak{l}_{\overline{0}}^{\omega}$ is reductive, the set $\mathscr{I}_{\overline{0}}^{\omega,0}$ is well-understood, see e.g. [10,21]. Furthermore, by [35, Proposition 3.3], the set

$$\mathscr{I}_{\overline{0}}^{\omega} := \left\{L(\omega,V)|V \in \mathscr{I}_{\overline{0}}^{\omega,0}\right\}$$

is a full set of representatives of isomorphism classes of simple objects in $\mathcal{O}^{\omega}$ (and in this sense $\mathscr{I}_{\overline{0}}^{\omega,0}$ and $\mathscr{I}_{\overline{0}}^{\omega}$ are canonically identified). Denote by $\mathscr{I}_{\overline{1}}^{\omega}$ an odd copy of $\mathscr{I}_{\overline{0}}^{\omega}$. Now a rough description of simple objects in $\widetilde{\mathcal{O}}^{\omega}$ is given by the following:

Proposition 2 *Let L be a simple object in $\widetilde{\mathcal{O}}^{\omega}$. Then there is $V \in \mathscr{I}_{\overline{0}}^{\omega,0}$ such that L is a quotient of* $\mathrm{Ind}_{\mathfrak{g}_{\overline{0}}}^{\mathfrak{g}} L(\omega,V)$ *up to parity change.*

Proof. As $\mathrm{Res}_{\mathfrak{g}_{\overline{0}}}^{\mathfrak{g}} L \in \mathcal{O}^{\omega}$ and each object in $\mathcal{O}^{\omega}$ has finite length, $\mathrm{Res}_{\mathfrak{g}_{\overline{0}}}^{\mathfrak{g}} L$ has a simple subobject, which is isomorphic to $L(\omega,V)$ for some $V \in \mathscr{I}_{\overline{0}}^{\omega,0}$ by the above (and which we may assume to be even up to parity change). By adjunction, we have

$$0 \neq \mathrm{Hom}_{\mathfrak{g}_{\overline{0}}}\left(L(\omega,V),(\mathrm{Res}_{\mathfrak{g}_{\overline{0}}}^{\mathfrak{g}} L)_{\overline{0}}\right) = \mathrm{Hom}_{\mathfrak{g}}\left(\mathrm{Ind}_{\mathfrak{g}_{\overline{0}}}^{\mathfrak{g}} L(\omega,V),L\right).$$

and the claim follows. □

We denote by $\mathscr{I}^{\omega}$ the set of isomorphism classes of simple objects in $\widetilde{\mathcal{O}}^{\omega}$. Define a binary relation $\Omega \subset \mathscr{I}^{\omega} \times (\mathscr{I}_{\overline{0}}^{\omega} \cup \mathscr{I}_{\overline{1}}^{\omega})$ (here the last union is automatically disjoint) by $(L,L(\omega,V)) \in \Omega$ if $L(\omega,V)$ is isomorphic to a submodule of $\mathrm{Res}_{\mathfrak{g}_{\overline{0}}}^{\mathfrak{g}} L$. Then Ω is *finitary* in the sense that for each $L \in \mathscr{I}^{\omega}$ the set

$$\{L(\omega,V) \in \mathscr{I}_{\overline{0}}^{\omega} \cup \mathscr{I}_{\overline{1}}^{\omega}|(L,L(\omega,V)) \in \Omega\}$$

is non-empty and finite, moreover, for each $L(\omega,V) \in \mathscr{I}_{\overline{0}}^{\omega} \cup \mathscr{I}_{\overline{1}}^{\omega}$ the set

$$\{L \in \mathscr{I}^{\omega}|(L,L(\omega,V)) \in \Omega\}$$

is non-empty and finite. Unfortunately, in the general case Ω is not a function in any direction. Anyway, $\mathscr{I}^{\omega}$ can be considered as a "finite cover" of $\mathscr{I}_{\overline{0}}^{\omega}$ in some sense. Put differently, the set $\mathscr{I}^{\omega}$ is only "finitely more complicated" than the very well-understood set $\mathscr{I}_{\overline{0}}^{\omega}$. An alternative description of $\mathscr{I}^{\omega}$ which uses $\mathfrak{l}^{\omega}$ will be given in Proposition 6.

3.4 Blocks of $\widetilde{\mathcal{O}}^{\omega}$

Let $\sim$ be the minimal equivalence relation on $\mathcal{I}_{\overline{0}}^{\omega}$ which contains all pairs $(L,L') \in \mathcal{I}_{\overline{0}}^{\omega} \times \mathcal{I}_{\overline{0}}^{\omega}$ such that $\mathrm{Ext}^1_{\mathcal{O}^{\omega}}(L,L') \neq 0$. For an equivalence class $\mathcal{X} \in \mathcal{I}_{\overline{0}}^{\omega}/\sim$ let $\mathcal{O}^{\omega}(\mathcal{X})$ denote the Serre subcategory of $\mathcal{O}^{\omega}$ generated by simples in $\mathcal{X}$. Then we have the usual decomposition

$$\mathcal{O}^{\omega} \cong \bigoplus_{\mathcal{X} \in \mathcal{I}_{\overline{0}}^{\omega}/\sim} \mathcal{O}^{\omega}(\mathcal{X})$$

into a direct sum of indecomposable subcategories, called *blocks* of $\mathcal{O}^{\omega}$. Every equivalence class $\mathcal{X}$ is finite as there are only finitely many (up to isomorphism) simple highest weight $\mathfrak{g}_{\overline{0}}$-modules for each given central character, see [10, Chap. 7] for details.

Let $\approx$ be the minimal equivalence relation on $\mathcal{I}^{\omega}$ which contains all pairs $(L,L') \in \mathcal{I}^{\omega} \times \mathcal{I}^{\omega}$ such that $\mathrm{Ext}^1_{\widetilde{\mathcal{O}}^{\omega}}(L,L') \neq 0$. For an equivalence class $\mathcal{X} \in \mathcal{I}^{\omega}/\approx$ let $\widetilde{\mathcal{O}}^{\omega}(\mathcal{X})$ denote the Serre subcategory of $\widetilde{\mathcal{O}}^{\omega}$ generated by simples in $\mathcal{X}$. Then we have the decomposition

$$\widetilde{\mathcal{O}}^{\omega} \cong \bigoplus_{\mathcal{X} \in \mathcal{I}^{\omega}/\approx} \widetilde{\mathcal{O}}^{\omega}(\mathcal{X})$$

into a direct sum of indecomposable subcategories, called *blocks* of $\widetilde{\mathcal{O}}^{\omega}$. For example, an explicit description of blocks for $\mathfrak{g} = \mathfrak{gl}(m|n)$ can be found in [6].

Proposition 3 *Each $\mathcal{X} \in \mathcal{I}^{\omega}/\approx$ is at most countable.*

Proof. Let $L \hookrightarrow N \twoheadrightarrow L'$ be a non-split extension in $\widetilde{\mathcal{O}}^{\omega}(\mathcal{X})$ with L, L' simple. Take some $\lambda \in \mathrm{supp}(N)$ such that $N_\lambda \neq L_\lambda$. Then $N = U(\mathfrak{g})N_\lambda$ and it follows that $\mathrm{supp}(N) \subset \lambda + \mathbb{Z}R'$, in particular, we have both $\mathrm{supp}(L) \subset \lambda + \mathbb{Z}R'$ and $\mathrm{supp}(L') \subset \lambda + \mathbb{Z}R'$. Therefore $\mathrm{supp}(N') \subset \lambda + \mathbb{Z}R'$ for any $N' \in \widetilde{\mathcal{O}}^{\omega}(\mathcal{X})$. Note that $\lambda + \mathbb{Z}R'$ is an at most countable set.

Simples in $\mathcal{O}^{\omega}$ are classified by their highest weight (see [35, Proposition 3.3]), in particular, there are only at most countably many simple objects in $\mathcal{O}^{\omega}$ with support in $\lambda + \mathbb{Z}R'$. Now each simple in $\widetilde{\mathcal{O}}^{\omega}(\mathcal{X})$ has, as a $\mathfrak{g}_{\overline{0}}$-module and up to parity change, some simple submodule $L(\omega, V)$ from $\mathcal{O}^{\omega}$ with support in $\lambda + \mathbb{Z}R'$ and hence is a quotient of $\mathrm{Ind}_{\mathfrak{g}_{\overline{0}}}^{\mathfrak{g}} L(\omega, V)$ (see Proposition 2). The module $\mathrm{Ind}_{\mathfrak{g}_{\overline{0}}}^{\mathfrak{g}} L(\omega, V)$ has finite length by Proposition 1(b). Since we have only at most countably many $L(\omega, V)$ to start with, the claim follows. □

For each $L \in \mathcal{I}^{\omega}$ fix an indecomposable projective cover $P(L)$ of L (which exists by Proposition 1(c)). For $\mathcal{X} \in \mathcal{I}^{\omega}/\approx$ let $\widetilde{\mathcal{P}}^{\omega}(\mathcal{X})$ denote the full subcategory of $\widetilde{\mathcal{O}}^{\omega}$ with objects $P(L)$, $L \in \mathcal{X}$. From Proposition 1 it follows that the $\mathbb{C}$-linear category $\widetilde{\mathcal{P}}^{\omega}(\mathcal{X})$ has the following properties:

- for each $L \in \mathscr{X}$ we have $\mathrm{Hom}_{\widetilde{\mathscr{P}}^{\omega}(\mathscr{X})}(P(L), P(L')) \neq 0$ for at most finitely many $L' \in \mathscr{X}$;
- for each $L \in \mathscr{X}$ we have $\mathrm{Hom}_{\widetilde{\mathscr{P}}^{\omega}(\mathscr{X})}(P(L'), P(L)) \neq 0$ for at most finitely many $L' \in \mathscr{X}$.

We also set $\widetilde{\mathscr{P}}^{\omega} = \bigcup_{\mathscr{X} \in \mathscr{I}^{\omega}/\approx} \widetilde{\mathscr{P}}^{\omega}(\mathscr{X})$.

Let $\widetilde{\mathscr{P}}^{\omega}(\mathscr{X})^{\mathrm{op}}$ be the category, opposite to $\widetilde{\mathscr{P}}^{\omega}(\mathscr{X})$. Consider the category $\widetilde{\mathscr{P}}^{\omega}(\mathscr{X})^{\mathrm{op}}$-fmod of finite dimensional $\widetilde{\mathscr{P}}^{\omega}(\mathscr{X})^{\mathrm{op}}$-modules, that is the category of $\mathbb{C}$-linear functors

$$\mathrm{F} : \widetilde{\mathscr{P}}^{\omega}(\mathscr{X})^{\mathrm{op}} \to \mathbb{C}\text{-mod}$$

satisfying the condition $\sum_{L \in \mathscr{X}} \dim(\mathrm{F}(P(L))) < \infty$. Now from the standard abstract nonsense (see e.g. [17]), we have:

Proposition 4 *For $\mathscr{X} \in \mathscr{I}^{\omega}/\approx$ the categories $\widetilde{\mathscr{O}}^{\omega}(\mathscr{X})$ and $\widetilde{\mathscr{P}}^{\omega}(\mathscr{X})^{\mathrm{op}}$-fmod are equivalent.*

3.5 Duality

The category $\mathscr{O}^{\omega}$ has the standard *simple preserving duality* $\star$, that is a contravariant anti-equivalence which preserves isomorphism classes of simple objects (see [21, Sect. 3.2] for details). This duality lifts to $\widetilde{\mathscr{O}}^{\omega}$ in the obvious way (and will also be denoted by $\star$), however, because of (2), simple modules are preserved by the lifted duality $\star$ only up to a possible parity change. We note also that some simple modules in $\widetilde{\mathscr{O}}^{\omega}$ might be stable under Π (that is, isomorphic, in $\widetilde{\mathscr{O}}^{\omega}$, to their parity changed counterparts). We will not need any explicit criterion for when $\star$ preserves simples strictly or only up to parity change, we refer the reader to [13] for the $\mathfrak{q}_n$-example.

4 Stratification

4.1 Standard and proper standard objects

The superalgebra $\mathfrak{l}^{\omega}$ is a classical Lie superalgebra in the sense of Sect. 2.1. The category ${}^{\mathfrak{l}^{\omega}_{\bar{0}}}\mathscr{O}^{0}$ is just the category of semi-simple finite dimensional $\mathfrak{l}^{\omega}_{\bar{0}}$-modules. Consider now the category ${}^{\mathfrak{l}^{\omega}}\widetilde{\mathscr{O}}^{0}$ which has all the properties described in Proposition 1. Furthermore, we have:

Proposition 5 *Projective and injective modules in ${}^{\mathfrak{l}^{\omega}}\widetilde{\mathscr{O}}^{0}$ coincide.*

Proof. It is enough to show that the indecomposable projective cover of each simple object in ${}^{\mathfrak{l}^{\omega}}\widetilde{\mathscr{O}}^{0}$ is injective and that the indecomposable injective envelope of each simple object in ${}^{\mathfrak{l}^{\omega}}\widetilde{\mathscr{O}}^{0}$ is projective. We will prove the first claim an the second is proved similarly. From the proof of Proposition 1(c) it follows that the indecomposable projective cover of each simple object in ${}^{\mathfrak{l}^{\omega}}\widetilde{\mathscr{O}}^{0}$ is a direct summand of a module of the form $\mathrm{Ind}_{\mathfrak{g}_{\overline{0}}}^{\mathfrak{g}} P$, where P is projective in ${}^{\mathfrak{l}^{\omega}_{\overline{0}}}\mathcal{O}^0$. The latter category is semi-simple and hence P is also injective. Now (2) implies that, up to parity change, the module $\mathrm{Ind}_{\mathfrak{g}_{\overline{0}}}^{\mathfrak{g}} P$ is isomorphic to $\mathrm{Coind}_{\mathfrak{g}_{\overline{0}}}^{\mathfrak{g}} P$ where P is injective. Then the module $\mathrm{Coind}_{\mathfrak{g}_{\overline{0}}}^{\mathfrak{g}} P$ is injective as $\mathrm{Coind}_{\mathfrak{g}_{\overline{0}}}^{\mathfrak{g}}$ is right adjoint to an exact functor (and thus sends injective modules to injective modules). The claim follows. □

Denote by ${}^{\mathfrak{l}}\mathscr{I}^{\omega}$ the set of isomorphism classes of simple objects in ${}^{\mathfrak{l}^{\omega}}\widetilde{\mathcal{O}}^0$. For $V \in {}^{\mathfrak{l}}\mathscr{I}^{\omega}$ denote by $\hat{V}$ the indecomposable projective cover of V in ${}^{\mathfrak{l}^{\omega}}\widetilde{\mathcal{O}}^0$ (note that the module $\hat{V}$ is also injective by Proposition 5 but it does not have to coincide with the indecomposable injective envelope of V). Set $\mathfrak{n}^{\omega,+}V = \mathfrak{n}^{\omega,+}\hat{V} = 0$. Define the *proper standard* or *generalized Verma* $\mathfrak{g}$-module

$$\overline{\Delta}(V) := U(\mathfrak{g}) \bigotimes_{U(\mathfrak{l}^{\omega}\oplus\mathfrak{n}^{\omega,+})} V$$

and the *standard* $\mathfrak{g}$-module

$$\Delta(V) := U(\mathfrak{g}) \bigotimes_{U(\mathfrak{l}^{\omega}\oplus\mathfrak{n}^{\omega,+})} \hat{V}.$$

Since the parabolic induction from $\mathfrak{l}^{\omega} \oplus \mathfrak{n}^{\omega,+}$ to $\mathfrak{g}$ is exact, Proposition 1(b) implies that each standard module has a finite filtration whose subquotients are proper standard modules (these subquotients do **not** have to be isomorphic one to the other).

Proposition 6 *Let $V \in {}^{\mathfrak{l}}\mathscr{I}^{\omega}$.*

(a) *The module $\overline{\Delta}(V)$ has simple top denoted by $L(V)$.*
(b) *The module $L(V)$ is also the simple top of $\Delta(V)$.*
(c) *The set $\{L(V)|V \in {}^{\mathfrak{l}}\mathscr{I}^{\omega}\}$ is a full set of representatives of isomorphism classes of simple objects in $\widetilde{\mathcal{O}}^{\omega}$.*

Proof. For $\lambda,\mu \in \mathfrak{h}_{\overline{0}}^*$ write $\lambda \leq_{\omega} \mu$ if and only if $\mu - \lambda \in \mathbb{Z}_+\mathrm{supp}(\mathfrak{l}^{\omega} \oplus \mathfrak{n}^{\omega,+})$. Now for $\lambda \in \mathrm{supp}(V)$ the unique maximal submodule of $\overline{\Delta}(V)$ is the sum of all submodules M of $\overline{\Delta}(V)$ which satisfy the following condition: $\mu \in \mathrm{supp}(M)$ implies $\mu <_{\omega} \lambda$. This implies claim (a) and claim (c) follows from the definition of $\widetilde{\mathcal{O}}^{\omega}$ and the universal property of induced modules. Claim (b) follows from claim (a) and definitions. We refer the reader to [10, Chap. 7] for similar properties of the usual Verma modules written with all details. □

Proposition 6(c) allows us to canonically identify ${}^{\mathfrak{l}}\mathscr{I}^{\omega}$ and $\mathscr{I}^{\omega}$. From the proof of Proposition 6(a) it follows that the simple top $L(V)$ has composition multiplicity one in $\overline{\Delta}(V)$.

4.2 Stratified structure

Theorem 2 *Each projective module in $\widetilde{\mathcal{O}}^{\omega}$ has a* standard filtration, *that is a filtration whose subquotients are isomorphic to standard modules.*

Proof. This claim is proved similarly to e.g. [16, Proposition 3] or [13, Theorem 12]. As mentioned above, each projective object in $\widetilde{\mathcal{O}}^{\omega}$ is a direct summand of a module of the form $\mathrm{Ind}_{\mathfrak{g}_{\bar{0}}}^{\mathfrak{g}} P$ where P is projective in $\mathcal{O}^{\omega}$. Similarly to [2] one shows that existence of a standard filtration is and additive property, that is inherited by all direct summands. Each projective in $\mathcal{O}^{\omega}$ has a Verma filtration, that is a filtration whose subquotients are isomorphic to Verma modules. Hence it is enough to show that each module of the form $\mathrm{Ind}_{\mathfrak{g}_{\bar{0}}}^{\mathfrak{g}} M(\lambda)$, where $M(\lambda)$ is a usual Verma module, has a standard filtration. The induction from $\mathfrak{g}_{\bar{0}}$ to $\mathfrak{g}$ can be factorized via $\mathfrak{l}^{\omega}$. From Proposition 5 it thus follows that the module $\mathrm{Ind}_{\mathfrak{g}_{\bar{0}}}^{\mathfrak{g}} M(\lambda)$, when considered as an $\mathfrak{l}^{\omega}$-module, is a direct sum of projective-injective modules in ${}^{\mathfrak{l}^{\omega}}\widetilde{\mathcal{O}}^{0}$. Moreover, by construction, the module $\mathrm{Ind}_{\mathfrak{g}_{\bar{0}}}^{\mathfrak{g}} M(\lambda)$ is free of finite rank over $U(\mathfrak{n}^{\omega,-})$. Take an $\mathfrak{l}^{\omega}$-direct summand N of $\mathrm{Ind}_{\mathfrak{g}_{\bar{0}}}^{\mathfrak{g}} M(\lambda)$ of maximal possible weight (with respect to the order $\leq_{\omega}$ introduced in the previous subsection). From the universal property of induced modules and the fact that $\mathrm{Ind}_{\mathfrak{g}_{\bar{0}}}^{\mathfrak{g}} M(\lambda)$ is free over $U(\mathfrak{n}^{\omega,-})$ it follows that $U(\mathfrak{g})N$ is a direct sum of standard modules. Furthermore, $U(\mathfrak{g})N$, when considered as an $\mathfrak{l}^{\omega}$-module, is a direct sum of projective-injective objects in ${}^{\mathfrak{l}^{\omega}}\widetilde{\mathcal{O}}^{0}$. This implies that $U(\mathfrak{g})N$ is a direct summand of $\mathrm{Ind}_{\mathfrak{g}_{\bar{0}}}^{\mathfrak{g}} M(\lambda)$ as an $\mathfrak{l}^{\omega}$-module. Now the proof is completed by induction with respect to $\leq_{\omega}$. □

Combination of Proposition 6 and Theorem 2 means that each $\widetilde{\mathcal{P}}^{\omega}(\mathcal{X})$ is weakly properly stratified in the sense of [11] (in particular, it is standardly stratified in the sense of [8]).

Using $\star$ we define *proper costandard* modules as $\star$-duals of proper standard modules. We also define *costandard* modules as $\star$-duals of standard modules. Then every costandard module has a *proper costandard filtration*, that is a filtration whose subquotients are isomorphic to proper costandard modules. The $\star$-dual of Theorem 2 says that each injective module in $\widetilde{\mathcal{O}}^{\omega}$ has a *costandard filtration*, that is a filtration whose subquotients are isomorphic to costandard modules.

The $\star$-dual of Proposition 6 says that all costandard and proper costandard modules have simple socle. For $V \in {}^{\mathfrak{l}}\mathcal{J}^{\omega}$ we denote by $\nabla(V)$ and $\overline{\nabla}(V)$ the costandard and proper costandard modules with simple socle $L(V)$, respectively. We denote by $\mathcal{F}(\Delta)$ the full subcategory of $\widetilde{\mathcal{O}}^{\omega}$ consisting of all modules having a standard filtration and define $\mathcal{F}(\overline{\Delta})$, $\mathcal{F}(\nabla)$ and $\mathcal{F}(\overline{\nabla})$ similarly.

By standard arguments (see e.g. [11, 12]), the fact that $\widetilde{\mathcal{P}}^{\omega}(\mathcal{X})$ is weakly properly stratified is equivalent to the following homological orthogonality:

Corollary 1 *For $V, V' \in {}^{\mathfrak{l}}\mathcal{J}^{\omega}$ we have*

$$\mathrm{Ext}^{i}_{\mathcal{O}}(\Delta(V), \overline{\nabla}(V')) \cong \begin{cases} \mathbb{C}, & \text{if } V \cong V' \text{ and } i = 0; \\ 0, & \text{otherwise.} \end{cases}$$

Using $\star$ one obtains a similar homological orthogonality between proper standard and costandard modules. As a consequence of this, for $N \in \mathcal{F}(\Delta)$ and $V \in {}^{\mathfrak{l}}\mathcal{J}^{\omega}$ the number of occurrences of $\Delta(V)$ as a subquotient of a standard filtration of N does not depend on the choice of the filtration and will be denoted by $(N : \Delta(V))$.

Weakly proper stratification of $\widetilde{\mathscr{P}}^{\omega}(\mathscr{X})$ also implies the following standard characterization of modules with (proper) (co)standard filtration (see [11,34]):

Corollary 2

(a) $\mathscr{F}(\Delta) = \{M \in \widetilde{\mathscr{O}}^{\omega} | \mathrm{Ext}^i_{\widetilde{\mathscr{O}}^{\omega}}(M, \overline{\nabla}(V)) = 0$ *for any* $V \in {}^{\mathfrak{l}}\mathscr{I}^{\omega}$ *and* $i > 0\}$.
(b) $\mathscr{F}(\nabla) = \{M \in \widetilde{\mathscr{O}}^{\omega} | \mathrm{Ext}^i_{\widetilde{\mathscr{O}}^{\omega}}(\overline{\Delta}(V), M) = 0$ *for any* $V \in {}^{\mathfrak{l}}\mathscr{I}^{\omega}$ *and* $i > 0\}$.
(c) $\mathscr{F}(\overline{\Delta}) = \{M \in \widetilde{\mathscr{O}}^{\omega} | \mathrm{Ext}^i_{\widetilde{\mathscr{O}}^{\omega}}(M, \nabla(V)) = 0$ *for any* $V \in {}^{\mathfrak{l}}\mathscr{I}^{\omega}$ *and* $i > 0\}$.
(d) $\mathscr{F}(\overline{\nabla}) = \{M \in \widetilde{\mathscr{O}}^{\omega} | \mathrm{Ext}^i_{\widetilde{\mathscr{O}}^{\omega}}(\Delta(V), M) = 0$ *for any* $V \in {}^{\mathfrak{l}}\mathscr{I}^{\omega}$ *and* $i > 0\}$.

For a simple L we denote by $[N : L]$ the composition multiplicity of L in N. Another standard corollary is the following *BGG-reciprocity*:

Corollary 3 *For* $V, V' \in {}^{\mathfrak{l}}\mathscr{I}^{\omega}$ *we have*

$$(P(V) : \Delta(V')) = [\overline{\nabla}(V') : L(V)].$$

For example, description of the stratified structure of the category $\mathscr{O}$ for the queer Lie superalgebra $\mathfrak{q}_n$ can be found with all details in [13].

4.3 Tilting modules

An object in $\widetilde{\mathscr{O}}^{\omega}$ is called a *tilting* or *cotilting* module if it belongs to $\mathscr{F}(\Delta) \cap \mathscr{F}(\overline{\nabla})$ or $\mathscr{F}(\nabla) \cap \mathscr{F}(\overline{\Delta})$, respectively. We have the following standard description of (co)tilting modules (see [11,27,31,34]):

Proposition 7

(a) *Each (co)tilting module is a direct sum of indecomposable (co)tilting modules.*
(b) *For every* $V \in {}^{\mathfrak{l}}\mathscr{I}^{\omega}$ *there is a unique (up to isomorphism) indecomposable tilting module* $T(V)$ *such that* $\Delta(V) \hookrightarrow T(V)$ *and the cokernel of this embedding has a standard filtration.*
(c) $T(V)$ *is also cotilting.*

Proof. From Corollary 2 it follows that all categories $\mathscr{F}(\Delta)$, $\mathscr{F}(\overline{\nabla})$, $\mathscr{F}(\nabla)$ and $\mathscr{F}(\overline{\Delta})$ are fully additive. This implies claim (a). Uniqueness of $T(V)$ follows from Corollary 1 by standard arguments, e.g. as in [34]. To prove existence, recall that it is well-known, see e.g. [21], that $\mathscr{O}^{\omega}$ has tilting modules, that is $\star$-self-dual modules with (generalized) Verma flag. Inducing these up to $\mathfrak{g}$ gives $\star$-self-dual (up to parity change) modules with standard filtration. Now existence of $T(V)$ follows by tracking the highest weight (with respect to $\leq_{\omega}$), which proves claim (b). Furthermore, by construction, all these induced modules belong to the category $\mathscr{F}(\Delta) \cap \mathscr{F}(\overline{\nabla}) \cap \mathscr{F}(\nabla) \cap \mathscr{F}(\overline{\Delta})$, which proves claim (c). □

The standard useful property of tilting modules (see [11,34]) is that every module with standard filtration has a finite coresolution by tilting modules. Furthermore, every module with a proper costandard filtration has a (possibly infinite) resolution by tilting modules.

For each $V \in {}^{\mathfrak{l}}\mathscr{I}^{\omega}$ we fix some $T(V)$ as given by Proposition 7(b). Denote by $\widetilde{\mathscr{T}^{\omega}}$ the full subcategory of $\widetilde{\mathscr{O}}^{\omega}$ with objects $T(V)$, $V \in {}^{\mathfrak{l}}\mathscr{I}^{\omega}$. Similarly, for $\mathscr{X} \in {}^{\mathfrak{l}}\mathscr{I}^{\omega}/\approx$ we denote by $\widetilde{\mathscr{T}^{\omega}}(\mathscr{X})$ the full subcategory of $\widetilde{\mathscr{O}}^{\omega}$ with objects $T(V)$, $V \in \mathscr{X}$.

4.4 Finitistic dimension

As an application of tilting module in the subsection we obtain a bound for the finitistic dimension of $\widetilde{\mathscr{O}}^{\omega}$ which, in particular, implies Theorem 1. Let $\mathscr{C}$ be an abelian category with enough projectives. The *global* dimension gl.dim$(\mathscr{C})$ is defined as the supremum of projective dimensions p.dim(X) taken over all objects $X \in \mathscr{C}$. The *finitistic* dimension fin.dim$(\mathscr{C})$ is defined as the supremum of p.dim(X) taken over all objects in $X \in \mathscr{C}$ for which p.dim$(X) < \infty$. It is known that the global dimension of the category $\mathscr{O}^{\omega}$ is finite and hence coincides with the finitistic dimension of $\mathscr{O}^{\omega}$ (finiteness follows from [35] and [37] and explicit bounds and different interpretations can be found in [14,24,26,28,30]). For $\widetilde{\mathscr{O}}^{\omega}$ we have:

Theorem 3

$$\text{fin.dim}(\widetilde{\mathscr{O}}^{\omega}) = 2 \cdot \max_{V \in {}^{\mathfrak{l}}\mathscr{I}^{\omega}} \text{p.dim}(T(V)) \leq \text{gl.dim}(\mathscr{O}^{\omega}).$$

Proof. Let us prove the inequality first. We start with the claim that all injective modules in $\widetilde{\mathscr{O}}^{\omega}$ have finite projective dimension. Indeed, this is obviously true for injectives in $\mathscr{O}^{\omega}$. Given a finite projective resolution of an injective I in $\mathscr{O}^{\omega}$, we can induce this resolution up to $\widetilde{\mathscr{O}}^{\omega}$ and obtain a finite projective resolution of $\text{Ind}_{\mathfrak{g}_{\bar{0}}}^{\mathfrak{g}} I$ in $\widetilde{\mathscr{O}}^{\omega}$. As any injective in $\widetilde{\mathscr{O}}^{\omega}$ is a direct summand of some $\text{Ind}_{\mathfrak{g}_{\bar{0}}}^{\mathfrak{g}} I$, we have our claim. Moreover, as a bonus we even have that $\max\limits_{V \in {}^{\mathfrak{l}}\mathscr{I}^{\omega}} \text{p.dim}(I(V)) \leq \text{gl.dim}(\mathscr{O}^{\omega})$.

Similarly one shows that all projective modules in $\widetilde{\mathscr{O}}^{\omega}$ have finite injective dimension. Now we claim that every $M \in \widetilde{\mathscr{O}}^{\omega}$ has finite projective dimension if and only if it has finite injective dimension. By symmetry, it is enough to prove the "if" statement. Let

$$0 \to I_0 \to I_1 \to \cdots \to I_k \to 0$$

be an injective resolution of M. Each I_s, $s = 0, \ldots, k$, has a finite projective resolution by the above. Substituting each I_s by its projective resolution (using the iterated cone construction, see e.g. [30]), we get a finite complex of projective modules with unique non-zero homology M concentrated in position 0. Deleting all trivial direct summands we obtain a finite projective resolution of M and hence M has finite projective dimension.

Next we claim that fin.dim$(\widetilde{\mathscr{O}}^{\omega}) = \max\limits_{V \in {}^{\mathfrak{l}}\mathscr{I}^{\omega}} \text{p.dim}(I(V))$. Note that the right hand side is bounded by gl.dim$(\mathscr{O}^{\omega})$ by the above. Set $N := \max\limits_{V \in {}^{\mathfrak{l}}\mathscr{I}^{\omega}} \text{p.dim}(I(V))$. Assume that $X \in \widetilde{\mathscr{O}}^{\omega}$ is such that p.dim$(X) > N$. Consider a short exact sequence

$$X \hookrightarrow I \twoheadrightarrow Y \tag{6}$$

Then, because of the dimension shift in the long exact sequence obtained by applying to (6) the functor $\mathrm{Hom}(\_,L(V))$, $V \in {}^{\mathfrak{l}}\mathscr{I}^{\omega}$, we get $\mathrm{p.dim}(Y) > \mathrm{p.dim}(X)$. At the same time, X has finite injective dimension by the above and hence the injective dimension of Y is strictly smaller then the injective dimension of X. Proceeding inductively we get an injective module of projective dimension greater than N, a contradiction. This completes the proof of the inequality $\mathrm{fin.dim}(\widetilde{\mathscr{O}}^{\omega}) \leq \mathrm{gl.dim}(\mathscr{O}^{\omega})$.

Now let us prove the equality

$$\max_{V\in{}^{\mathfrak{l}}\mathscr{I}^{\omega}} \mathrm{p.dim}(I(V)) = 2\cdot \max_{V\in{}^{\mathfrak{l}}\mathscr{I}^{\omega}} \mathrm{p.dim}(T(V)). \tag{7}$$

First we claim that the right hand side of (7) is finite. Indeed, it is finite in the case of $\mathscr{O}^{\omega}$. Having a projective resolution of a tilting module in $\mathscr{O}^{\omega}$, we can induce this resolution up to $\widetilde{\mathscr{O}}^{\omega}$ and get a projective resolution of the induced tilting module. Since, by the highest weight argument, each tilting module is a direct summand of an induced tilting module, we have our claim. Now the proof is completed as in [30, Theorem 1]. □

We note one important difference between $\widetilde{\mathscr{O}}^{\omega}$ and $\mathscr{O}^{\omega}$, namely the fact addressed in Sect. 3.4 that blocks of $\mathscr{O}^{\omega}$ are described by finite dimensional associative algebras while blocks of $\widetilde{\mathscr{O}}^{\omega}$ are described, in general, only by infinite dimensional associative algebras with local units. This fact makes Theorem 3 non-trivial and, to some extend, surprising.

5 Projective-injective modules and their applications

5.1 Irving-type theorems

Category $\mathscr{O}^{\omega}$ has a lot of projective-injective modules with remarkable properties, see [22] and [37]. These extend to $\widetilde{\mathscr{O}}^{\omega}$ as follows. For $V \in {}^{\mathfrak{l}}\mathscr{I}^{\omega}$ let us denote by $I(V)$ the indecomposable injective envelope of $L(V)$ in $\widetilde{\mathscr{O}}^{\omega}$.

Theorem 4 *Let $V \in {}^{\mathfrak{l}}\mathscr{I}^{\omega}$. Then the following assertions are equivalent:*

(i) *$P(V)$ is injective.*
(ii) *$P(V)$ is isomorphic to $I(V)$ up to parity change.*
(iii) *$L(V)$ occurs in the socle of a projective-injective module in $\widetilde{\mathscr{O}}$.*
(iv) *$L(V)$ occurs in the top of a projective-injective module in $\widetilde{\mathscr{O}}$.*
(v) *$L(V)$ occurs in the socle of some standard module.*
(vi) *$L(V)$ occurs in the socle of some proper standard module.*

Proof. Equivalence of claims (4) and (4) follows by applying $\star$. Equivalence of claims (4) and (4) follows from Proposition 5 and the fact that every standard module has a proper standard filtration. Claim (4) obviously implies claim (4). Each projective-injective module has a standard filtration and hence claim (4) implies claim (4). That claim (4) implies claim (4) is obvious and the reverse application would follow from the fact that claim (4) implies claim (4). It is left to prove that claim (4) implies claim (4).

Assume claim (4). The module $\Delta(V)$, when restricted to $\mathfrak{g}_{\overline{0}}$, has a Verma flag. Therefore, by the main result of [22], we have $\mathrm{Res}^{\mathfrak{g}}_{\mathfrak{g}_{\overline{0}}}\Delta(V) \hookrightarrow I$, where I is projective-injective in $\mathcal{O}^{\omega}$. Adjunction gives a non-zero map $\Delta(V) \to \mathrm{Ind}^{\mathfrak{g}}_{\mathfrak{g}_{\overline{0}}} I$. This map is injective as it is non-zero when restricted to the socle (which is not annihilated by the induction). The module $\mathrm{Ind}^{\mathfrak{g}}_{\mathfrak{g}_{\overline{0}}} I$ is projective since I is projective and $\mathrm{Ind}^{\mathfrak{g}}_{\mathfrak{g}_{\overline{0}}}$ is left adjoint to an exact functor. The module $\mathrm{Ind}^{\mathfrak{g}}_{\mathfrak{g}_{\overline{0}}} I$ is injective since I is injective and $\mathrm{Ind}^{\mathfrak{g}}_{\mathfrak{g}_{\overline{0}}}$ is right adjoint to an exact functor by (2). This proves claim (4). Note that every projective module is tilting and hence self-dual with respect to $\star$. Now claim (4) follows applying $\star$. □

5.2 Dominance dimension

The next application of projective modules is the dominance dimension property, described for $\mathcal{O}^{\omega}$ in [25, 38].

Proposition 8 *Every projective P in $\widetilde{\mathcal{O}}^{\omega}$ admits a two step coresolution*

$$0 \to P \to X_1 \to X_2,$$

where both X_1 and X_2 are projective-injective.

Proof. This property is obviously additive. From the proof of Proposition 1 it follows that every projective in $\widetilde{\mathcal{O}}^{\omega}$ is a direct summand of a module induced from a projective module in $\mathcal{O}^{\omega}$. Induction is exact and preserves both projective and injective modules (the latter because of (2)). Hence the claim follows from the corresponding property of $\mathcal{O}^{\omega}$, see [25, 38]. □

5.3 Soergel's Struktursatz

Denote by $\widetilde{\mathcal{Q}}^{\omega}$ the full subcategory of $\widetilde{\mathcal{P}}^{\omega}$ whose objects are both projective and injective in $\widetilde{\mathcal{O}}^{\omega}$. As a corollary from Proposition 8, we have the following analogue of Soergel's *Struktursatz*, see [37], for $\widetilde{\mathcal{O}}^{\omega}$.

Theorem 5 *The bifunctor*

$$\Phi := \mathrm{Hom}_{\widetilde{\mathcal{P}}^{\omega}}(-,-) : (\widetilde{\mathcal{P}}^{\omega})^{\mathrm{op}} \times \widetilde{\mathcal{Q}}^{\omega} \to \mathbb{C}\text{-mod}$$

induces a functor $\tilde{\Phi} : (\widetilde{\mathcal{P}}^{\omega})^{\mathrm{op}} \to \widetilde{\mathcal{Q}}^{\omega}$-mod and the latter functor is full and faithful.

Proof. Mutatis mutandis the proof of [1, Theorem 4.4]. □

5.4 Ringel self-duality

Our final application of the above is the following statement about Ringel self-duality of $\widetilde{\mathcal{O}}^{\omega}$:

Theorem 6 *For any $\mathscr{X} \in {}^{\mathfrak{l}}\mathscr{I}^{\omega}/\approx$ the categories $\widetilde{\mathscr{T}}^{\omega}(\mathscr{X})$ and $\widetilde{\mathscr{P}}^{\omega}(\mathscr{X})$ are canonically isomorphic.*

Proof. For simplicity, we prove the claim in the case when ω is such that the decomposition (3) is a triangular decomposition. The general case can be dealt with using e.g. the approach of [29, Sect. 10.4].

Simples in the category $\mathscr{O}^{\omega}$ which occur as socles of projective-injective modules are exactly the simple objects of maximal Gelfand-Kirillov dimension (this follows from [22, Proposition 4.3] and [23, Chap. 8]). Furthermore, for ω as described above simple modules in $\mathscr{O}^{\omega}$ of maximal Gelfand-Kirillov dimension are tilting. Recall that the functor $\mathrm{Ind}_{\mathfrak{g}_{\overline{0}}}^{\mathfrak{g}}$ is given by a tensor product (over $\mathbb{C}$) with a finite dimensional vector space and hence preserves Gelfand-Kirillov dimension. Therefore, applying $\mathrm{Ind}_{\mathfrak{g}_{\overline{0}}}^{\mathfrak{g}}$ and taking into account that it maps projective-injective modules to projective-injective modules, we get that simples which occur as socles of projective-injective modules $\widetilde{\mathscr{O}}^{\omega}$ are again exactly the simple objects of maximal Gelfand-Kirillov dimension.

Given two modules M and N, the *trace* of M in N is the sum of images of all homomorphisms from M to N. For a projective module $P \in \mathscr{O}^{\omega}$ denote by P' the trace in P of all projective injective modules in $\mathscr{O}^{\omega}$. As mentioned in the previous paragraph, for our choice of ω the socle of a dominant Verma module in $\mathscr{O}^{\omega}$ (which is also projective) is a tilting module. Using translation functors we obtain that P' is a tilting module for any projective P. Applying $\mathrm{Ind}_{\mathfrak{g}_{\overline{0}}}^{\mathfrak{g}}$ and using the previous paragraph, we get the same property for $\widetilde{\mathscr{O}}^{\omega}$.

For any $P_1, P_2 \in \widetilde{\mathscr{P}}^{\omega}(\mathscr{X})$ any homomorphism from P_1 to P_2 restricts to a homomorphism from P_1' to P_2'. From Theorem 5 it follows that this restriction is, in fact, an isomorphism. This implies that P_1' is an indecomposable tilting module and that $P_1' \cong P_2'$ if and only if $P_1 \cong P_2$. Taking into account the previous paragraph, renaming P into P' defines an isomorphism from $\widetilde{\mathscr{P}}^{\omega}(\mathscr{X})$ to $\widetilde{\mathscr{T}}^{\omega}(\mathscr{X})$. This completes the proof. □

Acknowledgements The research is partially supported by the Swedish Research Council and the Royal Swedish Academy of Sciences.

References

1. Andersen, H.H., Mazorchuk, V.: Category $\mathscr{O}$ for quantum groups. Preprint arxiv:1105.5500, to appear in JEMS.
2. Bernstein, I., Gelfand, I., Gelfand, S.: A certain category of $\mathfrak{g}$-modules. Funkcional. Anal. i Prilozen. **10**(2), 1–8 (1976)
3. Brundan, J.: Kazhdan-Lusztig polynomials and character formulae for the Lie superalgebra $\mathfrak{gl}(m|n)$. J. Amer. Math. Soc. **16**(1), 185–231 (2003)
4. Brundan, J.: Kazhdan-Lusztig polynomials and character formulae for the Lie superalgebra $\mathfrak{q}(n)$. Adv. Math. **182**(1), 28–77 (2004)
5. Cheng, S.-J., Lam, N., Wang, W.: Brundan-Kazhdan-Lusztig conjecture for general linear Lie superalgebras. Preprint arXiv:1203.0092

6. Cheng, S.-J., Mazorchuk, V., Wang, W.: Equivalence of blocks for the general linear Lie superalgebra. Preprint arXiv:1301.1204
7. Cheng, S.-J., Wang, W.: Dualities and Representations of Lie Superalgebras. Graduate Studies in Mathematics, Vol. 144. American Mathematical Society (2012)
8. Cline, E., Parshall, B., Scott, L.: Stratifying endomorphism algebras. Mem. Amer. Math. Soc. **124**(591) (1996)
9. Dimitrov, I., Mathieu, O., Penkov, I.: On the structure of weight modules. Trans. Amer. Math. Soc. **352**(6), 2857–2869 (2000)
10. Dixmier, J.: Enveloping algebras. Revised reprint of the 1977 translation. Graduate Studies in Mathematics, Vol. 11. American Mathematical Society, Providence, RI (1996)
11. Frisk, A.: Two-step tilting for standardly stratified algebras. Algebra Discrete Math. **3**, 38–59 (2004)
12. Frisk, A.: Dlab's theorem and tilting modules for stratified algebras. J. Algebra **314**(2), 507–537 (2007)
13. Frisk, A.: Typical blocks of the category $\mathcal{O}$ for the queer Lie superalgebra. J. Algebra Appl. **6**(5), 731–778 (2007)
14. Frisk, A., Mazorchuk, V.: Properly stratified algebras and tilting. Proc. London Math. Soc. A (3) **92**(1), 29–61 (2006)
15. Frisk, A., Mazorchuk, V.: Regular strongly typical blocks of $\mathcal{O}_{\mathfrak{q}}$. Comm. Math. Phys. **291** (2009), .
16. Futorny, V., König, S., Mazorchuk, V.: Categories of induced modules and standardly stratified algebras. Algebr. Represent. Theory **5**(3), 259–276 (2002)
17. Gabriel, P.: Indecomposable representations. II. Symposia Mathematica, Vol. XI (Convegno di Algebra Commutativa, INDAM, Rome, 1971), pp. 81–104. Academic Press, London (1973)
18. M. Gorelik; On the ghost centre of Lie superalgebras. Ann. Inst. Fourier (Grenoble) **50**(6), 1745–1764 (2000)
19. Gorelik, M.: Strongly typical representations of the basic classical Lie superalgebras. J. Amer. Math. Soc. **15**(1), 167–184 (2002)
20. Greenstein, J., Mazorchuk, V.: On Koszul duality for generalized Takiff algebras and superalgebras. To appear.
21. Humphreys, J.E.: Representations of semisimple Lie algebras in the BGG category $\mathcal{O}$. Graduate Studies in Math., Vol. 94. American Mathematical Society, Providence, RI (2008)
22. Irving, R.: Projective modules in the category $\mathcal{O}_S$: self-duality. Trans. Amer. Math. Soc. **291**(2), 701–732 (1985)
23. Jantzen, J.C.: Einhüllende Algebren halbeinfacher Lie-Algebren. Ergebnisse der Mathematik und ihrer Grenzgebiete, Vol. 3. Springer-Verlag, Berlin Heidelberg (1983)
24. Khomenko, O., Koenig, S., Mazorchuk, V.: Finitistic dimension and tilting modules for stratified algebras. J. Algebra **286**(2), 456–475 (2005)
25. König, S., Slungård, I., Xi, C.: Double centralizer properties, dominant dimension, and tilting modules. J. Algebra **240**(1), 393–412 (2001)
26. Mazorchuk, V.: On finitistic dimension of stratified algebras. Algebra Discrete Math. **2004**(3), 77–88 (2004)
27. Mazorchuk, V.: Koszul duality for stratified algebras II. Standardly stratified algebras. J. Aust. Math. Soc. **89**(1), 23–49 (2010)
28. Mazorchuk, V., Parker, A.: On the relation between finitistic and good filtration dimensions. Comm. Algebra **32**(5), 1903–1916 (2004)
29. Mazorchuk, V., Stroppel, C.: Categorification of (induced) cell modules and the rough structure of generalised Verma modules. Adv. Math. **219**(4), 1363–1426 (2008)
30. Mazorchuk, V., Ovsienko, S.: Finitistic dimension of properly stratified algebras. Adv. Math. **186**(1), 251–265 (2004)
31. Miemietz, V., Turner, W.: Homotopy, homology, and GL_2. Proc. Lond. Math. Soc. (3) **100**(2), 585–606 (2010)
32. Moody, R., Pianzola, A.: Lie algebras with triangular decompositions. Canadian Mathematical Society Series of Monographs and Advanced Texts. A Wiley-Interscience Publication. John Wiley & Sons, Inc., New York (1995)

33. Musson, I.: Lie superalgebras and enveloping algebras. Graduate Studies in Mathematics, Vol. 131. American Mathematical Society, Providence, RI (2012)
34. Ringel, C.M.: The category of modules with good filtrations over a quasi-hereditary algebra has almost split sequences. Math. Z. **208**(2), 209–223 (1991)
35. Rocha-Caridi, A.: Splitting criteria for $\mathfrak{g}$-modules induced from a parabolic and the Bernstein-Gelfand-Gelfand resolution of a finite-dimensional, irreducible $\mathfrak{g}$-module. Trans. Amer. Math. Soc. **262**(2), 335–366 (1980)
36. Saito, K.: Polyhedra Dual to the Weyl Chamber. Publ. RIMS, Kyoto Univ. **40**, 1337–1384 (2004)
37. Soergel, W.: Kategorie $\mathscr{O}$, perverse Garben und Moduln über den Koinvarianten zur Weylgruppe. J. Amer. Math. Soc. **3**(2), 421–445 (1990)
38. Stroppel, C.: Category $\mathscr{O}$: quivers and endomorphism rings of projectives. Represent. Theory **7**, 322–345 (2003)

On Kostant's theorem for Lie superalgebras

Elena Poletaeva

Abstract We study finite W-algebras for even regular (principal) nilpotent elements for classical Lie superalgebras. We give the precise description of the principal finite W-algebra for the exceptional simple Lie superalgebra $D(2,1;\alpha)$ and obtain partial results for $\mathfrak{osp}(1|2n)$. We show that the principal finite W-algebra for the Lie superalgebra $Q(n)$ is isomorphic to a factor algebra of the super-Yangian of $Q(1)$.

1 Introduction

The *finite W-algebras* are certain associative algebras associated to a complex semi-simple Lie algebra $\mathfrak{g}$ and a nilpotent element $e \in \mathfrak{g}$. They are quantizations of Poisson algebras of functions on the Slodowy slice at e to the orbit $Ad(G)e$, where $\mathfrak{g} = Lie(G)$ [5, 11]. Due to recent results of I. Losev, A. Premet and others, W-algebras play a very important role in description of primitive ideals [8, 11].

It is a result of B. Kostant that for a regular nilpotent element e, the finite W-algebra coincides with the center of $U(\mathfrak{g})$ [7].

Finite W-algebras for semi-simple Lie algebras were introduced by A. Premet [11] (see also [8]). We adopt A. Premet's definition of finite W-algebra for simple Lie superalgebras, and study the case when e is an even regular (principal) nilpotent element. Kostant's result does not hold in this case.

In [10] we obtained the precise description of the principal finite W-algebras for regular e for classical Lie superalgebras of Type I and defect one. In this work we describe the principal finite W-algebra for the exceptional Lie superalgebra $D(2,1;\alpha)$ and obtain partial results for $\mathfrak{osp}(1|2n)$.

It was observed by C. Briot and E. Ragoucy that certain finite W-algebras based on $\mathfrak{gl}(m|n)$ can be realized as truncations of the super-Yangian of $\mathfrak{gl}(m|n)$ [1]. J. Brown,

E. Poletaeva (✉)
Department of Mathematics, University of Texas-Pan American, Edinburg, TX 78539, USA
e-mail: elenap@utpa.edu

M. Gorelik, P. Papi (eds.): *Advances in Lie Superalgebras.* Springer INdAM Series 7,
DOI 10.1007/978-3-319-02952-8_10,

J. Brundan and S. Goodwin have recently described the principal finite W-algebras for $\mathfrak{g} = \mathfrak{gl}(m|n)$ as certain truncations of a shifted version of the super-Yangian of $\mathfrak{gl}(1|1)$ [3].

In this work we show that the principal finite W-algebra for the Lie superalgebra $Q(n)$ is isomorphic to a factor algebra of the super-Yangian of $Q(1)$.
All results for $Q(n)$, $D(2,1;\alpha)$ and $\mathfrak{osp}(1|2n)$ are joint work with V. Serganova.

2 Preliminaries

Let $\mathfrak{g}$ be a finite-dimensional semi-simple or reductive Lie algebra over $\mathbb{C}$, $(\cdot|\cdot)$ be a non-degenerate invariant symmetric bilinear form.

Definition 1 An element $e \in \mathfrak{g}$ is *nilpotent* if and only if ade is a nilpotent endomorphism of $\mathfrak{g}$.

Example 1 $\mathfrak{g} = \mathfrak{gl}(n)$.

$e \in \mathfrak{gl}(n)$ is nilpotent if and only if e is an $n \times n$-matrix with eigenvalues zero.

Definition 2 A nilpotent element $e \in \mathfrak{g}$ is called *regular nilpotent* if and only if $\mathfrak{g}^e = Ker(ade)$ attains the minimal dimension, which is equal to rank$\mathfrak{g}$.

Theorem 1 (Jacobson-Morozov) *Associated to a nonzero nilpotent element $e \in \mathfrak{g}$, there always exists $\{e,h,f\}$ which satisfy*

$$[e,f] = h, \quad [h,e] = 2e, \quad [h,f] = -2f.$$

Proof. Induction on $\dim \mathfrak{g}$, see [4].

Definition 3 (A Dynkin $\mathbb{Z}$-grading) Let $\mathfrak{sl}(2) =< e,h,f >$. The eigenspace decomposition of the adjoint action

$$adh : \mathfrak{g} \longrightarrow \mathfrak{g}$$

provides a $\mathbb{Z}$-grading:

$$\mathfrak{g} = \oplus_{j \in \mathbb{Z}} \mathfrak{g}_j, \quad \mathfrak{g}_j = \{x \in \mathfrak{g} \mid adh(x) = jx\}.$$

Properties

(1) $e \in \mathfrak{g}_2$,
(2) $ade : \mathfrak{g}_j \longrightarrow \mathfrak{g}_{j+2}$ is injective for $j \leq -1$,
(3) $ade : \mathfrak{g}_j \longrightarrow \mathfrak{g}_{j+2}$ is surjective for $j \geq -1$,
(4) $\mathfrak{g}^e \subset \oplus_{j \geq 0} \mathfrak{g}_j$,
(5) $(\mathfrak{g}_i|\mathfrak{g}_j) = 0$ unless $i + j = 0$,
(6) $\dim \mathfrak{g}^e = \dim \mathfrak{g}_0 + \dim \mathfrak{g}_1$.

Definition 4 (A good $\mathbb{Z}$-grading) A $\mathbb{Z}$-grading $\mathfrak{g} = \oplus_{j\in\mathbb{Z}}\mathfrak{g}_j$ for a semi-simple Lie algebra $\mathfrak{g}$ is called *a good* $\mathbb{Z}$-grading for e, if it satisfies the conditions (1)–(3).

For a reductive $\mathfrak{g}$, there is an additional condition: the center of $\mathfrak{g}$ is in $\mathfrak{g}_0$.

A good $\mathbb{Z}$-grading $\mathfrak{g} = \oplus_{j\in\mathbb{Z}}\mathfrak{g}_j$ is called *even*, if $\mathfrak{g}_j = 0$ unless j is an even integer.

Remark 1 Properties (4)–(6) remain to be valid for every good $\mathbb{Z}$-grading of $\mathfrak{g}$.

Proof. See [14].

3 Definition of finite *W*-algebras

Finite W-algebras for semi-simple or reductive Lie algebras were introduced by A. Premet [11] (see also [8]).

Let $\mathfrak{g}$ be a reductive Lie algebra, $(\cdot|\cdot)$ be a non-degenerate invariant symmetric bilinear form, e be a nilpotent element in $\mathfrak{g}$. Let $\mathfrak{g} = \oplus_{j\in\mathbb{Z}}\mathfrak{g}_j$ be a good $\mathbb{Z}$-grading for e, and let $\chi \in \mathfrak{g}^*$ be defined by $\chi(x) := (x|e)\ \forall x \in \mathfrak{g}$.

Define a bilinear form on $\mathfrak{g}_{-1}$ as follows

$$(x,y) := ([x,y]|e) = \chi([x,y]) \quad \forall x,y \in \mathfrak{g}_{-1}.$$

Remark 2 The bilinear form on $\mathfrak{g}_{-1}$ is skew-symmetric and non-degenerate.

Proof. The skew-symmetry follows by definition. The non-degeneracy follows from the bijection

$$ade : \mathfrak{g}_{-1} \longrightarrow \mathfrak{g}_1$$

and the identity

$$(x,y) = (x|[y,e]).$$

Hence $\dim \mathfrak{g}_{-1}$ is even.

□

Pick a Lagrangian (i.e. a maximal isotropic) subspace $\mathfrak{l}$ of $\mathfrak{g}_{-1}$ with respect to the form $(\cdot,\cdot)$. Then $\dim \mathfrak{l} = \frac{1}{2}\dim \mathfrak{g}_{-1}$. Let $\mathfrak{m} = (\oplus_{j\le -2}\mathfrak{g}_j) \oplus \mathfrak{l}$. The restriction of χ to $\mathfrak{m}$

$$\chi : \mathfrak{m} \longrightarrow \mathbb{C}$$

defines a one-dimensional representation $\mathbb{C}_\chi = <v>$ of $\mathfrak{m}$ thanks to the Lagrangian condition on $\mathfrak{l}$. Let I_χ be the left ideal of $U(\mathfrak{g})$ generated by $a - \chi(a)$ for $a \in \mathfrak{m}$.

Definition 5 *The generalized Whittaker module* is

$$Q_\chi := U(\mathfrak{g}) \otimes_{U(\mathfrak{m})} \mathbb{C}_\chi \cong U(\mathfrak{g})/I_\chi.$$

Definition 6 *The finite W-algebra* associated to the nilpotent element e is

$$W_\chi := \mathrm{End}_{U(\mathfrak{g})}(Q_\chi)^{op}.$$

Remark 3 W_χ can be identified as the space of *Whittaker vectors* in $U(\mathfrak{g})/I_\chi$. Let $\pi : U(\mathfrak{g}) \to U(\mathfrak{g})/I_\chi$ be the natural projection, and let $y \in U(\mathfrak{g})$. Then

$$W_\chi = (Q_\chi)^{ad\mathfrak{m}} = \{\pi(y) \in U(\mathfrak{g})/I_\chi \mid [a,y] \in I_\chi \quad \forall a \in \mathfrak{m}\}.$$

The multiplication is given by

$$\pi(y_1)\pi(y_2) = \pi(y_1 y_2)$$

for $y_i \in U(\mathfrak{g})$ such that $[a,y_i] \in I_\chi$ $\forall a \in \mathfrak{m}$ and $i = 1,2$.

Remark 4 The isoclasses of finite W-algebras do not depend on Lagrangian subspace $\mathfrak{l}$ ([5]) and good $\mathbb{Z}$-grading ([2]).

Example 2 Let $e = 0$. Then $\chi = 0$, $\mathfrak{g}_0 = \mathfrak{g}, \mathfrak{m} = 0$.

$$Q_\chi = U(\mathfrak{g}), \quad W_\chi = U(\mathfrak{g}).$$

Theorem 2 (B. Kostant (1978)) *For a regular nilpotent element $e \in \mathfrak{g}$, $W_\chi \cong Z(\mathfrak{g})$, the center of $U(\mathfrak{g})$, see [7].*

Definition 7 (Kazhdan filtration on W_χ) Let $\mathfrak{g}$ be a reductive Lie algebra with a Dynkin $\mathbb{Z}$-grading, let $\mathfrak{m} = (\oplus_{j\le -2}\mathfrak{g}_j) \oplus \mathfrak{l}$.
Let $\mathfrak{n} \subset \mathfrak{g}$ be an *adh*-invariant subspace such that $\mathfrak{g} = \mathfrak{m} \oplus \mathfrak{n}$. Then

$$W_\chi = \{X \in U(\mathfrak{g})/U(\mathfrak{g})\mathfrak{m} \cong S(\mathfrak{n}) \mid aXv = \chi(a)Xv \quad \forall a \in \mathfrak{m}\}.$$

For any $y \in \mathfrak{n}$, let $\mathrm{wt}(y)$ be the weight of y with respect to *adh* and

$$\deg(y) = \mathrm{wt}(y) + 2.$$

The degree function deg induces a $\mathbb{Z}$-grading on $S(\mathfrak{n})$. This grading defines a filtration on W_χ.

Theorem 3 (A. Premet) *The associated graded algebra $Gr(W_\chi)$ is isomorphic to $S(\mathfrak{g}^e)$.*

Idea of Proof. Introduce the map

$$P : W_\chi \longrightarrow S(\mathfrak{g}^e).$$

For $X \in W_\chi \subset S(\mathfrak{n})$ the term $P(X)$ of highest degree and highest weight belongs to $S(\mathfrak{g}^e)$, see [11].

Example 3 $\mathfrak{g} = \mathfrak{gl}(n)$.

Form: $(a|b) = \mathrm{tr}(ab)$.

$$e = \begin{pmatrix} 0 & 1 & 0 & 0 & 0 & 0 \\ 0 & 0 & 1 & 0 & 0 & 0 \\ 0 & 0 & 0 & \cdots & 0 & 0 \\ 0 & 0 & 0 & 0 & 1 & 0 \\ 0 & 0 & 0 & 0 & 0 & 1 \\ 0 & 0 & 0 & 0 & 0 & 0 \end{pmatrix}, \quad f = \begin{pmatrix} 0 & 0 & 0 & 0 & 0 & 0 \\ n-1 & 0 & 0 & 0 & 0 & 0 \\ 0 & 2(n-2) & 0 & \cdots & 0 & 0 \\ 0 & 0 & \cdots & 0 & 0 & 0 \\ 0 & 0 & 0 & \cdots & 0 & 0 \\ 0 & 0 & 0 & 0 & n-1 & 0 \end{pmatrix},$$

$h = \mathrm{diag}(n-1, n-3, \ldots, 3-n, 1-n)$.

e is a regular nilpotent element, h defines an even Dynkin $\mathbb{Z}$-grading of $\mathfrak{g}$ whose degrees on the elementary matrices E_{ij} are

$$\begin{pmatrix} 0 & 2 & 4 & 6 & \cdots & 2n-2 \\ -2 & 0 & 2 & 4 & \cdots & 2n-4 \\ -4 & -2 & 0 & 2 & \cdots & 2n-6 \\ -6 & -4 & -2 & 0 & \cdots & 2n-8 \\ \cdots & \cdots & \cdots & \cdots & \cdots & \cdots \\ 2-2n & \cdots & -6 & -4 & -2 & 0 \end{pmatrix}$$

$z = \mathrm{diag}(1, \ldots, 1)$ is the center of $\mathfrak{gl}(n)$.

$$\mathfrak{g}^e = < z, e, e^2, e^3, \ldots, e^{n-1} >, \quad \dim \mathfrak{g}^e = n.$$

$$\mathfrak{m} = \bigoplus_{j \geq 2}^{n} \mathfrak{g}_{2-2j}, \quad \chi(E_{i+1,i}) = 1, \quad \chi(E_{i+k,i}) = 0 \text{ if } k \geq 2.$$

W_χ is a polynomial algebra generated by n elements:

$$\pi(z), \pi(\Omega_2), \pi(\Omega_3), \ldots, \pi(\Omega_n),$$

where Ω_k is the k-th Casimir element of $\mathfrak{gl}(n)$:

$$\Omega_k = \sum_{i_1, i_2, \ldots, i_k} E_{i_1 i_2} E_{i_2 i_3} \ldots E_{i_k i_1}.$$

The generators of W_χ can be identified with elements of $\mathfrak{g}^e$ as follows:

$$\pi(z) \xrightarrow{P} z,$$

$$\frac{1}{k}\pi(\Omega_k) \xrightarrow{P} e^{k-1} \quad \textit{for} \quad k = 2, \ldots, n.$$

4 Finite W-algebras for Lie superalgebras

In the case of Lie superalgebras, finite W-algebras were studied by mathematicians and physicists in the following works [1, 3, 10, 14, 15].

Let $\mathfrak{g}$ be a classical simple Lie superalgebra, i.e. $\mathfrak{g} = \mathfrak{g}_{\bar{0}} \oplus \mathfrak{g}_{\bar{1}}$, $\mathfrak{g}_{\bar{0}}$ is a reductive Lie algebra, and $\mathfrak{g}$ has an invariant supersymmetric bilinear form. Let $e \in \mathfrak{g}_{\bar{0}}$ be an even nilpotent element, and we fix $\mathfrak{sl}(2) =< e,h,f >$.

The definition of W_χ given in section 3 makes sense, however the Theorem of Kostant does not hold in this case since W_χ must have a non-trivial odd part, and the center of $U(\mathfrak{g})$ is even. Kazhdan filtration on W_χ can be defined exactly as in the Lie algebra case.

Proposition 1 *$Gr(W_\chi)$ is supercommutative.*

Remark 5 If $\dim(\mathfrak{g}_{-1})_{\bar{1}}$ is even, then one can construct the similar map

$$P : W_\chi \longrightarrow S(\mathfrak{g}^e)$$

by taking the monomials of the highest degree and the highest weight.
If $\dim(\mathfrak{g}_{-1})_{\bar{1}}$ is odd, then there exists an odd element θ in $\mathfrak{g}_{-1} \cap \mathfrak{l}^\perp$ such that $\pi(\theta) \in W_\chi$ and $\pi(\theta)^2 = 1$.

In this work we study the principal finite W-algebras, which are the finite W-algebras associated to even regular nilpotent elements.

5 The case of $\mathfrak{g} = Q(n)$

In this section, we consider the principal finite W-algebra W_χ for the Lie superalgebra $Q(n)$. We construct a complete set of generators of W_χ. We also make a conjecture about the principal finite W-algebra for $\mathfrak{g}$ for the case when $\dim(\mathfrak{g}_{-1})_{\bar{1}}$ is even.

By definition

$$Q(n) = \{ \left(\begin{array}{c|c} A & B \\ \hline B & A \end{array} \right) \mid A, B \text{ are } n \times n \text{ matrices} \}.$$

Let $e_{i,j}$ and $f_{i,j}$ be standard bases in A and B respectively.
$\mathfrak{g}$ admits an odd non-degenerate $\mathfrak{g}$-invariant supersymmetric bilinear form

$$(x|y) := otr(xy) \text{ for } x, y \in \mathfrak{g},$$

where $otr \left(\begin{array}{c|c} A & B \\ \hline B & A \end{array} \right) = trB$.
Let $\mathfrak{sl}(2) =< e,h,f >$, where

$$e = \sum_{i=1}^{n-1} e_{i,i+1}, \quad h = \mathrm{diag}(n-1, n-3, \ldots, 3-n, 1-n), \quad f = \sum_{i=1}^{n-1} i(n-i) e_{i+1,i}.$$

e is a regular nilpotent element, h defines an even Dynkin $\mathbb{Z}$-grading of $\mathfrak{g}$, whose degrees on the elementary matrices are

$$\left(\begin{array}{cccc|cccc} 0 & 2 & \cdots & 2n-2 & 0 & 2 & \cdots & 2n-2 \\ -2 & 0 & \cdots & 2n-4 & -2 & 0 & \cdots & 2n-4 \\ \cdots & \cdots & \cdots & \cdots & \cdots & \cdots & \cdots & \cdots \\ 2-2n & \cdots & \cdots & 0 & 2-2n & \cdots & \cdots & 0 \\ \hline 0 & 2 & \cdots & 2n-2 & 0 & 2 & \cdots & 2n-2 \\ -2 & 0 & \cdots & 2n-4 & -2 & 0 & \cdots & 2n-4 \\ \cdots & \cdots & \cdots & \cdots & \cdots & \cdots & \cdots & \cdots \\ 2-2n & \cdots & \cdots & 0 & 2-2n & \cdots & \cdots & 0 \end{array}\right).$$

Let $E = \sum_{i=1}^{n-1} f_{i,i+1}$, thus E is an odd element. Let $\chi \in \mathfrak{g}^*$ be defined by $\chi(x) := (x|E)$. Note that

$$\mathfrak{g}^E = \{z, e, e^2, \dots, e^{n-1} \mid H_0, H_1, \dots, H_{n-1}\}, \quad \dim(\mathfrak{g}^E) = (n|n),$$

where $H_0 = \sum_{i=1}^n (-1)^{i+1} f_{i,i}$, $H_1 = \sum_{i=1}^{n-1} (-1)^i f_{i,i+1}, \dots H_{n-1} = (-1)^{n+1} f_{1,n}$, $z = \sum_{i=1}^n e_{i,i}$, $\mathfrak{m} = \oplus_{j=2}^n \mathfrak{g}_{2-2j}$, and it is generated by $e_{i+1,i}, f_{i+1,i}$ where $i = 1, \dots, n-1$; $\chi(e_{i+1,i}) = 1$, $\chi(f_{i+1,i}) = 0$.

In [13] A. Sergeev defined by induction the elements $e_{i,j}^{(m)}$ and $f_{i,j}^{(m)}$ belonging to $U(Q(n))$:

$$e_{i,j}^{(m)} = \sum_{k=1}^n e_{i,k} e_{k,j}^{(m-1)} + (-1)^{m+1} \sum_{k=1}^n f_{i,k} f_{k,j}^{(m-1)},$$
$$f_{i,j}^{(m)} = \sum_{k=1}^n e_{i,k} f_{k,j}^{(m-1)} + (-1)^{m+1} \sum_{k=1}^n f_{i,k} e_{k,j}^{(m-1)}.$$

Then

$$[e_{i,j}, e_{k,l}^{(m)}] = \delta_{j,k} e_{i,l}^{(m)} - \delta_{i,l} e_{k,j}^{(m)}, \quad [e_{i,j}, f_{k,l}^{(m)}] = \delta_{j,k} f_{i,l}^{(m)} - \delta_{i,l} f_{kj}^{(m)}, \tag{1}$$

$$[f_{i,j}, e_{k,l}^{(m)}] = (-1)^{m+1} \delta_{j,k} f_{i,l}^{(m)} - \delta_{i,l} f_{k,j}^{(m)}, \quad [f_{i,j}, f_{k,l}^{(m)}] = (-1)^{m+1} \delta_{j,k} e_{i,l}^{(m)} + \delta_{i,l} e_{k,j}^{(m)}.$$

Proposition 2 (A. Sergeev) *The elements* $\sum_{i=1}^n e_{i,i}^{(2m+1)}$ *generate* $Z(Q(n))$, *see* [13].

Lemma 1 $\pi(e_{n,1}^{(m)})$ *and* $\pi(f_{n,1}^{(m)})$ *are Whittaker vectors.*

Proof. By (1) we have

$$[e_{i,j}, e_{n,1}^{(m)}] = [f_{i,j}, e_{n,1}^{(m)}] = [e_{i,j}, f_{n,1}^{(m)}] = [f_{i,j}, f_{n,1}^{(m)}] = 0$$

for all $i > j$. In other words $e_{n,1}^{(m)}, f_{n,1}^{(m)} \in U(\mathfrak{g})^{ad\,\mathfrak{m}}$. □

Theorem 4 W_χ *has* n *even generators:* $\pi(e_{n,1}^{(n+k-1)})$ *and* n *odd generators:* $\pi(f_{n,1}^{(n+k-1)})$, $k = 1, \dots, n$.

Corollary 1 *The natural homomorphism* $U(\mathfrak{g})^{ad\,\mathfrak{m}} \longrightarrow W_\chi$ *is surjective.*

Let $\mathfrak{p} := \oplus_{j\geq 0}\mathfrak{g}_j$. Let $\mathfrak{f} =< e_{i,i}, f_{i,i} \mid i = 1,\dots,n >$, and let $\vartheta : U(\mathfrak{p}) \longrightarrow U(\mathfrak{f})$ be the Harish-Chandra homomorphism.

Proposition 3 *The Harish-Chandra homomorphism is injective.*

Denote

$$x_i = e_{i,i}, \quad \xi_i = (-1)^{i+1} f_{i,i}.$$

Theorem 5 *Under the Harish-Chandra homomorphism:*

$$\vartheta(\pi(e_{n,1}^{(n+k-1)})) = [\Sigma_{i_1\geq i_2\geq\dots\geq i_k}(x_{i_1} + (-1)^{k+1}\xi_{i_1})\dots(x_{i_{k-1}} - \xi_{i_{k-1}})(x_{i_k} + \xi_{i_k})]_{even},$$

$$\vartheta(\pi(f_{n,1}^{(n+k-1)})) = [\Sigma_{i_1\geq i_2\geq\dots\geq i_k}(x_{i_1} + (-1)^{k+1}\xi_{i_1})\dots(x_{i_{k-1}} - \xi_{i_{k-1}})(x_{i_k} + \xi_{i_k})]_{odd}.$$

Theorem 6

$$\pi(e_{n,1}^{(n+1)}) = \pi(\frac{1}{2}\sum_{i=1}^{n} e_{i,i}^2 + \sum_{i=1}^{n-1} e_{i,i+1} + \sum_{i<j}(-1)^{i-j} f_{i,i}f_{j,j} + \frac{1}{2}z^2 - z).$$

One can define odd generators $\Phi_0,\dots,\Phi_{n-1}$ *of* W_χ *as follows:*

$$\Phi_0 = \pi(f_{n,1}^{(n)}) = \pi(H_0),$$
$$\Phi_1 = [\pi(e_{n,1}^{(n+1)}), \Phi_0],$$
$$\dots$$
$$\Phi_{n-1} = [\pi(e_{n,1}^{(n+1)}), \Phi_{n-2}].$$

Then

$$[\Phi_m, \Phi_p] = 0, \text{ if } m+p \text{ is odd},$$
$$[\Phi_m, \Phi_p] \in Z(Q(n)), \text{ if } m+p \text{ is even}.$$

Lemma 2 *For odd m and p we have*

$$[\pi(e_{n,1}^{(n+m)}), \pi(e_{n,1}^{(n+p)})] = 0.$$

We set

$$z_i = \pi(e_{n,1}^{(n+i)}) \quad \text{for odd } i,$$
$$z_i = [\Phi_0, \Phi_i] \quad \text{for even } i.$$

Theorem 7 *Elements* $z_0,\dots,z_{n-1}$ *are algebraically independent in* W_χ. *Together with* $\Phi_0,\dots,\Phi_{n-1}$ *they form a complete set of generators in* W_χ.

Conjecture 1 In the case when $\dim(\mathfrak{g}_{-1})_{\bar{1}}$ is even, it is possible to find a set of generators of the principal finite W-algebra for $\mathfrak{g}$ such that even generators commute, and the commutators of odd generators are in $Z(\mathfrak{g})$.

6 Super-Yangian of $Q(n)$

In this section, we describe the principal finite W-algebra for $Q(n)$ as a factor algebra of the super-Yangian of $Q(1)$.

The super-Yangian $Y(Q(n))$ was studied by M. Nazarov and A. Sergeev in [9]. $Y(Q(n))$ is the associative unital superalgebra over $\mathbb{C}$ with the countable set of generators

$$T_{ij}^{(m)} \text{ where } m = 1,2,\dots \text{ and } i,j = \pm 1, \pm 2, \dots, \pm n.$$

The $\mathbb{Z}_2$-grading of the algebra $Y(Q(n))$ is defined as follows:

$$p(T_{ij}^{(m)}) = p(i) + p(j), \text{ where } p(i) = 0 \text{ if } i > 0, \text{ and } p(i) = 1 \text{ if } i < 0.$$

To write down defining relations for these generators we employ the formal series in $Y(Q(n))[[u^{-1}]]$:

$$T_{i,j}(u) = \delta_{ij} \cdot 1 + T_{i,j}^{(1)} u^{-1} + T_{i,j}^{(2)} u^{-2} + \dots.$$

Then for all possible indices i,j,k,l we have the relations

$$\begin{aligned}(u^2 - v^2)[T_{i,j}(u), T_{k,l}(v)] \cdot (-1)^{p(i)p(k)+p(i)p(l)+p(k)p(l)} \\ = (u+v)(T_{k,j}(u)T_{i,l}(v) - T_{k,j}(v)T_{i,l}(u)) \\ -(u-v)(T_{-k,j}(u)T_{-i,l}(v) - T_{k,-j}(v)T_{i,-l}(u)) \cdot (-1)^{p(k)+p(l)},\end{aligned} \tag{2}$$

where v is a formal parameter independent of u, so that (2) is an equality in the algebra of formal Laurent series in u^{-1}, v^{-1} with coefficients in $Y(Q(n))$.
For all indices i,j we also have the relations

$$T_{i,j}(-u) = T_{-i,-j}(u). \tag{3}$$

Note that the relations (2) and (3) are equivalent to the following defining relations:

$$\begin{aligned}&([T_{i,j}^{(m+1)}, T_{k,l}^{(r-1)}] - [T_{i,j}^{(m-1)}, T_{k,l}^{(r+1)}]) \cdot (-1)^{p(i)p(k)+p(i)p(l)+p(k)p(l)} && (2')\\ &= T_{k,j}^{(m)} T_{i,l}^{(r-1)} + T_{k,j}^{(m-1)} T_{i,l}^{(r)} - T_{k,j}^{(r-1)} T_{i,l}^{(m)} - T_{k,j}^{(r)} T_{i,l}^{(m-1)} \\ &\quad + (-1)^{p(k)+p(l)} (-T_{-k,j}^{(m)} T_{-i,l}^{(r-1)} + T_{-k,j}^{(m-1)} T_{-i,l}^{(r)} + T_{k,-j}^{(r-1)} T_{i,-l}^{(m)} - T_{k,-j}^{(r)} T_{i,-l}^{(m-1)}) \\ &T_{-i,-j}^{(m)} = (-1)^m T_{i,j}^{(m)} && (3')\end{aligned}$$

where $m, r = 1, \dots$ and $T_{ij}^{(0)} = \delta_{ij}$.

Theorem 8 *There exists a surjective homomorphism:*

$$\varphi : Y(Q(1)) \longrightarrow W_\chi$$

defined as follows:

$$\varphi(T_{1,1}^{(k)}) = (-1)^k \pi(e_{n,1}^{(n+k-1)}), \quad \varphi(T_{-1,1}^{(k)}) = (-1)^k \pi(f_{n,1}^{(n+k-1)}), \textit{ for } k = 1,2,\dots.$$

7 The case of $\mathfrak{g} = D(2,1;\alpha)$

In this section, we describe the principal finite W-algebra for the exceptional Lie superalgebra $D(2,1;\alpha)$ in terms of generators and relations. We follow the construction of this Lie superalgebra given by M. Scheunert [12].

Let $\sigma_1, \sigma_2, \sigma_3$ be complex numbers such that $\sigma_1 + \sigma_2 + \sigma_3 = 0$.
By definition, $\Gamma(\sigma_1, \sigma_2, \sigma_3) = \Gamma_{\bar{0}} \oplus \Gamma_{\bar{1}}$, where

$$\Gamma_{\bar{0}} = \mathfrak{sl}(2)_{\bar{1}} \oplus \mathfrak{sl}(2)_{\bar{2}} \oplus \mathfrak{sl}(2)_{\bar{3}},$$
$$\Gamma_{\bar{1}} = V_1 \otimes V_2 \otimes V_3,$$

$\mathfrak{sl}(2)_i = <X_i, H_i, Y_i>, \quad V_i = <e_i, f_i>, \quad i = 1,2,3,$
$P_i : V_i \times V_i \longrightarrow \mathfrak{sl}(2)_i$ is $\mathfrak{sl}(2)_i$-invariant bilinear mapping:
$P_i(e_i, e_i) = 2X_i, \quad P_i(f_i, f_i) = -2Y_i, \; P_i(e_i, f_i) = P_i(f_i, e_i) = -H_i.$
ψ_i is a non-degenerate skew-symmetric form on V_i:

$$\psi_i(e_i, f_i) = 1.$$

$[\Gamma_{\bar{0}}, \Gamma_{\bar{1}}]$ is the natural representation, $[\Gamma_{\bar{1}}, \Gamma_{\bar{1}}]$ is given by

$$[x_1 \otimes x_2 \otimes x_3, y_1 \otimes y_2 \otimes y_3] = \sigma_1 \psi_2(x_2, y_2)\psi_3(x_3, y_3)P_1(x_1, y_1) +$$
$$\sigma_2 \psi_1(x_1, y_1)\psi_3(x_3, y_3)P_2(x_2, y_2) + \sigma_3 \psi_1(x_1, y_1)\psi_2(x_2, y_2)P_3(x_3, y_3),$$

where $x_i, y_i \in V_i$.

Remark 6 The superalgebra $\Gamma(\sigma_1, \sigma_2, \sigma_3)$ is simple if and only if $\sigma_i \neq 0$ for $i = 1,2,3$. $\Gamma(\sigma_1, \sigma_2, \sigma_3) \cong \Gamma(\sigma_1', \sigma_2', \sigma_3')$ if and only if the sets $\{\sigma_i'\}$ and $\{\sigma_i\}$ are obtained from each other by a permutation and multiplication of all elements of one set by a nonzero complex number (see [12]).
$\Gamma(\sigma_1, \sigma_2, \sigma_3)$ is a one-parameter family of deformations of $osp(4|2)$.
$\Gamma(1, -1-\alpha, \alpha) \cong D(2,1;\alpha)$, where $\alpha \neq 0, -1$ (see [6]).

We consider the non-degenerate invariant supersymmetric bilinear form on $\mathfrak{g}$ given as follows:

$$(X_i, Y_i) = \tfrac{1}{\sigma_i}, \quad (H_i, H_i) = \tfrac{2}{\sigma_i},$$
$$(e_1 \otimes e_2 \otimes e_3, f_1 \otimes f_2 \otimes f_3) = -2, \quad (e_1 \otimes e_2 \otimes f_3, f_1 \otimes f_2 \otimes e_3) = 2,$$
$$(e_1 \otimes f_2 \otimes e_3, f_1 \otimes e_2 \otimes f_3) = 2, \quad (f_1 \otimes e_2 \otimes e_3, e_1 \otimes f_2 \otimes f_3) = 2.$$

Let $\mathfrak{sl}(2) = <e, h, f>$, where

$$e = X_1 + X_2 + X_3, \quad h = H_1 + H_2 + H_3, \quad f = Y_1 + Y_2 + Y_3.$$

Then e is a regular nilpotent element, and h defines a Dynkin $\mathbb{Z}$-grading of $\mathfrak{g}$:

$\mathfrak{g} = \oplus_{j=-3}^{3} \mathfrak{g}_j$, where

$$\mathfrak{g}_3 =< e_1 \otimes e_2 \otimes e_3 >, \quad \mathfrak{g}_2 =< X_1, X_2, X_3 >,$$
$$\mathfrak{g}_1 =< e_1 \otimes e_2 \otimes f_3, e_1 \otimes f_2 \otimes e_3, f_1 \otimes e_2 \otimes e_3 >,$$
$$\mathfrak{g}_0 =< H_1, H_2, H_3 >, \quad \mathfrak{g}_{-1} =< e_1 \otimes f_2 \otimes f_3, f_1 \otimes e_2 \otimes f_3, e_1 \otimes e_2 \otimes f_3 >,$$
$$\mathfrak{g}_{-2} =< Y_1, Y_2, Y_3 >, \quad \mathfrak{g}_{-3} =< f_1 \otimes f_2 \otimes f_3 > .$$

Note that $\dim(\mathfrak{g}^e) = (3|3)$, $\mathfrak{g}^e = (\mathfrak{g}^e)_{\bar{0}} \oplus (\mathfrak{g}^e)_{\bar{1}}$, where

$$(\mathfrak{g}^e)_{\bar{0}} =< X_1, X_2, X_3 >,$$
$$(\mathfrak{g}^e)_{\bar{1}} =< e_1 \otimes f_2 \otimes e_3 - e_1 \otimes e_2 \otimes f_3, f_1 \otimes e_2 \otimes e_3 - e_1 \otimes e_2 \otimes f_3, e_1 \otimes e_2 \otimes e_3 > .$$

Note that

$$\mathfrak{m} = \mathfrak{g}_{-3} \oplus \mathfrak{g}_{-2} \oplus \mathfrak{l}, \quad \mathfrak{l} =< e_1 \otimes f_2 \otimes f_3 >,$$

$\mathfrak{m}$ is generated by Y_1, Y_2, Y_3 and $e_1 \otimes f_2 \otimes f_3$. Also,

$$\chi(Y_i) = \frac{1}{\sigma_i} \text{ for } i = 1,2,3, \text{ and } \chi(e_1 \otimes f_2 \otimes f_3) = 0.$$

$\theta = f_1 \otimes e_2 \otimes f_3 - f_1 \otimes f_2 \otimes e_3 \in \mathfrak{g}_{-1} \cap \mathfrak{l}^{\perp}$, $\pi(\theta) \in W_\chi$, $\pi(\theta)^2 = -2$.

Even generators of W_χ are
$C_1 = \pi(2X_1 + \sigma_1(\frac{1}{2}H_1^2 - H_1))$,
$C_2 = \pi(2X_2 + \frac{1}{2}\sigma_2 H_2^2 + (f_1 \otimes e_2 \otimes f_3)(e_1 \otimes e_2 \otimes f_3))$,
$C_3 = \pi(2X_3 + \frac{1}{2}\sigma_3 H_3^2 + (f_1 \otimes f_2 \otimes e_3)(e_1 \otimes f_2 \otimes e_3))$.

Odd generators of W_χ are
$R_1 = \pi(2(e_1 \otimes f_2 \otimes e_3 - e_1 \otimes e_2 \otimes f_3) + \sigma_1 H_1(f_1 \otimes e_2 \otimes f_3 - f_1 \otimes f_2 \otimes e_3))$,
$R_2 = \pi(2(f_1 \otimes e_2 \otimes e_3 - e_1 \otimes e_2 \otimes f_3)$
$\quad + (\sigma_1 H_1 - \sigma_3 H_3)(f_1 \otimes e_2 \otimes f_3) - \sigma_2 H_2(f_1 \otimes f_2 \otimes e_3))$,
$R_3 = \pi(4(e_1 \otimes e_2 \otimes e_3) - \sigma_1 H_1 R_2 - 4\sigma_1(f_1 \otimes e_2 \otimes f_3)X_1$
$\quad -2(\sigma_1 H_1(e_1 \otimes e_2 \otimes f_3) + \sigma_2 H_2(e_1 \otimes f_2 \otimes e_3) + \sigma_3 H_3(e_1 \otimes e_2 \otimes f_3)))$,
and $\pi(\theta)$.

Note that the quadratic Casimir element of $\mathfrak{g}$ is

$$\Omega = \Sigma_{i=1}^{3}(\frac{\sigma_i}{2} H_i^2 + 2X_i) - (e_1 \otimes e_2 \otimes f_3)(f_1 \otimes f_2 \otimes e_3) - (e_1 \otimes f_2 \otimes e_3)(f_1 \otimes e_2 \otimes f_3).$$

Hence

$$\pi(\Omega) = C_1 + C_2 + C_3 - \frac{1}{2} R_1 \pi(\theta).$$

Theorem 9 *The principal finite W-algebra W_χ is generated by even elements $\pi(\Omega)$, C_1 and C_2, and odd element $\pi(\theta)$. The relations are*

$$
\begin{gathered}
[C_1,C_2]=0, \quad [\pi(\theta),C_i]=R_i \mp \tfrac{\sigma_i}{2}\pi(\theta), \quad i=1,2,\\
[C_2,R_1]=-\tfrac{\sigma_2}{2}R_1+R_3, \quad [C_1,R_2]=\tfrac{\sigma_1}{2}R_2+R_3,\\
[R_i,R_i]=8\sigma_i C_i-2\sigma_i R_i\pi(\theta), \quad i=1,2,\\
[R_1,R_2]=-4(\sigma_1C_2+\sigma_2C_1+\sigma_3\pi(\Omega))+(\sigma_1R_2+\sigma_2R_1)\pi(\theta),\\
[R_i,\pi(\theta)]=\mp 2\sigma_i, \quad i=1,2, \quad [\pi(\theta),\pi(\theta)]=-4,\\
[\pi(\Omega),\pi(\theta)]=0, \quad [\pi(\Omega),C_i]=0,\ i=1,2, \quad [\pi(\Omega),R_i]=0,\ i=1,2,3.
\end{gathered}
$$

8 The case of $\mathfrak{g}=\mathfrak{osp}(1|2n)$

In this section, we present partial results for the principal finite W-algebra for $\mathfrak{osp}(1|2n)$ and make a conjecture for this case.

Form: $(a|b)=-str(ab)$.
We use the following notations for some elementary matrices in $\mathfrak{osp}(1|2n)$:

$$
\left(\begin{array}{c|cccc|cccc}
0 & s_1 & s_2 & \dots & \mathbf{s_n} & r_1 & r_2 & \dots & r_n\\
\hline
r_1 & h_1 & x_1 & \dots & \dots & p_1 & \dots & \dots & \dots\\
\dots & y_1 & h_2 & x_2 & \dots & .. & p_2 & \dots & \dots\\
\dots & \dots & \dots & \dots & x_{n-1} & \dots & \dots & \dots & \dots\\
r_n & \dots & \dots & y_{n-1} & h_n & \dots & \dots & \dots & p_n\\
\hline
s_1 & q_1 & \dots & \dots & \dots & h_1 & y_1 & \dots & \dots\\
\dots & \dots & q_2 & \dots & \dots & x_1 & h_2 & y_2 & \dots\\
\dots & \dots & \dots & \dots & \dots & \dots & \dots & \dots & y_{n-1}\\
\mathbf{s_n} & \dots & \dots & \dots & q_n & \dots & \dots & x_{n-1} & h_n
\end{array}\right).
$$

Let $\mathfrak{sl}(2)=<e,h,f>$, where $e=(x_1+\dots x_{n-1})+p_n$,
$h=\mathrm{diag}(0|2n-1,2n-3,\dots,3,1;-2n+1,-2n+3,\dots,-3,-1)$,
$f=(\sum_{k=1}^{n-1}k(2n-k)y_k)+n^2q_n$.
Note that e is a regular nilpotent element, and h defines a Dynkin $\mathbb{Z}$-grading of $\mathfrak{g}$ whose degrees on the elementary matrices are

$$
\left(\begin{array}{c|cccc|cccc}
0 & -2n+1 & \dots & -3 & \mathbf{-1} & 2n-1 & \dots & 3 & 1\\
\hline
2n-1 & 0 & 2 & \dots & \dots & 4n-2 & \dots & 2n+2 & 2n\\
\dots & -2 & 0 & 2 & \dots & 4n-4 & \dots & 2n & 2n-2\\
3 & \dots & \dots & 0 & 2 & \dots & \dots & 6 & 4\\
1 & \dots & -4 & -2 & 0 & 2n & \dots & 4 & 2\\
\hline
-2n+1 & -4n+2 & \dots & -2n-2 & -2n & 0 & -2 & \dots & \dots\\
\dots & -4n+4 & \dots & -2n & -2n+2 & 2 & 0 & \dots & -4\\
-3 & \dots & \dots & -6 & -4 & \dots & \dots & 0 & -2\\
\mathbf{-1} & -2n & \dots & -4 & -2 & \dots & \dots & 2 & 0
\end{array}\right).
$$

Note that $\dim \mathfrak{g}^e = (n|1)$, $(\mathfrak{g}^e)_{\bar{1}} = < r_1 >$, $\mathfrak{g}_{-1} = < \theta >$, where $\theta = s_n$, $\dim \mathfrak{g}_{-1} = 1$.
$\mathfrak{m} = \oplus_{j \leq -2} \mathfrak{g}_j$, where $\mathfrak{g}_{-2} = < y_i, q_n >$, $i = 1, \ldots, n-1$.
$\mathfrak{m}$ is generated by y_i, q_n; $\chi(y_i) = 2$, for $i = 1, \ldots, n-1$, and $\chi(q_n) = 1$.
Note that $\pi(\theta) \in W_\chi$, $\pi(\theta)^2 = -1$.

Conjecture 2 The principal finite W-algebra W_χ is generated by the first n Casimir elements in $Z(\mathfrak{g})$ and odd elements $\pi(\theta)$ and R, where R is induced by r_1 so that

$$[R,R] \in Z(\mathfrak{g}), \quad [R, \pi(\theta)] \in Z(\mathfrak{g}), \quad [\pi(\theta), \pi(\theta)] = -2.$$

Acknowledgements The author would like to thank the organizers of the Conference "Lie superalgebras", Roma, Istituto Nazionale di Alta Matematica, December 10-14, 2012 for the very interesting conference and for the hospitality.

The author thanks V. Serganova for very helpful discussions and A. Sergeev for pointing out the reference [9].

References

1. Briot C., Ragoucy, E.: W-superalgebras as truncations of super-Yangians. J. Phys. A **36**(4), 1057–1081 (2003)
2. Brundan J. and Goodwin S.: Good grading polytopes. Proc. London Math. Soc. **94**, 155–180 (2007)
3. Brown, J., Brundan, J., Goodwin, S.: Principal W-algebras for $GL(m|n)$. arXiv:1205.0992
4. Carter, R.: Finite groups of Lie type: Conjugacy classes and complex characters. Pure and Applied Math. John Wiley & Sons, Inc., New York (1985)
5. Gan, W.L., Ginzburg, V.: Quantization of Slodowy slices. Internat. Math. Res. Notices **5**, 243–255 (2002)
6. Kac, V.G.: Lie superalgebras. Adv. Math. **26**, 8–96 (1977)
7. Kostant, B.: On Whittaker vectors and representation theory. Invent. Math. **48**, 101–184 (1978)
8. Losev, I.: Finite W-algebras. In: Proceedings of the International Congress of Mathematicians. Vol. III, pp. 1281–1307. Hindustan Book Agency, New Delhi (2010). arXiv:1003.5811v1
9. Nazarov, M., Sergeev, A.: Centralizer construction of the Yangian of the queer Lie superalgebra. Studies in Lie Theory. Progr. Math. **243**, 417–441 (2006)
10. Poletaeva, E., Serganova, V.: On finite W-algebras for Lie superalgebras in the regular case. In: Dobrev, V. (ed.) Lie Theory and Its Applications in Physics. IX International Workshop. 20–26 June 2011, Varna, Bulgaria. Springer Proceedings in Mathematics and Statistics, Vol. **36**, pp. 487–497. Springer, Tokyo (2013)
11. Premet, A.: Special transverse slices and their enveloping algebras. Adv. Math. **170**, 1–55 (2002)
12. Scheunert, M.: The Theory of Lie Superalgebras. Lecture Notes in Mathematics, Vol. 716. Springer-Verlag, Berlin Heidelberg (1979)
13. Sergeev, A.: The centre of enveloping algebra for Lie superalgebra $Q(n, \mathbb{C})$. Letters in Math. Phys. **7**, 177–179 (1983)
14. Wang, W.: Nilpotent orbits and finite W-algebras. Fields Institute Communications Series, Vol. 59, pp. 71–105 (2011)
15. Zhao, L.: Finite W-superalgebras for queer Lie superalgebras. arXiv:1012.2326v2

Classical Lie superalgebras at infinity

Vera Serganova

Abstract We study certain categories of modules over direct limits of classical Lie superalgebras. In many cases these categories are equivalent to similar categories for classical Lie algebras. The functors establishing this equivalence can be used to obtain a new result for representation theory of direct limits of Lie algebras.

1 Introduction

There are several generalizations of simple Lie algebras and superalgebras in the infinite-dimensional case. In this paper, we discuss representations of locally simple Lie algebras, i.e. Lie algebras and superalgebras we consider are the direct limits $\mathfrak{g} = \varinjlim \mathfrak{g}_i$ of finite-dimensional simple Lie algebras (or superalgebras) $\mathfrak{g}_i$. In particular, we are interested in the cases when $\mathfrak{g} = \mathfrak{sl}(\infty), \mathfrak{so}(\infty)$ or $\mathfrak{sp}(\infty)$.

In [5] we tried to define a nice analogue of the category of finite-dimensional modules for $\mathfrak{g}$. The most obvious analogue, the category of integrable modules, is rather difficult to study. Even the problem of classifying simple modules involves infinitely many continuous parameters.

On the other hand, $\mathfrak{g}$ has a very natural class of representations in the tensor powers of the standard and costandard modules. In [2] we give an intrinsic definition of a category $\mathbb{T}_\mathfrak{g}$ that contains all such representations. It turns out that $\mathbb{T}_\mathfrak{g}$ has many remarkable properties. Although it is not semi-simple, it is a Koszul category in the sense of [1]. That allows one to calculate extensions between simple modules and their injective resolutions. We also prove that the categories $T_\mathfrak{g}$ for $\mathfrak{g} = \mathfrak{so}(\infty)$ and $\mathfrak{sp}(\infty)$ are equivalent. In the recent preprint [8] the same categories are studied from slightly different point of view.

V. Serganova (✉)
Department of Mathematics, University of California, Berkeley, CA 94720, USA
e-mail: serganov@math.berkeley.edu

M. Gorelik, P. Papi (eds.): *Advances in Lie Superalgebras.* Springer INdAM Series 7,
DOI 10.1007/978-3-319-02952-8_11,

The goal of the present paper is to define and study analogues of $\mathbb{T}_{\mathfrak{g}}$ for direct limits of classical Lie superalgebras. As follows from the Kac classification, [4], there are four such superalgebras $\mathfrak{sl}(\infty,\infty)$, $\mathfrak{osp}(\infty,\infty)$, $P(\infty)$ and $Q(\infty)$. We will see that in the first three cases we do not obtain new categories. Namely, $\mathbb{T}_{\mathfrak{sl}(\infty,\infty)}$ is equivalent to $\mathbb{T}_{\mathfrak{sl}(\infty)}$, and $\mathbb{T}_{\mathfrak{osp}(\infty,\infty)}$ and $\mathbb{T}_{P(\infty)}$ are equivalent to $\mathbb{T}_{\mathfrak{o}(\infty)}$. The latter fact can be used to construct a direct equivalence functor $\mathbb{T}_{\mathfrak{o}(\infty)} \to \mathbb{T}_{\mathfrak{sp}(\infty)}$. This result is somewhat surprising, since it appears that these categories are easier than the corresponding categories for finite-dimensional superalgebras. The rather complicated matter of atypical representations disappears after going to direct limits.

In the case of $Q(\infty)$ we obtain a completely new category. It is interesting to study it in detail.

2 Direct limits of classical Lie algebras

2.1 General and special Lie algebras at infinity

Let V and V_* be countable-dimensional vector spaces with non-degenerate pairing $tr : V \otimes V_* \to \mathbb{C}$.

Definition 1 $\mathfrak{gl}(\infty) = V \otimes V_*$ has a natural Lie algebra structure given by

$$[v_1 \otimes u_1, v_2 \otimes u_2] = tr(v_2 \otimes u_1) v_1 \otimes u_2 - tr(v_1 \otimes u_2) v_2 \otimes u_1.$$

$\mathrm{Ker}(tr) = \mathfrak{sl}(\infty)$ is a simple Lie subalgebra of $\mathfrak{gl}(\infty)$.

One can also realize $\mathfrak{g}$ as a direct limit

$$\mathfrak{sl}(\infty) = \varinjlim \mathfrak{sl}(n), \ \mathfrak{gl}(\infty) = \varinjlim \mathfrak{gl}(n),$$

and identify $\mathfrak{gl}(\infty)$ with the space of infinite matrices $(a_{ij})_{i,j \in \mathbb{N}}$ with finitely many non-zero entries and $\mathfrak{sl}(\infty)$ with the subalgebra of traceless matrices in $\mathfrak{gl}(\infty)$.

It is clear that V and V_* are simple $\mathfrak{g}$-modules. Furthermore, the classical Schur-Weyl duality works in the infinite-dimensional case.

Theorem 1 (Schur-Weyl duality) *Let* $\mathfrak{g} = \mathfrak{gl}(\infty)$ *or* $\mathfrak{sl}(\infty)$. *Then*

$$V^{\otimes n} = \bigoplus_{|\lambda|=n} V^{\lambda} \otimes Y_{\lambda}, \ V_*^{\otimes n} = \bigoplus_{|\lambda|=n} V_*^{\lambda} \otimes Y_{\lambda},$$

where λ runs the set of partitions of size n and Y_λ denotes the corresponding irreducible representation of S_n and $V^{\lambda} = \pi_{\lambda}(V^{\otimes n})$, where π_λ is a Young projector with the Young diagram λ.

However, the representation of $\mathfrak{g}$ in the space of mixed tensors $V^{\otimes n} \otimes V_*^{\otimes m}$ is not completely reducible in contrast with finite-dimensional case. Indeed, for instance, the exact sequence of $\mathfrak{sl}(\infty)$-modules

$$0 \to \mathfrak{sl}(\infty) \to V \otimes V_* \to \mathbb{C} \to 0$$

does not split. The following result gives a description of the $\mathfrak{g}$-module structure on $V^{\otimes n} \otimes V_*^{\otimes m}$.

Theorem 2 ([6]) *Let $\mathfrak{g} = \mathfrak{gl}(\infty)$ or $\mathfrak{sl}(\infty)$. Then*

$$V^{\otimes n} \otimes V_*^{\otimes m} = \bigoplus_{|\lambda|=n, |\mu|=m} \tilde{V}^{\lambda,\mu} \otimes (Y_\lambda \boxtimes Y_\mu),$$

where each $\tilde{V}^{\lambda,\mu} = V^\lambda \otimes V_*^\mu$ is an indecomposable $\mathfrak{g}$-module with irreducible socle $V^{\lambda,\mu}$ and $Y_\lambda \boxtimes Y_\mu$ is the exterior tensor product of irreducible S_n and S_m-modules.

The socle filtration of $\tilde{V}^{\lambda,\mu}$ is given by

$$\mathrm{soc}^k(\tilde{V}^{\lambda,\mu})/\mathrm{soc}^{k-1}(\tilde{V}^{\lambda,\mu}) = \bigoplus_{|\gamma|=k} N^\lambda_{\gamma,\lambda'} N^\mu_{\gamma,\mu'} V^{\lambda',\mu'}.$$

Here $N^\lambda_{\gamma,\lambda'}$ stand for Littlewood–Richardson coefficients.

The proof is based on the results of Howe, Tan and Willenbring [3] who calculated asymptotic decomposition of mixed tensor products in the finite-dimensional case.

2.2 Orthogonal and symplectic Lie algebras

Assume now that a countable-dimensional vector space V has a non-degenerate symmetric (resp. skew-symmetric) form $\omega : V \otimes V \to \mathbb{C}$.

Definition 2 $\mathfrak{so}(\infty)$ (resp. $\mathfrak{sp}(\infty)$) is the Lie subalgebra of finite rank linear operators in V preserving ω.

One can use identification

$$\mathfrak{so}(\infty) = \Lambda^2(V), \ \mathfrak{sp}(\infty) = S^2(V)$$

given by

$$X_{v\wedge w}(u) = \omega(v,u)w - \omega(u,w)v, \forall v,w,u \in V. \tag{1}$$

Another way to define $\mathfrak{g}$ is via direct limits

$$\mathfrak{so}(\infty) = \lim_{\rightarrow} \mathfrak{so}(n), \ \mathfrak{sp}(\infty) = \lim_{\rightarrow} \mathfrak{sp}(n).$$

The representation of $\mathfrak{g}$ in the tensor algebra $T(V)$ were also described by Penkov and Styrkas.

Theorem 3 *([6]) Let $\mathfrak{g} = \mathfrak{so}(\infty)$ or $\mathfrak{sp}(\infty)$. Then*

$$V^{\otimes n} = \bigoplus_{|\lambda|=n} \tilde{V}^\lambda \otimes Y_\lambda,$$

where each $\tilde{V}^\lambda$ is an indecomposable $\mathfrak{g}$-module with irreducible socle V^λ.

The socle filtration of $\tilde{V}^{\lambda}$ is given by

$$\mathrm{soc}^k(\tilde{V}^{\lambda})/\mathrm{soc}^{k-1}(\tilde{V}^{\lambda}) = \bigoplus_{|\gamma|=k} N^{\lambda}_{2\gamma,\lambda'} V^{\lambda'},$$

for $\mathfrak{g} = \mathfrak{so}(\infty)$, *and*

$$\mathrm{soc}^k(\tilde{V}^{\lambda})/\mathrm{soc}^{k-1}(\tilde{V}^{\lambda}) = \bigoplus_{|\gamma|=k} N^{\lambda}_{2\gamma^{\perp},\lambda'} V^{\lambda'},$$

for $\mathfrak{g} = \mathfrak{sp}(\infty)$.

3 The category $\mathbb{T}_{\mathfrak{g}}$

Definition 3 Let $\mathfrak{g} = \mathfrak{gl}(\infty)$ or $\mathfrak{sl}(\infty)$. A subalgebra $\mathfrak{k} \subset \mathfrak{g}$ is a finite corank subalgebra if there exist finite dimensional subspaces $W \subset V$ and $W_* \subset V_*$ such that the restriction of the canonical pairing to $W \otimes W_* \to \mathbb{C}$ is non-degenerate and $\mathfrak{k} \supset \mathfrak{g} \cap (W_*^{\perp} \otimes W^{\perp})$.

If $\mathfrak{g} = \mathfrak{so}(\infty)$ or $\mathfrak{sp}(\infty)$, then $\mathfrak{k} \subset \mathfrak{g}$ is a finite corank subalgebra if there exists a finite dimensional subspace $W \subset V$ such that the restriction of ω on W is non-degenerate and $\mathfrak{k} \supset \Lambda^2(W^{\perp})$ or $S^2(W^{\perp})$ respectively.

We define $\mathbb{T}_{\mathfrak{g}}$ as a full subcategory of $\mathfrak{g}$-modules whose objects M satisfy the following conditions

- M is integrable.
- For every $m \in M$ the annihilator of m in $\mathfrak{g}$ is a finite corank subalgebra.
- M has finite length.

It is not difficult to see that $\mathbb{T}_{\mathfrak{g}}$ is closed under tensor product, hence it is a monoidal category. However, it is not rigid since there is no a reasonable duality functor on $\mathbb{T}_{\mathfrak{g}}$. The following results relate tensor representations of $\mathfrak{g}$ with $\mathbb{T}_{\mathfrak{g}}$.

Theorem 4 ([2])

- *For* $\mathfrak{g} = \mathfrak{gl}(\infty)$ *or* $\mathfrak{sl}(\infty)$ *all (up to isomorphism) simple objects of* $\mathbb{T}_{\mathfrak{g}}$ *are* $V^{\lambda,\mu}$.
- *For* $\mathfrak{g} = \mathfrak{so}(\infty)$ *or* $\mathfrak{sp}(\infty)$ *all (up to isomorphism) simple objects of* $\mathbb{T}_{\mathfrak{g}}$ *are* V^{λ}.
- $\tilde{V}^{\lambda,\mu}$ *(respectively* $\tilde{V}^{\lambda}$*) are all up to isomorphism indecomposable injective in* $\mathbb{T}_{\mathfrak{g}}$.

To prove injectivity of $\tilde{V}^{\lambda,\mu}$ we use the fact (proven in [5]) that for any integrable $\mathfrak{g}$-module M, the integrable part of M^* is injective in the category of integrable modules. From this it is easy to see that the functor Γ from the category of all integrable $\mathfrak{g}$-modules to $\mathbb{T}_{\mathfrak{g}}$ defined by

$$\Gamma(M) = \bigcup M^{\mathfrak{k}}$$

(where the union is taken over all finite corank $\mathfrak{k} \subset \mathfrak{g}$) maps an injective module in the category of integrable modules to an injective module in $\mathbb{T}_{\mathfrak{g}}$. On the other hand, by a direct calculation done in [2]

$$\Gamma(V^{\otimes m} \otimes V_*^{\otimes n}) = \bigoplus_{k \le m, l \le n} (V^{\otimes k} \otimes V_*^{\otimes l})^{\oplus c(k,l)}.$$

Note that Γ can be considered as a certain version of the Zuckerman functor [9].

There exists a Borel subalgebra $\mathfrak{b} \subset \mathfrak{g}$ such that all simple modules of $\mathbb{T}_\mathfrak{g}$ are highest weight modules. For instance, if $\mathfrak{g} = \mathfrak{sl}(\infty)$, considered as the algebra of matrices $(a_{ij})_{i,j \in \mathbb{N}}$, we define a nonstandard total order on $\mathbb{N}$ by $1 < 3 < 5 < \cdots < 6 < 4 < 2$ and positive roots $\varepsilon_i - \varepsilon_j$ for all $i < j$. The corresponding infinite "Dynkin diagram" is

$$\circ - \circ - \cdots - \circ - \circ.$$

Lemma 1 *If S is a non-zero quotient of the Verma module and $S \in \mathbb{T}_\mathfrak{g}$, then S is simple.*

To show that the the category $\mathbb{T}_\mathfrak{g}$ is Koszul we use the following

Lemma 2 ([2])

- *If $\mathfrak{g} = \mathfrak{gl}(\infty)$ or $\mathfrak{sl}(\infty)$, then*

$$\mathrm{Ext}^k(V^{\lambda,\mu}, V^{\nu,\kappa}) \neq 0$$

implies $|\lambda| - |\nu| = |\mu| - |\kappa| = k$.
- *If $\mathfrak{g} = \mathfrak{so}(\infty)$ or $\mathfrak{sp}(\infty)$, then*

$$\mathrm{Ext}^k(V^{\lambda}, V^{\nu}) \neq 0$$

implies $|\lambda| - |\nu| = 2k$.

Let $T = T(V)$ for $\mathfrak{g} = \mathfrak{so}(\infty)$ or $\mathfrak{sp}(\infty)$ and $T = \bigoplus_{m,n \ge 0} V^{\otimes m} \otimes V_*^{\otimes n}$ for $\mathfrak{g} = \mathfrak{sl}(\infty)$ or $\mathfrak{gl}(\infty)$. For $\mathfrak{g} = \mathfrak{gl}(\infty)$ or $\mathfrak{sl}(\infty)$ we set

$$\mathscr{A}_\mathfrak{g}^k = \bigoplus_{m,n \ge 0} \mathrm{Hom}_\mathfrak{g}(V^{\otimes m} \otimes V_*^{\otimes n}, V^{\otimes m-k} \otimes V_*^{\otimes n-k}),\ \mathscr{A}_\mathfrak{g} = \bigoplus_{k \ge 0} \mathscr{A}_\mathfrak{g}^k.$$

For $\mathfrak{g} = \mathfrak{so}(\infty)$ or $\mathfrak{sp}(\infty)$ set

$$\mathscr{A}_\mathfrak{g}^k = \bigoplus_{n \ge 0} \mathrm{Hom}_\mathfrak{g}(V^{\otimes n}, V^{\otimes n-2k}),\ \mathscr{A}_\mathfrak{g} = \bigoplus_{k \ge 0} \mathscr{A}_\mathfrak{g}^k.$$

Note that $\mathscr{A}_\mathfrak{g}$ is by definition a graded algebra. It does not have the identity but it is a direct limit of unital algebras.

Theorem 5 ([2])

- *The category $\mathbb{T}_\mathfrak{g}$ is antiequivalent to the category $A_\mathfrak{g}$-fmod of locally unitary finite-dimensional $A_\mathfrak{g}$-modules.*
- *$A_\mathfrak{g}$ is a direct limit of Koszul rings.*

For $\mathfrak{g} = \mathfrak{gl}(\infty)$ and $\mathfrak{sl}(\infty)$ the corresponding algebras $\mathscr{A}_{\mathfrak{g}}$ are the same. It is shown in [2] that $\mathscr{A}_{\mathfrak{g}}^0 = \bigoplus_{m,n\geq 0} \mathbb{C}[S_n \times S_m]$ and $\mathscr{A}_{\mathfrak{g}}^1$ is generated by contractions.

In this case one can prove that $\mathscr{A}_{\mathfrak{g}}$ is Koszul self-dual, i.e. $(A_{\mathfrak{g}}^!)^{op} \simeq \mathscr{A}_{\mathfrak{g}}$. That implies the following formulas for extensions between simple modules

$$\dim \mathrm{Ext}^k(V^{\lambda',\mu'}, V^{\lambda,\mu}) = \sum_{|\gamma|=k} N^{\lambda}_{\gamma,\lambda'} N^{\mu}_{\gamma^{\perp},\mu'}.$$

It is also shown in [2] that for $\mathfrak{g} = \mathfrak{so}(\infty)$ and $\mathfrak{sp}(\infty)$ $\mathscr{A}_{\mathfrak{g}}^0 = \bigoplus_{n\geq 0} \mathbb{C}[S_n]$ and $\mathscr{A}_{\mathfrak{g}}^1$ is generated by contractions. Knowing this it is easy to obtain an isomorphism

$$\mathscr{A}_{\mathfrak{so}(\infty)} \simeq \mathscr{A}_{\mathfrak{sp}(\infty)}.$$

The latter implies an equivalence of abelian categories $\mathbb{T}_{\mathfrak{sp}(\infty)}$ and $\mathbb{T}_{\mathfrak{so}(\infty)}$ by Theorem 5. Under this equivalence V^{λ} goes to $V^{\lambda^{\perp}}$. It is proven in [8] that this is an equivalence of monoidal categories. A different proof of this fact is given in the next section.

4 Superalgebras

4.1 Direct limits of classical Lie superalgebras

Let $V = V_0 \oplus V_1$ be a superspace, both V_0 and V_1 are countable-dimensional. Below we consider the following possibilities.

- There is a countable-dimensional V_* and a non-degenerate pairing $str : V \otimes V_* \to \mathbb{C}$. Then we set $\mathfrak{gl}(\infty,\infty) = V \otimes V_*$ and $\mathfrak{sl}(\infty,\infty) = \mathrm{Ker}(str)$ with the commutator defined in the same way as in the purely even case (with the usual sign convention). Note that $\mathfrak{g} = \mathfrak{gl}(\infty,\infty)$ has a $\mathbb{Z}$-grading (compatible with $\mathbb{Z}_2$-grading) $\mathfrak{g} = \mathfrak{g}_{-1} \oplus \mathfrak{g}_0 \oplus \mathfrak{g}_1$, where

$$\mathfrak{g}_0 = V_0 \otimes (V_0)_* \oplus (V_1)_* \otimes V_1 \simeq (\mathfrak{gl}(\infty)) \oplus (\mathfrak{gl}(\infty)),$$
$$\mathfrak{g}_1 = V_0 \otimes (V_1)_*, \quad \mathfrak{g}_{-1} = (V_0)_* \otimes V_1.$$

  This grading naturally can be restricted to $\mathfrak{sl}(\infty,\infty)$. The Lie superalgebra $\mathfrak{sl}(\infty,\infty)$ is simple since it can be obtained as a direct limit of simple Lie superalgebras.

- We fix an *even* non-degenerate symmetric form $\omega : S^2(V) \to \mathbb{C}$ and define $\mathfrak{osp}(\infty,\infty)$ as the subalgebra of operators in V of finite rank preserving ω. One can identify $\mathfrak{osp}(\infty,\infty)$ with $\Lambda^2(V)$ using (1). In this case

$$\mathfrak{g}_0 = \mathfrak{so}(\infty) \oplus \mathfrak{sp}(\infty), \ \mathfrak{g}_1 = V_0 \otimes V_1.$$

  The Lie superalgebra $\mathfrak{osp}(\infty,\infty)$ is simple because it is isomorphic to a direct limit of finite-dimensional simple Lie superalgebras.

- We fix an *odd* non-degenerate form $\omega : S^2(V) \to \mathbb{C}$ and define the Lie superalgebra $P(\infty)$ as the subalgebra of linear operators of finite rank in V preserving ω. The superalgebra $P(\infty)$ has a $\mathbb{Z}$-grading $\mathfrak{g} = \mathfrak{g}_{-1} \oplus \mathfrak{g}_0 \oplus \mathfrak{g}_1$, with

$$\mathfrak{g}_0 = V_0 \otimes V_1 \; \mathfrak{g}_1 = S^2(V_0), \; \mathfrak{g}_{-1} = \Lambda^2(V_1).$$

Note that $\mathfrak{g}_0$ is isomorphic to $\mathfrak{gl}(\infty)$, V_0 and V_1 are standard and costandard representation of $\mathfrak{g}_0$. Furthermore, $\mathfrak{g}_1$ (resp. $\mathfrak{g}_{-1}$) is identified with $S^2(V_0)$ (resp. $\Lambda^2(V_1)$) by setting

$$X_{u,w}(v) = (u,v)w + (w,v)u,$$

for any $u, w \in V_0$, $v \in V$ and

$$X_{u,w}(v) = (u,v)w - (w,v)u,$$

for any $u, w \in V_1$, $v \in V$. Observe that $P(\infty) = \varinjlim P(n)$ is not a simple Lie algebra. Its commutator is a simple ideal $SP(\infty)$ of all traceless matrices in $P(\infty)$.

- Let $J : V \to V$ be an odd operator such that $J^2 = -1$. The Lie superalgebra $Q(\infty)$ is the centralizer of J in $\mathfrak{gl}(\infty, \infty)$. As in the finite-dimensional case $\mathfrak{g}_0 = \mathfrak{gl}(\infty)$ and $\mathfrak{g}_1$ is the adjoint representation of $\mathfrak{g}_0$. Note that $Q(\infty) = \varinjlim Q(n)$ is not simple. It contains a simple ideal $SQ(\infty)$ of odd codimension 1 consisting of operators X such that $str(XJ) = 0$. Note that in contrast with all other cases, $SQ(\infty)$ is the direct limit of $SQ(n)$, but $SQ(n)$ are not simple.

We leave to the reader the definition of finite corank subalgebras in this case.

4.2 $\mathbb{T}_{\mathfrak{g}}$ *for Lie superalgebras*

Now let $\mathfrak{g}$ denote one of the Lie superalgebras defined in the previous section. Let $\mathbb{T}_{\mathfrak{g}}$ be a full subcategory of $\mathfrak{g}$-modules M satisfying the following three conditions:

(1) M is integrable over $\mathfrak{g}_0$, and therefore over $\mathfrak{g}$,
(2) the annihilator of every vector in M is a finite corank subalgebra in $\mathfrak{g}$,
(3) M has finite length over $\mathfrak{g}_0$.

It is clear that $\mathbb{T}_{\mathfrak{g}}$ is an abelian monoidal category. If $\mathfrak{g} \neq Q(\infty)$, in order to avoid the annoying but not essential parity chasing we allow morphisms which change parity, i. e. the standard functor Π changing the parity is an isomorphism in our category. In fact, it is not difficult to show that if $\mathfrak{g} \neq Q(\infty)$, then

$$\mathbb{T}_{\mathfrak{g}} = \mathbb{T}^+_{\mathfrak{g}} \oplus \mathbb{T}^-_{\mathfrak{g}}$$

with $\Pi : \mathbb{T}^+_{\mathfrak{g}} \to \mathbb{T}^-_{\mathfrak{g}}$ defining an equivalence of categories. Threfore, admitting odd isomorphisms in the category $\mathbb{T}_{\mathfrak{g}}$ is the same as studying $\mathbb{T}^+_{\mathfrak{g}}$ instead of $\mathbb{T}_{\mathfrak{g}}$.

4.3 Orthosymplectic superalgebra

Let $\mathfrak{g} = \mathfrak{osp}(\infty,\infty)$. The goal of this subsection is to prove the following

Theorem 6 *The monoidal categories $\mathbb{T}_{\mathfrak{g}}$, $\mathbb{T}_{\mathfrak{sp}(\infty)}$ and $\mathbb{T}_{\mathfrak{so}(\infty)}$ are equivalent.*

We start with studying tensor powers of the standard representation V. If M is a $\mathfrak{g}$-module and $\mathfrak{k} \subset \mathfrak{g}$ is a subalgebra, then $M^{\mathfrak{k}}$ denotes the space of $\mathfrak{k}$-invariants in M.

Lemma 3 (a) *$(V^{\otimes n})^{\mathfrak{so}(\infty)} = V_1^{\otimes n}$ and $(V^{\otimes n})^{\mathfrak{sp}(\infty)} = V_0^{\otimes n}$.*
(b) *$V_0^{\otimes n}$ or $V_1^{\otimes n}$ generates $V^{\otimes n}$ over $\mathfrak{g}$.*

Proof. We have the obvious isomorphism of $\mathfrak{g}_0$-modules

$$V^{\otimes n} \simeq \bigoplus_{p+q=n} (V_0^{\otimes p} \otimes V_1^{\otimes q})^{\oplus C(n,p)}.$$

The identity

$$(V_0^{\otimes p} \otimes V_1^{\otimes q})^{\mathfrak{so}(\infty)} = (V_0^{\otimes p})^{\mathfrak{so}(\infty)} \otimes V_1^{\otimes q}$$

together with the fact that $(V_0^{\otimes p})^{\mathfrak{so}(\infty)} = 0$ for $p \neq 0$ imply $(V^{\otimes n})^{\mathfrak{so}(\infty)} = V_1^{\otimes n}$. The second statement of (a) is similar.

Now we prove that $V_0^{\otimes n}$ generates $V^{\otimes n}$. Assume that the statement is not true. Define the grading on $V^{\otimes n}$ by setting the degree of a homogeneous indecomposable tensor $u = u_1 \otimes \cdots \otimes u_n$ equal the number of $u_i \in V_1$. Pick up an indecomposable u of minimal degree that does not belong to $U(\mathfrak{g})V_0^{\otimes n}$. Then $k = \deg(u) > 0$ and $u_i \in V_1$ for some i. Pick up $e, e' \in V_0$ such that $(e,e') = 1$, $(e,u_1) = \cdots = (e,u_n) = 0$. Then

$$u = \pm X_{e\wedge u_i}(u_1 \otimes \cdots \otimes u_{i-1} \otimes e' \otimes u_{i+1} \otimes \ldots u_n) + v,$$

for some v of degree $k-2$. Note that $\deg(u_1 \otimes \cdots \otimes u_{i-1} \otimes e' \otimes u_{i+1} \otimes \ldots u_n) = k-1$. Hence both v and $(u_1 \otimes \cdots \otimes u_{i-1} \otimes e' \otimes u_{i+1} \otimes \ldots u_n)$ belong to $U(\mathfrak{g})V_0^{\otimes n}$. Therefore $u \in U(\mathfrak{g})V_0^{\otimes n}$. Contradiction. In the same way one can prove that $V_1^{\otimes n}$ generates $V^{\otimes n}$ over $\mathfrak{g}$.

Let

$$T(V) = \bigoplus_{n\geq 0} V^{\otimes n},\ T^{\geq m}(V) = \bigoplus_{n\geq m} V^{\otimes n},\ T^{\leq m}(V) = \bigoplus_{n\leq m} V^{\otimes n}.$$

A linear operator $X \in \mathrm{End}_k(V)$ is called *bounded* if the there are n and m such that $T^{\geq n}(V) \subset \mathrm{Ker} X$ and $\mathrm{Im} X \subset T^{\leq m}(V)$.

We denote by $\mathscr{A}_{\mathfrak{g}}$ the subalgebra of all bounded operators in $\mathrm{End}_{\mathfrak{g}}(T(V))$. Note that $\mathscr{A}_{\mathfrak{g}} = \bigoplus_{m,n\geq 0} \mathrm{Hom}_{\mathfrak{g}}(V^{\otimes m}, V^{\otimes n})$. By $\mathscr{A}_{\mathfrak{so}(\infty)}$ (resp. $\mathscr{A}_{\mathfrak{sp}(\infty)}$) we denote the algebras of bounded operators in $\mathrm{End}_{\mathfrak{so}(\infty)}(T(V_0))$ (resp. $\mathrm{End}_{\mathfrak{sp}(\infty)}(T(V_0))$).

Lemma 3 (a) implies that there are natural homomorphisms

$$\rho_{\mathfrak{so}} : \mathscr{A}_{\mathfrak{g}} \to \mathscr{A}_{\mathfrak{so}(\infty)},\ \rho_{\mathfrak{sp}} : \mathscr{A}_{\mathfrak{g}} \to \mathscr{A}_{\mathfrak{sp}(\infty)}$$

given by the restriction to $T(V)^{\mathfrak{sp}(\infty)}$ and $T(V)^{\mathfrak{so}(\infty)}$ respectively.

Lemma 4 *Both $\rho_{\mathfrak{so}}$ and $\rho_{\mathfrak{sp}}$ are isomorphisms.*

Proof. Injectivity of $\rho_{\mathfrak{so}}$ and $\rho_{\mathfrak{sp}}$ follows from Lemma 3 (b). To prove surjectivity recall from [2] that $\mathscr{A}_{\mathfrak{so}(\infty)}$ (resp. $\mathscr{A}_{\mathfrak{sp}(\infty)}$) are generated by permutation groups acting on $V_0^{\otimes n}$ (resp. $V_1^{\otimes n}$) and contractions. Both permutations and contraction are defined on $T(V)$ and hence lie in the image of $\rho_{\mathfrak{so}}$ (resp. $\rho_{\mathfrak{sp}}$).

Let λ be a partition and π_λ be the corresponding Young projector in $\mathbb{C}[S_n]$. We define

$$\tilde{V}^\lambda = \pi_\lambda(V^{\otimes n}),\ \tilde{V_0}^\lambda = \pi_\lambda(V_0^{\otimes n}),\ \tilde{V_1}^\lambda = \pi_\lambda(V_1^{\otimes n}).$$

Recall that the socles of $\tilde{V_0}^\lambda$ and $\tilde{V_1}^\lambda$ are simple $\mathfrak{g}_0$-modules. We denote them V_0^λ and V_1^λ respectively.

Lemma 5 (a) *$V_0^\lambda \otimes V_1^\mu$ is a simple $\mathfrak{g}_0$-module.*
(b) *Every simple module in $\mathbb{T}_{\mathfrak{g}_0}$ is isomorphic to $V_0^\lambda \otimes V_1^\mu$ for some partitions λ and μ.*
(c) *$\tilde{V}_0^\lambda \otimes \tilde{V}_1^\mu$ is indecomposable injective in $\mathbb{T}_{\mathfrak{g}_0}$ with socle equal to $V_0^\lambda \otimes V_1^\mu$.*

Proof. (a) can be proven by the standard argument using the Jacobson density theorem. Let $v = \sum_{j=1}^m e_j \otimes f_j \neq 0$ for some linearly independent $e_j \in V_0^\lambda$ and $f_j \in V_1^\lambda$. Let $u = e \otimes f$. Since V_0^λ and V_1^μ are simple there exist $X \in U(\mathfrak{so}(\infty))$ and $Y \in U(\mathfrak{sp}(\infty))$ such that $X(e_1) = e$, $Y(f_1) = f$ and $X(e_j) = Y(f_j) = 0$ for $j > 1$. Hence $u = X \otimes Y(v)$. Thus any non-zero v generates the whole $V_0^\lambda \otimes V_1^\mu$.

(b) Let $M \in \mathbb{T}_{\mathfrak{g}_0}$ be simple. Then M as an $\mathfrak{so}(\infty)$-module lies in a slightly bigger category $\hat{T}_{\mathfrak{so}(\infty)}$ of modules satisfying (1) and (2). Therefore M contains a subquotient isomorphic to V_0^λ for some λ. Since M is simple $M = V_0^\lambda \otimes W$, where W is some simple module in $\hat{T}_{\mathfrak{sp}(\infty)}$. Hence $W = V_1^\mu$ for some partition μ.

(c) Injectivity of $\tilde{V}_0^\lambda$ in $\mathbb{T}_{\mathfrak{so}(\infty)}$ implies injectivity of $\tilde{V}_0^\lambda \otimes \tilde{V}_1^\mu$ in $\mathbb{T}_{\mathfrak{so}(\infty)}$. By the same reason $\tilde{V}_0^\lambda \otimes \tilde{V}_1^\mu$ is injective in $\mathbb{T}_{\mathfrak{sp}(\infty)}$. For any simple $V_0^{\lambda'} \otimes V_1^{\mu'}$ we have that an exact sequence

$$0 \to \tilde{V}_0^\lambda \otimes \tilde{V}_1^\mu \to M \to V_0^{\lambda'} \otimes V_1^{\mu'} \to 0$$

splits over $\mathfrak{so}(\infty)$ and $\mathfrak{sp}(\infty)$. Hence it splits over $\mathfrak{g}_0$. That proves injectivity of $\tilde{V}_0^\lambda \otimes \tilde{V}_1^\mu$. Irreducibility of the socle follows from the identity

$$\mathrm{Hom}_{\mathfrak{g}_0}(V_0^{\lambda'} \otimes V_1^{\mu'}, \tilde{V}_0^\lambda \otimes \tilde{V}_1^\mu) \simeq \mathrm{Hom}_{\mathfrak{so}(\infty)}(V_0^{\lambda'}, \tilde{V}_0^\lambda) \otimes \mathrm{Hom}_{\mathfrak{sp}(\infty)}(V_1^{\mu'}, \tilde{V}_1^\mu).$$

We define functors $R_{\mathfrak{so}} : \mathbb{T}_{\mathfrak{g}} \to \mathbb{T}_{\mathfrak{so}(\infty)}$ and $R_{\mathfrak{sp}} : \mathbb{T}_{\mathfrak{g}} \to \mathbb{T}_{\mathfrak{sp}(\infty)}$ by

$$R_{\mathfrak{so}}(M) = M^{\mathfrak{sp}(\infty)},\ R_{\mathfrak{sp}}(M) = M^{\mathfrak{so}(\infty)}.$$

Lemma 6 *If $M \in \mathbb{T}_{\mathfrak{g}}$ and $M \neq 0$, then $R_{\mathfrak{so}}(M) \neq 0$ and $R_{\mathfrak{sp}}(M) \neq 0$.*

Proof. Let $L \simeq V_0^\lambda \otimes V_1^\mu$ be a simple submodule in $\mathrm{soc}_{\mathfrak{g}_0}(M)$ with maximal $|\lambda|$. Consider the natural morphism $\theta : L \otimes \mathfrak{g}_1 \to M$ of $\mathfrak{g}_0$-modules given by $\theta(u \otimes X_w) = X_w(u)$. Then $\mathrm{soc}_{\mathfrak{g}_0}(\mathrm{Im}\theta)$ has only constituents $V_0^\kappa \otimes V_1^\nu$ with $|\kappa| < |\lambda|$. Therefore if $u \otimes w \in L$ is such that $u = \pi_\lambda(u_1 \otimes \cdots \otimes u_n)$, $w \in V^{\otimes|\mu|}$ and $e \otimes f \in V_0 \otimes V_1$ is such that $(e, u_i) = 0$ for all $i = 1, \ldots, n$, then

$$\theta(u \otimes w \otimes e \otimes f) = X_{e\wedge f}(u \otimes w) = 0. \tag{2}$$

Pick up $e, e' \in V_0, f, f' \in V_1$ such that $(e, u_i) = (e', u_i) = 0$ for all $i = 1, \ldots, n$, $(e, e') = 1$ and $(f, f') = 0$. Then $X_{f\wedge f'} = [X_{e\wedge f}, X_{e'\wedge f'}]$. By (2) $X_{f\wedge f'}(u \otimes w) = 0$. It is easy to see that $X_{f\wedge f'}$ for all orthogonal f, f' generate $\mathfrak{sp}(\infty)$ we obtain $\mathfrak{sp}(\infty)w = 0$. Hence $\mu = 0$, i.e. $L \subset R_{\mathfrak{so}}(M)$.

The proof that $R_{\mathfrak{sp}}(M) \neq 0$ is similar.

Now we are ready to describe simple objects in $\mathbb{T}_\mathfrak{g}$. Let λ be a partition with $|\lambda| = n$. Choose a Cartan subalgebra $\mathfrak{h}$ such that the roots of $\mathfrak{g}$ are as in $D(\infty, \infty)$. The even roots of $\mathfrak{g}$ are $\{\pm\varepsilon_i \pm \varepsilon_j | i, j > 0, i \neq j\} \cup \{\pm\delta_i \pm \delta_j | i, j > 0\}$ and the odd roots are $\{\pm(\varepsilon_i \pm \delta_j | i, j > 0\}$. Let $\mathfrak{b}$ be the Borel subalgebra defined by the set of positive roots

$$\{\varepsilon_i \pm \varepsilon_j | i < j\} \cup \{\delta_i \pm \delta_j | i < j\} \cup \{2\delta_i | i > 0\} \cup \{\varepsilon_i \pm \delta_j | i, j > 0\}.$$

Let V^λ denote the simple highest weight module with highest weight $\lambda = \sum \lambda_i \varepsilon_i$. We introduce the standard order on weights by setting $\lambda \leq \mu$ if $\mu - \lambda$ is a non-negative integral linear combination of positive roots.

Lemma 7 $V^\lambda \in \mathbb{T}_\mathfrak{g}$. *Any simple object in* $\mathbb{T}_\mathfrak{g}$ *is isomorphic to* V^λ *for some partition* λ.

Proof. Recall that V_0^λ is a highest weight module over $\mathfrak{so}(\infty)$. Hence there exists a unique up to proportionality $v \in V_0^\lambda \subset \tilde{V}^\lambda$ of weight $\lambda = \sum \lambda_i \varepsilon_i$. An easy calculation shows that $\mathfrak{n}(v) = 0$. By Frobenius reciprocity there exists a non-zero homomorphism ψ from the Verma module M^λ to $\tilde{V}^\lambda$. The image of ψ has a unique simple quotient isomorphic to V^λ. Thus $V^\lambda \in \mathbb{T}_\mathfrak{g}$.

Now let $M \in \mathbb{T}_\mathfrak{g}$ be a simple module. Pick up a $\mathfrak{g}_0$-submodule $L \simeq V_0^\lambda$ in $R_{\mathfrak{so}}(M)$ with maximal $|\lambda|$. Then a non-zero $v \in L$ of weight λ is annihilated by $\mathfrak{n}$. Therefore $M \simeq V^\lambda$.

Lemma 8 *Let* $M \in \mathbb{T}_\mathfrak{g}$ *be a non-zero quotient of the Verma module* M^λ *for some partition* λ. *Then* $M \simeq V^\lambda$.

Proof. Suppose that the statement is false. Then there exists $V^\mu \subset M$ with $\mu < \lambda$. Let $v_\mu \in V^\mu$ be a highest vector of weight μ and v_λ be a non-zero vector of weight λ. Then $v_\mu \in U(\mathfrak{n}^-)v_\lambda$ and therefore $v_\mu \in U(\mathfrak{n}^- \cap \mathfrak{g}')v_\lambda$ for some finite-dimensional $\mathfrak{g}' = \mathfrak{osp}(2p, 2q) \subset \mathfrak{g}$. Without loss of generality we may assume $p > 2q + |\mu|$. Therefore the quadratic Casimir element in $U(\mathfrak{g}')$ has the same eigenvalue on v_μ and v_λ. This is impossible since $(\lambda + \rho, \lambda + \rho) \neq (\mu + \rho, \mu + \rho)$ if $p > 2q + |\mu|$, here ρ is the half sum of even positive roots minus the half sum of odd positive roots of $\mathfrak{g}'$.

Lemma 9 *Let*

$$0 \to V^\lambda \to M \to V^\mu \to 0$$

be a non-split exact sequence in $\mathbb{T}_{\mathfrak{g}}$, *then* $\mu < \lambda$ *in the standard order. Furthermore,* $|\mu| < |\lambda|$.

Proof. If μ and λ are not comparable then a vector of weight μ spans in M a submodule isomorphic to V^μ and the sequence splits. Since $M \in \mathbb{T}_{\mathfrak{g}_0}$, M is semisimple over $\mathfrak{h}$, the sequence also splits in the case $\mu = \lambda$. If $\mu > \lambda$, then M is a quotient of the Verma module M^μ. By Lemma 8 $M \simeq V^\mu$. Therefore the only possibility is $\mu < \lambda$. Note that this implies $|\mu| \leq |\lambda|$. We claim that $|\mu| < |\lambda|$. Indeed, let $|\mu| = |\lambda|$. For any module N set N^+ be the span of all weight spaces of weights $\sum a_i\varepsilon_i + \sum b_j\delta_j$ with $\sum a_i = |\lambda|$. If $\mathfrak{k} \simeq \mathfrak{gl}(\infty)$ be the subalgebra in $\mathfrak{g}$ generated by the roots of the form $\varepsilon_i - \varepsilon_j$, then N^+ is obviously $\mathfrak{k}$-stable. It is easy to see that $(V^\lambda)^+$ and $(V^\mu)^+$ are simple $\mathfrak{k}$-submodules in the tensor algebra of the standard $\mathfrak{k}$-module. Therefore

$$0 \to (V^\lambda)^+ \to M^+ \to (V^\mu)^+ \to 0$$

splits over $\mathfrak{k}$. But then the original sequence must split as well by Lemma 8.

Lemma 10 $\tilde{V}^\lambda$ *is injective in* $\mathbb{T}_{\mathfrak{g}}$.

Proof. It is sufficient to prove that for any μ an exact sequence

$$0 \to \tilde{V}^\lambda \to M \to V^\mu \to 0$$

of modules in $\mathbb{T}_{\mathfrak{g}}$ splits. If $|\mu| \geq |\lambda|$, then the sequence splits by Lemma 9, as $|\mu| \geq |\nu|$ for any simple subquotient V^ν in $\tilde{V}^\lambda$. Hence we may assume $|\mu| < |\lambda|$.

By Lemma 5 $\tilde{V}^\lambda$ is injective in $\mathbb{T}_{\mathfrak{g}_0}$. Therefore the sequence splits over $\mathfrak{g}_0$. Thus, we can write $M = V^\mu \oplus \tilde{V}^\lambda$ as a $\mathfrak{g}_0$-module. The action of $\mathfrak{g}_1$ is given by $X(u,w) = (Xu, c(u \otimes X) + Xw)$ for any $u \in V^\mu, w \in \tilde{V}^\lambda$, $X \in \mathfrak{g}_1$ and some $c \in \mathrm{Hom}_{\mathfrak{g}_0}(V^\mu \otimes \mathfrak{g}_1, \tilde{V}^\lambda)$. By Lemma 5

$$\mathrm{soc}_{\mathfrak{g}_0}\tilde{V}^\lambda = \bigoplus_{(\nu,\nu')} V_0^\nu \otimes V_1^{\nu'}$$

for some set of pairs (ν,ν') such that $|\nu| + |\nu'| = |\lambda|$. Therefore we have

$$c(V_0^\mu \otimes \mathfrak{g}_1) \cap \mathrm{soc}_{\mathfrak{g}_0}\tilde{V}^\lambda = \bigoplus_{\nu} V_0^\nu \otimes V_1,$$

where the summation is taken over some set of partitions ν such that $|\nu| = |\lambda| - 1$. If $u \in V_0^\mu$, then

$$c(u \otimes X_{e\wedge f}) = \sum a_\nu \pi_\nu(u \otimes e) \otimes f,$$

for some $a_\nu \in \mathbb{C}$. We claim that in fact all $a_\nu = 0$. Indeed, assume $a_\nu \neq 0$. Let $u = \pi_\lambda(u_1 \otimes \cdots \otimes u_n) \in V_0^\lambda$ for some linearly independent isotropic mutually orthogonal $u_1, \dots, u_n$. For any $e \in V_0$ linearly independent of $u_1, \dots, u_n$ and any non-zero $f \in V_1$, we have $c(u \otimes X_{e\wedge f}) \neq 0$. But then the annihilator of $u \in M$ is not a finite corank subalgebra, hence M is not in $\mathbb{T}_{\mathfrak{g}}$. Contradiction.

Thus, $c(V_0^\mu \otimes \mathfrak{g}_1) = 0$. Let $v_\mu \in V_0^\mu$ be the highest vector. Then $(v_\mu, 0) \in M$ is $\mathfrak{n}$-invariant. Consider the submodule $N \subset M$ generated by $(v_\mu, 0)$. By Lemma 8 $N \simeq V^\mu$. Therefore the exact sequence splits.

Corollary 1 *The socle of* $\tilde{V}^\lambda$ *coincides with* V^λ.

Proof. Follows from Lemma 6 and 7 since $R_{\mathfrak{so}}(\tilde{V}^\lambda) = \tilde{V}_0^\lambda$.

Corollary 2

$$R_{\mathfrak{so}}(V^\lambda) = V_0^\lambda,\ R_{\mathfrak{sp}}(V^\lambda) = V_1^{\lambda^\perp}. \tag{3}$$

Proof. By Corollary 1 and Lemma 10

$$V^\lambda = \bigcap_{\varphi \in \mathrm{Hom}_{\mathfrak{g}}(\tilde{V}^\lambda, T^{\leq |\lambda|-1}(V))} \mathrm{Ker}\varphi.$$

Therefore Lemma 4 implies the statement.

One can prove as in [2] that $\mathbb{T}_{\mathfrak{g}}$ is antiequvivalent to the category of locally unitary finite-dimensional $\mathscr{A}_{\mathfrak{g}}$- modules. Therefore Lemma 4 implies that $\mathbb{T}_{\mathfrak{osp}(\infty,\infty)}$ and $\mathbb{T}_{\mathfrak{so}(\infty)}$ are equivalent abelian categories.

Define now the functors $S_{\mathfrak{so}} : \mathbb{T}_{\mathfrak{so}(\infty)} \to \mathbb{T}_{\mathfrak{g}}$ and $S_{\mathfrak{sp}} : \mathbb{T}_{\mathfrak{sp}(\infty)} \to \mathbb{T}_{\mathfrak{g}}$ as follows. Let $M \in \mathbb{T}_{\mathfrak{so}(\infty)}$ (resp. $\mathbb{T}_{\mathfrak{sp}(\infty)}$). By $I(M)$ we denote the induced module $U(\mathfrak{g}) \otimes_{U(\mathfrak{g}_0)} M$, where we define the action of $\mathfrak{sp}(\infty)$ (resp. $\mathfrak{so}(\infty)$) on M to be trivial. We set

$$S_{\mathfrak{so}}(M) = I(M)/(\bigcap_{\varphi \in \mathrm{Hom}_{\mathfrak{g}}(I(M), T(V))} \mathrm{Ker}\varphi),$$

respectively

$$S_{\mathfrak{sp}}(M) = I(M)/(\bigcap_{\varphi \in \mathrm{Hom}_{\mathfrak{g}}(I(M), T(V))} \mathrm{Ker}\varphi).$$

Observe that $\mathrm{Hom}_{\mathfrak{g}}(I(M), T^{\geq n}(V)) = \mathrm{Hom}_{\mathfrak{g}_0}(M, T^{\geq n}(V)) = 0$ for sufficiently large n. Thus, $S_{\mathfrak{so}}(M)$ (resp. $S_{\mathfrak{sp}}(M)$) have finite length over $\mathfrak{g}_0$ and hence lie in $\mathbb{T}_{\mathfrak{g}}$. It is not hard to see that $S_{\mathfrak{so}}(M)$ (resp. $S_{\mathfrak{sp}}(M)$) is the maximal quotient of $I(M)$ belonging to $\mathbb{T}_{\mathfrak{g}}$. Hence by Frobenius reciprocity

$$\begin{aligned} \mathrm{Hom}_{\mathfrak{g}}(S_{\mathfrak{so}}(M), N) &\simeq \mathrm{Hom}_{\mathfrak{so}(\infty)}(M, R_{\mathfrak{so}}(N)), \\ \mathrm{Hom}_{\mathfrak{g}}(S_{\mathfrak{sp}}(M), N) &\simeq \mathrm{Hom}_{\mathfrak{sp}(\infty)}(M, R_{\mathfrak{sp}}(N)). \end{aligned} \tag{4}$$

Proposition 1 *The functors* $S_{\mathfrak{so}}$ *(resp.* $S_{\mathfrak{sp}}$*) and* $R_{\mathfrak{so}}$ *(resp.* $R_{\mathfrak{sp}}$*) are mutually inverse equivalences between* $\mathbb{T}_{\mathfrak{so}(\infty)}$ *(resp.* $\mathbb{T}_{\mathfrak{sp}(\infty)}$*) and* $\mathbb{T}_{\mathfrak{g}}$.

Proof. A functor $F : \mathbb{T}_{\mathfrak{g}} \to \mathbb{T}_{\mathfrak{so}(\infty)}$ establishing an equivalence can be taken as a composition of $F_1 : \mathscr{A}_{\mathfrak{g}}\text{-fmod} \to \mathbb{T}_{\mathfrak{so}(\infty)}$ and $F_2 : \mathbb{T}_{\mathfrak{g}} \to \mathscr{A}_{\mathfrak{g}}\text{-fmod}$, where $F_1 = \mathrm{Hom}_{\mathscr{A}_{\mathfrak{g}}}(\cdot, T(V_0))$ and $F_2 = \mathrm{Hom}_{\mathfrak{g}}(\cdot, T(V))$. Lemma 4 implies $R_{\mathfrak{so}} = F_1 \circ F_2$. Since $S_{\mathfrak{so}}$ is left adjoint to $R_{\mathfrak{so}}$ by (4), $S_{\mathfrak{so}}$ must be inverse of $R_{\mathfrak{so}}$ by general nonsense.

The case of $\mathfrak{sp}(\infty)$ is similar.

To prove Theorem 6 we claim the following.

Proposition 2 *The functors $S_{\mathfrak{so}}$ (resp. $S_{\mathfrak{sp}}$) and $R_{\mathfrak{so}}$ (resp. $R_{\mathfrak{sp}}$) are equivalences of monoidal categories.*

The proof of the above Proposition amounts to showing that $R_{\mathfrak{so}}$ and $R_{\mathfrak{sp}}$ preserve tensor products. It is an easy consequence of the following curious fact. We leave its proof to the reader.

Lemma 11 *Let $\mathfrak{k} = \mathfrak{gl}(\infty), \mathfrak{sl}(\infty), \mathfrak{so}(\infty), \mathfrak{sp}(\infty)$. In the category $\mathbb{T}_{\mathfrak{k}}$ the functor of invariants preserves tensor product. In other words, $(M \otimes N)^{\mathfrak{k}} = M^{\mathfrak{k}} \otimes N^{\mathfrak{k}}$.*

4.4 The case of $\mathfrak{gl}(\infty,\infty)$

The case $\mathfrak{g} = \mathfrak{gl}(\infty,\infty)$ is similar to the case of $\mathfrak{osp}(\infty,\infty)$. Therefore we only state the results omitting the proofs. Recall that the even part $\mathfrak{g}_0$ is a direct sum of two copies of $\mathfrak{gl}(\infty)$. Let $\mathfrak{k} = V_0 \otimes (V_0)_*$ and $\mathfrak{l} = V_1 \otimes (V_1)_*$. We define the functors $R_{\mathfrak{k}} : \mathbb{T}_{\mathfrak{g}} \to \mathbb{T}_{\mathfrak{k}}$, $R_{\mathfrak{l}} : \mathbb{T}_{\mathfrak{g}} \to \mathbb{T}_{\mathfrak{l}}$ by $R_{\mathfrak{k}}(M) = M^{\mathfrak{l}}$, $R_{\mathfrak{l}}(M) = M^{\mathfrak{k}}$ and $S : \mathbb{T}_{\mathfrak{gl}(\infty)} \to \mathbb{T}_{\mathfrak{gl}(\infty|\infty)}$ by

$$S(M) = I(M)/(\bigcap_{\varphi \in \mathrm{Hom}(I(M), T(V \oplus V_*))} \mathrm{Ker}\varphi,$$

where $I(M) = U(\mathfrak{g}) \otimes_{U(\mathfrak{g}_0)} M$.

It is not hard to see that S is left adjoint to $R_{\mathfrak{k}}$ and $R_{\mathfrak{l}}$.

Theorem 7 *The functors $R_{\mathfrak{k}}$ and S establish an equivalence of monoidal categories $\mathbb{T}_{\mathfrak{gl}(\infty,\infty)}$ and $\mathbb{T}_{\mathfrak{gl}(\infty)}$.*

Note that the composition $S \circ R_{\mathfrak{l}} : \mathbb{T}_{\mathfrak{k}} \to \mathbb{T}_{\mathfrak{l}}$ defines an autoequivalence of the category $\mathbb{T}_{\mathfrak{gl}(\infty)}$ that sends $V^{\lambda,\mu} \to V^{\lambda^{\perp},\mu^{\perp}}$.

4.5 The case of $P(\infty)$

It turns out that for $\mathfrak{g} = P(\infty)$ the category $\mathbb{T}_{\mathfrak{g}}$ is also equivalent to $\mathbb{T}_{\mathfrak{so}(\infty)}$ and hence to $\mathbb{T}_{\mathfrak{sp}(\infty)}$ and $\mathbb{T}_{\mathfrak{osp}(\infty|\infty)}$. However, the proof is different in this case.

We claim that any module M in $\mathbb{T}_{\mathfrak{g}}$ can be equipped with $\mathbb{Z}$-grading $M = \bigoplus M^k$ such that $\mathfrak{g}_i M^k \subset M^{k+2i}$. Indeed, note that any simple $\mathfrak{g}_0$-subquotient of M is isomorphic $V_0^{\lambda,\mu}$, and we assign to it degree $|\lambda| - |\mu|$.

Define a functor $R : \mathbb{T}_{\mathfrak{g}} \to \mathbb{T}_{\mathfrak{g}_0}$ by

$$R(M) = M^{\mathfrak{g}_1}.$$

Lemma 12 (a) *For any $M \in \mathbb{T}_{\mathfrak{g}}$, $R(M) \neq 0$.*
(b) *If M is simple then $R(M)$ is simple.*
(c) *If $R(M) \simeq R(L)$ for two simple $M, L \in \mathbb{T}_{\mathfrak{g}}$, then $M \simeq L$.*

Proof. Since M has finite length over $\mathfrak{g}_0$ there exists a maximal k such that $M^k \neq 0$. Then $\mathfrak{g}_1(M^k) \subset M^{k+2} = 0$. That proves (a).

(b) Let k be maximal such that $M^k \neq 0$. If N is a proper submodule in M^k, then $U(\mathfrak{g})N = U(\mathfrak{g}_{-1})N$ is a proper submodule in M since $(U(\mathfrak{g})N)^k = N$. Therefore M^k is a simple $\mathfrak{g}_0$-module. If $R(M)^i \neq 0$ for some $i < k$, then $U(\mathfrak{g})(R(M)^i)$ is a proper non-zero submodule in M. Hence if M is simple, then $R(M) = M^k$ is simple.

(c) It is clear that both M and L are simple quotients of the parabolically induced module $U(\mathfrak{g}) \otimes_{U(\mathfrak{g}_0 \oplus \mathfrak{g}_1)} R(M)$. But the latter has a unique simple quotient. Hence (c).

Lemma 13 *Let* $M \in \mathbb{T}_{\mathfrak{g}}$.
(a) *If* $V_0^\lambda \subset M$ *is an embedding of* $\mathfrak{g}_0$*-modules, then* $V_0^\lambda \subset R(M)$.
(b) *Any simple submodule in* $R(M)$ *is isomorphic to* V_0^λ *for some partition* λ.

Proof. (a) Consider the map $\psi : \mathfrak{g}_1 \otimes V_0^\lambda \to M$ defined by $\psi(X \otimes v) = Xv$. Suppose $|\lambda| = k$. Consider linearly independent $u_1, \dots, u_k \in V_0$. Then $u = \pi_\lambda(u_1 \otimes \cdots \otimes u_k) \in V_0^\lambda$ is not zero. Moreover, for any $w, v \in V_0$ we have

$$\psi(X_{w,v} \otimes u) = \sum_\nu a_\nu \pi_\nu(w \otimes v \otimes u)$$

for some ν obtained from λ by adding two boxes not in the same column and some $a_\nu \in \mathbb{C}$. Note that if some $a_\nu \neq 0$, then for any w, v such that $u_1, \dots, u_k, v, w$ are linearly independent, we have $\psi(X_{w,v} \otimes u) \neq 0$. But then the annihilator of $v \in V_0^\lambda$ is not of finite corank. Hence $\psi = 0$, i.e. $V_0^\lambda \subset R(M)$.

(b) Let L be a simple $\mathfrak{g}_0$-submodule in $R(M)$. Then $L \simeq \mathrm{soc}_{\mathfrak{g}_0}(V_0^\lambda \otimes V_1^\mu)$. We want to show that $\mu = 0$. Assume the opposite. Consider the $\mathfrak{g}_0$-homorphism $\psi : L \otimes \mathfrak{g}_{-1} \to M$ given by $\psi(w \otimes X) = X(w)$ for all $X \in \mathfrak{g}_{-1}, w \in L$. Since the annihilator of any vector in M has finite corank we have

$$\psi(\mathrm{soc}_{\mathfrak{g}_0}(L \otimes \mathfrak{g}_{-1})) = 0. \tag{5}$$

Let $w \in L$ be of the form

$$w = \pi_\lambda(v_1 \otimes \cdots \otimes v_n) \otimes \pi_\mu(u_1 \otimes \cdots \otimes u_m),$$

where $v_i \in V_0, u_j \in V_1$ are linearly independent, $n = |\lambda|, m = |\mu|$, $(v_i, u_j) = 0$ for all $i \leq n, j \leq m$. Let $f_1, f_2 \in V_1$ be orthogonal to $v_1, \dots, v_n$ and linearly independent with $u_1, \dots, u_m$. By (5)

$$\psi(w \otimes X_{f_1,f_2}) = X_{f_1,f_2}(w) = 0. \tag{6}$$

Let $e \in V_0$. Then for any $v \in V_0, u \in V_1$ we have

$$\begin{aligned} [X_{e,e}, X_{f_1,f_2}](v) &= 2((f_1,v)(f_2,e) - (f_2,v)(f_1,e))e, \\ [X_{e,e}, X_{f_1,f_2}](u) &= 2(e,u)((f_1,e)f_2 - (f_2,e)f_1). \end{aligned} \tag{7}$$

Pick up $e \in V_0, f_1, f_2 \in V_1$ such that $(e, f_1) = 1, (e, f_2) = 0$, $(e, u_1) = 1$ and $(e, u_i) = 0$ for $i > 1$. Then by (6) $[X_{e,e}, X_{f_1,f_2}](w) = 0$ and by (7)

$$[X_{e,e}, X_{f_1,f_2}](w) = 2\pi_\lambda(v_1 \otimes \cdots \otimes v_n) \otimes \pi_\mu(f_2 \otimes u_2 \otimes \cdots \otimes u_m) \neq 0.$$

Contradiction.

Corollary 3 *For any simple $M \in \mathbb{T}_{\mathfrak{g}}$, $R(M) \simeq V_0^{\lambda}$ for some λ.*

We use the notation V^{λ} for the simple $M \in \mathbb{T}_{\mathfrak{g}}$ with $R(M) = V_0^{\lambda}$.

We will prove now that $V^{\otimes n}$ is injective in $\mathbb{T}_{\mathfrak{g}}$. Consider the action of S_n on $V^{\otimes n}$ such that an adjacent transposition $\sigma_{i,i+1}$ acts by

$$\sigma_{i,i+1}(v_1 \otimes \cdots \otimes v_n) = (-1)^{p(v_i)p(v_{i+1})} v_1 \otimes \cdots \otimes v_{i+1} \otimes v_i \otimes \cdots \otimes v_n.$$

This action commutes with the action of $\mathfrak{g}$.

Proposition 3 (a) $R(V^{\otimes n}) = (V_0)^{\otimes n}$;
(b) $\mathrm{soc} V^{\otimes n} = U(\mathfrak{g}_0)(V_0^{\otimes n})$;
(c) $\mathrm{End}_{\mathfrak{g}}(V^{\otimes n}) = \mathrm{End}_{\mathfrak{g}_0}(V_0^{\otimes n}) = \mathbb{C}[S_n]$;
(d) $V^{\otimes n} = \bigoplus_{|\lambda|=n} \tilde{V}^{\lambda} \otimes \mathbb{Y}_{\lambda}$, *where $\mathbb{Y}_{\lambda}$ is the irreducible S_n-module associated to λ and $\tilde{V}^{\lambda}$ is an indecomposable module with socle V^{λ}.*

Proof. (a) Consider $V^{\otimes n}$ as a $\mathfrak{g}_0$-module. It splits into indecomposable sumands ${V_0}^{\mu} \otimes {V_1}^{\nu}$ with $|\mu| + |\nu| = n$. Hence the statement follows from Lemma 13(b)

(b) By Lemma 12(a) and (a) each $V_0^{\mu} \subset V_0^{\otimes n}$ generates a simple submodule. Therefore (b) follows from (a).

(c) The restriction map: $\mathrm{End}_{\mathfrak{g}}(V^{\otimes n}) \to \mathrm{End}_{\mathfrak{g}_0}(V_0^{\otimes n}) = \mathbb{C}[S_n]$; is obviously surjective. We claim that the quotient of $V^{\otimes n}$ by the socle can have only simple subquotients V^{μ} for $|\mu| < n$. Indeed, if V^{μ} is a simple subquotient of $V^{\otimes n}$, then V_0^{μ} is a simple $\mathfrak{g}_0$-subquotient of $V^{\otimes n}$. Hence $|\mu| \leq n$. On the other hand, if $|\mu| = n$, then $V_0^{\mu} \subset V_0^{\otimes n} \subset \mathrm{soc} V^{\otimes n}$. Therefore any $\varphi \in \mathrm{End}_{\mathfrak{g}}(V^{\otimes n})$ that kills the socle must be zero. That implies injectivity of the restriction map.

(d) is a consequence of (c) and (b).

Note that (b) also implies

Corollary 4 *If $\mathrm{Hom}_{\mathfrak{g}}(V^{\otimes n}, V^{\otimes m}) \neq 0$, then $n - m$ is non-negative even.*

Lemma 14 *Let $M \in \mathbb{T}_{\mathfrak{g}}$, $V_0^{\lambda} \subset R(M)$ generates M, then $M \simeq V^{\lambda}$.*

Proof. Let $|\lambda| = n$. Consider the parabolically induced module $K = U(\mathfrak{g}) \otimes_{U(\mathfrak{g}_0 \oplus \mathfrak{g}_1)} V_0^{\lambda}$. We have an isomorphism $K \simeq \Lambda(\mathfrak{g}_{-1}) \otimes V_0^{\lambda}$ of $\mathfrak{g}_0$-modules. Let $N_0 = \mathrm{soc}_{\mathfrak{g}_0}(\mathfrak{g}_{-1} \otimes V_0^{\lambda})$. First, we show that $\mathfrak{g}_1(N_0) = 0$. For this we fix the Borel subalgebra of $\mathfrak{g}_0$ such that all tensor modules are highest weight modules (see [2] and Sect. 3). If $v \in V_0^{\lambda}$ is a highest vector, $Y \in \mathfrak{g}_{-1}$ is a highest vector with respect to the adjoint action of $\mathfrak{g}_0$, then $Y \otimes v$ generates N_0 (as a $\mathfrak{g}_0$-module). Let $X \in \mathfrak{g}_1$, then $X(Y \otimes v) = 1 \otimes [X, Y]v$. By a straightforward check $[X, Y]v = 0$. Thus, $\mathfrak{g}_1(Y \otimes v) = 0$ and hence the whole N_0 is annihilated by $\mathfrak{g}_1$.

Now let $N \subset K$ be the submodule generated by N_0 and $Q = K/N$. Let $\pi : K \to Q$ denote the natural projection. Note that the annihilator of $\pi(1 \otimes v)$ is of finite corank. Since $U(\mathfrak{g})\pi(1 \otimes v) = Q$, Q satisfies (1) and (2). Note that M is a quotient of K. Consider the natural projection $\sigma : K \to M$. If $\sigma(N_0) \neq 0$, then $Z\sigma(1 \otimes v) \neq 0$ for any

$Z \in \mathfrak{g}_{-1}$, that contradicts our assumption $M \in \mathbb{T}_{\mathfrak{g}}$, as $\sigma(1 \otimes v)$ does not have the annihilator of finite corank. Hence $\sigma(N_0) = 0$ and therefore M is a quotient of Q.

Note that although $K \notin \mathbb{T}_{\mathfrak{g}}$, it is still equipped with the $\mathbb{Z}$-grading such that $K^{n-2k} = \Lambda^k(\mathfrak{g}_{-1}) \otimes V_0^{\lambda}$. Hence both N and Q are also graded. We claim that for all μ and $k > 0$

$$\mathrm{Hom}_{\mathfrak{g}_0}(V_0^{\mu}, Q^{n-2k}) = 0. \tag{8}$$

Indeed, N^{n-2k} is generated by $\Lambda^{k-1}(\mathfrak{g}_{-1})(Y \otimes v)$ over $\mathfrak{g}_0$. Any weight vector in $\Lambda^{k-1}(\mathfrak{g}_{-1})(Y \otimes v)$ has weight $\sum a_i \varepsilon_i$ with at least two negative a_i and hence belongs to $\mathrm{soc}_{\mathfrak{g}_0}^{2k-1} K^{n-2k}$. But then

$$N^{n-2k} \subset \mathrm{soc}_{\mathfrak{g}_0}^{2k-1} K^{n-2k}.$$

Therefore we have an embedding of $\mathfrak{g}_0$-modules

$$\mathrm{soc}_{\mathfrak{g}_0} Q^{n-2k} \subset (\mathrm{soc}_{\mathfrak{g}_0}^{2k} K^{n-2k})/N^{n-2k}.$$

All $\mathfrak{g}_0$-simple subquotients of $\mathrm{soc}_{\mathfrak{g}_0}^{2k} K^{n-2k}$ are of the form $\mathrm{soc}_{\mathfrak{g}_0}(V_0^{\mu} \otimes V_1^{\nu})$ with $|\nu| \geq 1$. Hence (8).

Now we can prove that Q is simple and hence isomorphic to V^{λ}. Indeed, it is equivalent to proving that $R(Q) = V_0^{\lambda}$. Suppose the latter is false, i.e. $R(Q)^{n-2k} \neq 0$ for some $k > 0$. By Lemma 13(a) there exists μ such that $V_0^{\mu} \subset R(Q)^{n-2k}$. But this contradicts (8).

Theorem 8 *$\tilde{V}^{\lambda}$ is the injective hull of V^{λ} in $\mathbb{T}_{\mathfrak{g}}$.*

Proof. It suffices to prove that any exact sequence in $\mathbb{T}_{\mathfrak{g}}$ of the form

$$0 \to \tilde{V}^{\lambda} \to M \to V^{\mu} \to 0$$

splits. Since $\tilde{V}^{\lambda}$ is injective in $\mathbb{T}_{\mathfrak{g}_0}$ the sequence splits over $\mathfrak{g}_0$. In particular, we have an embedding $V_0^{\mu} \subset M$ of $\mathfrak{g}_0$-modules. By Lemma 13(a) $V_0^{\mu} \subset R(M)$. By Lemma 14 V_0^{μ} generates $V^{\mu} \subset M$. Hence the statement.

The above theorem and Proposition 3(d) imply

Corollary 5 *$V^{\otimes n}$ is injective in $\mathbb{T}_{\mathfrak{g}}$.*

Recall that $\mathscr{A}_{\mathfrak{g}}$ denotes the subalgebra of all bounded operators in $\mathrm{End}_{\mathfrak{g}}(T(V))$. Theorem 8 implies

Corollary 6 *$\mathbb{T}_{\mathfrak{g}}$ is antiequivalent to the category of locally unitary finite-dimensional modules over $\mathscr{A}_{\mathfrak{g}}$.*

By Corollary 4 we have a $\mathbb{Z}$-grading

$$\mathscr{A}_{\mathfrak{g}} = \bigoplus_{i \geq 0} \mathscr{A}_{\mathfrak{g}}^i, \ \mathscr{A}_{\mathfrak{g}}^i = \bigoplus_{n \geq 0} \mathrm{Hom}_{\mathfrak{g}}(V^{\otimes n}, V^{\otimes n-2i}).$$

By Proposition 3

$$\mathscr{A}^0_{\mathfrak{g}} = \bigoplus_{n\geq 0} \mathbb{C}[S_n].$$

For any $1 \leq i < j \leq n$ define $\tau^n_{ij} \in \mathrm{Hom}_{\mathfrak{g}}(V^{\otimes n}, V^{\otimes n-2})$ by the formula

$$\tau^n_{ij}(v_1 \otimes \cdots \otimes v_n) = (-1)^s (v_i, v_j) v_1 \otimes \cdots \otimes \hat{v}_i \otimes \cdots \otimes \hat{v}_j \otimes \cdots \otimes v_n,$$

where $s = (p(v_i) + p(v_j))(p(v_1) + \cdots + p(v_{i-1})) + p(v_j)(p(v_{i+1}) + \cdots + p(v_{j-1}))$.

Lemma 15 *$\{\tau^n_{ij}\}$ for all $n > 1$ and $1 \leq i < j \leq n$ form a basis in $\mathscr{A}^1_{\mathfrak{g}}$.*

Proof. It is sufficient to show that for a fixed n, the set $\{\tau^n_{ij}\}$ for all $1 \leq i < j \leq n$ is a basis in $\mathrm{Hom}_{\mathfrak{g}}(V^{\otimes n}, V^{\otimes n-2})$. Linear independence is straightforward. To prove that $\{\tau^n_{ij}\}$ span $\mathrm{Hom}_{\mathfrak{g}}(V^{\otimes n}, V^{\otimes n-2})$ consider the homomorphism

$$\rho : \mathrm{Hom}_{\mathfrak{g}}(V^{\otimes n}, V^{\otimes n-2}) \to \mathrm{Hom}_{\mathfrak{g}_0}((V^{\otimes n})^{n-2}, (V^{\otimes n-2})^{n-2})$$

defined by restriction to the $n-2$-th graded component. Since $(V^{\otimes n-2})^{n-2} = V_0^{\otimes n-2}$ generates the socle of $V^{\otimes n-2}$, this homomorphism is injective. Write

$$(V^{\otimes n})^{n-2} = \bigoplus_{k=1}^{n} M_k,$$

where $M_k = V_0^{\otimes k-1} \otimes V_1 \otimes V_0^{\otimes n-k}$. Any $\psi \in \mathrm{Hom}_{\mathfrak{g}_0}((V^{\otimes n})^{n-2}, V_0^{\otimes n-2})$ can be written in the form $\sum a_{kl}\theta^l_k$ where $\theta^l_k : M_k \to V_0^{\otimes n-2}$ is the restriction of τ^n_{kl} to M_k if $k < l$ or τ^n_{lk} if $k > l$. If ψ lies in the image of ρ, then $\psi = \rho(\phi)$ and therefore

$$\psi(X_{f_1,f_2}(v_1 \otimes \cdots \otimes v_n)) = X_{f_1,f_2}\phi(v_1 \otimes \cdots \otimes v_n) = 0 \tag{9}$$

for any $v_1, \ldots, v_n, f_1, f_2 \in V_0$. Choose $v_1, \ldots, v_n \in V_0, f_1, f_2 \in V_1$ so that $(v_i, f_1) = 0$ for any $i \neq k$, $(v_i, f_2) = 0$ for any $i \neq l$, $(v_k, f_1) = (v_l, f_2) = 1$. Then (9) implies $a_{kl} = a_{lk}$. Therefore $\psi = \rho(\sum_{k<l} a_{kl}\tau^n_{kl})$. The result follows now from injectivity of ρ.

Lemma 16 *Let λ^+ (resp. λ^-) denote the set of all μ obtained from λ by adding (resp. removing) one box. Then we have the following exact sequence*

$$0 \to \bigoplus_{\nu\in\lambda^+} V^\nu \to V^\lambda \otimes V \to \bigoplus_{\mu\in\lambda^-} V^\mu \to 0.$$

Proof. Assume $|\lambda| = n$. From embedding $V^\lambda \otimes V \subset V^{\otimes n+1}$ we have

$$R(V^\lambda \otimes V) = V_0^\lambda \otimes V_0 = \bigoplus_{\nu\in\lambda^+} V_0^\nu.$$

Let M be the submodule in $V^\lambda \otimes V$ generated by $R(V^\lambda \otimes V)$. Then $M = \bigoplus_{\nu\in\lambda^+} V^\nu$ by Lemma 14 and Pierri rule. Let $S = (V^\lambda \otimes V)/M$ and $\pi : V^\lambda \otimes V \to S$ be the natural projection. Then S is generated by $\pi(V_0^\lambda \otimes V_1)$. Moreover, $\pi(V_0^\lambda \otimes V_1) \subset R(S)$. By Lemma 13(b) and [5]

$$\pi(V_0^\lambda \otimes V_1) \subset \bigoplus_{\mu \in \lambda^-} V_0^\mu.$$

To see that

$$\pi(V_0^\lambda \otimes V_1) = \bigoplus_{\mu \in \lambda^-} V_0^\mu$$

observe that the set $\{\tau_{j,n+1}^{n+1} | 1 \le j \le n\}$ spans $\mathrm{Hom}_{\mathfrak{g}_0}(V_0^\lambda \otimes V_1, V_0^{n-1})$. Since $\tau_{j,n+1}^{n+1}(M) = 0$ for all $j \le n$ we have

$$M \cap (V_0^\lambda \otimes V_1) \subset \mathrm{soc}_{\mathfrak{g}_0}(V_0^\lambda \otimes V_1)$$

Now the statement follows from Lemma 14.

Lemma 17

$$\mathrm{soc} V^{\otimes n} = \bigcap_{\varphi \in \mathrm{Hom}_{\mathfrak{g}}(V^{\otimes n}, V^{\otimes n-2})} \mathrm{Ker}\varphi = \bigcap_{1 \le i < j \le n} \mathrm{Ker}\tau_{ij}^n.$$

Proof. The inclusion

$$\mathrm{soc} V^{\otimes n} \subset \bigcap_{1 \le i < j \le n} \mathrm{Ker}\tau_{ij}^n$$

is trivial since $V_0^{\otimes n} \subset \mathrm{Ker}\tau_{ij}^n$ for all i, j and $\mathrm{soc} V^{\otimes n}$ is generated by $V_0^{\otimes n}$.

We prove equality by induction in n. Let $X_n = \bigcap_{1 \le i < j \le n} \mathrm{Ker}\tau_{ij}^n$. By induction assumption we have

$$X_n \subset X_{n-1} \otimes V = (\mathrm{soc} V^{\otimes n-1}) \otimes V.$$

Using the previous lemma one can easily see that there is an exact sequence

$$0 \to \mathrm{soc} V^{\otimes n} \to X_{n-1} \otimes V \to Z \to 0,$$

where Z is a direct sum of some V^μ with $|\mu| = n-2$. By the above exact sequence it is sufficient to check

$$(\mathrm{soc} V^{\otimes n})^{n-2} = \bigcap_{1 \le i < j \le n} \mathrm{Ker}\tau_{ij}^n \cap (V^{\otimes n})^{n-2}.$$

Our calculation in the proof of Lemma 15 implies that

$$\bigcap_{1 \le i < j \le n} \mathrm{Ker}\tau_{ij}^n \cap (V^{\otimes n})^{n-2} = \mathfrak{g}_{-1} V_0^{\otimes n}.$$

Hence the statement.

Lemma 18 *$\mathscr{A}_{\mathfrak{g}}$ is generated by $\mathscr{A}_{\mathfrak{g}}^0$ and $\mathscr{A}_{\mathfrak{g}}^1$.*

Proof. Let $\varphi \in \mathrm{Hom}_{\mathfrak{g}}(V^{\otimes n}, V^{\otimes n-2k})$ for $k > 1$. Then $\mathrm{soc}V^{\otimes n} \subset \mathrm{Ker}\varphi$. Let $M = V^{\otimes n}/\mathrm{soc}V^{\otimes n}$. By Lemma 17

$$\bigoplus_{1\leq i<j\leq n} \tau_{ij}^n : V^{\otimes n} \to (V^{\otimes n-2})^{\oplus n(n-1)/2}$$

defines the embedding $M \subset (V^{\otimes n-2})^{\oplus n(n-1)/2}$. By injectivity of $V^{\otimes n-2k}$ there exists $\psi \in \mathrm{Hom}_{\mathfrak{g}}((V^{\otimes n-2})^{\oplus n(n-1)/2}, V^{\otimes n-2k})$ such that

$$\varphi = \psi \circ (\bigoplus_{1\leq i<j\leq n} \tau_{ij}^n).$$

Hence

$$\varphi = \sum_{1\leq i<j\leq n} \psi_{ij} \circ \tau_{ij}^n$$

for some $\psi_{ij} \in \mathrm{Hom}_{\mathfrak{g}}(V^{\otimes n-2}, V^{\otimes n-2k})$. Now the statement easily follows by induction in k.

Theorem 9 *The graded algebras $\mathscr{A}_{P(\infty)}$ and $\mathscr{A}_{\mathfrak{so}(\infty)}$ are isomorphic. Hence the categories $\mathbb{T}_{P(\infty)}$, $\mathbb{T}_{\mathfrak{so}(\infty)}$, $\mathbb{T}_{\mathfrak{sp}(\infty)}$ and $\mathbb{T}_{\mathfrak{osp}}$ are all equivalent.*

It is an open problem to construct directly a functor $\mathbb{T}_{P(\infty)} \to \mathbb{T}_{\mathfrak{so}(\infty)}$ that preserves tensor product.

Finally, let us observe that we know very little about finite-dimensional representations of $P(n)$. In particular, characters of simple modules and extensions between simple modules are unknown. On the other hand, in $\mathbb{T}_{P(\infty)}$ these questions are easy to answer. Is it possible to use information about representations of $P(\infty)$ to make some progress in the finite-dimensional case?

5 The case of the queer Lie superalgebra $Q(n)$

5.1 Sergeev duality

In this section we assume $\mathfrak{g} = Q(\infty)$. Let us recall the analogue of Schur–Weyl duality result in this case. It is due to Sergeev [7].

Let H_n be the semidirect product of $\mathbb{C}[S_n]$ and the Clifford algebra Cliff_n with generators $p_i, i = 1, \ldots, n$ satisfying the relations

$$p_i^2 = -1, p_i p_j + p_j p_i = 0.$$

Define the action of H_n on $V^{\otimes n}$ as follows

$$s_{i,i+1}(v_1 \otimes \cdots \otimes v_i \otimes v_{i+1} \otimes \cdots \otimes v_n) = (-1)^{p(v_i)p(v_{i+1})}(v_1 \otimes \cdots \otimes v_{i+1} \otimes v_i \otimes \cdots \otimes v_n),$$

$$p_i(v_1 \otimes \cdots \otimes v_i \otimes \cdots \otimes v_n) = (-1)^{p(v_1)+\cdots+p(v_{i-1})}(v_1 \otimes \cdots \otimes J(v_i) \otimes \cdots \otimes v_n).$$

One can show (see [7]) that $H_n \simeq U_n \otimes \mathrm{Cliff}_n$ for a certain finite-dimensional algebra U_n. The category of finite-dimensional representations of U_n is equivalent to the category of projective representations of S_n. Irreducible representations of U_n are enumerated by strict partitions.

Theorem 10 **([7])** *H_n is the centralizer of $Q(\infty)$ in $V^{\otimes n}$. There is a decomposition*

$$V^{\otimes n} = \bigoplus 2^{-\delta(\lambda)} V^{\lambda} \otimes T_{\lambda},$$

here summation is taken over all strict partitions of size n, T_λ is the irreducible H_n-module corresponding to the strict partition λ, $\delta(\lambda)$ is the parity of the length of λ.

The coefficient $2^{-\delta(\lambda)}$ appears in the case when $\dim \mathrm{End}_{Q(\infty)}(V_\lambda) = (1|1)$. For example, the second tensor power of the standard representation V has a decomposition

$$V \otimes V = S^2(V) \oplus \Lambda^2(V).$$

But $S^2(V) \simeq \Lambda^2(V)$ as $Q(\infty)$-modules because $p_1 p_2(S^2(V)) = \Lambda^2(V)$. There is only one strict partition $\lambda = (2,0,\dots,0)$ of size 2, $\delta(\lambda) = 1$. The reader will check that H_2 has an irreducible 4-dimensional module (in the category of $\mathbb{Z}_2$-graded modules).

5.2 *The category $\mathbb{T}_{Q(\infty)}$*

Let us consider the mixed tensor module

$$T = \bigoplus_{m,n \geq 0} V^{\otimes n} \otimes V_*^{\otimes m}$$

We claim that T is an injective cogenerator in the category $\mathbb{T}_{Q(\infty)}$.

Lemma 19

$$V^{\otimes n} \otimes V_*^{\otimes m} = \bigoplus_{|\lambda|=n, |\mu|=m} 2^{-max(\delta(\lambda),\delta(\mu))} \tilde{V}^{\lambda,\mu} \otimes (T_\lambda \boxtimes T_\mu),$$

where $\tilde{V}^{\lambda,\mu}$ is an indecomposable injective module in $\mathbb{T}_{\mathfrak{g}}$ with simple socle $V^{\lambda,\mu}$.

Define the graded subalgebra $\mathscr{A}_{Q(\infty)} \subset \mathrm{End}_{Q(\infty)}(T)$ generated by $\bigoplus_{n,m \geq 0} H_n \otimes H_m$ and all contractions in $\mathrm{Hom}_{Q(\infty)}(V^{\otimes n} \otimes V_*^{\otimes m}, V^{\otimes n-1} \otimes V_*^{\otimes m-1})$.

Conjecture Let $\mathfrak{g} = Q(\infty)$.

- $V^{\lambda,\mu}$ (for pairs of strict partitions (λ,μ)) are all up to isomorphism simple objects in $\mathbb{T}_{\mathfrak{g}}$.
- $\tilde{V}^{\lambda,\mu}$ are all up to isomorphism indecomposable injective objects in $\mathbb{T}_{\mathfrak{g}}$.
- $\mathscr{A}_{\mathfrak{g}}$ is a direct limit of self-dual Koszul rings.
- $\mathbb{T}_{\mathfrak{g}}$ is antiequivalent to the category of finite dimensional (locally unitary) $\mathscr{A}_{\mathfrak{g}}$-modules.

- The socle filtration of $\tilde{V}^{\lambda,\mu}$ is given by

$$\mathrm{soc}^k(\tilde{V}^{\lambda,\mu}) = \bigoplus_{|\gamma|=k} R^{\lambda}_{\gamma,\lambda'} R^{\mu}_{\gamma,\mu'} V^{\lambda',\mu'}.$$

Here $R^{\lambda}_{\gamma,\lambda'}$ stand for the Littlewood–Richardson coefficients for the Lie superalgebra $Q(\infty)$.

Note that an analogue of Howe, Tan, Willenbring result for $Q(N)$ for $N >> 0$ is difficult to get since there is no complete reducibility. On the other hand, even if we had such result, it is unclear how to proceed to ∞, because we "loose" some simple constituents at ∞.

For instance, $SQ(n)$ has a one dimensional center for every n but

$$SQ(\infty) = \lim_{\rightarrow} SQ(n)$$

is simple.

Acknowledgements This work was supported by the NSF grants 0901554 and 1303301.

References

1. Beilinson, A., Ginzburg, V., Soergel, V.: Koszul duality patterns in representation theory. J. Amer. Math. Soc. **9**, 473–527 (1996)
2. Dan-Cohen, E., Penkov, I., Serganova, V.: A Koszul category of representations of finitary Lie algebras. preprint 2011, arXiv:1105.3407
3. Howe, R., Tan, E.-C., Willenbring, J.: Stable branching rules for classical symmetric pairs. Trans. Amer. Math. Soc. **357**(4), 1601–1626 (2005)
4. Kac, V.: Lie superalgebras. Advances in Math. *26*(1), 8–96 (1977)
5. Penkov, I., Serganova, V.: Categories of integrable $sl(\infty)$-, $o(\infty)$-, $sp(\infty)$-modules. In: Representation Theory and Mathematical Physics. Contemporary Mathematics **557**, 335–357 (2011)
6. Penkov, I., Styrkas, K.: Tensor representations of infinite-dimensional root-reductive Lie algebras. In: Developments and Trends in Infinite-Dimensional Lie Theory, pp. 127-150. Progress in Mathematics, Vol. 288. Birkhäuser, Boston (2011)
7. Sergeev, A.: The Howe duality and the projective representations of symmetric groups. Represent. Theory (electronic) *3*, 416–434 (1999)
8. Sam, S., Snowden, A.: Stability patterns in representation theory. Preprint 2013, arXiv: 1302.5859
9. Zuckerman, G.: Generalized Harish-Chandra Modules. In: Highlights in Lie Algebraic Methods, pp. 123–143. Progress in Mathematics, Vol. 295. Birkhäuser, New York (2012)

Classical $\mathcal{W}$-algebras within the theory of Poisson vertex algebras

Daniele Valeri

Abstract We review the Poisson vertex algebra theory approach to classical $\mathcal{W}$-algebras. First, we provide a description of the Drinfeld-Sokolov Hamiltonian reduction for the construction of classical $\mathcal{W}$-algebras within the framework of Poisson vertex algebras and we establish, under certain sufficient conditions, the applicability of the Lenard-Magri scheme of integrability and the existence of the corresponding integrable hierarchy of bi-Hamiltonian equations. Then we provide a Poisson vertex algebra analogue of the Gelfand-Dickey construction of classical $\mathcal{W}$-algebras and we show the relations with the Drinfeld-Sokolov Hamiltonian reduction. It will be also shown that classical $\mathcal{W}$-algebras are the Poisson vertex algebras which are of interest from the conformal field theory point of view.

1 Introduction

Classical (also quantum) $\mathcal{W}$-algebras appear in many areas of mathematics and mathematical physics. One of the first appearance of $\mathcal{W}$-algebras as mathematical objects is related to the conformal field theory. The main problem of the conformal field theory is a description of fields having conformal symmetry. Only in dimension two, the group of conformal diffeomorphisms is rich enough to give rise to a meaningful theory.

After the fundamental paper by Belavin, Polyakov and Zamolodchikov [3] it was realized by Zamolodchikov [19] that extended symmetries in two dimensional conformal field theory in general do not give rise to algebras with linear defining relations. He constructed the so-called $\mathcal{W}_3$-algebra, which is an extension of the Virasoro algebra obtained adding one primary field of conformal weight 3. Later, this construction was generalized by Fateev and Lukyanov [12] to construct what are

D. Valeri (✉)
SISSA, via Bonomea 265, 34136 Trieste, Italy
e-mail: dvaleri@sissa.it

M. Gorelik, P. Papi (eds.): *Advances in Lie Superalgebras*. Springer INdAM Series 7,
DOI 10.1007/978-3-319-02952-8_12,

known as $\mathscr{W}_n$-algebras. Roughly speaking, these algebras are non-linear extensions of the Virasoro algebra obtained by adding primary fields.

The key point in the construction of the algebras $\mathscr{W}_n$ by Fateev and Lukyanov was the relation between $\mathscr{W}$-algebras and integrable systems. They identified the $\mathscr{W}_n$-algebras with the so-called second Poisson structure of the n-th (or generalized) Korteweg-de Vries (KdV) type equation (the KdV equation corresponds to $n = 2$).

These Poisson structures are known as Gelfand-Dickey algebras [15] and are obtained as Poisson algebras of local functionals on a suitable space of differential operators. More general, one can consider the larger space of pseudodifferential operators. The corresponding Poisson structure is related to the Kadomtsev-Petviashvili (KP) equation and the n-th KdV equations can be obtained with a reduction procedure from the equations of the KP hierarchy (see the lecture notes [10] for a good reference).

The fact that classical $\mathscr{W}$-algebras have in general non-linear defining relations puts them outside of the scope of the Lie algebra theory. However, the deep connection with Lie algebras reveals in the seminal paper by Drinfeld and Sokolov [11]: a classical $\mathscr{W}$-algebra is associated to a simple Lie algebra and its principal nilpotent element slightly modifying the general scheme of the Hamiltonian reduction by a group action. Moreover, they constructed an integrable hierarchy of bi-Hamiltonian equations related to this classical $\mathscr{W}$-algebra.

In [2], Barakat, De Sole and Kac explained the relation between Poisson vertex algebras and Hamiltonian equations and proved that Poisson vertex algebras provide a very convenient framework to study Hamiltonian systems (in particular bi-Hamiltonian systems).

These tools were used in [18] to develop the theory of classical $\mathscr{W}$-algebras in the Poisson vertex algebra framework. This leads to a better understanding of the Poisson structure underlying classical $\mathscr{W}$-algebras and to a generalization of results, both in Gelfand-Dickey and Drinfeld-Sokolov approach (see [7]), to the study of integrability of Hamiltonian equations. The aim of this paper is to review the Poisson vertex algebra theory approach to classical $\mathscr{W}$-algebras.

The paper is organized as follows. In Sect. 2 we review, following [2], the basic definitions and notations of the Poisson vertex algebra theory and its application to the theory of integrable bi-Hamiltonian equations. As the main application, we show how to apply successfully the Lenard-Magri scheme of integrability for the affine Poisson vertex algebras.

Section 3 is mainly devoted to present the definition of classical $\mathscr{W}$-algebras in the Poisson vertex algebra framework. First, we review the results of [7] about the description of the Drinfeld and Sokolov Hamiltonian reduction using the Poisson vertex algebra theory. We will show that the Drinfeld-Sokolov Hamiltonian group action on the phase space is encoded by (the exponential of) a Lie conformal algebra action on a suitable differential subalgebra of the affine Poisson vertex algebra and we will give sufficient conditions in order to apply successfully the Lenard-Magri scheme of integrability. Finally, we show how to define a Poisson vertex algebra analogue of the Gelfand-Dickey approach to classical $\mathscr{W}$-algebras and we show

how this construction is related to the Drinfeld-Sokolov Hamiltonian reduction approach.

2 Poisson vertex algebras and Hamiltonian equations

In this section we review the connection between Poisson vertex algebras and the theory of Hamiltonian equations as laid down in [2]. As the main application we explain how to use the Lenard-Magri scheme of integrability in order to get an integrable hierarchy of the Hamiltonian equations for affine Poisson vertex algebras.

2.1 Poisson vertex algebras

By a *differential algebra* we mean a unital commutative associative algebra $\mathcal{V}$ over $\mathbb{C}$ with a derivation ∂, that is a $\mathbb{C}$-linear map from $\mathcal{V}$ to itself such that, for $a,b\in\mathcal{V}$

$$\partial(ab)=\partial(a)b+a\partial(b)\,.$$

In particular $\partial 1=0$.

Definition 1 Let $\mathcal{V}$ be a differential algebra. A *λ-bracket* on $\mathcal{V}$ is a $\mathbb{C}$-linear map $\mathcal{V}\otimes\mathcal{V}\to\mathbb{C}[\lambda]\otimes\mathcal{V}$, denoted by $f\otimes g\to\{f_\lambda g\}$, satisfying *sesquilinearity* ($f,g\in\mathcal{V}$):

$$\{\partial f_\lambda g\}=-\lambda\{f_\lambda g\},\qquad \{f_\lambda \partial g\}=(\lambda+\partial)\{f_\lambda g\}, \tag{1}$$

and the *left and right Leibniz rules* ($f,g,h\in\mathcal{V}$):

$$\begin{aligned}\{f_\lambda gh\}&=\{f_\lambda g\}h+\{f_\lambda h\}g,\\ \{fh_\lambda g\}&=\{f_{\lambda+\partial}g\}_\to h+\{h_{\lambda+\partial}g\}_\to f,\end{aligned}$$

where we use the following notation: if $\{f_\lambda g\}=\sum_{n\in\mathbb{Z}_+}\lambda^n c_n$, then

$$\{f_{\lambda+\partial}g\}_\to h=\sum_{n\in\mathbb{Z}_+}c_n(\lambda+\partial)^n h\in\mathcal{V}[\lambda]\,.$$

We say that the λ-bracket is *skew-symmetric* if

$$\{g_\lambda f\}=-\{f_{-\lambda-\partial}g\}, \tag{2}$$

where, now, $\{f_{-\lambda-\partial}g\}=\sum_{n\in\mathbb{Z}_+}(-\lambda-\partial)^n c_n$ (if there is no arrow we move ∂ to the left).

Definition 2 A *Poisson vertex algebra* is a differential algebra $\mathcal{V}$ endowed with a λ-bracket which is skew-symmetric and satisfies the following *Jacobi identity* in $\mathcal{V}[\lambda,\mu]$ ($f,g,h\in\mathcal{V}$):

$$\{f_\lambda\{g_\mu h\}\}-\{g_\mu\{f_\lambda h\}\}=\{\{f_\lambda g\}_{\lambda+\mu}h\}\,. \tag{3}$$

In this paper we consider Poisson vertex algebra structures on an algebra of differential polynomials $\mathscr{V}$ in the variables $\{u_i\}_{i\in I}$:

$$\mathscr{V} = \mathbb{C}[u_i^{(n)} \mid i \in I, n \in \mathbb{Z}_+],$$

where ∂ is the derivation defined by $\partial(u_i^{(n)}) = u_i^{(n+1)}$, $i \in I, n \in \mathbb{Z}_+$. In this case, thanks to sesquilinearity and Leibniz rules, the λ-brackets $\{u_{i\lambda}u_j\}$, $i,j \in I$, completely determine the λ-bracket on the whole differential algebra $\mathscr{V}$ as stated by the following theorem.

Theorem 1 ([2, Theorem 1.15]) *Let $\mathscr{V}$ be an algebra of differential polynomials in the variables $\{u_i\}_{i\in I}$, and choose $\{u_{i\lambda}u_j\} \in \mathbb{C}[\lambda]\otimes\mathscr{V}$, $i,j \in I$.*

(a) The Master Formula

$$\{f_\lambda g\} = \sum_{i,j\in I m,n\in\mathbb{Z}_+} \frac{\partial g}{\partial u_j^{(n)}} (\lambda+\partial)^n \{u_{i\lambda+\partial}u_j\}_\to (-\lambda-\partial)^m \frac{\partial f}{\partial u_i^{(m)}} \tag{4}$$

defines a λ-bracket on $\mathscr{V}$ with given $\{u_{i\lambda}u_j\}$, $i,j \in I$.

(b) The λ-bracket (4) *on $\mathscr{V}$ satisfies the skew-symmetry condition* (2) *provided that the same holds on generators $(i,j \in I)$:*

$$\{u_{i\lambda}u_j\} = -\{u_{j-\lambda-\partial}u_i\}, \tag{5}$$

(c) Assuming that the skew-symmetry condition (5) *holds, the λ-bracket* (4) *satisfies the Jacobi identity* (3)*, thus making $\mathscr{V}$ a Poisson vertex algebra, provided that the Jacobi identity holds on any triple of generators $(i,j,k \in I)$:*

$$\{u_{i_\lambda}\{u_{j_\mu}u_k\}\} - \{u_{j_\mu}\{u_{i_\lambda}u_k\}\} = \{\{u_{i\lambda}u_j\}_{\lambda+\mu}u_k\}. \tag{6}$$

In Sect. 3 we will define classical $\mathscr{W}$-algebras in terms of representations of Lie conformal algebras. Let us recall here the definitions [16].

Definition 3 (a) A *Lie conformal algebra* is a $\mathbb{C}[\partial]$-module R with a $\mathbb{C}$-linear map $\{\cdot_\lambda\cdot\}: R\otimes R \to \mathbb{C}[\lambda]\otimes R$ satisfying (1), (2) and (3).

(b) A *representation* of a Lie conformal algebra R on a $\mathbb{C}[\partial]$-module $\mathscr{V}$ is a λ-action $R\otimes\mathscr{V} \to \mathbb{C}[\lambda]\otimes\mathscr{V}$, denoted $a\otimes g \mapsto a_\lambda^\rho g$, satisfying sesquilinearity, $(\partial a)_\lambda^\rho g = -\lambda a_\lambda^\rho g$, $a_\lambda^\rho(\partial g) = (\lambda+\partial)a_\lambda^\rho g$, and Jacobi identity $a_\lambda^\rho(b_\mu^\rho g) - b_\mu^\rho(a_\lambda^\rho g) = \{a_\lambda b\}_{\lambda+\mu}^\rho g$ $(a,b \in R, g \in \mathscr{V})$.

2.2 *Affine Poisson vertex algebras*

Let $\mathfrak{g}$ be a Lie algebra over $\mathbb{C}$ with a non-degenerate symmetric invariant bilinear form κ, and let s be an element of $\mathfrak{g}$. The *affine Poisson vertex algebra* $\mathscr{V}(\mathfrak{g},\kappa,s)$, associated to the triple $(\mathfrak{g},\kappa,s)$, is the algebra of differential polynomials $\mathscr{V} = S(\mathbb{C}[\partial]\mathfrak{g})$ (where $\mathbb{C}[\partial]\mathfrak{g}$ is the free $\mathbb{C}[\partial]$-module generated by $\mathfrak{g}$ and $S(R)$ denotes the symmetric algebra over the $\mathbb{C}$-vector space R) together with the λ-bracket given by

$$\{a_\lambda b\}_z = [a,b] + \kappa(a \mid b)\lambda + z\kappa(s \mid [a,b]) \qquad \text{for } a,b \in \mathfrak{g}, z \in \mathbb{C}. \tag{7}$$

It is easy to check that skew-symmetry condition (2) holds for any pair of elements of $\mathfrak{g}$ and Jacobi identity holds for any triple of elements of $\mathfrak{g}$. Hence, by Theorem 1, we can extend the λ-bracket (7) to the whole differential algebra $\mathcal{V}$ using the Master Formula (4). We want to emphasize the fact that we have a one-parameter family of λ-brackets if we think at z as a parameter (this will be useful in the applications to the theory of integrable bi-Hamiltonian equations). Indeed, for any $a,b \in \mathfrak{g}$, we can write $\{a_\lambda b\}_z = \{a_\lambda b\}_0 - z\{a_\lambda b\}_\infty$, where

$$\{a_\lambda b\}_0 = [a,b] + \kappa(a \mid b)\lambda \qquad \text{and} \qquad \{a_\lambda b\}_\infty = -\kappa(s|[a,b]) . \tag{8}$$

Let $\{u_i\}_{i\in I}$ be a basis of $\mathfrak{g}$ (hence, as a differential algebra, $\mathcal{V} = \mathbb{C}[u_i^{(n)} \mid i \in I, n \in \mathbb{Z}_+]$) and let $\{u^i\}_{i\in I}$ be the dual basis with respect to κ, namely $\kappa(u_i \mid u^j) = \delta_{ij}$, for all $i,j \in I$. Let us consider the following element of $\mathcal{V}$:

$$L = \frac{1}{2}\sum_{i\in I} u_i u^i + \partial x \in \mathcal{V} , \tag{9}$$

where $x \in \mathfrak{g}$, and let $a \in \mathfrak{g}$ be such that $[x,a] = ma$, for $m \in \mathbb{C}$. Then, it is an easy exercise in λ-bracket calculus to prove that

$$\{L_\lambda a\}_z = (\partial + (1-m)\lambda)a - \kappa(a \mid x)\lambda^2 - z([s,a] + m\kappa(s \mid a)\lambda) , \tag{10}$$
$$\{L_\lambda L\}_z = (\partial + 2\lambda)L - \kappa(x \mid x)\lambda^3 - z(\partial + 2\lambda)[s,x] . \tag{11}$$

2.3 Poisson vertex algebras of conformal field theory type

Definition 4 Let $\mathcal{V}$ be a Poisson vertex algebra. An element $L \in \mathcal{V}$ is called a *Virasoro element* of *central charge* $c \in \mathbb{C}$ if

$$\{L_\lambda L\} = (\partial + 2\lambda)L + c\lambda^3 .$$

An element $a \in \mathcal{V}$ is called an *L-eigenvector* of *conformal weight* $\Delta_a \in \mathbb{C}$ if

$$\{L_\lambda a\} = (\partial + \Delta_a\lambda)a + O(\lambda^2) .$$

It is called a *primary element* of conformal weight Δ_a if $\{L_\lambda a\} = (\partial + \Delta_a\lambda)a$.

Definition 5 A Poisson vertex algebra is of *conformal field theory type* if, as a differential algebra, it is an algebra of differential polynomials in a Virasoro element and some primary elements.

Let us assume that the element $x \in \mathfrak{g}$ appearing in (9) is semisimple. Then, we have the ad$x$-eigenspace decomposition $\mathfrak{g} = \oplus_{m\in\mathbb{C}}\mathfrak{g}_m$. By (11), we see that L is a Virasoro element (for the $z = 0$ λ-bracket) and by (10) that any element $a \in \mathfrak{g}_m$ is a primary element (for the $z = 0$ λ-bracket) of conformal weight $\Delta_a = 1 - m$, provided that $\kappa(a \mid x) = 0$.

2.4 Hamiltonian equations

The relation between Poisson vertex algebras and systems of Hamiltonian equations is based on the following simple observation.

Proposition 1 *Let $\mathscr{V}$ be a Poisson vertex algebra. The 0-th product on $\mathscr{V}$ induces a well defined Lie algebra bracket on the quotient space $\mathscr{V}/\partial\mathscr{V}$:*

$$\{\textstyle\int f, \int g\} = \int \{f_\lambda g\}|_{\lambda=0} , \tag{12}$$

where $\int : \mathscr{V} \to \mathscr{V}/\partial\mathscr{V}$ is the canonical quotient map. Moreover, we have a well defined Lie algebra action of $\mathscr{V}/\partial\mathscr{V}$ on $\mathscr{V}$ by derivations of the commutative associative product on $\mathscr{V}$, commuting with ∂, given by

$$\{\textstyle\int f, g\} = \{f_\lambda g\}|_{\lambda=0} .$$

This motivates the following definition.

Definition 6 Let $\mathscr{V}$ be a Poisson vertex algebra.

(a) Elements of $\mathscr{V}/\partial\mathscr{V}$ are called *local functionals*.
(b) Given a local functional $\int h \in \mathscr{V}/\partial\mathscr{V}$, the corresponding *Hamiltonian equation* is

$$\frac{du}{dt} = \{\textstyle\int h, u\} \tag{13}$$

(c) A local functional $\int f \in \mathscr{V}/\partial\mathscr{V}$ is called an *integral of motion* of equation (13) if $\int h$ and $\int f$ are *in involution*:

$$\{\textstyle\int h, \int f\} = 0 .$$

(d) Equation (13) is called *integrable* if there exists an infinite sequence $\int f_0 = \int h, \int f_1, \int f_2, \ldots,$ of linearly independent integrals of motion in involution, namely such that

$$\{\textstyle\int f_n, \int f_m\} = 0, \quad \text{for all } n, m \in \mathbb{Z}_+ .$$

The corresponding *integrable hierarchy of Hamiltonian equations* is

$$\frac{du}{dt_n} = \{\textstyle\int f_n, u\}, \quad n \in \mathbb{Z}_+ . \tag{14}$$

2.5 Compatible Poisson vertex algebras and Lenard-Magri scheme of integrability

Definition 7 Two λ-brackets on a differential algebra $\mathscr{V}$, say $\{\cdot_\lambda\cdot\}_0$ and $\{\cdot_\lambda\cdot\}_\infty$, are *compatible* if any their $\mathbb{C}$-linear combination defines a Poisson vertex algebra structure on $\mathscr{V}$.

Remark 1 For affine Poisson vertex algebras defined in Sect. 2.2, the λ-brackets $\{\cdot_\lambda\cdot\}_0$ and $\{\cdot_\lambda\cdot\}_\infty$ given by (8) are compatible.

Let $\mathcal{V}$ be an algebra of differential polynomials in the variables $\{u_i\}_{i\in I}$ endowed with two compatible λ-brackets $\{\cdot_\lambda\cdot\}_0$ and $\{\cdot_\lambda\cdot\}_\infty$. According to the *Lenard-Magri scheme of integrability* [17] (see also [2]), in order to obtain an infinite sequence of integrals of motion, one needs to find a sequence $\{f_n\}_{n\in\mathbb{Z}_+}\subset\mathcal{V}$ such that

$$\{f_{n\lambda}u_i\}_0|_{\lambda=0}=\{f_{n+1\lambda}u_i\}_\infty|_{\lambda=0}\,,\qquad \text{for all } n\in\mathbb{Z}_+, i\in I\,. \tag{15}$$

Indeed, if this is the case, it follows easily that the elements $\int f_n$, $n\in\mathbb{Z}_+$, form an infinite sequence of local functionals in involution: $\{\int f_m,\int f_n\}_0=\{\int f_m,\int f_n\}_\infty=0$, for all $m,n\in\mathbb{Z}_+$. According to Definition 6(d), we get the corresponding integrable hierarchy of Hamiltonian equations (14) (in fact bi-Hamiltonian) provided that the local functionals $\int f_n$, $n\in\mathbb{Z}_+$, span an infinite dimensional subspace of $\mathcal{V}/\partial\mathcal{V}$.

2.6 Application of the Lenard-Magri scheme of integrability for affine Poisson vertex algebras

As an application of the Lenard-Magri scheme of integrability, we construct integrable hierarchies of bi-Hamiltonian equations (13) for affine Poisson vertex algebras defined in Sect. 2.2

Let $\mathfrak{g}$ be a finite-dimensional Lie algebra with a non-degenerate symmetric invariant bilinear form κ, let s be an element of $\mathfrak{g}$, and let $\mathcal{V}=S(\mathbb{C}[\partial]\mathfrak{g})$. Recall from Remark 1 that we have two compatible λ-brackets on $\mathcal{V}$:

$$\{a_\lambda b\}_0=[a,b]+\kappa(a\,|\,b)\lambda \qquad\text{and}\qquad \{a_\lambda b\}_\infty=-\kappa(s|[a,b])\,,$$

for $a,b\in\mathfrak{g}$ and extended to the whole $\mathcal{V}$ using the Master Formula (4). Let $\{u_i\}_{i\in I}$ be a basis of $\mathfrak{g}$, hence $\mathcal{V}=\mathbb{C}[u_i^{(n)}\mid i\in I,n\in\mathbb{Z}_+]$. According to the previous section our goal is to find a sequence $f_n\in\mathcal{V}$, $n\in\mathbb{Z}_+$, such that the Lenard-Magri recursion relation (15) holds. We will do this under the assumption that $s\in\mathfrak{g}$ is a semisimple element and denote $\mathfrak{h}=\ker(\operatorname{ad}s)\subset\mathfrak{g}$ (it is clearly a subalgebra). By invariance of the bilinear form κ we have that $\mathfrak{h}^\perp=\operatorname{im}(\operatorname{ad}s)$, and that $\mathfrak{g}=\ker(\operatorname{ad}s)\oplus\operatorname{im}(\operatorname{ad}s)$.

Let us introduce the operator of *variational derivative* $\frac{\delta}{\delta u}:\mathcal{V}/\partial\mathcal{V}\to\mathfrak{g}\otimes\mathcal{V}$, defined by $\frac{\delta f}{\delta u}=\sum_{i\in I}u_i\otimes\frac{\delta f}{\delta u_i}$, where

$$\frac{\delta f}{\delta u_i}=\sum_{n\in\mathbb{Z}_+}(-\partial)^n\frac{\partial f}{\partial u_i^{(n)}}\,,$$

for any $i\in I$ and $f\in\mathcal{V}$ (it is easy to prove that $\frac{\delta}{\delta u}\circ\partial=0$). Then, using the generating series

$$F(z)=\sum_{n\in\mathbb{Z}_+}\frac{\delta f_n}{\delta u}z^{-n}$$

it is shown in [7] that solving (15) is equivalent to solve the following equation:

$$[L(z),F(z)]=0,\qquad [s\otimes 1,F_0]=0\,, \tag{16}$$

(here $F_0 = \frac{\delta f_0}{\delta u}$) where $L(z) = \partial + \sum_{i\in I} u^i \otimes u_i + zs \otimes 1$, with $\{u^i\}_{i\in I}$ being the dual basis with respect to κ, and the commutator is extended to $\mathfrak{g} \otimes \mathscr{V}$ in the obvious way (and componentwise to $(\mathfrak{g} \otimes \mathscr{V})((z^{-1}))$).

The solution of the above problem will be achieved in Propositions 2 and 3 below (due to Drinfeld and Sokolov [11]) which are a short reformulation of Propositions 2.3 and 2.4 of [7] (where one can find further details). Proposition 2 shows us how to construct a series $F(z) = \sum_{n\in\mathbb{Z}_+} F_n z^{-n} \in (\mathfrak{g} \otimes \mathscr{V})[[z^{-1}]]$ which satisfies (16) and Proposition 3 shows that, for any $n \in \mathbb{Z}_+$, $F_n = \frac{\delta f_n}{\delta u}$, for some $f_n \in \mathscr{V}$.

Proposition 2 *There exist formal series $U(z) \in (\mathfrak{g} \otimes \mathscr{V})[[z^{-1}]]z^{-1}$ and $h(z) \in (\mathfrak{h} \otimes \mathscr{V})[[z^{-1}]]$ such that*

$$L_0(z) = e^{\operatorname{ad} U(z)}(L(z)) = \partial + zs \otimes 1 + h(z) . \tag{17}$$

Let $a \in Z(\mathfrak{h})$ (the center of $\mathfrak{h}$), and let $U(z) \in (\mathfrak{g} \otimes \mathscr{V})[[z^{-1}]]z^{-1}$, $h(z) \in (\mathfrak{h} \otimes \mathscr{V})[[z^{-1}]]$ solve equation (17). Then

$$F(z) = e^{-\operatorname{ad} U(z)}(a \otimes 1) \in (\mathfrak{g} \otimes \mathscr{V})[[z^{-1}]] \tag{18}$$

solves (16), namely $[s \otimes 1, F_0] = 0$ and $[L(z), F(z)] = 0$.

Proposition 3 *Let $U(z) \in (\mathfrak{g} \otimes \mathscr{V})[[z^{-1}]]z^{-1}$ and $h(z) \in (\mathfrak{h} \otimes \mathscr{V})[[z^{-1}]]$ be a solution of equation* (17). *Then the formal power series $F(z) \in (\mathfrak{g} \otimes \mathscr{V})[[z^{-1}]]$ defined in* (18) *(where $a \in Z(\mathfrak{h})$) satisfies $F(z) = \frac{\delta f(z)}{\delta u}$, where*

$$\textstyle\int f(z) = \int \kappa(a \otimes 1 \mid h(z)) \in \mathscr{V}/\partial\mathscr{V}[[z^{-1}]] .$$

According to the discussion of Sect. 2.5 we found a sequence of local functionals $\int f_n \in \mathscr{V}/\partial\mathscr{V}$ in involution. In order to get an integrable hierarchy of Hamiltonian equations (14) we need to show that these local functionals span an infinite dimensional subspace in $\mathscr{V}/\partial\mathscr{V}$. This is true provided that $a \in Z(\mathfrak{h})\backslash Z(\mathfrak{g})$. We refer to Corollary 2.9 in [7] for the proof of this fact.

Example 1 (The N-wave equation) Let $\mathfrak{g} = \mathfrak{gl}_N$, with the bilinear form $\kappa(A \mid B) = \operatorname{tr}(AB)$, and let $s = \operatorname{diag}(s_1, \dots, s_N)$ be a diagonal matrix with distinct eigenvalues. Then $\mathfrak{h} = \ker(\operatorname{ad} s)$ is the abelian subalgebra of diagonal $N \times N$ matrices, and $\mathfrak{h}^\perp = \operatorname{im}(\operatorname{ad} s)$ consists of $N \times N$ matrices with zeros along the diagonal. Let $\{E_{ij}\}_{i,j=1}^N$ be the basis of $\mathfrak{gl}_N$ given by elementary matrices.

It is possible to compute recursively the Hamiltonian functionals in involution associated to the non-scalar element $a = \operatorname{diag}(a_1, \dots, a_N) \in Z(\mathfrak{h}) = \mathfrak{h}$. The first three of them are (the terms with zero denominator are dropped from the sums):

$$\int f_0 = \int \sum_k a_k E_{kk}, \qquad \int f_1 = \int \sum_{k,\alpha} \frac{a_k E_{k\alpha} E_{\alpha k}}{s_k - s_\alpha},$$
$$\int f_2 = \int \sum_{\alpha,\beta,k} \frac{a_k E_{k\alpha} E_{\alpha\beta} E_{\beta k}}{(s_k - s_\alpha)(s_k - s_\beta)} - \sum_{\alpha,k} \frac{a_k \left(E_{k\alpha} E'_{\alpha k} + E_{k\alpha} E_{\alpha k} E_{kk}\right)}{(s_k - s_\alpha)^2} .$$

The first two equations of the corresponding hierarchy of bi-Hamiltonian equations (14) are ($1 \le i, j \le N$):

$$\frac{dE_{ij}}{dt_0} = (a_i - a_j)E_{ij}, \qquad \frac{dE_{ij}}{dt_1} = \gamma_{ij}E'_{ij} + \sum_k (\gamma_{ik} - \gamma_{kj})E_{ik}E_{kj},$$

where $\gamma_{ij} = \frac{a_i - a_j}{s_i - s_j}$ for $i \neq j$ and $\gamma_{ij} = 0$ for $i = j$. The last equation is known as the *N-wave equation.*

3 Classical $\mathcal{W}$-algebras

In this section we provide a description of the Drinfeld-Sokolov Hamiltonian reduction [11] for the construction of classical $\mathcal{W}$-algebras within the framework of Poisson vertex algebras following [7]. We show how this approach is related to the original definition of Drinfeld and Sokolov. Indeed, in this context, the gauge group action on the phase space is translated in terms of (the exponential of) a Lie conformal algebra action on the space of functions. We thus obtain two compatible λ-brackets for classical $\mathcal{W}$-algebras, that we will use to apply successfully (under some futher assumptions) the Lenard-Magri scheme of integrability.

Finally, following [18], we want to define the Poisson vertex algebra analogue of the Gelfand-Dickey (first and second) Hamiltonian structure on the space of (pseudo) differential operators [10]. We show that we have an isomorphism with some classical $\mathcal{W}$-algebras obtained with the Drinfeld-Sokolov Hamiltonian reduction.

3.1 Setup

Throughout the rest of the paper we make the following assumptions.

Let $\mathfrak{g}$ be a finite-dimensional Lie algebra over the field $\mathbb{C}$ with a non-degenerate symmetric invariant bilinear form κ, and let us assume that $(f, h = 2x, e) \subset \mathfrak{g}$ is an $\mathfrak{sl}_2$-triple. Then we have the $\operatorname{ad}x$-eigenspace decomposition

$$\mathfrak{g} = \bigoplus_{i \in \frac{1}{2}\mathbb{Z}} \mathfrak{g}_i . \tag{19}$$

Clearly, $f \in \mathfrak{g}_{-1}$, $h \in \mathfrak{g}_0$ and $e \in \mathfrak{g}_1$.

There is a well-defined skew-symmetric bilinear form ω on $\mathfrak{g}_{\frac{1}{2}}$ defined by

$$\omega(a,b) = \kappa(f \mid [a,b]), \qquad a, b \in \mathfrak{g}_{\frac{1}{2}},$$

which is non-degenerate since $\operatorname{ad} f : \mathfrak{g}_{\frac{1}{2}} \to \mathfrak{g}_{-\frac{1}{2}}$ is bijective. We fix an isotropic subspace $\mathfrak{l} \subset \mathfrak{g}_{\frac{1}{2}}$ (with respect to ω) and we denote by $\mathfrak{l}^{\perp\omega} = \{a \in \mathfrak{g}_{\frac{1}{2}} \mid \omega(a,b) = 0 \text{ for all } b \in \mathfrak{l}\} \subset \mathfrak{g}_{\frac{1}{2}}$ its symplectic complement with respect to ω. We consider the following nilpotent subalgebras of $\mathfrak{g}$:

$$\mathfrak{m} = \mathfrak{l} \oplus \mathfrak{g}_{\geq 1} \subset \mathfrak{n} = \mathfrak{l}^{\perp\omega} \oplus \mathfrak{g}_{\geq 1},$$

where $\mathfrak{g}_{\geq 1} = \bigoplus_{i \geq 1} \mathfrak{g}_i$.

3.2 The Poisson vertex algebra structure

Let us fix $s \in \mathfrak{g}$ and let us consider the affine Poisson vertex algebra $\mathscr{V}(\mathfrak{g}, \kappa, s)$ (see Sect. 2.2). We recall that, as a differential algebra, it is $\mathscr{V}(\mathfrak{g}) = S(\mathbb{C}[\partial]\mathfrak{g})$, with λ-bracket on generators given by (7).

We let $\widetilde{\mathscr{J}} = \langle m - \kappa(f \mid m) \mid m \in \mathfrak{m}\rangle_{\mathscr{V}(\mathfrak{g})}$ be the differential ideal generated by the elements $m - \kappa(f \mid m)$, for $m \in \mathfrak{m}$. Note that this is not a Poisson vertex algebra ideal. Indeed, for example, if $a \in \mathfrak{g}_1$ is such that $\kappa(f \mid a) = 1$, then $a - 1$ lies in $\widetilde{\mathscr{J}}$, but the coefficient of λ in $\{f_\lambda a - 1\}$ is 1, which does not lie in $\widetilde{\mathscr{J}}$. Let us define

$$\widetilde{\mathscr{W}_z} = \left\{ p \in \mathscr{V}(\mathfrak{g}) \,\middle|\, \{a_\lambda p\}_z \in \widetilde{\mathscr{J}}[\lambda] \; \forall a \in \mathfrak{n} \right\} \subset \mathscr{V}(\mathfrak{g}) .$$

Lemma 1 *(a) If $p \in \widetilde{\mathscr{W}_z}$ and $q \in \widetilde{\mathscr{J}}$, then $\{p_\lambda q\}_z, \{q_\lambda p\}_z \in \widetilde{\mathscr{J}}[\lambda]$.*
(b) $\widetilde{\mathscr{W}_z} \subset \mathscr{V}(\mathfrak{g})$ is a Poisson vertex subalgebra.
(c) If $s \in \ker(\operatorname{ad}\mathfrak{n})$, then the differential subalgebra $\widetilde{\mathscr{W}_z} \subset \mathscr{V}(\mathfrak{g})$ does not depend on z (while its Poisson vertex algebra structure depends) and $\widetilde{\mathscr{J}} \subset \widetilde{\mathscr{W}_z}$.

Proof. To prove (a) we note that any element of $\widetilde{\mathscr{J}}$ is a finite sum of elements of the form $r\partial^i(m - \kappa(f \mid m))$, with $r \in \mathscr{V}(\mathfrak{g})$, $m \in \mathfrak{m}$, and $i \in \mathbb{Z}_+$. Then we get, by the Leibniz rule and sesquilinearity,

$$\begin{aligned}\{p_\lambda q\}_z &= \sum \{p_\lambda r\partial^i(m - \kappa(f \mid m))\}_z = \\ &= \sum \{p_\lambda r\}_z \partial^i(m - \kappa(f \mid m)) + \sum r(\lambda + \partial)^i \{p_\lambda m\}_z \in \widetilde{\mathscr{J}}[\lambda] ,\end{aligned}$$

since $p \in \widetilde{\mathscr{W}_z}$ and $m \in \mathfrak{m} \subset \mathfrak{n}$. From the fact that $\widetilde{\mathscr{J}}$ is a differential ideal, using skew-commutativity we get also $\{q_\lambda p\}_z \in \widetilde{\mathscr{J}}[\lambda]$.

For (b), first we prove that $\widetilde{\mathscr{W}_z} \subset \mathscr{V}(\mathfrak{g})$ is a differential subalgebra. Indeed, if $p, q \in \widetilde{\mathscr{W}_z}$ and $a \in \mathfrak{n}$, we have $\{a_\lambda pq\}_z = p\{a_\lambda q\}_z + q\{a_\lambda p\}_z \in \widetilde{\mathscr{J}}[\lambda]$, proving that pq lies in $\widetilde{\mathscr{W}}$, and $\{a_\lambda \partial p\}_z = (\lambda + \partial)\{a_\lambda p\}_z \in \widetilde{\mathscr{J}}[\lambda]$, proving that ∂p lies in $\widetilde{\mathscr{W}_z}$. Next, we show that $\widetilde{\mathscr{W}_z}$ is closed for the λ-bracket. Let $a \in \mathfrak{n}$ and $p, q \in \widetilde{\mathscr{W}_z}$. By the Jacobi identity we have

$$\{a_\lambda \{p_\mu q\}_z\}_z = \{p_\mu \{a_\lambda q\}_z\}_z + \{\{a_\lambda p\}_{z\lambda+\mu} q\}_z .$$

By (a) both terms in the right hand side lie in $\widetilde{\mathscr{J}}[\lambda, \mu]$. Hence $\{p_\lambda q\}_z \in \widetilde{\mathscr{W}_z}[\lambda]$.

Finally, let us prove (c). The first assertion follows from the fact that, under the assumption $s \in \ker(\operatorname{ad}\mathfrak{n})$, we have $\{a_\lambda p\}_z = \{a_\lambda p\}_0$, for all $a \in \mathfrak{n}$ and $p \in \mathscr{V}(\mathfrak{g})$. Hence, we are left to prove that $\widetilde{\mathscr{J}} \subset \widetilde{\mathscr{W}_z}$. By the Leibniz rule and sesquilinearity this reduces to the fact that elements of the form $m - \kappa(f \mid m)$, with $m \in \mathfrak{m}$, lie in $\widetilde{\mathscr{W}_z}$. Take $a \in \mathfrak{n}$, then

$$\{a_\lambda m - \kappa(f \mid m)\}_z = \{a_\lambda m\}_z = [a, m] ,$$

since, by assumption, $s \in \ker(\operatorname{ad}\mathfrak{n})$. On the other hand, we have $[a, m] \in [\mathfrak{l}, \mathfrak{l}^{\perp\omega}] \oplus (\oplus_{j>1}\mathfrak{g}_j) \subset \widetilde{\mathscr{J}}$, proving that $m - \kappa(f \mid m) \in \widetilde{\mathscr{W}_z}$. □

If $s \in \ker(\operatorname{ad}\mathfrak{n})$, by the above lemma, $\widetilde{\mathscr{J}} \subset \widetilde{\mathscr{W}_z}$ is a Poisson vertex algebra ideal. Hence, the quotient space has an induced Poisson vertex algebra structure.

Definition 8 The quotient Poisson vertex algebra

$$\mathscr{W} = \widetilde{\mathscr{W}_z} / \widetilde{\mathscr{J}}$$

is called *classical $\mathscr{W}$-algebra.*

Remark 2 It follows from Lemma 1(c) that we have a one-parameter family of λ-brackets on the quotient differential algebra $\mathscr{W}$, if we think at z as a parameter. In literature, the name classical $\mathscr{W}$-algebra is referred to the Poisson vertex algebra structure corresponding to the case $z = 0$. As we will see, though, the λ-bracket corresponding to the case $z = \infty$, plays an important role in obtaining an integrable hierarchy of bi-Hamiltonian equations associated to the classical $\mathscr{W}$-algebra.

In order to prove the equivalence with the Drinfeld-Sokolov Hamiltonian reduction and to explore better the structure of the classical $\mathscr{W}$-algebras we need to be able to explicitly compute the induced λ-bracket in $\mathscr{W}$.

3.3 Equivalence with Drinfeld-Sokolov Hamiltonian reduction

Note that, since, by assumption, $[s,\mathfrak{n}] = 0$, $\mathbb{C}[\partial]\mathfrak{n} \subset \mathscr{V}(\mathfrak{g})$ is a Lie conformal subalgebra (see Definition 3), with the λ-bracket $\{a_\lambda b\}_z = [a,b]$, $a,b \in \mathfrak{n}$ (it is independent on z).

Let $\mathfrak{g} = \mathfrak{m} \oplus \mathfrak{p}$, where $\mathfrak{p} \subset \mathfrak{g}_{\leq \frac{1}{2}}$ is an arbitrary subspace of $\mathfrak{g}$ complementary to $\mathfrak{m}$. Consider the differential subalgebra $\mathscr{V}(\mathfrak{p}) = S(\mathbb{C}[\partial]\mathfrak{p})$ of $\mathscr{V}(\mathfrak{g})$, and denote by $\rho : \mathscr{V}(\mathfrak{g}) \twoheadrightarrow \mathscr{V}(\mathfrak{p})$, the differential algebra homomorphism defined on generators by

$$\rho(a) = \pi_{\mathfrak{p}}(a) + \kappa(f \mid a), \quad a \in \mathfrak{g}, \tag{20}$$

where $\pi_{\mathfrak{p}} : \mathfrak{g} \twoheadrightarrow \mathfrak{p}$ is the projection map with kernel $\mathfrak{m}$.

Lemma 2 (a) *We have a representation of the Lie conformal algebra $\mathbb{C}[\partial]\mathfrak{n}$ on the differential subalgebra $\mathscr{V}(\mathfrak{p}) \subset \mathscr{V}(\mathfrak{g})$ given by ($a \in \mathfrak{n}$, $g \in \mathscr{V}(\mathfrak{p})$):*

$$a^{\rho}_{\lambda} g = \rho\{a_\lambda g\}_z \tag{21}$$

(note that the RHS is independent of z since, by assumption, $s \in \ker(\operatorname{ad}\mathfrak{n})$).

(b) *Let $\mathscr{W} \subset \mathscr{V}(\mathfrak{p})$ be the subspace killed by the Lie conformal algebra action of $\mathbb{C}[\partial]\mathfrak{n}$:*

$$\mathscr{W} = \mathscr{V}(\mathfrak{p})^{\mathbb{C}[\partial]\mathfrak{n}} = \left\{ g \in \mathscr{V}(\mathfrak{p}) \,\middle|\, a^{\rho}_{\lambda} g = 0 \textit{ for all } a \in \mathfrak{n} \right\}. \tag{22}$$

The map $\{\cdot\,{}_\lambda\,\cdot\}_{z,\rho} : \mathscr{W} \otimes \mathscr{W} \to \mathbb{C}[\lambda] \otimes \mathscr{W}$ given by

$$\{g_\lambda h\}_{z,\rho} = \rho\{g_\lambda h\}_z \tag{23}$$

defines a Poisson vertex algebra structure on $\mathscr{W}$ (which is equivalent to Definition 8).

Thus we can give the following equivalent definition.

Definition 9 The classical $\mathscr{W}$-algebra is the differential algebra $\mathscr{W}$ defined by (22) with the Poisson vertex algebra structure given by (23).

Remark 3 The Poisson vertex algebra $\mathscr{W}$ can be constructed in the same way for an arbitrary choice of s in $\mathfrak{g}$ (taking the Lie conformal subalgebra $\mathbb{C}[\partial]\mathfrak{n}\oplus\mathbb{C}$ of $\mathscr{V}(\mathfrak{g})$. However the differential algebra $\mathscr{W}$ is independent of the choice of $z\in\mathbb{C}$ if and only if $[s,\mathfrak{n}]=0$. As already stated in Remark 2, this independence of z is very important in constructing integrable hierarchies of bi-Hamiltonian equations, since there we need to view z as a formal parameter.

Let us fix a basis $\{q_i\}_{i\in P}$ of $\mathfrak{p}$. We can find an explicit formula for the λ-bracket in $\mathscr{W}$ as follows. Recalling the Master Formula (4) and using (20) and the definition (7) of the λ-bracket in $\mathscr{V}$, we get ($g,h\in\mathscr{W}$):

$$\{g_\lambda h\}_{z,\rho}=\{g_\lambda h\}_{0,\rho}-z\{g_\lambda h\}_{\infty,\rho}\,,$$

where

$$\{g_\lambda h\}_{\varepsilon,\rho}=\sum_{\substack{i,j\in P\\ m,n\in\mathbb{Z}_+}}\frac{\partial h}{\partial q_j^{(n)}}(\lambda+\partial)^n\{q_{i\lambda+\partial}q_j\}_{\varepsilon,\rho\to}(-\lambda-\partial)^m\frac{\partial g}{\partial q_i^{(m)}}\,,$$

with $\varepsilon=\{0,\infty\}$ and

$$\{q_{i\lambda}q_j\}_{0,\rho}=\pi_{\mathfrak{p}}[q_i,q_j]+\kappa(q_i\mid q_j)\lambda+\kappa(f\mid[q_i,q_j]),\qquad(24)$$

$$\{q_{i\lambda}q_j\}_{\infty,\rho}=-\kappa(s\mid[q_i,q_j])\,,\qquad(25)$$

for $i,j\in P$.

In [11], Drinfeld and Sokolov defined the classical $\mathscr{W}$-algebra as the subspace $\mathscr{W}\subset\mathscr{V}(\mathfrak{p})$ consisting of *gauge invariant* differential polynomials. A *gauge transformation* is, by definition, a change of variables formula $q=(q_i)_{i\in P}\mapsto q^a=(q_i^a)_{i\in P}$, for $a\in\mathfrak{n}\otimes\mathscr{V}(\mathfrak{p})$ (see [11] or [7] for a precise definition). Hence, $\mathfrak{g}\in\mathscr{V}(\mathfrak{p})$ belongs to $\mathscr{W}$ if and only if $g(q^a)=g(q)$ for all $a\in\mathfrak{n}\otimes\mathscr{V}(\mathfrak{p})$. Here and further we use the following notation: for $g\in\mathscr{V}(\mathfrak{p})$ and $r=(r_i)_{i\in P}\subset\mathscr{V}(\mathfrak{p})$, we let $g(r)$ be the differential polynomial in q_i obtained replacing $q_i^{(m)}$ by $\partial^m r_i$ in the differential polynomial g.

We claim that the space of gauge invariant differential polynomials coincides with the space $\mathscr{W}$ defined in (22). The key observation is that the action of the gauge group $g\mapsto g(q^a)\in\mathscr{V}(\mathfrak{p})$ is obtained by exponentiating the Lie conformal algebra action of $\mathbb{C}[\partial]\mathfrak{n}$ on $\mathscr{V}(\mathfrak{p})$ given by (21). This is given by the following result.

Theorem 2 *For every $a\otimes h\in\mathfrak{n}\otimes\mathscr{V}(\mathfrak{p})$ and $g\in\mathscr{V}(\mathfrak{p})$, we have*

$$g(q^{a\otimes h})=\sum_{n\in\mathbb{Z}_+}\frac{(-1)^n}{n!}(a^\rho_{\lambda_1}\dots a^\rho_{\lambda_n}g)\left(\big|_{\lambda_1=\partial}h\right)\dots\left(\big|_{\lambda_n=\partial}h\right),$$

where, for a polynomial $p(\lambda_1,\dots,\lambda_n)=\sum c\lambda_1^{i_1}\dots\lambda_n^{i_n}$, we denote

$$p(\lambda_1,\dots,\lambda_n)\left(\big|_{\lambda_1=\partial}h_1\right)\dots\left(\big|_{\lambda_n=\partial}h_n\right)=\sum c(\partial^{i_1}h_1)\dots(\partial^{i_n}h_n)\,.$$

Finally, it follows by a straightforward computation, using (24) and (25), that the Lie algebra structure on the quotient space $\mathscr{W}/\partial\mathscr{W}$, defined by (12), coincides with the Lie algebra structure defined on the space of gauge invariant local functionals in [11].

3.4 Structure of classical $\mathscr{W}$-algebras

Using the description of the classical $\mathscr{W}$-algebras in terms of gauge invariance, it is possible to prove, following the ideas of Drinfeld and Sokolov, that the differential algebra $\mathscr{W}$ is an algebra of differential polynomials in $r = \dim\ker(\operatorname{ad} f)$ variables, and it is possible to provide an algorithm to find explicit generators.

Theorem 3 *Assume that $\mathfrak{g}^f \subset \mathfrak{p}$ (this is the case, for example, if $\mathfrak{p} \subset \mathfrak{g}$ is compatible with the* ad$x$*-eigenspace decomposition* (19)*), and let $\{v_j\}_{j\in J}$ is a basis of $\mathfrak{g}^f$. Then the differential algebra $\mathscr{W} \subset \mathscr{V}(\mathfrak{p})$ is the algebra of differentials polynomials in the variables w_j, $j \in J$, where $w_j = v_j + g_j \in \mathscr{V}(\mathfrak{p})$.*

The differential polynomials $g_j \in \mathscr{V}(\mathfrak{p})$ satisfy some properties of homogeneity with respect to the *conformal weight grading* of $\mathscr{V}(\mathfrak{p})$ (see [7]). Furthermore, it is proved that classical $\mathscr{W}$-algebras contain a Virasoro element L and a basis (as differential algebras) of L-eigenvectors.

Proposition 4 *Consider the Poisson vertex algebra $\mathscr{W}$ with the λ-bracket $\{\cdot\,{}_\lambda\,\cdot\}_{z,\rho}$ defined by Eq.* (23)*.*

(a) *There exists an element $L \in \mathscr{W}$ such that the λ-bracket of L with itself is*

$$\{L_\lambda L\}_{z,\rho} = (\partial + 2\lambda)L - \kappa(x \mid x)\lambda^3 + 2\kappa(f \mid s)z\lambda\,.$$

In particular, $L \in \mathscr{W}$ is Virasoro element for $z = 0$ (or for arbitrary z provided that $\kappa(f \mid s) = 0$) of central charge $c = -\kappa(x \mid x)$.

(b) *Assume that $\mathfrak{p} \subset \mathfrak{g}$ is compatible with the* ad$x$*-eigenspace decomposition* (19)*, and consider the generators $w_j = v_j + g_j \in \mathscr{W}$, $j \in J$ provided by Theorem 3, where $\{v_j\}_{j\in J}$ is a basis of $\mathfrak{g}^f$ consisting of* ad$x$*-eigenvectors: $[x, v_j] = (1 - \Delta_j)v_j$, $j \in J$ (with $\Delta_j \geq 1$). Then w_j is an L-eigenvector of conformal weight Δ_j for $z = 0$.*

(c) *The Poisson vertex algebra $\mathscr{W}$, with the λ-bracket $\{g_\lambda h\}_{z,\rho}$, is independent of the choice of the isotropic subspace $\mathfrak{l} \subset \mathfrak{g}_{\frac{1}{2}}$ for $z = 0$, and for arbitrary z, provided that $s \in \ker(\operatorname{ad}\mathfrak{g}_{\geq\frac{1}{2}})$ is fixed.*

If we assume $\mathfrak{l} \subset \mathfrak{g}_{\frac{1}{2}}$ to be a maximal isotropic subspace, then the Virasoro element $L \in \mathscr{W}$ is the image of the Virasoro element of affine Poisson vertex algebras defined in (9) via the map ρ in (20).

3.5 Integrable hierarchies of Hamiltonian equations for classical $\mathscr{W}$-algebras

Slightly modifying the argument used to prove applicability of the Lenard-Magri scheme of integrability for affine Poisson vertex algebras it is possible (under some assumptions on the nilpotent element $f \in \mathfrak{g}$) to apply successfully the Lenard-Magri scheme of integrability for classical $\mathscr{W}$-algebras endowed with the compatible λ-brackets described in (24) and (25). The first result in this direction appeared in the original paper of Drinfeld and Sokolov.

Theorem 4 ([11]) *If $f \in \mathfrak{g}$ is principal nilpotent, then there exists an infinite sequence of linearly independent local functionals $\int f_0, \int f_1, \ldots \in \mathscr{W}/\partial\mathscr{W}$ such that*

$$\{\textstyle\int f_n, \int f_m\}_{0,\rho} = \{\int f_n, \int f_m\}_{\infty,\rho} = 0,$$

for all $n, m \in \mathbb{Z}_+$.

Later this result was improved in the papers [4–6, 13, 14] using the theory of Heisenberg subalgebras of Kac-Moody algebras. In [7] it is proved the following result.

Theorem 5 *If $f + zs$ is semisimple in $\mathfrak{g}((z^{-1}))$, then there exists an infinite sequence of linearly independent local functionals $\int f_0, \int f_1, \ldots \in \mathscr{W}/\partial\mathscr{W}$ such that*

$$\{\textstyle\int f_n, \int f_m\}_{0,\rho} = \{\int f_n, \int f_m\}_{\infty,\rho} = 0,$$

for all $n, m \in \mathbb{Z}_+$.

We note that if f is principal nilpotent, then $e_\theta \in \ker(\operatorname{ad}\mathfrak{n})$ (e_θ being the highest root vector) and $f + ze_\theta$ is semisimple in $\mathfrak{g}((z^{-1}))$ thus recovering Theorem 4. Moreover, we point out that the freedom in the choice of $\mathfrak{l} \subset \mathfrak{g}_{\frac{1}{2}}$ (see Proposition 4(c)) gives more chances to find $s \in \ker(\operatorname{ad}\mathfrak{n})$ such that the hypotheses of Theorem 5 are satisfied.

3.6 Examples of classical $\mathscr{W}$-algebras and corresponding integrable hierarchies of Hamiltonian equations

3.6.1 The KdV hierarchy

Let $\mathfrak{g} = \mathfrak{sl}_2$ with standard generators $f, h = 2x, e$ With respect to the $\operatorname{ad}x$-eigenspace decomposition (19), we have $\mathfrak{n} = \mathfrak{m} = \mathbb{C}e$. We fix the subspace $\mathfrak{p} = \mathbb{C}h \oplus \mathbb{C}f \subset \mathfrak{sl}_2$ complementary to $\mathfrak{m}$. Since $\mathfrak{g}^f = \mathbb{C}f$, by Theorem 3, $\mathscr{W} \subset \mathscr{V}(\mathfrak{p})$ is the algebra of differential polynomials in one generator, say L, of the form $L = f + g$, where $g \in \mathscr{V}(\mathfrak{p})$. By an explicit computation, we find

$$L = f - \frac{1}{2c}x^2 + \partial x \in \mathscr{V}(\mathfrak{p}),$$

where $c = -\kappa(x \mid x)$. Let us assume $s = e \in \ker(\operatorname{ad}\mathfrak{n})$. Then the λ-bracket of L with itself is given by (see Proposition 4)

$$\{L_\lambda L\}_{z,\rho} = (2\lambda + \partial)L + c\lambda^3 - 4c\lambda z.$$

The $z = 0$ λ-bracket is known as the *Virasoro-Magri* λ-bracket, while the $z = \infty$ λ-bracket is known as the *Gardner-Faddeev-Zakharov* λ-bracket (up to the factor $4c$).

It is also possible to compute explicitly the first terms of the series of integrals of motion. We get $\int f_0 = \int L$ and the corresponding Hamiltonian equation is $\frac{dL}{dt_0} = L'$. The next integral of motion is $\int f_1 = \int \frac{1}{8c} L^2$ and the corresponding Hamiltonian equation is the *Korteweg-de Vries equation*

$$\frac{dL}{dt_1} = \frac{1}{4}(L''' + \frac{3}{c}LL').$$

3.6.2 The Boussinesq hierarchy

Let $\mathfrak{g} = \mathfrak{sl}_3$ and fix $\kappa(a \mid b) = c\,\mathrm{tr}(ab)$, $c \in \mathbb{C}^*$, for $a, b \in \mathfrak{sl}_3$. Let $f \in \mathfrak{sl}_3$ be its principal nilpotent element and let us choose $s = e_\theta$, where θ is the highest weight root. Then $\mathscr{W} = S(\mathbb{C}[\partial]L \oplus \mathbb{C}[\partial]w_3)$ with λ-bracket

$$\begin{aligned}
\{L_\lambda L\}_{z,\rho} &= (\partial + 2\lambda)L - 2c\lambda^3, \\
\{L_\lambda w_3\}_{z,\rho} &= (\partial + 3\lambda)w_3 + 3cz\lambda, \\
\{w_{3\lambda} w_3\}_{z,\rho} &= \frac{1}{3c}(2\lambda + \partial)L^2 - \frac{1}{6}(\lambda + \partial)^3 L - \frac{1}{6}\lambda^3 L \\
&\quad - \frac{1}{4}\lambda(\lambda + \partial)(2\lambda + \partial)L + \frac{c}{6}\lambda^5.
\end{aligned}$$

The first integral of motion is $\int f_0 = \int w_3$ and the corresponding Hamiltonian equation is

$$\begin{cases} L_t = 2w_3' \\ w_{3t} = -\frac{1}{6}L''' + \frac{2}{3c}LL' \end{cases}$$

from which we can eliminate w_3 and get the Boussinesq equation

$$L_{tt} = -\frac{1}{3}\left(L^{(4)} - \frac{4}{c}(LL')'\right).$$

3.6.3 Minimal nilpotent element case in $\mathfrak{sl}_3$

In $\mathfrak{g} = \mathfrak{sl}_3$, we can also consider $f = e_{-\theta}$ to be the lowest root vector. In this case $\mathfrak{g}_{\frac{1}{2}} \neq 0$. Let us assume $\mathfrak{l} = \mathbb{C}(e_\alpha + e_\beta) \subset \mathfrak{g}_{\frac{1}{2}}$ ($\mathfrak{l}$ is maximal isotropic), where α and β are the simple roots of $\mathfrak{sl}_3$, and let us choose $s = e_\alpha + e_\beta$. We get $\mathscr{W} = S(\mathbb{C}[\partial]L \oplus \mathbb{C}[\partial]\varphi \oplus \mathbb{C}[\partial]\psi_\pm)$ with λ-bracket

$$\begin{aligned}
\{L_\lambda L\}_{z,\rho} &= (2\lambda + \partial)L - \frac{c}{2}\lambda^3, \\
\{L_\lambda \psi_\pm\}_{z,\rho} &= (\frac{3}{2}\lambda + \partial)\psi_\pm + \frac{3}{2}cz\lambda, \\
\{L_\lambda \varphi\}_{z,\rho} &= (\lambda + \partial)\varphi, \\
\{\psi_{\pm\lambda} \psi_\pm\}_{z,\rho} &= 0, \\
\{\psi_{+\lambda} \psi_-\}_{z,\rho} &= -L + \frac{1}{3c}\varphi^2 - \frac{1}{2}(2\lambda + \partial)\varphi + c\lambda^2, \\
\{\psi_{\pm\lambda} \varphi\}_{z,\rho} &= \pm 3\psi_\pm \pm 3cz, \\
\{\varphi_\lambda \varphi\}_{z,\rho} &= 6c\lambda.
\end{aligned}$$

Since $f+zs$ satisfies the hypothesis of Theorem 5 we can find a sequence of linearly independent integrals of motion in involution. The first integral of motion is $\int f_0 = \int(\psi_+ + \psi_-)$ and the corresponding Hamiltonian equation is

$$\begin{cases} L_t = \frac{1}{2}(\psi_+ + \psi_-)' \\ \varphi_t = 3(\psi_+ - \psi_-) \\ (\psi_+)_t = L - \frac{1}{3c}\varphi^2 - \frac{1}{2}\varphi' \\ (\psi_-)_t = -L + \frac{1}{3c}\varphi^2 - \frac{1}{2}\varphi'. \end{cases}$$

The corresponding hierarchy of Hamiltonian equations is known as *fractional KdV hierarchy* (see [1]). We note that we can eliminate L and $\psi_\pm$ from the system and get the equation

$$\varphi'' = -\frac{1}{3}\left(\varphi_{tttt} + \frac{4}{c}(\varphi\varphi_t)_t\right),$$

which, up to rescaling, is the Boussinesq equation with derivatives in x and t exchanged.

We can also consider $\mathfrak{l} = 0$ and $s = e_\theta$. Then we get the following λ- bracket for the generators of $\mathscr{W}$:

$$\begin{aligned} \{L_\lambda L\}_{z,\rho} &= (2\lambda+\partial)L - \frac{c}{2}\lambda^3 + 2cz\lambda, \\ \{L_\lambda \psi_\pm\}_{z,\rho} &= (\frac{3}{2}\lambda+\partial)\psi_\pm, \\ \{L_\lambda \varphi\}_{z,\rho} &= (\lambda+\partial)\varphi, \\ \{\psi_{\pm\lambda}\psi_\pm\}_{z,\rho} &= 0, \\ \{\psi_{+\lambda}\psi_-\}_{z,\rho} &= -L + \frac{1}{3c}\varphi^2 - \frac{1}{2}(2\lambda+\partial)\varphi + c\lambda^2 - cz, \\ \{\psi_{\pm\lambda}\varphi\}_{z,\rho} &= \pm 3\psi_\pm, \\ \{\varphi_\lambda\varphi\}_{z,\rho} &= 6c\lambda\,. \end{aligned}$$

As stated in Proposition 4 (c) we note that λ-bracket corresponding to $z=0$ does not change for different choices of the isotropic subspace $\mathfrak{l} \subset \mathfrak{g}_{\frac{1}{2}}$ (but it does change for arbitrary z with the change of s). Also in this case, $f+zs$ satisfies the hypothesis of Theorem 5 and we get the corresponding integrable hierarchy of Hamiltonian equations. The first two integrals of motion are

$$\int f_0 = \int\left(L - \frac{1}{12c}\varphi^2\right) \text{ and } \int f_1 = \int\left(\frac{1}{2c}(\psi_+\psi_-' - \psi_+'\psi_-) - \frac{1}{2c^2}\varphi\psi_+\psi_- - \frac{1}{4c}f_0^2\right).$$

After performing a Dirac reduction (see [9]) we can set $\varphi = 0$ and the first non-trivial equation is

$$\begin{cases} \dfrac{d\psi_\pm}{dt_1} = \psi_\pm''' - \alpha L\psi_\pm' - \dfrac{\alpha}{2}\psi_\pm L' \pm \dfrac{2}{3}\alpha^2\psi_\pm\psi_+\psi_- \\ \dfrac{dL}{dt_1} = \dfrac{1}{4}L''' - \alpha LL' + \alpha(\psi_+\psi_-'' - \psi_+''\psi_-) \end{cases}$$

where $\alpha = \frac{3}{2c} \in \mathbb{C}^*$.

Remark 4 The explicit formula for the λ-bracket on the generators of the classical $\mathscr{W}$-algebra corresponding to a simple Lie algebra $\mathfrak{g}$ and a minimal nilpotent element $f \in \mathfrak{g}$ is computed in [8]. In the same paper there is the classification of all the elements $s \in \ker(\mathrm{ad}\,\mathfrak{n})$ such that $f + zs \in \mathfrak{g}((z^{-1}))$ is semisimple and the computation of the first non-trivial equations of the corresponding integrable hierarchies (which exist by Theorem 5).

3.7 Gelfand-Dickey algebras

Let $\mathscr{V}$ be a differential algebra with a derivation ∂. We consider the algebra $\mathscr{V}((\partial^{-1}))$ of formal pseudodifferentials operators with coefficients in $\mathscr{V}$, with multiplication defined by

$$\partial^n \circ a = \sum_{k \in \mathbb{Z}_+} \binom{n}{k} a^{(k)} \partial^{n-k} \tag{26}$$

for any $n \in \mathbb{Z}$ and $a \in \mathscr{A}$. It can easily be verified that multiplication given by (26) is associative, thus making the space of formal pseudodifferential operators an associative algebra with unity.

The algebra of formal pseudodifferential operators has a natural anti-homomorphism, which we denote by *, called *formal adjoint*, defined by $f^* = f$, for $f \in \mathscr{A}$, and $\partial^* = -\partial$.

The *symbol* of a pseudodifferential operator $A(\partial) = \sum a_n \partial^n \in \mathscr{V}((\partial^{-1}))$ is the formal Laurent series $A(z) = \sum a_n z^n \in \mathscr{V}((z^{-1}))$, obtained replacing ∂ with z, where z is an indeterminate commuting with $\mathscr{V}$. This gives us a bijective map $\mathscr{V}((\partial^{-1})) \longrightarrow \mathscr{V}((z^{-1}))$ which is not an algebra homomorphism. If we consider the symbol of the multiplication rule (26) we get, for any $n \in \mathbb{Z}$ and $a \in \mathscr{V}$, the well known binomial formula

$$(z+\partial)^n a = \sum_{k \in \mathbb{Z}_+} \binom{n}{k} a^{(k)} z^{n-k} . \tag{27}$$

This allows us to write a closed formula for the associative product in $\mathscr{V}((\partial^{-1}))$ in terms of the corresponding symbols. For $A(\partial), B(\partial) \in \mathscr{V}((\partial^{-1}))$, it is given by the following

$$(A \circ B)(z) = A(z+\partial) B(z) ,$$

where, for any $n \in \mathbb{Z}$, we expand $(z+\partial)^n$ as in (27), namely in non-negative powers of ∂.

Let us fix $N \in \mathbb{Z}_+$ and let us consider $\mathscr{V} = S(\oplus_{i \geq -N} \mathbb{C}[\partial] u_i)$ be the algebra of differential polynomials in infinitely many variables u_i, $i \geq -N$. We let, for $c \in \mathbb{C}$,

$$L(\partial) = \partial^N + u_{-N} \partial^{N-1} + \ldots + u_{-1} + c + u_0 \partial^{-1} + \ldots \in \mathscr{V}((\partial^{-1})) .$$

Proposition 5 *For all $i, j \geq -N$ let us define $\{u_{i\,\lambda} u_j\}_c \in \mathscr{V}[\lambda]$ via the generating series*

$$\begin{aligned} \{L(z)_\lambda L(w)\}_c = & \, L(z) i_z (z-w-\lambda-\partial)^{-1} L(w) \\ & - L(w+\lambda+\partial) i_z (z-w-\lambda-\partial)^{-1} L^*(\lambda - z) , \end{aligned} \tag{28}$$

(where i_z means we should expand in negative powers of z, and the expansions of $L(w+\lambda+\partial)$ and $L^*(\lambda-z)$ are done according to the binomial formula (27)*, namely we expand in non-negative powers of $\lambda+\partial$). Then* (28) *defines a Poisson vertex algebra structure on $\mathcal{V}$.*

Proof. We note that by Theorem 1(b) and (c), in order to prove the proposition, it suffices to check equations (5) and (6) for all $i,j,k \geq -N$. Using the symbol $L(z)$ these equations become, respectively,

$$\{L(z)_\lambda L(w)\}_c = -\{L(w)_{-\lambda-\partial} L(z)\}_c \tag{29}$$

and

$$\{L(z)_\lambda \{L(w)_\mu L(t)\}_c\}_c - \{L(w)_\mu \{L(z)_\lambda L(t)\}_c\}_c = \{\{L(z)_\lambda L(w)\}_{c\,\lambda+\mu} L(t)\}_c\,. \tag{30}$$

By an easy computation we get

$$\begin{aligned}-\{L(w)_{-\lambda-\partial}L(z)\}_c = {} & L(z)i_w(z-w-\lambda-\partial)^{-1}L(w) \\ & -L(w+\lambda+\partial)i_w(z-w-\lambda-\partial)^{-1}L^*(\lambda-z)\,.\end{aligned}$$

Hence, (29) is equivalent to the identity

$$L(w+\lambda+\partial)\delta(z-w-\lambda-\partial)L^*(\lambda-z) = L(z)\delta(z-w-\lambda-\partial)L(w)\,, \tag{31}$$

where

$$\delta(z-w) = \sum_{k\in\mathbb{Z}} z^k w^{-k-1} \in \mathbb{C}[[z,z^{-1},w,w^{-1}]]$$

is the *formal δ-function*. The identity (31) follows easily applying the properties of the formal δ-function (which can be found in [16]). The Jacobi identity (30) follows by a straightforward computation using again the properties of the formal δ-function (see [18]). □

We remark that we have a one-parameter family of λ-brackets if we think at c as a parameter. For $c=0$ we recover the so-called *second Gelfand-Dickey Poisson structure*, while for $c=\infty$ we recover the *first Gelfand-Dickey Poisson structure* (see [18] for further details).

Let $\mathcal{I} = \langle u_i \mid i \in \mathbb{Z}_+\rangle$ be the differential ideal in $\mathcal{V}$ generated by u_i, for $i\in\mathbb{Z}_+$. It can be proved it is also a Poisson vertex algebra ideal. Thus we can consider the quotient Poisson vertex algebra $\mathcal{W}_N \cong \mathcal{V}/\mathcal{I}$.

Theorem 6 *There is a Poisson vertex algebra isomorphism between $\mathcal{W}_N$ and the classical $\mathcal{W}$-algebra defined in Definition 9 for $\mathfrak{g}=\mathfrak{gl}_N$, $f\in\mathfrak{g}$ principal nilpotent, $\kappa(a \mid b) = -\mathrm{tr}(ab)$, for all $a,b\in\mathfrak{g}$, and $s=e_\theta$.*

A similar result can be proved for any classical Lie algebra (see [18]).

Acknowledgements I wish to thank Maria Gorelik and Paolo Papi for inviting me to give a talk at the "Lie superalgebras" conference, and INDAM for the kind hospitality.

References

1. Bakas, I., Depireux, D.: A fractional KdV hierarchy, Modern Phys. Lett. A **6**(17), 1561–1573 (1991)
2. Barakat, A., De Sole, A., Kac, V.: Poisson vertex algebras in the theory of Hamiltonian equations, Jpn. J. Math. **4**(2), 141–252 (2009)
3. Belavin, A., Polyakov, A., Zamolodchikov, A.: Infinite conformal symmetry in two dimensional quantum field theory. Nucl. Phys. B **241**, 333–380 (1984)
4. Burruoughs, N., de Groot, M., Hollowood, T., Miramontes, L.: Generalized Drinfeld-Sokolov hierarchies II: the Hamiltonian structures. Comm. Math. Phys. **153**, 187–215 (1993)
5. de Groot, M., Hollowood, T., Miramontes, L.: Generalized Drinfeld-Sokolov hierarchies. Comm. Math. Phys. **145**, 57–84 (1992)
6. Delduc, F., Fehér, L.: Regular conjugacy classes in the Weyl group and integrable hierarchies. J. Phys. A **28**(20), 5843–5882 (1995)
7. De Sole, A., Kac, V., Valeri, D.: Classical $\mathscr{W}$-algebras and generalized Drinfeld-Sokolov bi-Hamiltonian systems within the theory of Poisson vertex algebras. Comm. Math. Phys. **323**(2), 663–711 (2013)
8. De Sole, A., Kac, V., Valeri, D.: Classical $\mathscr{W}$-algebras and generalized Drinfeld-Sokolov hierarchies for minimal and short nilpotents. arXiv:1306.1684 [math-ph]
9. De Sole, A., Kac, V., Valeri, D.: Dirac reduction for Poisson vertex algebras. arXiv:1306.6589 [math-ph]
10. Dickey, L.A.: Lectures on classical $\mathscr{W}$-algebras. Acta Appl. Math. **47**, 243–321 (1997)
11. Drinfeld, V., Sokolov, V.: Lie algebras and equations of KdV type. Soviet J. Math. **30**, 1975–2036 (1985)
12. Fateev, V., Lukyanov, S.: The models of two dimensional conformal quantum field theory with $\mathbb{Z}_n$ symmetry. Int. J. Mod. Phys. A **3**, 507–520 (1988)
13. Fehér, L., Harnad, J., Marshall, I.: Generalized Drinfeld-Sokolov reductions and KdV type hierarchies. Comm. Math. Phys. **154**(1), 181–214 (1993)
14. Fernández-Pousa, C., Gallas, M., Miramontes, L., Sánchez Guillén, J.: $\mathscr{W}$-algebras from soliton equations and Heisenberg subalgebras. Ann. Physics **243**(2), 372–419 (1995)
15. Gelfand, I., Dickey, L.: Family of Hamiltonian structures connected with integrable non-linear equations. Preprint, IPM, Moscow (in Russian) (1978). English version in: Collected papers of I.M. Gelfand, Vol. 1, pp. 625–646. Springer-Verlag, Berlin New York (1987)
16. Kac, V.: Vertex algebras for beginners. University Lecture Series, AMS, Vol. 10 (1996); 2nd ed. AMS (1998)
17. Magri, F.: A simple model of the integrable Hamiltonian equation. J. Math. Phys. **19**(5), 1156–1162 (1978)
18. Valeri, D.: Classical $\mathscr{W}$-algebras. Ph.D. thesis (2012)
19. Zamolodchikov, A.: Infinite additional symmetries in two dimensional conformal quantum field theory. Theor. Math. Phys. **65**, 1205–1213 (1985)

References

Vertex operator superalgebras and odd trace functions

Jethro van Ekeren

Abstract We begin by reviewing Zhu's theorem on modular invariance of trace functions associated to a vertex operator algebra, as well as a generalisation by the author to vertex operator superalgebras. This generalisation involves objects that we call 'odd trace functions'. We examine the case of the $N = 1$ superconformal algebra. In particular we compute an odd trace function in two different ways, and thereby obtain a new representation theoretic interpretation of a well known classical identity due to Jacobi concerning the Dedekind eta function.

1 Introduction

One of the most significant theorems in the theory of vertex operator algebras (VOAs) is the modular-invariance theorem of Zhu [17]. The theorem states that under favourable circumstances the graded dimensions of certain modules over a VOA are modular forms for the group $SL_2(\mathbb{Z})$. The favourable circumstances are that the VOA be rational, C_2-cofinite, and be graded by integer conformal weights (we define all terms in Sect. 2 and state Zhu's theorem fully in Sect. 3 below).

Numerous generalisations of Zhu's theorem have appeared in the literature: to twisted modules over VOAs [2], to vertex operator superalgebras (VOSAs) and their twisted modules [3, 4] (see also [8]), to intertwining operators for VOAs [6, 13], to twisted intertwining operators [16], and to non rational VOAs [14].

In [15] the present author relaxed the assumption of integer conformal weights of V to allow arbitrary rational conformal weights. This work was carried out in the setting of twisted modules over a rational C_2-cofinite VOSA. Actually it is worth noting that in that paper the condition of C_2-cofiniteness was also relaxed slightly, allowing applications to some interesting examples such as affine VOAs at admissible level.

J. van Ekeren (✉)
IMPA, Rio de Janeiro, RJ 22460-320, Brasil
e-mail: jethrovanekeren@gmail.com

M. Gorelik, P. Papi (eds.): *Advances in Lie Superalgebras.* Springer INdAM Series 7,
DOI 10.1007/978-3-319-02952-8_13,

One of the features of [15] is the appearance of odd trace functions (see Sect. 4 for the definition) which are to be included alongside the more usual (super)trace functions in order to achieve modular invariance. Although similar in many ways, these odd traces differ from (super)traces in that they act nontrivially on odd elements of a vector superspace, whereas the (super)trace must always vanish on such elements. The results of [15] are reviewed in Sect. 4 (for simplicity in the special case of Ramond twisted modules).

In the present work, in Sect. 6, we compute an odd trace function for a particular example: the $N = 1$ superconformal minimal model of central charge $c = -21/4$. We evaluate the odd trace function on the superconformal generator (which is an odd element of conformal weight $3/2$), using the strong constraint of its modular invariance. The odd trace function in question equals the weight $3/2$ modular form $\eta(\tau)^3$, where $\eta(\tau)$ is the well known Dedekind eta function.

We then give a different proof of this equality (up to an ambiguity of signs) using a BGG resolution and some simple combinatorics. The result is a representation theoretic interpretation of the classical identity

$$\eta(\tau)^3 = q^{1/8} \sum_{n\in\mathbb{Z}} (4n+1) q^{n(2n+1)}$$

similar in spirit, but a little different, to the celebrated proof coming from the affine Weyl-Kac denominator identity [9, Chap. 12].

2 Definitions

For us a *vertex superalgebra* [5, 10] is a quadruple $V, |0\rangle, T, Y$ where V is a vector superspace, $|0\rangle \in V$ an even vector, $T : V \to V$ an even linear map, and $Y : V \otimes V \to V((z))$, denoted $u \otimes v \mapsto Y(u,z)v = \sum_{n\in\mathbb{Z}} u_{(n)} v z^{-n-1}$, is also even. These data are to satisfy the following axioms.

- The unit identities $Y(|0\rangle, z) = I_V$ and $Y(u,z)|0\rangle|_{z=0} = u$.
- The translation invariance identity $Y(Tu,z) = \partial_z Y(u,z)$.
- The Cousin property that the three expressions

  $$Y(u,z)Y(v,w)x \qquad p(u,v)Y(v,w)Y(u,z)x, \quad \text{and} \quad Y(Y(u,z-w)v,w)x,$$

  which are elements of $V((z))((w))$, $V((w))((z))$, and $V((w))((z-w))$, are images of a single element of $V[[z,w]][z^{-1}, w^{-1}, (z-w)^{-1}]$ under natural inclusions into those three spaces.

An equivalent definition, more convenient for some applications, is the following. A vertex superalgebra is a triple $V, |0\rangle, Y$ where these data are as above, but satisfy the following axioms.

- The unit identities $|0\rangle_{(n)} u = \delta_{n,-1} u$, $u_{(-1)}|0\rangle = u$ and $u_{(n)}|0\rangle = 0$ for $n > 0$.
- The Borcherds identity (also known, in a different notation, as the Jacobi identity)

  $$B(u,v,x;m,k,n) = 0 \quad \text{for all } u,v,x \in V,\ m,k,n \in \mathbb{Z},$$

where

$$B(u,v,x;m,k,n) = \sum_{j\in\mathbb{Z}_+} \binom{m}{j} (u_{(n+j)}v)_{(m+k-j)}x$$
$$- \sum_{j\in\mathbb{Z}_+} (-1)^j \binom{n}{j} \left[u_{(m+n-j)}v_{(k+j)} - (-1)^n p(u,v) v_{(k+n-j)}u_{(m+j)}\right] x.$$

A *vertex algebra* is a purely even vertex superalgebra.

Let V be a vertex superalgebra. A V-module is a vector superspace M with a vertex operation $Y^M : V\otimes M \to M((z))$ such that

$$Y^M(|0\rangle, z) = I_M, \quad \text{and} \quad B(u,v,x;m,k,n) = 0 \tag{1}$$

for all $u,v\in V$, $x\in M$, $m,k,n\in\mathbb{Z}$.

For the present theory we require the extra structure of a *conformal vector*. This is a vector $\omega\in V$ such that its modes $L_n = \omega_{(n-1)} \in \operatorname{End} V$ furnish V with a representation of the Virasoro algebra, i.e.,

$$[L_m, L_n] = (m-n)L_{m+n} + \delta_{m,-n}\frac{m^3-m}{12}c$$

(here $c\in\mathbb{C}$ is an invariant of V called the central charge), L_0 is diagonal with real eigenvalues bounded below, and $L_{-1} = T$. We call a vertex superalgebra with conformal vector a *vertex operator superalgebra* or VOSA, and we use the term VOA to distinguish the purely even case.

A V-module M is called a *positive energy* module if $L_0 \in \operatorname{End} M$ acts diagonally with eigenvalues bounded below. In particular V is a positive energy V-module. The eigenvalues of $L_0 \in \operatorname{End} V$ are called *conformal weights*, and if $L_0 u = \Delta u$ we write

$$Y(u,z) = \sum_{n\in\mathbb{Z}} u_{(n)} z^{-n-1} = \sum_{n\in-\Delta+\mathbb{Z}} u_n z^{-n-\Delta}$$

(so that $u_n = u_{(n-\Delta+1)}$). The *zero mode* $u_0 \in \operatorname{End} M$ attached to $u\in V$ is special because it commutes with L_0 and thus preserves the eigenspaces of the latter. A VOSA V is said to be *rational* if its category of positive energy modules is semisimple, i.e., it contains finitely many irreducible objects, and any object is isomorphic to a direct sum of irreducible ones.

The condition of C_2-cofiniteness is an important finiteness condition of vertex (super)algebras introduced by Zhu. We say that V is *C_2-cofinite* if

$$\dim\left(V/V_{(-2)}V\right) < \infty.$$

3 The theorem of Zhu

Now we come to the theorem of Zhu [17].

Theorem 1 (Zhu) *Let V be a VOA (i.e., purely even VOSA) such that:*

- *V is rational;*
- *the conformal weights of V lie in* $\mathbb{Z}_+$*;*
- *V is* C_2*-cofinite.*

We associate to each irreducible positive energy module M, and $u \in V$*, the series*

$$S_M(u,\tau) = \mathrm{Tr}_M\, u_0 q^{L_0 - c/24},$$

convergent for $q = e^{2\pi i\tau}$ *of modulus less than 1. There is a grading* $V = \bigoplus_{\nabla \in \mathbb{Z}_+} V_{[\nabla]}$ *such that for* $u \in V_{[\nabla]}$ *the span* $\mathscr{C}(u)$ *of the (finitely many) functions* $S_M(u,\tau)$ *defined above is modular invariant of weight* ∇*, i.e.,*

$$(c\tau+d)^{\nabla} f\left(\frac{a\tau+b}{c\tau+d}\right) \in \mathscr{C}(u) \quad \text{for all } f(\tau) \in \mathscr{C}(u) \text{ and } \left(\begin{smallmatrix} a & b \\ c & d \end{smallmatrix}\right) \in SL_2(\mathbb{Z}).$$

Here is an outline of the proof of Zhu's theorem.

1. Introduce a space $\mathscr{C}$ of maps $S(u,\tau) : V \times \mathscr{H} \to \mathbb{C}$ linear in V and holomorphic in $\mathscr{H} = \{\tau \in \mathbb{C} | \mathrm{Im}\tau > 0\}$ satisfying certain axioms, this space is called the 'conformal block' of V or the space of conformal blocks of V. The definition of $\mathscr{C}$ can be understood in terms of elliptic curves and their moduli [5].
2. It is automatic from its definition that $\mathscr{C}$ admits an action of the group $SL_2(\mathbb{Z})$, namely,
$$[S \cdot A](u,\tau) = (c\tau+d)^{-\nabla_u} S(u, A\tau),$$
where ∇ is the grading mentioned above.
3. It is proved by direct calculation that $\mathrm{Tr}_M\, u_0^M q^{L_0 - c/24}$ is a conformal block (at least as a formal power series).
4. Using the C_2-cofiniteness condition, one shows that any fixed $S \in \mathscr{C}$ satisfies some differential equation, and consequently is expressible as a power series in q (whose coefficients are linear maps $V \to \mathbb{C}$).
5. The lowest order coefficient $C_0 : V \to \mathbb{C}$ in the series expansion factors to a certain quotient $\mathrm{Zhu}(V)$ of V. This quotient has the structure of a unital associative algebra, and C_0 is symmetric, i.e., $C_0(ab) = C_0(ba)$.
6. There is a natural bijection
$$\text{irreducible positive energy } V\text{-modules} \longleftrightarrow \text{irreducible } \mathrm{Zhu}(V)\text{-modules},$$
and so if V is rational, $\mathrm{Zhu}(V)$ is finite dimensional semisimple. Thus we can write $C_0 = \sum_N \alpha_N \mathrm{Tr}_N$ where the sum is over irreducible $\mathrm{Zhu}(V)$-modules and $\alpha_N \in \mathbb{C}$.
7. Write the corresponding sum $\sum_N \alpha_N S_M$ where M is the V-module associated to N. Subtract this conformal block from S.
8. One can repeat the process and show that S is exhausted by trace functions in a finite number of steps.

4 Generalisation to the supersymmetric case, and to rational conformal weights

Results described in this section are drawn from [15].

Many examples of interest, especially in the supersymmetric case, are graded by noninteger conformal weights. So it is first necessary (and of independent interest) to relax the condition of integer conformal weights in Theorem 1. Therefore let V be a VOA whose conformal weights lie in $\mathbb{Q}$ (and are bounded below) rather than in $\mathbb{Z}_+$, and which satisfies the other conditions of the theorem. Then the trace functions $S_M(u,\tau)$ and their span $\mathscr{C}(u)$ are defined as before. There exists a certain *rational* grading $V = \oplus_{\nabla\in\mathbb{Q}} V_{[\nabla]}$ in place of the usual integer grading. It is then true that for $u \in V_{[\nabla]}$ the space $\mathscr{C}(u)$ is invariant under the weight ∇ action[1], not of $SL_2(\mathbb{Z})$, but of its congruence subgroup

$$\Gamma_1(N) = \{\left(\begin{smallmatrix} a & b \\ c & d \end{smallmatrix}\right) \in SL_2(\mathbb{Z}) | b \equiv 0 \bmod N, \text{ and } a \equiv d \equiv 1 \bmod N\}.$$

Here N is the least common multiple of the denominators of conformal weights of vectors in V. This number is finite because of the condition of C_2-cofinitness.

It is possible to achieve invariance under the whole of $SL_2(\mathbb{Z})$ by altering our definition of V-module. Define a *Ramond twisted* V-module to be a vector superspace M together with fields

$$Y^M(u,z) = \sum_{n\in-\Delta_u+\mathbb{Z}} u_{(n)} z^{-n-1} = \sum_{n\in\mathbb{Z}} u_n z^{-n-\Delta}$$

satisfying (1) for all $m \in -\Delta_u + \mathbb{Z}$, $k \in -\Delta_v + \mathbb{Z}$, $n \in \mathbb{Z}$. Notice that the ranges of indices of the modes are modified so that $u \in V$ always possesses integrally graded modes $u_n \in \operatorname{End} M$, and in particular always possesses a zero mode. Let us call a VOA *Ramond rational* if its category of positive energy Ramond twisted modules is semisimple.

Let V be a Ramond rational, C_2-cofinite VOA with rational conformal weights bounded below. Attach the trace function

$$S_M(u,\tau) = \operatorname{Tr}_M u_0 q^{L_0 - c/24}$$

to $u \in V$ and M an irreducible positive energy Ramond twisted V-module, and let $\mathscr{C}(u)$ be the span of all such trace functions. Then for $u \in V_{[\nabla]}$ the space $\mathscr{C}(u)$ is invariant under the weight ∇ action of the full modular group $SL_2(\mathbb{Z})$.

The previous paragraph stated a result for VOAs. Upon passage from VOAs to VOSAs, one might expect the claim to hold with trace functions simply replaced by supertrace functions $S_M(u,\tau) = \operatorname{STr}_M u_0 q^{L_0 - c/24}$. This would be true, except for an interesting subtlety which can be traced to Step 5 of the proof outline given in Sect. 3.

---

[1] The precise definition of the action involves choices of roots of unity in general. See [15] for details.

In the present situation it is appropriate to replace the usual Zhu algebra with a certain superalgebra (which we also refer to as the Zhu algebra and denote $\text{Zhu}(V)$) introduced in the necessary level of generality in [1]. If V is Ramond rational then $\text{Zhu}(V)$ is finite dimensional and semisimple. The lowest coefficient C_0 of the series expansion of a conformal block now descends to a supersymmetric function on $\text{Zhu}(V)$, i.e., $C_0(ab) = p(a,b)C_0(ba)$.

The classification of pairs (A, φ), where A is a finite dimensional simple superalgebra (over $\mathbb{C}$) and φ is a supersymmetric function on A, is as follows:

- $A = \text{End}(N)$ for some finite dimensional vector superspace N, and φ is a scalar multiple of STr_N;
- $A = \text{End}(P)[\theta]/(\theta^2 - 1)$ where P is a vector space and θ is an odd indeterminate, and φ is a scalar multiple of $a \mapsto \text{Tr}_P(a\theta)$.

The first case is the analogue of the usual Wedderburn theorem. The superalgebra of the second case is known as the *queer superalgebra* and is often denoted Q_n (where $n = \dim P$). Clearly we have

$$Q_n \cong \left\{ \left(\begin{smallmatrix} X & Y \\ Y & X \end{smallmatrix}\right) | X, Y \in \text{Mat}_n(\mathbb{C}) \right\},$$

and the unique up to a scalar factor supersymmetric function on Q_n is $\left(\begin{smallmatrix} X & Y \\ Y & X \end{smallmatrix}\right) \mapsto \text{Tr}\, Y$, which is known as the *odd trace*.

Roughly speaking modular invariance will hold for $\mathscr{C}(u)$ defined as the span of supertrace functions together with apropriate analogues of odd trace functions. More precisely:

Definition 1 Let V be a Ramond rational, C_2-cofinite VOSA with rational conformal weights bounded below. Let A be a simple component of $\text{Zhu}(V)$, N the corresponding unique $\mathbb{Z}_2$-graded irreducible module, and M the corresponding irreducible positive energy Ramond twisted V-module. If $A \cong \text{End}(P)[\theta]/(\theta^2 - 1)$ is queer then let $\Theta : M \to M$ denote the lift to M of the map $\theta : N \to N$ of multiplication by θ. In this case we define the *odd trace function*

$$S_M(u, \tau) = \text{Tr}_M\, u_0 \Theta q^{L_0 - c/24}.$$

If A is not queer then we define the *supertrace function*

$$S_M(u, \tau) = \text{STr}_M\, u_0 q^{L_0 - c/24}.$$

Now we can state the main theorem: it is Theorem 1.3 of [15] applied to the special case of untwisted characters of Ramond twisted V-modules.

Theorem 2 ([15]) *Let V be a VOSA as in Definition 1, and let $\mathscr{C}(u)$ be the span of the supertrace functions and odd trace functions attached to all irreducible positive energy Ramond twisted V-modules. There exists a grading $V = \oplus_{\nabla \in \mathbb{Q}} V_{[\nabla]}$ such that for $u \in V_{[\nabla]}$ the space $\mathscr{C}(u)$ is invariant under the weight ∇ action of $SL_2(\mathbb{Z})$.*

In the next two sections we view some examples of odd trace functions.

5 Example: The neutral free fermion

This example is considered more fully in [15], the interested reader may refer there for further details.

We consider the Lie superalgebras A^{tw} (resp. A^{untw})

$$(\oplus_n \mathbb{C}\psi_n) \oplus \mathbb{C}1$$

where the direct sum ranges over $n \in 1/2 + \mathbb{Z}$ (resp. $n \in \mathbb{Z}$). Here the vector 1 is even, ψ_n is odd. The commutation relations in both cases are

$$[\psi_m, \psi_n] = \delta_{m,-n} 1.$$

We introduce the Fock module

$$V = U(A^{\text{tw}}) \otimes_{U(A_+^{\text{tw}})} \mathbb{C}|0\rangle,$$

where

$$A_+^{\text{tw}} = \mathbb{C}1 \oplus (\oplus_{n \geq 1/2} \mathbb{C}\psi_n)$$

and $\mathbb{C}|0\rangle$ is the A_+^{tw}-module on which 1 acts as the identity and ψ_n acts trivially.

It is well known [10] that V can be given the structure of a VOSA. The Virasoro element is $\omega = \frac{1}{2}\psi_{-3/2}\psi_{-1/2}|0\rangle$ with central charge $c = 1/2$. The vector $\psi = \psi_{-1/2}|0\rangle$ has conformal weight $1/2$ and associated vertex operator

$$Y(\psi, z) = \sum_{n \in 1/2 + \mathbb{Z}} \psi_n z^{-n-1/2}.$$

Ramond twisted V-modules are, in particular, modules over the untwisted Lie superalgebra A^{untw}. In fact the unique irreducible positive energy Ramond twisted V-module is

$$M = U(A^{\text{untw}}) \otimes_{U(A_+^{\text{untw}})} (\mathbb{C}v + \mathbb{C}\overline{v})$$

where

$$A_+^{\text{untw}} = \mathbb{C}1 \oplus (\oplus_{n \geq 0} \mathbb{C}\psi_n)$$

and $\mathbb{C}v + \mathbb{C}\overline{v}$ is the A_+^{untw}-module on which 1 acts as the identity, ψ_n acts trivially for $n > 0$, $\psi_0 v = \overline{v}$ and $\psi_0 \overline{v} = v/2$. We note that V is C_2-cofinite and Ramond rational (as well as being rational).

The (Ramond) Zhu algebra of V is explicitly isomorphic to the queer superalgebra $Q_1 = \mathbb{C}[\theta]/(\theta^2 = 1)$ via the map $[|0\rangle] \mapsto 1$, $[\psi] \mapsto \sqrt{2}\theta$. Thus the lowest graded piece $M_0 = \mathbb{C}v + \mathbb{C}\overline{v}$ of M is a Q_1-module.

The corresponding odd trace function is

$$S_M(u, \tau) = \operatorname{Tr}_M u_0 \Theta q^{L_0 - c/24}$$

where $\Theta : M \to M$ is as in Definition 1. By unwinding that definition we see that $\Theta : \mathbf{m}w \mapsto \mathbf{m}\psi_0 w$, where w is v or $\overline{v}$, and the monomial $\mathbf{m} \in U(A^{\text{untw}}/A_+^{\text{untw}})$.

The odd trace function $S_M(u,\tau)$ vanishes on $u = |0\rangle$ (indeed on all even vectors of V), but it acts nontrivially on the odd vector ψ (which is pure of Zhu weight $1/2$). Therefore $S_M(\psi,\tau)$ must be a modular form on $SL_2(\mathbb{Z})$ of weight $1/2$ (with possible multiplier system).

Indeed one may verify that $\psi_0\Theta$ acts as $(-1)^{\text{length}(\mathbf{m})}$ on the monomial vector $\mathbf{m}w$, and so

$$\begin{aligned} S_M(\psi,\tau) &= q^{-c/24} q^{L_0|M_0}(1-q^1)(1-q^2)\cdots \\ &= q^{1/24}\prod_{n=1}^{\infty}(1-q^n) = \eta(\tau) \end{aligned}$$

(here we have used that $c = 1/2$, and that $L_0|_{M_0} = 1/16$). We have recovered the well known Dedekind eta function $\eta(\tau)$ which is indeed a modular form on $SL_2(\mathbb{Z})$ of weight $1/2$.

6 Example: The $N = 1$ superconformal algebra

First we recall the definition of the Neveu-Schwarz Lie superalgebra $\mathrm{NS}^{\mathrm{tw}}$, and its Ramond-twisted variant $\mathrm{NS}^{\mathrm{untw}}$ (which is often called the Ramond superalgebra).

Definition 2 As vector superspaces the Lie superalgebras $\mathrm{NS}^{\mathrm{tw}}$ (resp. $\mathrm{NS}^{\mathrm{untw}}$) are

$$(\oplus_{n\in\mathbb{Z}}\mathbb{C}L_n)\oplus(\oplus_m\mathbb{C}G_m)\oplus\mathbb{C}C$$

where the direct sum ranges over $m \in \mathbb{Z}$ (resp. $m \in 1/2+\mathbb{Z}$). Here C and L_n are even, G_m is odd. The commutation relations in both cases are

$$\begin{aligned} [L_m, L_n] &= (m-n)L_{m+n} + \frac{m^3-m}{12}\delta_{m,-n}C, \\ [G_m, L_n] &= (m-\frac{n}{2})G_{m+n}, \\ [G_m, G_n] &= 2L_{m+n} + \frac{1}{3}(m^2-\frac{1}{4})\delta_{m,-n}C, \end{aligned} \tag{2}$$

with C central.

As usual we introduce the Verma $\mathrm{NS}^{\mathrm{tw}}$-module

$$M^{\mathrm{tw}}(c,h) = U(\mathrm{NS}^{\mathrm{tw}}) \otimes_{U(\mathrm{NS}^{\mathrm{tw}}_+)} \mathbb{C}v_{c,h}$$

where

$$\mathrm{NS}^{\mathrm{tw}}_+ = \mathbb{C}C + \oplus_{n\geq 0}\mathbb{C}L_n + \oplus_{m\geq 1/2}\mathbb{C}G_m,$$

and $\mathbb{C}v_{c,h}$ is the $\mathrm{NS}^{\mathrm{tw}}_+$-module on which C acts by $c \in \mathbb{C}$, L_0 acts by $h \in \mathbb{C}$, and higher modes act trivially. It is well known [10, 11] that the quotient $\mathrm{NS}^c = M^{\mathrm{tw}}(c,0)/U(\mathrm{NS}^{\mathrm{tw}})G_{-1/2}v_{c,0}$ is a VOSA of central change c, as is the irreducible quotient NS_c.

We shall also require the (generalised) Verma $\mathrm{NS}^{\mathrm{untw}}$-modules

$$M(c,h) = U(\mathrm{NS}^{\mathrm{untw}}) \otimes_{U(\mathrm{NS}_{+}^{\mathrm{untw}})} S_{c,h}$$

and their irreducible quotients $L(c,h)$ (we omit the superscript untw) where

$$\mathrm{NS}_{+}^{\mathrm{untw}} = \mathbb{C}C + \oplus_{n\geq 0}\mathbb{C}L_n + \oplus_{m\geq 0}\mathbb{C}G_m,$$

and $S_{c,h}$ is the $\mathrm{NS}_{+}^{\mathrm{untw}}$-module characterised by:

$$\begin{aligned} S_{c,h} &= \mathbb{C}v_{c,h} \text{ with } G_0 v_{c,h} = 0 \text{ if } h = c/24, \\ S_{c,h} &= \mathbb{C}v_{c,h} + \mathbb{C}G_0 v_{c,h} \qquad \text{if } h \neq c/24, \end{aligned}$$

with $C = c$, $L_0 = h$, and positive modes acting trivially in both cases.

A clear summary of the representations of NS^c and NS_c can be found in [12]. Here we focus on the Ramond twisted representations and shall often omit the adjective 'Ramond twisted'. Generically $\mathrm{NS}_c = \mathrm{NS}^c$ is irreducible and all the $\mathrm{NS}^{\mathrm{untw}}$-modules $L(c,h)$ acquire the structure of positive energy NS_c-modules. For certain values of c though, NS_c is a nontrivial quotient of NS^c and the irreducible positive energy NS_c-modules are finite in number and are all of the form $L(c,h)$. In fact, NS_c is a (Ramond) rational VOSA when

$$c = c_{p,p'} = \frac{3}{2}\left(1 - \frac{2(p'-p)^2}{pp'}\right)$$

for $p, p' \in \mathbb{Z}_{>0}$ with $p < p'$, $p' - p \in 2\mathbb{Z}$ and $\gcd(\frac{p'-p}{2}, p) = 1$. In this case the irreducible positive energy NS_c-modules are precisely the $\mathrm{NS}^{\mathrm{untw}}$-modules $L(c,h)$ where

$$h = h_{r,s} = \frac{(rp' - sp)^2 - (p'-p)^2}{8pp'} + \frac{1}{16}$$

for $1 \leq r \leq p-1$ and $1 \leq s \leq p'-1$ with $r-s$ odd.

Let c be one of these special values from now on. The irreducible $\mathrm{Zhu}(\mathrm{NS}_c)$-modules are precisely the lowest graded pieces of the modules $L(c,h)$ introduced above. The lowest graded piece is of dimension 1 if $h = c/24$ (there is clearly at most one such module for any fixed value of c), and is of dimension $1|1$ if $h \neq c/24$. It is known that $\mathrm{Zhu}(\mathrm{NS}_c)$ is supercommutative (it is a quotient of $\mathrm{Zhu}(\mathrm{NS}^c) \cong \mathbb{C}[x,\theta]/(\theta^2 - x + c/24)$ where x is even and θ odd). Therefore the simple components of $\mathrm{Zhu}(\mathrm{NS}_c)$ with the $1|1$-dimensional modules are all copies of the queer superalgebra Q_1, and the component with the 1-dimensional module (if it exists) is $\mathbb{C}$.

Let us consider the case $c = -21/4$ (so $p = 2$, $p' = 8$) for which the two irreducible positive energy modules are $M_i = L(c,h_i)$ where $h_1 = -3/32$ and $h_2 = -7/32 = c/24$. The first of these is the unique queer module. Theorem 2 tells us that the supertrace function $S_{M_2}(u,\tau)$ and the odd trace function $S_{M_1}(u,\tau)$ together span an $SL_2(\mathbb{Z})$-invariant space whose weight is the Zhu weight of u. Assume further that $u \in V$ is odd. Then $S_{M_2}(u,\tau)$ vanishes as the supertrace of an odd element. But $S_{M_1}(u,\tau)$ need not vanish, and it will be a modular form (with multiplier system).

Unwinding Definition 1 we see that $\Theta : \mathbf{m}v_{c,h} \mapsto \mathbf{m}G_0 v_{c,h}$ where $\mathbf{m}$ is an element of $U(\mathrm{NS}^{\mathrm{untw}}/\mathrm{NS}^{\mathrm{untw}}_+)$, and

$$S_{M_1}(u,\tau) = \mathrm{Tr}_{M_1} u_0 \Theta q^{L_0 - c/24}.$$

The VOSA NS_c possesses a distinguished element $v = G_{-3/2}|0\rangle$ of conformal weight $3/2$, it satisfies $v_0 = G_0$. It turns out that v is of pure Zhu weight $3/2$ and so by the above remarks

$$F(\tau) := \mathrm{Tr}_{M_1} G_0 \Theta q^{L_0 - c/24}$$

is a modular form of weight $3/2$. On the top level of M_1, $G_0\Theta = G_0^2 = h - c/24 = 1/8$, so the top level contribution to $F(\tau)$ is $\frac{1}{4}q^{1/8}$. This is already enough information to determine $F(\tau)$ completely. The cube of the Dedekind eta function is $q^{1/8}$ times an ordinary power series in q, so the quotient $f(\tau) = F(\tau)/\eta(\tau)^3$ is a holomorphic modular form of weight 0 for $SL_2(\mathbb{Z})$, possibly with a multiplier system. Since the q-series of f has integer powers of q we have $f(T\tau) = f(\tau)$, and since $S^2 = 1$ we have only the possibilities $f(S\tau) = \pm f(\tau)$. But in $SL_2(\mathbb{Z})$ we have the relation $(ST)^3 = 1$, so if S acted by -1 on f we would have $f(\tau) = f(-T^3\tau) = -f(\tau)$. Hence $f(S\tau) = f(\tau)$ and, since it is a genuine holomorphic modular form on $SL_2(\mathbb{Z})$, we have $f(\tau) = 1$. Thus

$$F(\tau) = \eta(\tau)^3/4. \tag{3}$$

We next compute $F(\tau)$ using representation theory. We obtain (up to some undetermined signs) the following well known classical identity of Jacobi

$$\eta(\tau)^3 = q^{1/8} \sum_{n\in\mathbb{Z}} (4n+1) q^{n(2n+1)}. \tag{4}$$

We begin by considering the trace of $G_0\Theta q^{L_0}$ on the Verma module $M(c,h)$. The action of $G_0\Theta$ on the monomial

$$\mathbf{m}v = L_{m_1} \cdots L_{m_s} G_{n_1} \cdots G_{n_t} v,$$

where $m_1 \leq \ldots \leq m_s \leq -1$, $n_1 < \ldots < n_t \leq -1$, and v is $v_{c,h}$ or $G_0 v_{c,h}$, looks like

$$\mathbf{m}G_0 v \mapsto G_0 \mathbf{m} G_0 v = (-1)^t \mathbf{m} G_0^2 v + \text{reduced terms}.$$

Reduced terms resulting from a single use of the commutation relations are of the same length as $\mathbf{m}$, but contain different numbers of the symbols L and G. Reduced terms resulting from more than one use of the commutation relations are strictly shorter than $\mathbf{m}$. Therefore none of these terms contribute to the trace. Consider monomials $\mathbf{m}$ as above with a fixed value of $N = \sum_{i=1}^s m_i + \sum_{j=1}^t n_j$. A simple generating function argument shows that if $N > 0$ then the number of such monomials with t even is the same as the number with t odd. Thus the only nonzero term in $\mathrm{Tr}\, G_0\Theta q^{L_0}$ is the leading term.

It is known that $L(c=-21/4, h_0=-3/32)$ has a BGG resolution

$$0 \leftarrow L(c,h_0) \leftarrow M(c,h_0) \leftarrow M(c,h_1) \oplus M(c,h_{-1}) \leftarrow M(c,h_2) \oplus M(c,h_{-2}) \leftarrow \cdots,$$

where $h_n = -3/32 - n(2n+1)$ for all $n \in \mathbb{Z}$, and that all $M(c,h_k)$ are naturally embedded in $M(c,h_0)$ [7]. We therefore identify each Verma module with its image in $M(c,h_0)$. The trace we seek is given as an alternating sum over the terms in the resolution.

From this we see already that the only nonzero coefficients in the q-expansion of $\eta(\tau)^3$ must be for powers $q^{1/8-n(2n+1)}$. We can also easily determine the coefficients up to a sign. Indeed we have $\Theta^2 = h_0 - c/24 = 1/8$, while the operator $G_0|_{M(c,h_k)}$ preserves the top piece S_k of $M(c,h_k)$ and squares to $h_k^2 - c/24$. Therefore the operator $(G_0\Theta)|_{S_k}$ (which is diagonal on the $1|1$-dimensional space S_k) squares to

$$(h_k - c/24)^2/8 = [(4k+1)/8]^2.$$

This matches perfectly with (4). To determine the signs of the coefficients directly it seems to be necessary to know some further information about the singular vectors, it would be nice to find a simpler derivation.

Of course similar arguments may be applied to other rational NS_c and their modules. If L_0 happens to take the value $c/24$ on one of the levels of a module then the arguments potentially become more intricate.

Acknowledgements I would like to warmly thank the organisers of the conference 'Lie Superalgebras' at Università di Roma Sapienza where this work was presented. Also to express my gratitude to IMPA and to the IHES where the writing of this paper was completed.

References

1. De Sole, A., Kac, V.: Finite vs affine W-algebras. Jpn. J. Math. **1**, 137–261 (2006)
2. Dong, C., Li, H., Mason, G.: Modular-invariance of trace functions in orbifold theory and generalized Moonshine. Comm. Math. Phys. **214**, 1–56 (2000)
3. Dong, C., Zhao, Z.: Modularity in Orbifold Theory for Vertex Operator Superalgebras. Comm. Math. Phys. **260**, 227–256 (2005)
4. Dong, C., Zhao, Z.: Modularity of trace functions in orbifold theory for Z-graded vertex operator superalgebras. In: Moonshine: the first quarter century and beyond. London Math. Soc. Lecture Note Ser., Vol. 372, pp. 128–143. Cambridge University Press, Cambridge (2010)
5. Frenkel, E., Ben Zvi, D.: Vertex Algebras and Algebraic Curves. Mathematical Surveys and Monographs, Vol. 88, 2nd edn. Amer. Math. Soc., Providence (2004)
6. Huang, Y.: Differential equations, duality and modular invariance. Commun. Contemp. Math. **7**, 649–706 (2005)
7. Iohara, K., Koga, Y.: Representation theory of Neveu-Schwarz and Ramond algebras I: Verma modules. Adv. in Math. 178, 1–65 (2003)
8. Jordan, A.: A super version of Zhu's theorem. Ph.D. Thesis, University of Oregon (2008)
9. Kac, V.: Infinite Dimensional Lie Algebras. Progress in Mathematics, Vol. 44. Birkhäuser, Boston (1983)
10. Kac, V.: Vertex algebras for beginners. University Lecture Series, Vol. 10, 2nd edn. Amer. Math. Soc., Providence (1998)

11. Kac, V., Wang, W.: Vertex operator superalgebras and their representations. In: Mathematical aspects of conformal and topological field theories and quantum groups. Contemp. Math., Vol. 175, pp. 161–191. Amer. Math. Soc., Providence (1994)
12. Milas, A.: Characters, Supercharacters and Weber Modular Functions. J. Reine Angew. Math. **608**, 35–64 (2007)
13. Miyamoto, M.: Intertwining operators and modular invariance. arXiv:math/0010180
14. Miyamoto, M.: Modular invariance of vertex operator algebras satisfying C_2-cofiniteness. Duke Math. J. **122**, 51–91 (2004)
15. Van Ekeren, J.: Modular Invariance for Twisted Modules over a Vertex Operator Superalgebra. Comm. Math. Phys. To appear. See arXiv:1111.0682
16. Yamauchi, H.: Orbifold Zhu theory associated to intertwining operators. J. Algebra **265**, 513–538 (2003)
17. Zhu, Y.: Modular invariance of characters of vertex operator algebras. J. Amer. Math. Soc. **9**, 237–301 (1996)

Serre presentations of Lie superalgebras

Ruibin Zhang

Abstract We prove an analogue of Serre's theorem, which describes presentations in terms of Chevalley generators and Serre type relations for the finite dimensional simple contragredient Lie superalgebras relative to all possible choices of Borel subalgebras.

1 Introduction

A well known theorem of Serre gave presentations of finite dimensional semi-simple Lie algebras in terms of Chevalley generators and Serre relations. It was generalised to Kac-Moody algebras with symmetrisable Cartan matrices by Gabber and Kac [7]. The theorem and its generalisations now provide the standard method to present simple Lie algebras and Kac-Moody algebras [13], as well as the associated quantised universal enveloping algebras [4, 11, 20].

A natural question is how to present simple Lie superalgebras with Cartan matrices in a similar way. Surprisingly this did not receive much attention until quantised universal enveloping superalgebras [2, 25, 29, 34] became popular in the early 90s because of their applications in a variety of areas such as low dimensional topology [19, 27, 30], integrable models in physics [2, 33, 34] and noncommutative geometry [21, 31, 32]. It turned out that beside the usual Serre relations, higher order relations are required to present simple Lie superalgebras. These higher order relations still remain somewhat mysterious.

It was Kac who first realised the necessity of higher order Serre relations. Leites and Serganova [18] treated the relations for $\mathfrak{sl}_{m|n}$ in the distinguished root system (i.e., with simple roots defined relative to the distinguished Borel subalgebra). The corresponding quantum relations for $U_q(\mathfrak{sl}_{m|n})$ were constructed in [5, 23]. Yamane

R.B. Zhang (✉)
School of Mathematics and Statistics, University of Sydney, Sydney, New South Walse 2006, Australia
e-mail: ruibin.zhang@sydney.edu.au

M. Gorelik, P. Papi (eds.): *Advances in Lie Superalgebras.* Springer INdAM Series 7,
DOI 10.1007/978-3-319-02952-8_14, © Springer International Publishing Switzerland 2014

[25] wrote down higher order quantum Serre relations for quantised universal enveloping superalgebras of finite dimensional simple Lie superalgebras in the distinguished root systems and some other root systems. In the ensuing years, further work was done to find defining relations of contragredient Lie superalgebras (i.e., with Cartan matrices) by Leites and collaborators [1,8,9] and by Yamane [26]. Leites and coworkers used computers to construct relations, while Yamane used odd reflections. The relations (in the classical limit) in [25, 26] and in [8] look very different, and it is not clear how to prove their equivalence.

The key problem is whether the relations constructed so far are complete, that is, whether they are sufficient to define the Lie superalgebras. The methods used in [1, 8, 9, 25, 26] to find relations does not automatically address the problem; a separate treatment is required. As discussed in [8, §1], completeness of the relations in [8] was checked by computers for finite dimensional simple contragredient Lie superalgebras, but remained an open problem in the infinite dimensional case. A proof for the completeness of the classical analogues of the relations in [25, 26] is still lacking.

Therefore there is the need of a systematic treatment of Serre presentations for the finite dimensional simple contragredient Lie superalgebras relative to all possible choices of Borel subalgebras. We give such a treatment here, and present the analogue of Serre's theorem, Theorem 3.3. The completeness of the relations in Theorem 3.3 is established in Theorem 3.2.

The approach used here is different from those of [1, 8, 9] and [25, 26] both conceptually and technically. It is quite elementary and has the advantage of automatically generating a complete and minimal set of relations. Conceptually the approach is quite transparent in the sense that one can see how the defining relations arise. It also provides an alternative approach to Serre's theorem for finite dimensional semi-simple Lie algebras, see Remark 5.2.

Let us briefly explain our method. Given a realisation of the Cartan matrix $A = (a_{ij})$ of a simple contragredient Lie superalgebra with the set of simple roots $\Pi_{\mathfrak{b}} = \{\alpha_1, \ldots, \alpha_r\}$, we introduce an auxiliary Lie superalgebra $\tilde{\mathfrak{g}}$, which is generated by Chevalley generators $\{e_i,\ f_i,\ h_i \mid i = 1, 2, \ldots, r\}$ subject to quadratic relations only (see Definition 3.1, where more informative notation is used). Let $\mathfrak{r}$ be the $\mathbb{Z}_2$-graded maximal ideal of $\tilde{\mathfrak{g}}$ that intersects trivially the Cartan subalgebra spanned by all h_i. Then $L := \tilde{\mathfrak{g}}/\mathfrak{r}$ is the simple Lie superalgebra which we started with in all cases except in type $A(n,n)$ where L is $\mathfrak{sl}_{n+1|n+1}$ (see Theorem 3.1).

We introduce a $\mathbb{Z}_2$-graded ideal $\mathfrak{s}$ of the auxiliary Lie superalgebra, which is generated by explicitly given generators. A main result proved in Theorem 3.2 states that $\mathfrak{s} = \mathfrak{r}$, or equivalently, $\mathfrak{g} := \tilde{\mathfrak{g}}/\mathfrak{s} \cong L$. From this result, we deduce a super analogue of Serre's theorem, Theorem 3.3, which gives presentations of the finite dimensional simple contragredient Lie superalgebras relative to all possible choices of Borel subalgebras.

The completeness of the relations in Theorem 3.3 is guaranteed by Theorem 3.2. The proof of Theorem 3.2 makes use of a $\mathbb{Z}$-grading of $\tilde{\mathfrak{g}}$, which descends to L and $\mathfrak{g}$ to give $\mathbb{Z}$-gradings to these Lie superalgebras. Write $L = \oplus_k L_k$ and $\mathfrak{g} = \oplus_k \mathfrak{g}_k$ with respect to the $\mathbb{Z}$-gradings. Lemma 3.3 states that $L_0 \cong \mathfrak{g}_0$ as Lie superalgebras and $L_k \cong \mathfrak{g}_k$ as $\mathfrak{g}_0$-modules for all $k \neq 0$. Then Theorem 3.2 follows from this lemma.

The unconventional Serre relations can now be understood as arising from two sources: the conditions for $\mathfrak{g}_{\pm 1}$ to be irreducible $\mathfrak{g}_0$-modules; and the requirement that $[\mathfrak{g}_{\pm 1}, \mathfrak{g}_{\pm 1}] = L_{\pm 2}$ and similar requirements at other degrees.

Compared with [25, 26] (in the $q \longrightarrow 1$ limit), we have more relations, which are needed for presenting exceptional Lie superalgebras with non-distinguished Borel subalgebras. The relations obtained from [25, 26] in the $q \longrightarrow 1$ limit agree with the corresponding subset of relations in the present paper. Our relations look quite different from those in [8].

The organisation of the paper is as follows. Section 2 reviews Kac's classification of finite dimensional simple classical Lie superalgebras [12], and also clarifies certain subtle points about Cartan matrices and Dynkin diagrams in this context. Section 3 contains the statements of the main results, Theorem 3.2 and Theorem 3.3, which give presentations of contragredient Lie superalgebras in arbitrary root systems. The proof of Theorem 3.2, which implies Theorem 3.3 as a corollary, is given by using the key lemma, Lemma 3.3. Sections 4 and 5 are devoted to the proof of the key lemma. An outline of the proof is given in Sect. 4.2 to explain its conceptual aspects. We end the paper with a discussion of possible generalisation of the method developed here to affine Kac-Moody superalgebras to construct Serre type presentations in Sect. 6.

Two appendices are also included. Appendix A gives the root systems and Dynkin diagrams of all simple contragredient Lie superalgebras [3, 6, 12]. The material is used throughout the paper, and is also necessary in order to make precise the description of Dynkin diagrams in non-distinguished root systems. Appendix B describes the structure of some generalised Verma modules of lowest weight type and their irreducible quotients, which enter the proof of Lemma 3.3.

2 Finite dimensional simple Lie superalgebras

In this section, we present some background material, and clarify some issues about Cartan matrices and Dynkin diagrams of Lie superalgebras.

2.1 Finite dimensional simple Lie superalgebras

We work over the field $\mathbb{C}$ of complex numbers throughout the paper.

2.1.1 Classification

A Lie superalgebra $\mathfrak{g}$ is a $\mathbb{Z}_2$-graded vector space $\mathfrak{g} = \mathfrak{g}_{\bar{0}} \oplus \mathfrak{g}_{\bar{1}}$ endowed with a bilinear map $[\ ,\ ] : \mathfrak{g} \times \mathfrak{g} \longrightarrow \mathfrak{g}$, $(X, Y) \mapsto [X, Y]$, called the Lie superbracket, which is homogeneous of degree 0, graded skew-symmetric and satisfies the super Jacobian identity. The even subspace $\mathfrak{g}_{\bar{0}}$ of $\mathfrak{g}$ is a Lie algebra in its own right, which is called the even subalgebra of $\mathfrak{g}$. The odd subspace $\mathfrak{g}_{\bar{1}}$ forms a $\mathfrak{g}_{\bar{0}}$-module under the restriction of the adjoint action defined by the Lie superbracket. If $\mathfrak{g}_{\bar{0}}$ is a reductive Lie algebra and $\mathfrak{g}_{\bar{1}}$ is a semi-simple $\mathfrak{g}_{\bar{0}}$-module, $\mathfrak{g}$ is called *classical* [12, 22].

The classification of the finite dimensional simple Lie superalgebras was completed in the late 70s. The theorem below is taken from the foundational paper [12] of Kac, which is still the best reference on Lie superalgebras. Historical information and further references on the classification can be found in [15, 16] (also see [22]).

Theorem 2.1 *The finite dimensional simple classical Lie superalgebras comprise of the simple contragredient Lie superalgebras*

$$\begin{aligned}&A(m,n), \quad B(0,n), \quad B(m,n),\ m>0, \quad C(n),\ n>2, \quad D(m,n),\ m>1,\\ &F(4), \quad G(3), \quad D(2,1;\alpha),\ \alpha\in\mathbb{C}\backslash\{0,-1\},\end{aligned}$$

and simple strange Lie superalgebras $P(n)$ and $Q(n)$ ($n \geq 1$).

The simple contragredient Lie superalgebras admit non-degenerate invariant bilinear forms, while the strange Lie superalgebras $P(n)$ and $Q(n)$ do not. In the remainder of the paper, we shall consider only contragredient simple Lie superalgebras.

The A, B, C and D series are essentially the special linear and orthosymplectic Lie superalgebras, which are familiar examples of Lie superalgebras. The exceptional Lie superalgebras $F(4), G(3)$ and $D(2,1;\alpha)$ are less well-known, but one can understand their structures given the description of their roots in Appendix A.1.

Let $\mathfrak{g} = \mathfrak{g}_{\bar{0}} \oplus \mathfrak{g}_{\bar{1}}$ be a simple contragredient Lie superalgebra, and choose a Cartan subalgebra $\mathfrak{h}$ for $\mathfrak{g}$, which by definition is just a Cartan subalgebra of $\mathfrak{g}_{\bar{0}}$. Denote by $\mathfrak{g}_\alpha$ the root space of the root α, and call α even (resp. odd) if $\mathfrak{g}_\alpha \subset \mathfrak{g}_{\bar{0}}$ (resp. $\mathfrak{g}_\alpha \subset \mathfrak{g}_{\bar{1}}$). Denote by Δ_0 and Δ_1 the sets of the even and odd roots respectively, and set $\Delta = \Delta_0 \cup \Delta_1$. Let $(\ ,\ ) : \mathfrak{h}^* \times \mathfrak{h}^* \longrightarrow \mathbb{C}$ denote the Weyl group invariant non-degenerate symmetric bilinear form on $\mathfrak{h}^*$, where the Weyl group of $\mathfrak{g}$ is by definition the Weyl group of $\mathfrak{g}_{\bar{0}}$. A root β will be called isotropic if $(\beta,\beta) = 0$. Note that all isotropic roots are odd.

A Borel subalgebra of $\mathfrak{g}$ is a maximal soluble Lie super subalgebra containing a Borel subalgebra of $\mathfrak{g}_{\bar{0}}$. A new feature in the present context is that Borel subalgebras are not always conjugate under the Weyl groups. All the conjugacy classes of Borel subalgebras were given in [12, pp. 51–52] and [13, Proposition 1.2]. In particular, Kac described a particularly convenient Borel subalgebra, which he called distinguished, for each simple contragredient Lie superalgebra. We shall call a root system with the set of simple roots determined by this Borel subalgebra the *distinguished root system*. In this case, there exists only one odd simple root.

2.1.2 Cartan matrices and Dynkin diagrams

The precise forms of the Cartan matrices and Dynkin diagrams will be crucial in Sect. 3. However, there do not exist canonical definitions for them in the Lie superalgebra setting, thus we spell out the details of our definitions here.

Let $\Pi_{\mathfrak{b}} = \{\alpha_1, \alpha_2, \ldots, \alpha_r\}$ be the set of simple roots of a simple contragrediant Lie superalgebra $\mathfrak{g}$ relative to a Borel subalgebra $\mathfrak{b}$. The Cartan matrix and Dynkin diagram provide a convenient way to describe $\Pi_{\mathfrak{b}}$. We define a Cartan matrix in the following way. Denote by $\Theta \subset \{1,2,\ldots,r\}$ the subset such that $\alpha_t \in \Delta_1$ for all $t \in \Theta$. Let l_m^2 be the minimum of $|(\beta,\beta)|$ for all non-isotropic $\beta \in \Delta$ if $\mathfrak{g} \neq D(2,1;\alpha)$. If $\mathfrak{g}$ is $D(2,1;\alpha)$, let l_m^2 be the minimum of all $|(\beta,\beta)| > 0$ $(\beta \in \Delta)$, which are independent of the arbitrary parameter α. Let

$$\kappa = \begin{cases} 0, \text{ if } \mathfrak{g} \text{ is of type } B, \\ 1, \text{ otherwise}; \end{cases} \qquad d_i = \begin{cases} \frac{(\alpha_i,\alpha_i)}{2}, & \text{if } (\alpha_i,\alpha_i) \neq 0, \\ \frac{l_m^2}{2^\kappa}, & \text{if } (\alpha_i,\alpha_i) = 0. \end{cases}$$

Introduce the matrices

$$B = (b_{ij})_{i,j=1}^r, \quad b_{ij} = (\alpha_i, \alpha_j),$$
$$D = \mathrm{diag}(d_1, \ldots, d_r),$$

then the Cartan matrix A associated to the set of simple roots $\Pi_{\mathfrak{b}}$ is defined by

$$A = D^{-1}B.$$

When it is necessary to indicate the dependence on Θ, we write (A,Θ) for the Cartan matrix.

Note that if α_i is non-isotropic, $a_{it} = \frac{2(\alpha_i,\alpha_t)}{(\alpha_i,\alpha_i)}$ is a non-positive integer for all t. However, if α_t is isotropic, then $a_{tj} = \frac{2}{l_m^2}(\alpha_t, \alpha_j)$ can be an integer of any sign or zero (except in type $D(2,1;\alpha)$). If $b_{ij} \neq 0$, we define

$$sgn_{ij} = \text{sign of } b_{ij}. \tag{2.1}$$

As we shall see in Sect. 2.2, these signs provide the additional information required to recover a Cartan matrix from its Dynkin diagram.

Remark 2.1 Our definition of the Cartan matrix differs from the usual one due to Kac [12]. In Kac's definition, if $b_{ss} = 0$, then $d_s = (\alpha_s, \alpha_{s+k})$ for the smallest k such that $d_s \neq 0$. Note that in our definition, none of the signs sgn_{ij} is lost.

The Dynkin diagram associated with (A,Θ) consists of r nodes, which are connected by lines. The i-th node is coloured white if $i \notin \Theta$, black if $i \in \Theta$ but α_i is not isotropic, and grey if α_i is isotropic.

If (A,Θ) is of type $D(2,1;\alpha)$, the Dynkin diagram is obtained by simply connecting the i-th and j-th nodes by one line if $a_{ij} \neq 0$ and write b_{ij} at the line.

In all other cases, we join the i-th and j-th nodes by n_{ij} lines, where

$$n_{ij} = \max(|a_{ij}|, |a_{ji}|), \qquad \text{if } a_{ii} + a_{jj} \geq 2;$$
$$n_{ij} = |a_{ij}|, \qquad \text{if } a_{ii} = a_{jj} = 0.$$

When the i-th and j-th nodes are not both grey, say, the i-th one is not grey, and connected by more than one lines, we draw an arrow pointing to the j-th node if $-a_{ij} = 1$ and pointing to the i-th node if $-a_{ij} > 1$.

The Dynkin diagrams of the simple contragredient Lie superalgebras are given in the tables in Appendix A.2.

2.2 Comments on Dynkin diagrams

From the Cartan matrices in our definition, one can recover the corresponding root systems. Dynkin diagrams also uniquely represent Cartan matrices, except in the cases of $\mathfrak{osp}_{4|2}$ and $\mathfrak{sl}_{2|2}$. The Dynkin diagrams of these superalgebras relative to the distinguished root systems are exactly the same, but the two Lie superalgebras are non-isomorphic.

This problem can be resolved by incorporating the signs sgn_{ij} defined by (2.1) in the Dynkin diagram, e.g., by placing sgn_{ij} at the line(s) connecting two grey nodes i and j. Then the modified Dynkin diagrams are respectively given by

$\mathfrak{sl}_{2|2}$: 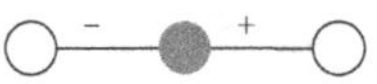 , $\mathfrak{osp}_{4|2}$:  . (2.2)

As we shall see, the signs enter the construction of higher order Serre relations.

In this paper we did not include the additional information of these signs in the definition of Dynkin diagrams, as they would make the diagrams look cumbersome. Also, there is no ambiguity about the signs in all the other Dynkin diagrams.

Similar signs were also discussed in [26].

Recall that if we remove a subset of vertices (i.e., nodes) and all the edges connected to these vertices from a Dynkin diagram of a semi-simple Lie algebra, we obtain the Dynkin diagram of another semi-simple Lie algebra of a smaller rank. This corresponds to taking regular subalgebras. In the context of Lie superalgebras, the notion of regular subalgebras still exists, but some explanation is required at the level of Dynkin diagrams.

Definition 2.1 Call a sub-diagram Γ' of a Dynkin diagram Γ *full* if for any two nodes i and j in Γ', the edges between them in Γ, the arrows on the edges, and also the b_{ij} labels of the edges when Γ is of type $D(2,1;\alpha)$, are all present in Γ'.

Consider for example the Dynkin diagram

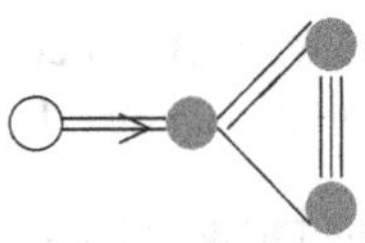

of $F(4)$, which has the following full sub-diagrams beside others:

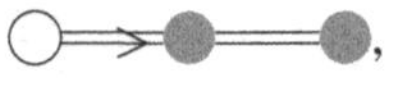 ,  . (2.3)

Note that none of these appears in Tables 1 and 2.

The reason is that the sub-matrices in the Cartan matrix of $F(4)$ associated with these full sub-diagrams are not Cartan matrices in the strict sense. The problem lies in the definition of a_{ij} when the node i is grey, which involves the number ℓ_m. The ℓ_m for $F(4)$ is not the correct ones for the full sub-diagrams. By properly renormalising the bilinear forms on the weight spaces associated with them, the full sub-diagrams can be cast into the form

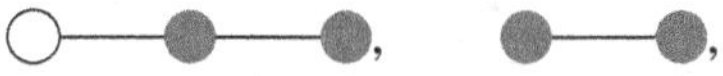

which are respectively Dynkin diagrams for $\mathfrak{sl}_{3|1}$ and $\mathfrak{sl}_{2|1}$.

We call the Dynkin diagrams in Table 1 and Table 2 standard, and the ones like those in (2.3) non-standard.

We mention that if a Lie superalgebra $\mathfrak{g}$ is contained as a regular subalgebra in another Lie superalgebra, defining relations of $\mathfrak{g}$ can in principle be extracted from relations of the latter by considering sub-diagrams of Dynkin diagrams. However, this involves subtleties, as we have just discussed, and requires more care than hitherto exercised in the literature.

3 Presentations of Lie superalgebras

In this section, we generalise Serre's theorem for simple Lie algebras to simple contragredient Lie superalgebras, obtaining presentations for such Lie superalgebras in terms of Chevalley generators and defining relations.

3.1 An auxiliary Lie superalgebra

We start by defining an auxiliary Lie superalgebra following the strategy of [14]. Let (A,Θ) with $A=(a_{ij})_{i,j=1}^r$ be the Cartan matrix of one of the simple contragredient Lie superalgebras relative to a given Borel subalgebra $\mathfrak{b}$. Let $\Pi_{\mathfrak{b}}$ be the set of simple roots relative to this Borel subalgebra.

Definition 3.1 Let $\tilde{\mathfrak{g}}(A,\Theta)$ be the Lie superalgebra generated by homogeneous generators e_i, f_i, h_i $(i=1,2,\dots,r)$, where e_s, f_s for all $s\in\Theta$ are odd while the rest are even, subject to the following relations

$$\begin{aligned}&[h_i,h_j]=0,\\ &[h_i,e_j]=a_{ij}e_j, \quad [h_i,f_j]=-a_{ij}f_j,\\ &[e_i,f_j]=\delta_{ij}h_i, \quad \forall i,j.\end{aligned} \tag{3.1}$$

Let $\tilde{\mathfrak{n}}^+$ (resp. $\tilde{\mathfrak{n}}^-$) be the subalgebra generated by all e_i (resp. all f_i) subject to the relevant relations, and $\mathfrak{h}=\oplus_{i=1}^r\mathbb{C}\mathfrak{h}_i$, the Cartan subalgebra. Then it is well known and easy to prove (following the reasoning of [14, §1]) that $\tilde{\mathfrak{g}}(A,\Theta)=\tilde{\mathfrak{n}}^+\oplus\mathfrak{h}\oplus\tilde{\mathfrak{n}}^-$. The Lie superalgebra is graded $\tilde{\mathfrak{g}}(A,\Theta)=\oplus_{\nu\in Q}\tilde{\mathfrak{g}}_\nu$ by $Q=\mathbb{Z}\Pi_{\mathfrak{b}}$, with $\tilde{\mathfrak{g}}_0=\mathfrak{h}$. Note that $\tilde{\mathfrak{n}}^+_\nu$ (rep. $\tilde{\mathfrak{n}}^-_{-\nu}$) is zero unless $\nu\in Q_{\mathbb{N}}$, where $\mathbb{N}=\{1,2,\dots\}$ and $Q_{\mathbb{N}}=\mathbb{N}\Pi_{\mathfrak{b}}$, that is,

$$\tilde{\mathfrak{n}}^+=\oplus_{\nu\in Q_{\mathbb{N}}}\tilde{\mathfrak{n}}^+_\nu, \quad \tilde{\mathfrak{n}}^-=\oplus_{\nu\in Q_{\mathbb{N}}}\tilde{\mathfrak{n}}^-_{-\nu}. \tag{3.2}$$

Let $\mathfrak{r}(A,\Theta)$ be the maximal $\mathbb{Z}_2$-graded ideal of $\tilde{\mathfrak{g}}(A,\Theta)$ that intersects $\mathfrak{h}$ trivially. Set $\mathfrak{r}^\pm=\mathfrak{r}(A,\Theta)\cap\tilde{\mathfrak{n}}^\pm$. Then $\mathfrak{r}(A,\Theta)=\mathfrak{r}^+\oplus\mathfrak{r}^-$. The following fact follows from the maximality of $\mathfrak{r}(A,\Theta)$.

Lemma 3.1 *Let $\Sigma = \Sigma^+ \cup \Sigma^-$ with $\Sigma^\pm \subset \tilde{\mathfrak{n}}^\pm$ be a subset of $\tilde{\mathfrak{g}}(A,\Theta)$ consisting of homogeneous elements. If $[f_i, \Sigma^+] \subset \mathbb{C}\Sigma^+$ and $[e_i, \Sigma^-] \subset \mathbb{C}\Sigma^-$ for all i, then $\Sigma \subset \mathfrak{r}(A,\Theta)$.*

Proof. It follows from the definition of Σ that the ideal generated by $\mathfrak{r}(A,\Theta) \cup \Sigma$ intersects $\mathfrak{h}$ trivially, hence must be equal to $\mathfrak{r}(A,\Theta)$ by the maximality of the latter. □

In particular, if $X^\pm \in \tilde{\mathfrak{n}}^\pm$ satisfy $[f_i, X^+] = 0$, and $[e_i, X^-] = 0$ for all i, then they belong to $\tilde{\mathfrak{n}}^\pm$ respectively.

Let us define the Lie superalgebra

$$L(A,\Theta) := \frac{\tilde{\mathfrak{g}}(A,\Theta)}{\mathfrak{r}(A,\Theta)}.$$

We have the following result.

Theorem 3.1 *Let $\mathfrak{g}$ be a finite dimensional simple contragredient Lie superalgebra, and let (A,Θ) be the Cartan matrix of $\mathfrak{g}$ relative to a given Borel subalgebra. Then $L(A,\Theta)$ is isomorphic to $\mathfrak{g}$ unless $\mathfrak{g} = A(n,n)$, and in the latter case $L(A,\Theta) \cong \mathfrak{sl}_{n+1|n+1}$.*

Proof. This follows from Kac's classification [12] of the simple contragredient Lie superalgebras (see Theorem 2.1) except in the case of $A(n,n)$. In the latter case, we have $\det A = 0$. Therefore, $L(A,\Theta)$ contains a 1-dimensional center, and the quotient of $L(A,\Theta)$ by the center is $A(n,n)$. Hence $L(A,\Theta)$ is isomorphic to $\mathfrak{sl}_{n+1|n+1}$. □

3.2 Main theorem

3.2.1 Standard and higher order Serre elements

Let us first define some elements of $\tilde{\mathfrak{g}}(A,\Theta)$, which will play a crucial role in studying the presentation of Lie superalgebras.

We shall call the following elements the *standard Serre elements*:

$$(ad_{e_i})^{1-a_{ij}}(e_j), \quad (ad_{f_i})^{1-a_{ij}}(f_j), \quad \text{for } i \neq j, \text{ with } a_{ii} \neq 0 \text{ or } a_{ij} = 0;$$
$$[e_s, e_s], \quad [f_s, f_s], \quad \text{for } a_{ss} = 0.$$

We also introduce *higher order Serre elements* if the Dynkin diagram of (A,Θ) contains full sub-diagrams of the following kind:

1. 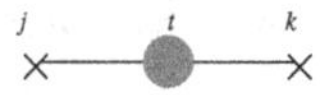

with $sgn_{jt} sgn_{tk} = -1$, the associated higher order Serre elements are

$$[e_t, [e_j, [e_t, e_k]]], \quad [f_t, [f_j, [f_t, f_k]]];$$

2. 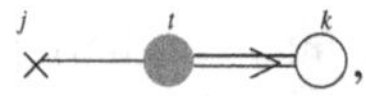

, the associated higher order Serre elements are

$$[e_t, [e_j, [e_t, e_k]]], \quad [f_t, [f_j, [f_t, f_k]]];$$

3. 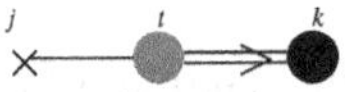

, the associated higher order Serre elements are

$$[e_t,[e_j,[e_t,e_k]]], \quad [f_t,[f_j,[f_t,f_k]]];$$

4. 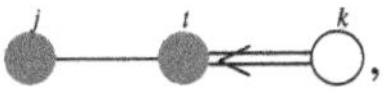

, the associated higher order Serre elements are

$$[[e_j,e_t],[[e_j,e_t],[e_t,e_k]]],$$
$$[[f_j,f_t],[[f_j,f_t],[f_t,f_k]]];$$

5. 

, the associated higher order Serre elements are

$$[[e_i,[e_j,e_t]],[[e_j,e_t],[e_t,e_k]]],$$
$$[[f_i,[f_j,f_t]],[[f_j,f_t],[f_t,f_k]]];$$

6. 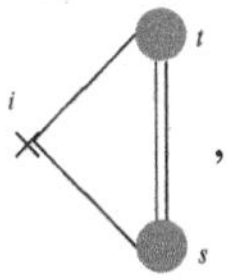

, the associated higher order Serre elements are

$$[e_t,[e_s,e_i]]-[e_s,[e_t,e_i]],$$
$$[f_t,[f_s,f_i]]-[f_s,[f_t,f_i]];$$

7. 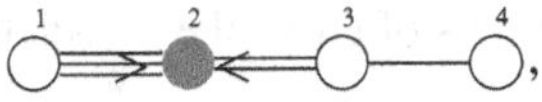

, which is a Dynkin diagram of $F(4)$, the associated higher order Serre elements are

$$[E,[E,[e_2,[e_3,e_4]]]],$$
$$[F,[F,[f_2,[f_3,f_4]]]],$$

where $E=[[e_1,e_2],[e_2,e_3]]$ and $F=[[f_1,f_2],[f_2,f_3]]$;

8. 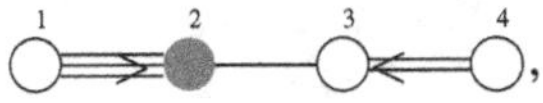

, which is a Dynkin diagram of $F(4)$, the associated higher order Serre elements are

$$[[e_1,e_2],[[e_2,e_3],[e_3,e_4]]-[[e_2,e_3],[[e_1,e_2],[e_3,e_4]],$$
$$[[f_1,f_2],[[f_2,f_3],[f_3,f_4]]-[[f_2,f_3],[[f_1,f_2],[f_3,f_4]];$$

9. 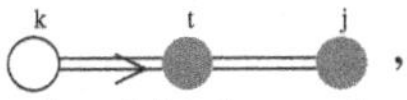

, which only appears in Dynkin diagrams of $F(4)$, the associated higher order Serre elements are

$$[e_t,[e_j,[e_t,e_k]]],$$
$$[f_t,[f_j,[f_t,f_k]]];$$

10. 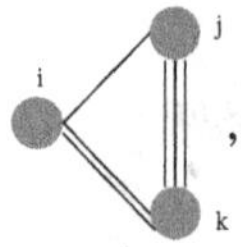

, which only appears in one of the Dynkin diagrams of $F(4)$, the associated higher order Serre elements are

$$2[e_i,[e_k,e_j]]+3[e_j,[e_k,e_i]],$$
$$2[f_i,[f_k,f_j]]+3[f_j,[f_k,f_i]];$$

11. 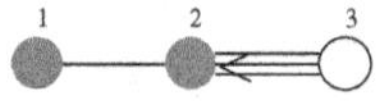

, which is one of the Dynkin diagrams of $G(3)$, the associated higher order Serre elements are

$$[[e_1,e_2],[[e_1,e_2],[[e_1,e_2],[e_2,e_3]]]],$$
$$[[f_1,f_2],[[f_1,f_2],[[f_1,f_2],[f_2,f_3]]]];$$

12. 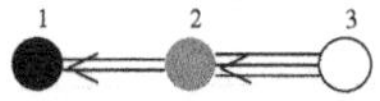

, which is one of the Dynkin diagrams of $G(3)$, the associated higher order Serre elements are

$$[[e_2,e_1],[e_3,[e_2,e_1]]]-[[e_2,e_3],[[e_1,e_1],e_2]],$$
$$[[f_2,f_1],[f_3,[f_2,f_1]]]-[[f_2,f_3],[[f_1,f_1],f_2]];$$

13. 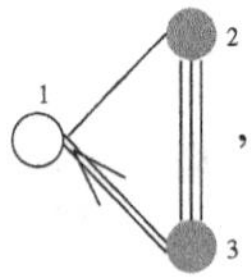

, which is one of the Dynkin diagrams of $G(3)$, the associated higher order Serre elements are

$$[e_2,[e_3,e_1]]-2[e_3,[e_2,e_1]],$$
$$[f_2,[f_3,f_1]]-2[f_3,[f_2,f_1]];$$

14. 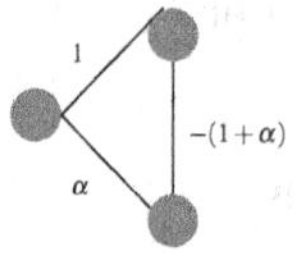

, which is one of the Dynkin diagram for $D(2,1;\alpha)$.

We label the left, top and bottom nodes by 1, 2 and 3 respectively. The higher order Serre elements are

$$\alpha[e_1,[e_2,e_3]]+(1+\alpha)[e_2,[e_1,e_3]],$$
$$\alpha[f_1,[f_2,f_3]]+(1+\alpha)[f_2,[f_1,f_3]].$$

Remark 3.1 The Dynkin diagrams of $D(2,1)$ and $D(2,1;\alpha)$ in their respective distinguished root systems are not among the full sub-diagrams listed above.

Remark 3.2 Diagram 9 above is a non-standard diagram of $\mathfrak{sl}_{3|1}$ (see Sect. 2.2).

Denote by $\mathscr{S}^+(A,\Theta)$ (resp. $\mathscr{S}^-(A,\Theta)$) the set of all the standard and higher order Serre elements (if defined) which involve generators e_k (resp. f_k) only. Set $\mathscr{S}(A,\Theta) = \mathscr{S}^+(A,\Theta) \cup \mathscr{S}^-(A,\Theta)$. We have the following result.

Lemma 3.2 *The set $\mathscr{S}(A,\Theta)$ is contained in the maximal ideal $\mathfrak{r}(A,\Theta)$ of $\tilde{\mathfrak{g}}$.*

Proof. Direct calculations show that

$$[f_i, \mathscr{S}^+(A,\Theta)] \subset \mathbb{C}\mathscr{S}^+(A,\Theta), \quad [e_i, \mathscr{S}^-(A,\Theta)] \subset \mathbb{C}\mathscr{S}^-(A,\Theta), \quad \forall i.$$

Hence $\mathscr{S}(A,\Theta) \subset \mathfrak{r}(A,\Theta)$ by Lemma 3.1. We leave out the details of the calculations. □

Definition 3.2 Let $\mathfrak{s}(A,\Theta)$ be the $\mathbb{Z}_2$-graded ideal of $\tilde{\mathfrak{g}}(A,\Theta)$ generated by the elements of $\mathscr{S}(A,\Theta)$.

Then $\mathfrak{s}(A,\Theta) \subset \mathfrak{r}(A,\Theta)$ by Lemma 3.2. Define the Lie superalgebra

$$\mathfrak{g}(A,\Theta) := \frac{\tilde{\mathfrak{g}}(A,\Theta)}{\mathfrak{s}(A,\Theta)}. \tag{3.3}$$

There exists a natural surjective Lie superalgebra map $\mathfrak{g}(A,\Theta) \longrightarrow L(A,\Theta)$. We shall show that it is in fact an isomorphism.

3.2.2 $\mathbb{Z}$-gradings

Let us discuss $\mathbb{Z}$-gradings for the Lie supealgebras $\mathfrak{g}(A,\Theta)$ and $L(A,\Theta)$. Fix a positive integer $d \leq r$, where r is the size of A. We assign degrees to the generators of $\tilde{\mathfrak{g}}(A,\Theta)$ as follows:

$$\begin{aligned} &deg(h_j) = 0, \quad \forall j, \\ &deg(e_i) = deg(f_i) = 0, \quad \forall i \neq d, \\ &deg(e_d) = -deg(f_d) = 1. \end{aligned} \tag{3.4}$$

This introduces a $\mathbb{Z}$-grading to the auxiliary Lie superalgebra $\tilde{\mathfrak{g}}(A,\Theta)$, which is not required to be compatible with the $\mathbb{Z}_2$-grading upon reduction modulo 2. In view of the Q-grading of $\tilde{\mathfrak{g}}(A,\Theta)$ and (3.2), the maximal ideal $\mathfrak{r}(A,\Theta)$ is $\mathbb{Z}$-graded. Since all elements in $\mathscr{S}(A,\Theta)$ are homogeneous with respect to the $\mathbb{Z}$-grading, $\mathfrak{s}(A,\Theta)$ is $\mathbb{Z}$-graded as well.

The Lie superalgebra $L(A,\Theta)$ inherits a $\mathbb{Z}$-grading from $\tilde{\mathfrak{g}}(A,\Theta)$ and $\mathfrak{r}(A,\Theta)$. Write $L(A,\Theta) = \oplus_{k\in\mathbb{Z}} L_k$. Since the roots of $L(A,\Theta)$ are known, we have a detailed understanding of all L_k as L_0-modules.

The Lie superalgebra $\mathfrak{g}(A,\Theta)$ inherits a $\mathbb{Z}$-grading from $\tilde{\mathfrak{g}}(A,\Theta)$ and $\mathfrak{s}(A,\Theta)$. Write $\mathfrak{g}(A,\Theta) = \oplus_{k\in\mathbb{Z}} \mathfrak{g}_k$, where $\mathfrak{g}_k$ is the homogeneous component of degree k. Note that $\mathfrak{g}_1$ (resp. $\mathfrak{g}_{-1}$) generates $\mathfrak{g}_k$ (resp. $\mathfrak{g}_{-k}$) for all $k > 0$. Thus if $\mathfrak{g}_p = 0$ (resp. $\mathfrak{g}_{-p} = 0$) for some $p > 0$, then $\mathfrak{g}_q = 0$ (resp. $\mathfrak{g}_{-q} = 0$) for all $q > p$. Also each $\mathfrak{g}_k$ forms a $\mathfrak{g}_0$-module in the obvious way.

We have the following result.

Lemma 3.3 (Key lemma) *There exist $\mathbb{Z}$-gradings for $\mathfrak{g}(A,\Theta)$ and $L(A,\Theta)$ determined by some d such that $\mathfrak{g}_0 = L_0$ as Lie superalgebras and $\mathfrak{g}_k = L_k$ as $\mathfrak{g}_0$-modules for all nonzero $k \in \mathbb{Z}$.*

This lemma is essential for establishing Theorem 3.2 below. Its proof is elementary but very lengthy, thus we relegate it to later sections. Here we consider some general properties of the Lie superalgebras $\mathfrak{g}(A,\Theta)$ and $L(A,\Theta)$, which will significantly simplify the proof of Lemma 3.3.

Recall that an anti-involution ω of a Lie superalgebra $\mathfrak{a}$ is a linear map on $\mathfrak{a}$ satisfying $\omega([X,Y]) = [\omega(Y), \omega(X)]$ for all $X, Y \in \mathfrak{a}$, and $\omega^2 = \mathrm{id}_{\mathfrak{a}}$. The Lie superalgebra $\tilde{\mathfrak{g}}(A,\Theta)$ admits an anti-involution defined by

$$\omega(e_i) = f_i, \quad \omega(f_i) = e_i, \quad \omega(h_i) = h_i, \quad \forall i.$$

Note that $\omega(\mathscr{S}^+) \subset -\mathscr{S}^- \cup \mathscr{S}^-$ and $\omega(\mathscr{S}^-) \subset -\mathscr{S}^+ \cup \mathscr{S}^+$, where $\mathscr{S}^\pm = \mathscr{S}^\pm(A,\Theta)$ and $-\mathscr{S}^\pm$ are respectively the sets consisting of the negatives of the elements of $\mathscr{S}^\pm$. Therefore, ω descents to an anti-involution on $\mathfrak{g}(A,\Theta)$, which sends $\mathfrak{g}_k$ to $\mathfrak{g}_{-k}$ for all $k \in \mathbb{Z}$ and provides a $\mathfrak{g}_0$-module isomorphism between $\mathfrak{g}_{-k}$ and the dual space of $\mathfrak{g}_k$.

The anti-involution of $\tilde{\mathfrak{g}}(A,\Theta)$ also descends to an anti-involution of $L(A,\Theta)$, which maps L_k to L_{-k} for all $k \in \mathbb{Z}$, and provides an isomorphism between the L_0-module L_{-k} and the dual L_0-module of L_k.

Therefore, if $\mathfrak{g}_0 = L_0$ and $\mathfrak{g}_k = L_k$ for all $k > 0$ as $\mathfrak{g}_0$-modules, the existence of the anti-involutions immediately implies that $\mathfrak{g}_{-k} = L_{-k}$ for all $k > 0$. Hence in order to prove Lemma 3.3, we only need to show that it holds for all $k > 0$.

The arguments above may be summarised as follows.

Lemma 3.4 *If $\mathfrak{g}_0 = L_0$ as Lie superalgebras and $\mathfrak{g}_k = L_k$ for all $k > 0$ as $\mathfrak{g}_0$-modules, then Lemma 3.3 holds.*

This result will play an essential role in the proof of Lemma 3.3.

3.2.3 Main theorem

The following theorem is the main result of this paper.

Theorem 3.2 *The Lie superalgebra $\mathfrak{g}(A,\Theta)$ coincides with $L(A,\Theta)$, or equivalently, the ideal $\mathfrak{s}(A,\Theta)$ of $\tilde{\mathfrak{g}}(A,\Theta)$ is equal to the maximal ideal $\mathfrak{r}(A,\Theta)$.*

Proof. Note that Lemma 3.3 immediately implies the claim. Indeed, we have already shown in Lemma 3.2 that $\mathfrak{s}(A,\Theta) \subset \mathfrak{r}(A,\Theta)$, and this is an inclusion of $\mathbb{Z}$-graded ideals of $\tilde{\mathfrak{g}}(A,\Theta)$. If $\mathfrak{s}(A,\Theta) \neq \mathfrak{r}(A,\Theta)$, there would exist a surjective Lie superalgebra homomorphism $\mathfrak{g}(A,\Theta) \longrightarrow L(A,\Theta)$ with a nonzero kernel. Thus for some k, the degree-k homogeneous components of $L(A,\Theta)$ and $\mathfrak{g}(A,\Theta)$ are not equal. This contradicts Lemma 3.3. □

3.3 Presentations of Lie superalgebras

Since the generators of the $\mathbb{Z}_2$-graded ideal $\mathfrak{s}(A,\Theta)$ are known explicitly, Theorem 3.2 provides a presentation for each simple contragredient Lie superalgebra and $\mathfrak{sl}_{n+1|n+1}$ in an arbitrary root system. We have the following result for the Lie superalgebra $L(A,\Theta)$.

Theorem 3.3 *The Lie superalgebra $L(A,\Theta)$ is generated by the generators e_i, f_i and h_i ($1 \le i \le r$), where e_i and f_i are odd if $i \in \Theta$, and even otherwise, subject to the quadratic relations*

$$\begin{aligned} &[h_i,h_j]=0,\\ &[h_i,e_j]=a_{ij}e_j, \quad [h_i,f_j]=-a_{ij}f_j,\\ &[e_i,f_j]=\delta_{ij}h_i, \quad \forall i,j; \end{aligned} \tag{3.5}$$

standard Serre relations

$$\begin{aligned} &(ad_{e_i})^{1-a_{ij}}(e_j)=0,\\ &(ad_{f_i})^{1-a_{ij}}(f_j)=0, \quad \text{for } i\neq j, \text{ with } a_{ii}\neq 0 \text{ or } a_{ij}=0;\\ &[e_t,e_t]=0, \quad [f_t,f_t]=0, \quad \text{for } a_{tt}=0; \end{aligned} \tag{3.6}$$

and higher order Serre relations if the Dynkin diagram of (A,Θ) contains any of the following diagrams as full sub-diagrams:

1. 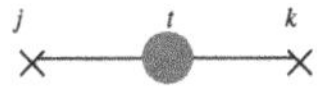 *with $sgn_{jt}sgn_{tk}=-1$, the associated higher order Serre relations are*

$$[e_t,[e_j,[e_t,e_k]]]=0, \quad [f_t,[f_j,[f_t,f_k]]]=0;$$

2. 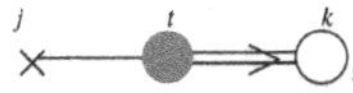*, the associated higher order Serre relations are*

$$[e_t,[e_j,[e_t,e_k]]=0, \quad [f_t,[f_j,[f_t,f_k]]]=0;$$

3. 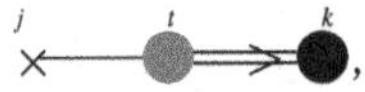*, the associated higher order Serre relations are*

$$[e_t,[e_j,[e_t,e_k]]]=0, \quad [f_t,[f_j,[f_t,f_k]]]=0;$$

4. *, the associated higher order Serre relations are*

$$\begin{aligned} &[[e_j,e_t],[[e_j,e_t],[e_t,e_k]]]=0,\\ &[[f_j,f_t],[[f_j,f_t],[f_t,f_k]]]=0; \end{aligned}$$

5. *, the associated higher order Serre relations are*

$$\begin{aligned} &[[e_i,[e_j,e_t]],[[e_j,e_t],[e_t,e_k]]]=0,\\ &[[f_i,[f_j,f_t]],[[f_j,f_t],[f_t,f_k]]]=0; \end{aligned}$$

6. 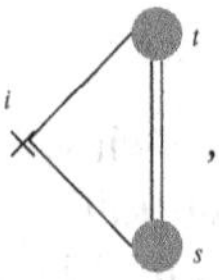

, *the associated higher order Serre relations are*

$$[e_t,[e_s,e_i]]-[e_s,[e_t,e_i]]=0,$$
$$[f_t,[f_s,f_i]]-[f_s,[f_t,f_i]]=0;$$

7. 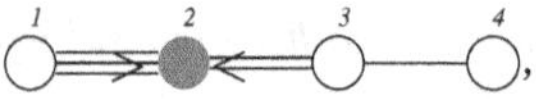

, *the associated higher order Serre relations are*

$$[E,[E,[e_2,[e_3,e_4]]]]=0,$$
$$[F,[F,[f_2,[f_3,f_4]]]]=0,$$

where $E=[[e_1,e_2],[e_2,e_3]]$ *and* $F=[[f_1,f_2],[f_2,f_3]]$;

8. 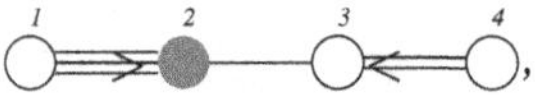

, *the associated higher order Serre relations are*

$$[[e_1,e_2],[[e_2,e_3],[e_3,e_4]]-[[e_2,e_3],[[e_1,e_2],[e_3,e_4]]=0,$$
$$[[f_1,f_2],[[f_2,f_3],[f_3,f_4]]-[[f_2,f_3],[[f_1,f_2],[f_3,f_4]]=0;$$

9. 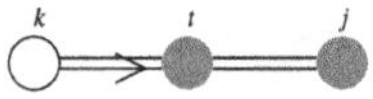

, *the associated higher order Serre relations are*

$$[e_t,[e_j,[e_t,e_k]]]=0,$$
$$[f_t,[f_j,[f_t,f_k]]]=0;$$

10. 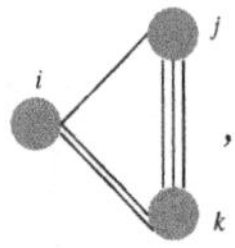

, *the associated higher order Serre relations are*

$$2[e_i,[e_k,e_j]]+3[e_j,[e_k,e_i]]=0,$$
$$2[f_i,[f_k,f_j]]+3[f_j,[f_k,f_i]]=0;$$

11. 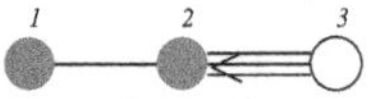

, *the higher order Serre relations are*

$$[[e_1,e_2],[[e_1,e_2],[[e_1,e_2],[e_2,e_3]]]]=0,$$
$$[[f_1,f_2],[[f_1,f_2],[[f_1,f_2],[f_2,f_3]]]]=0;$$

12. 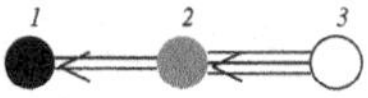

, *the higher order Serre relations are*

$$[[e_2,e_1],[e_3,[e_2,e_1]]]-[[e_2,e_3],[[e_1,e_1],e_2]]=0,$$
$$[[f_2,f_1],[f_3,[f_2,f_1]]]-[[f_2,f_3],[[f_1,f_1],f_2]]=0;$$

13. 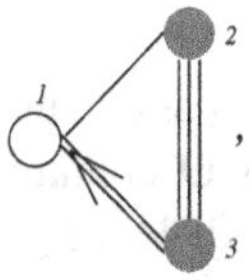

, *the higher order Serre relations are*

$$[e_2,[e_3,e_1]]-2[e_3,[e_2,e_1]]=0,$$
$$[f_2,[f_3,f_1]]-2[f_3,[f_2,f_1]]=0;$$

14. 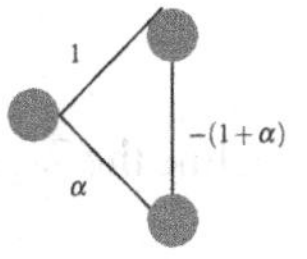

, *the higher order Serre relations are*

$$\alpha[e_1,[e_2,e_3]]+(1+\alpha)[e_2,[e_1,e_3]]=0,$$
$$\alpha[f_1,[f_2,f_3]]+(1+\alpha)[f_2,[f_1,f_3]]=0,$$

where the left node is labeled by 1, *the top node by* 2 *and bottom one by* 3.

When (A,Θ) is given in the distinguished root system, Theorem 3.3 simplifies considerably. We have the following result.

Theorem 3.4 *Let* (A,Θ) *with* $\Theta=\{s\}$ *be the Cartan matrix of a contragredient Lie superalgebra in the distinguished root system. Then* $L(A,\Theta)$ *is generated by* e_i, f_i, h_i $(i=1,2,\ldots,r)$*, where* e_s *and* f_s *are odd and the rest even, subject to the quadratic relations*

$$\begin{aligned}&[h_i,h_j]=0,\\ &[h_i,e_j]=a_{ij}e_j, \quad [h_i,f_j]=-a_{ij}f_j,\\ &[e_i,f_j]=\delta_{ij}h_i, \quad \forall i,j;\end{aligned} \tag{3.7}$$

standard Serre relations

$$\begin{aligned}&(ad_{e_i})^{1-a_{ij}}(e_j)=0,\\ &(ad_{f_i})^{1-a_{ij}}(f_j)=0, \quad for\ i\neq j,\ a_{ii}\neq 0;\\ &[e_s,e_s]=0, \quad [f_s,f_s]=0, \quad for\ a_{ss}=0;\end{aligned} \tag{3.8}$$

and higher order Serre relations

$$[e_s,[e_{s-1},[e_s,e_{s+1}]]]=0, \quad [f_s,[f_{s-1},[f_s,f_{s+1}]]]=0, \tag{3.9}$$

if the Dynkin diagram of A contains a full sub-diagram of the form

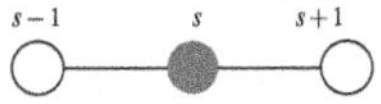

with $sgn_{s-1,s}sgn_{s,s+1}=-1$, *or* 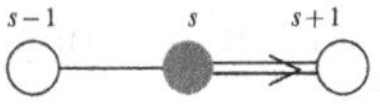

.

Remark 3.3 Note the importance of the signs sgn_{ij} in the above theorem. There are higher order Serre relations associated with the first Dynkin diagram in (2.2), but none with the second. The Dynkin diagrams in (2.2) are respectively those of $\mathfrak{sl}_{2|2}$ and $\mathfrak{osp}_{4|2}$ in their distinguished root systems. The Lie superalgebra $D(2,1;\alpha)$ in the distinguished root system has no higher order Serre relations either.

4 Proof of key lemma for distinguished root systems

Throughout this section, we assume that the Cartan matrix (A,Θ) is associated with the distinguished root system of a simple Lie superalgebra. Thus Θ contains only one element, which we denote by s. To simplify notation, we write $\tilde{\mathfrak{g}}(A)$ for $\tilde{\mathfrak{g}}(A,\Theta)$, $\mathfrak{g}(A)$ for $\mathfrak{g}(A,\Theta)$, and $L(A)$ for $L(A,\Theta)$.

4.1 The proof

The proof of Lemma 3.3 will make essential use of Lemma 3.4. Define the $\mathbb{Z}$-gradings for $\mathfrak{g}(A)$ and $L(A)$ as in Sect. 3.2.2 by taking $d = s$.

Lemma 4.1 *As reductive Lie algebras,* $\mathfrak{g}_0 = L_0$.

Proof. In this case, both $\mathfrak{g}_0$ and L_0 are generated by purely even elements. Let $\mathfrak{g}_0' = [\mathfrak{g}_0,\mathfrak{g}_0]$ and $L_0' = [L_0,L_0]$ be the derived algebras. Then by Serre's theorem for semi-simple Lie algebras $\mathfrak{g}_0' = L_0'$. Now the claim immediately follows. □

We now consider the $\mathfrak{g}_0$-modules $\mathfrak{g}_1$ and L_1.

Remark 4.1 For convenience, we continue to use e_i, h_i and f_i to denote the images of these elements in $\mathfrak{g}(A)$.

Examine the following relations in $\mathfrak{g}(A)$:

$$[h_i,e_s] = a_{is}e_s, \quad [f_i,e_s] = 0, \quad (ad_{e_i})^{1-a_{is}}e_s = 0, \quad \forall i \neq s. \tag{4.1}$$

The first two relations imply that e_s is a lowest weight vector of the $\mathfrak{g}_0$-module $\mathfrak{g}_1$, with weight α_s. Since a_{is} are non-positive integers for all $i \neq s$, by [10, Theorem 21.4], the third relation implies that $\mathfrak{g}_1$ is an irreducible finite dimensional $\mathfrak{g}_0$-module. The relations (4.1) also hold in $L(A)$. This immediately shows the following result.

Lemma 4.2 *Both* $\mathfrak{g}_1$ *and* L_1 *are irreducible* $\mathfrak{g}_0$*-modules, and* $\mathfrak{g}_1 = L_1$.

Note that $\mathfrak{g}_2$ is generated by $\mathfrak{g}_1$, that is $\mathfrak{g}_2 = [\mathfrak{g}_1,\mathfrak{g}_1]$. By induction one can show that $\mathfrak{g}_{k+1} = (ad_{\mathfrak{g}_1})^k(\mathfrak{g}_1)$ for all $k \geq 1$. If $\mathfrak{g}_i = 0$ for some $i > 1$, then $\mathfrak{g}_j = 0$ for all $j \geq i$. We have the $\mathfrak{g}_0$-module decomposition $\mathfrak{g}_1 \otimes \mathfrak{g}_1 = S_s^2(\mathfrak{g}_1) \oplus \wedge_s^2(\mathfrak{g}_2)$, where $S_s^2(\mathfrak{g}_1)$ denotes the second $\mathbb{Z}_2$-graded symmetric power, and $\wedge_s^2(\mathfrak{g}_1)$ the second $\mathbb{Z}_2$-graded skew power, of $\mathfrak{g}_1$.

Remark 4.2 Throughout the paper, we use $S_s^k(V)$ and $\wedge_s^k(V)$ to denote the $\mathbb{Z}_2$-graded symmetric and skew symmetric tensors of rank k in the $\mathbb{Z}_2$-graded vector space V, and $S^k(V)$ and $\wedge^k(V)$ to denote the usual symmetric and skew symmetric tensors of rank k, ignoring the $\mathbb{Z}_2$-grading of V.

We have the following result:

Lemma 4.3 *The Lie superbracket defines a surjective* $\mathfrak{g}_0$*-map* $\mathfrak{g}_1 \otimes \mathfrak{g}_1 \longrightarrow \mathfrak{g}_2$, $X \otimes Y \mapsto [X,Y]$. *The* $\mathfrak{g}_0$*-submodule* $S_s^2(\mathfrak{g}_1)$ *is in the kernel of this map, and* $\wedge_s^2(\mathfrak{g}_1)$ *is mapped surjectively onto* $\mathfrak{g}_2$.

Proof. For any $X,Y \in \mathfrak{g}_1$, an element $Z \in \mathfrak{g}_0$ acts on $X \otimes Y$ by

$$Z \cdot (X \otimes Y) = [Z,X] \otimes Y + X \otimes [Z,Y].$$

The Lie superbracket maps $Z \cdot (X \otimes Y)$ to $[[Z,X],Y] + [X,[Z,Y]] = [Z,[X,Y]]$. This proves the first claim. The second claim follows from the $\mathbb{Z}_2$-graded skew symmetry of the Lie superbracket. □

Therefore, the $\mathfrak{g}_0$-map $\Psi : \wedge_s^2(\mathfrak{g}_1) \longrightarrow \mathfrak{g}_2$ defined by the composition

$$\wedge_s^2(\mathfrak{g}_1) \hookrightarrow \mathfrak{g}_1 \otimes \mathfrak{g}_1 \xrightarrow{[\ ,\ ]} \mathfrak{g}_2$$

is also surjective, where the map on the left is the natural embedding. The structure of $\wedge_s^2(\mathfrak{g}_1)$ as a $\mathfrak{g}_0$-module can be understood; this enables us to understand the structure of $\mathfrak{g}_2$.

Recall that in the distinguished root systems, $L_2 = 0$ if $L(A)$ is of type I, and $L_2 \neq 0$ but $L_3 = 0$ if $L(A)$ is of type II. Thus in order to show that $\mathfrak{g}_k = L_k$ for all $k > 0$, it remains to prove that $\mathfrak{g}_2 = 0$ if the Cartan matrix A is of type I, and $\mathfrak{g}_2 = L_2$ and $\mathfrak{g}_3 = 0$ if A is of type II. In view of Lemma 3.4, the proof of Lemma 3.3 is done once this is accomplished.

The rest of the proof will be based on a case by case study. Let us start with the type I Lie superalgebras.

4.1.1 The case of $\mathfrak{sl}_{m|n}$

If the Cartan matrix A is that of $\mathfrak{sl}_{m|n}$, the Lie superalgebra $\mathfrak{g}(A)$ has $\mathfrak{g}_0 = \mathfrak{gl}_m \oplus \mathfrak{sl}_n$, and $\mathfrak{g}_1 \cong \mathbb{C}^m \otimes \overline{\mathbb{C}}^n$ up to parity change, where $\mathbb{C}^m$ denotes the natural module for $\mathfrak{gl}_m$, and $\overline{\mathbb{C}}^n$ denotes the dual of the natural module for $\mathfrak{sl}_n$. Assuming that both m and n are greater than 1. Then $\wedge_s^2(\mathfrak{g}_1) = S^2(\mathbb{C}^m) \otimes S^2(\overline{\mathbb{C}}^n) \oplus \wedge^2(\mathbb{C}^m) \otimes \wedge^2(\overline{\mathbb{C}}^n)$.

The lowest weight vectors of the irreducible submodules are respectively given by

$$\begin{aligned} v(2) :=& e_s \otimes e_s; \\ v(1^2) :=& e_{s-1,s+2} \otimes e_{s,s+1} + e_{s,s+1} \otimes e_{s-1,s+2} \\ & - (e_{s-1,s+1} \otimes e_{s,s+2} + e_{s,s+2} \otimes e_{s-1,s+1}), \end{aligned}$$

where $s = m$, and

$$\begin{aligned} & e_{s,s+1} = e_s, \quad e_{s,s+2} = [e_s, e_{s+1}], \\ & e_{s-1,s+1} = [e_{s-1}, e_s], \quad e_{s-1,s+2} = [e_{s-1}, e_{s,s+2}]. \end{aligned}$$

We have $\Psi(v(2)) = [e_s, e_s] = 0$ by one of the standard Serre relations. It follows that the entire irreducible $\mathfrak{g}_0$-submodule $S^2(\mathbb{C}^m) \otimes S^2(\overline{\mathbb{C}}^n)$ is mapped to zero. In particular, we have

$$[e_{s-1,s+2}, e_{s,s+1}] + [e_{s-1,s+1}, e_{s,s+2}] = 0. \tag{4.2}$$

The first term of (4.2) vanishes by the higher order Serre relation; this in turn forces the second term to vanish as well. Hence

$$\Psi(v(1^2)) = [e_{s-1,s+2}, e_{s,s+1}] - [e_{s-1,s+1}, e_{s,s+2}] = 0.$$

Therefore, $v(1^2)$ is in the kernel of Ψ, implying that the entire submodule $\wedge^2(\mathbb{C}^m) \otimes \wedge^2(\overline{\mathbb{C}}^n)$ is mapped to zero by Ψ. This shows that $\mathfrak{g}_2 = 0$, and hence $\mathfrak{g}_k = 0$ for all $k \geq 2$.

Note that if $min(m,n) = 1$, say, $n = 1$, $\wedge_s^2(\mathfrak{g}_1)$ is irreducible as $\mathfrak{g}_0$-module and is equal to $S^2(\mathbb{C}^m) \otimes \mathbb{C}$. The above proof obviously goes through but in a much simplified fashion.

Therefore, we have proved that $\mathfrak{g}_k = L_k$ for all $k \geq 2$ in the case $L(A) = \mathfrak{sl}_{m|n}$.

4.1.2 The case of $C(n+1)$ with $n > 1$

In this case, $\mathfrak{g}_0 = \mathfrak{sp}_{2n} \oplus \mathbb{C}$ and $\mathfrak{g}_1 = \mathbb{C}^{2n}$. The $\mathbb{Z}_2$-graded skew symmetric tensor $\wedge_s^2(\mathfrak{g}_1)$ is an irreducible $\mathfrak{g}_0$-module with the lowest weight vector $e_1 \otimes e_1$. Since $\Psi(e_1 \otimes e_1) = [e_1, e_1] = 0$ by the standard Serre relation, it immediately follows that $\mathfrak{g}_k = 0$ for all $k \geq 2$.

4.1.3 The case of $D(m,n)$ with $m > 2$

In this case, $\mathfrak{g}_0 = \mathfrak{gl}_n \oplus \mathfrak{so}_{2m}$, and $\mathfrak{g}_1$ is isomorphic to $\mathbb{C}^n \otimes \mathbb{C}^{2m}$ as $\mathfrak{g}_0$-module (up to parity) with e_n being the lowest weight vector. Let us first assume that $n > 1$. Then we have

$$\wedge_s^2(\mathfrak{g}_1) = S^2(\mathbb{C}^n) \otimes \frac{S^2(\mathbb{C}^{2m})}{\mathbb{C}} \oplus \wedge^2(\mathbb{C}^n) \otimes \wedge^2(\mathbb{C}^{2m}) \oplus S^2(\mathbb{C}^n) \otimes \mathbb{C}.$$

Lowest weight vectors of the first two irreducible submodules can be explicitly constructed in exactly the same way as in the case of $\mathfrak{sl}_{m|n}$. The same arguments used there also show that the Lie superbracket maps both submodules to zero. Hence $\mathfrak{g}_2 \cong S^2(\mathbb{C}^n) \otimes \mathbb{C}$. Inspecting the roots of $D(m,n)$ given in Appendix A.1, we can see that $\mathfrak{g}_2 = L_2$.

Let us examine $\mathfrak{g}_2$ in more detail. We use notation from Appendix A.1 for roots of the Lie superalgebra $D(m,n)$. Let $X_{\delta_i \pm \varepsilon_p}$, where $1 \leq i \leq n$ and $1 \leq p \leq m$, be a weight basis of $\mathfrak{g}_1$. Then in $\mathfrak{g}_2$, we have

$$\begin{aligned} &[X_{\delta_i - \varepsilon_p}, X_{\delta_j - \varepsilon_q}] = [X_{\delta_i + \varepsilon_p}, X_{\delta_j + \varepsilon_q}] = 0, \quad \forall i, j, p, q, \\ &[X_{\delta_i + \varepsilon_p}, X_{\delta_j - \varepsilon_q}] = 0, \quad \forall i, j, p \neq q, \end{aligned} \tag{4.3}$$

and there exist scalars $c_{ij,pq}$ such that

$$[X_{\delta_i-\varepsilon_p}, X_{\delta_j+\varepsilon_p}] = c_{ij,pq}[X_{\delta_i+\varepsilon_p}, X_{\delta_j-\varepsilon_p}] \neq 0, \quad \forall i,j,p,q.$$

By multiplying the elements $X_{\delta_i\pm\varepsilon_p}$ by appropriate scalars if necessary, we may assume

$$[X_{\delta_i-\varepsilon_p}, X_{\delta_j+\varepsilon_p}] = [X_{\delta_i-\varepsilon_q}, X_{\delta_j+\varepsilon_q}], \quad \forall i,j,p,q,$$

which we denote by $X_{\delta_i+\delta_j}$. Then the subset of $X_{\delta_i+\delta_j}$ with $1 \leq i \leq j \leq n$ forms a basis of $\mathfrak{g}_2$.

Now we consider $\mathfrak{g}_3$. It immediately follows from (4.3) that $[X_{\delta_i+\delta_j}, X_{\delta_k\pm\varepsilon_p}] = 0$ for all k,p and $i \leq j$, that is,

$$\mathfrak{g}_3 = [\mathfrak{g}_1, \mathfrak{g}_2] = 0. \tag{4.4}$$

Hence $\mathfrak{g}_k = 0$ for all $k \geq 3$.

When $n = 1$, the proof goes through much more simply. This completes the proof of Lemma 3.3 for the case of $D(m,n)$ with $m > 2$.

In contrast to the type I case, the complication here is that $\mathfrak{g}_3$ needs to be analysed separately as $\mathfrak{g}_2 \neq 0$.

4.1.4 The case of $D(2,n)$

In this case, $\mathfrak{g}_0 = \mathfrak{gl}_n \oplus \mathfrak{sl}_2 \oplus \mathfrak{sl}_2$, and $\mathfrak{g}_1 = \mathbb{C}^n \otimes \mathbb{C}^2 \otimes \mathbb{C}^2$. The $\mathbb{Z}_2$-graded skew symmetric rank two tensor $\wedge_s^2(\mathfrak{g}_1)$ decomposes into the direct sum of four irreducible $\mathfrak{g}_0$-modules if $n > 1$:

$$\begin{aligned}\wedge_s^2(\mathfrak{g}_1) =& L^n_{(2)} \otimes L^2_{(2)} \otimes L^2_{(2)} \oplus L^n_{(1,1)} \otimes L^2_{(2)} \otimes L^2_{(0)} \\ &\oplus L^n_{(1,1)} \otimes L^2_{(0)} \otimes L^2_{(2)} \oplus L^n_{(2)} \otimes L^2_{(0)} \otimes L^2_{(0)}.\end{aligned}$$

If $n = 1$, then $L^n_{(1,1)} = 0$, the two modules in the middle are absent.

The lowest weight vectors of the first three submodules can be easily worked out. Below we give the explicit formulae for their images under the Lie superbracket. Let

$$\begin{aligned}&e_{s;s+1} = [e_s, e_{s+1}], \quad e_{s;s+2} = [e_s, e_{s+2}], \quad e_{s-1;s} = [e_{s-1}, e_s], \\ &e_{s-1;s+1} = [e_{s-1}, e_{s;s+1}], \quad e_{s-1;s+2} = [e_{s-1}, e_{s;s+2}].\end{aligned}$$

Then the images of the lowest weight vectors are given by

$$[e_s, e_s], \quad [e_{s-1;s+1}, e_s] - [e_{s-1;s}, e_{s,s+1}], \quad [e_{s-1;s+2}, e_s] - [e_{s-1;s}, e_{s,s+2}]. \tag{4.5}$$

We have the Serre relation $[e_s, e_s] = 0$. This implies that the entire irreducible submodule $L^n_{(2)} \otimes L^2_{(2)} \otimes L^2_{(2)}$ is mapped to zero by the Lie superbracket.

In the case $n > 1$, this in particular implies

$$[e_{s-1;s+1}, e_s] + [e_{s-1;s}, e_{s,s+1}] = 0, \quad [e_{s-1;s+2}, e_s] + [e_{s-1;s}, e_{s,s+2}] = 0.$$

Note that $[e_{s-1;s+1}, e_s] = 0$ and $[e_{s-1;s+2}, e_s] = 0$ are the two higher order Serre relations involving e_s. Thus all the four terms on the left hand sides of the above equations should vanish separately. It then follows that the second and third elements in (4.5) are zero, that is, the lowest weight vectors of the irreducible submodules $L^n_{(1,1)} \otimes L^2_{(2)} \otimes L^2_{(0)}$ and $L^n_{(1,1)} \otimes L^2_{(0)} \otimes L^2_{(2)}$ are in the kernel of the Lie superbracket. Thus both irreducible submodules are mapped to zero by the Lie superbracket. The above analysis is vacuous if $n = 1$.

Therefore, $\mathfrak{g}_2 \cong L^n_{(2)} \otimes L^2_{(0)} \otimes L^2_{(0)}$, and this shows that $\mathfrak{g}_2 \cong L_2$.

To analyse $\mathfrak{g}_3$, we note that equation (4.4) still holds here as can be shown by adapting the arguments in the $m > 2$ case. This completes the proof in this case.

4.1.5 The case of $B(m,n)$

When $m \geq 1$, the proof is much the same as in the case of $D(m,n)$ with $m > 2$. We omit the details.

If $m = 0$, then $\mathfrak{g}_0 = \mathfrak{gl}_n$, $\mathfrak{g}_1 = \mathbb{C}^n$ and $\mathfrak{g}_2 \cong \wedge^2_s(\mathfrak{g}_1) \cong L_2$. Every root vector in $\mathfrak{g}_1$ is of the form $[X, e_s]$ for some positive root vector $X \in \mathfrak{g}_0$, where $s = n$. Thus it follows from the relation $(ad_{e_s})^3(e_{s-1}) = 0$ that $[\mathfrak{g}_1, [e_s, e_s]] = 0$. Since $[e_s, e_s]$ is a $\mathfrak{g}_0$ lowest weight vector of $\mathfrak{g}_2$, this implies $\mathfrak{g}_3 = 0$.

Remark 4.3 The Lie superalgebra $B(0,n)$ is essentially the same as the ordinary Lie algebra B_n. As a matter of fact, the corresponding quantum supergroup is isomorphic to the smash product of $U_q(B_n)$ with the group algebra of $\mathbb{Z}_2^n$ [17,28]. The usual proof of Serre presentations for semi-simple Lie algebras (see, e.g., [10]) works for $B(0,n)$. We gave the alternative proof here for the sake of uniformity.

4.1.6 The case of $F(4)$

Let us order the nodes in the Dynkin diagram from the right to left:

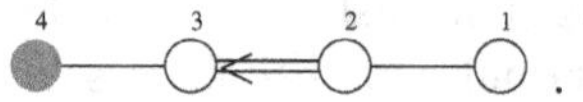

We may express the simple roots as $\alpha_1 = \varepsilon_1 - \varepsilon_2$, $\alpha_2 = \varepsilon_2 - \varepsilon_3$, $\alpha_2 = \varepsilon_3$ and $\alpha_4 = \frac{1}{2}\left(\delta - \varepsilon_1 - \varepsilon_2 - \varepsilon_3\right)$. The symmetric bilinear form on the weight space is defined in Appendix A.1, where further details about roots of $F(4)$ are given.

The first three simple roots are the standard simple roots of $\mathfrak{so}_7$, thus $\mathfrak{g}_0 = \mathfrak{so}_7 \oplus \mathfrak{gl}_1$. The subspace $\mathfrak{g}_1$ is an irreducible $\mathfrak{g}_0$-module, which has e_4 as a lowest weight vector, and restricts to the spinor module for $\mathfrak{so}_7$. Now $\wedge^2_s(\mathfrak{g}_1)$ decomposes into the direct sum of two irreducibles $\mathfrak{g}_0$-submodules, one of which is 1-dimensional, the other is 35-dimensional with lowest weight vector $e_4 \otimes e_4$.

The Serre relation $[e_4, e_4] = 0$ implies that the 35-dimensional submodule is in the kernel of the Lie superbracket, and hence $\mathfrak{g}_2$ is 1-dimensional. A basis element for $\mathfrak{g}_2$ is $E = [e_4, e_{\alpha_4+\varepsilon_1+\varepsilon_2+\varepsilon_3}]$.

For any weight β of $\mathfrak{g}_1$, we use $e_\beta \in \mathfrak{g}_1$ to denote a basis vector of the associated weight space, and set $e_{\alpha_4} = e_4$. Then we have

$$[e_\beta, E] = 0, \quad \text{for all odd positive root } \beta. \tag{4.6}$$

This is trivially true for $\beta = \alpha_4$ or $\alpha_4 + \varepsilon_1 + \varepsilon_2 + \varepsilon_3$. For $\beta = \alpha_4 + \varepsilon_1$ or $\alpha_4 + \varepsilon_i + \varepsilon_j$ $(i \neq j)$, we have $[e_\beta, E] = [[e_\beta, e_{\alpha_4}], e_{\alpha_4+\varepsilon_1+\varepsilon_2+\varepsilon_3}] - [e_{\alpha_4}, [e_\beta, e_{\alpha_4+\varepsilon_1+\varepsilon_2+\varepsilon_3}]]$, where both terms vanish as they involve images in $\mathfrak{g}_2$ of elements in the 35-dimensional submodule of $\wedge_s^2(\mathfrak{g}_1)$. Therefore, $\mathfrak{g}_k = \{0\}$ for all $k \geq 3$.

4.1.7 The case of $G(3)$

In this case, $\mathfrak{g}_0$ is isomorphic to the reductive Lie algebra $G_2 \oplus \mathfrak{gl}_1$, and $\mathfrak{g}_1$ is an irreducible $\mathfrak{g}_0$-module which restricts to the 7-dimensional irreducible G_2-module. The $\mathbb{Z}_2$-graded skew symmetric tensor $\wedge_s^2(\mathfrak{g}_1)$ decomposes into the direct sum $L(2\alpha_1) \oplus L(0)$ of two irreducible $\mathfrak{g}_0$-submodules. The submodule $L(2\alpha_1)$ has $e_1 \otimes e_1$ as lowest weight vector, thus its image under the Lie superbracket is zero by the Serre relation $[e_1, e_1] = 0$. The submodule $L(0)$ is 1-dimensional. Since the Lie superbracket maps $\wedge_s^2(\mathfrak{g}_1)$ surjectively to $\mathfrak{g}_2$, we immediately conclude that $\dim \mathfrak{g}_2 = 1$.

Let $X = e_{2\alpha_2+\alpha_3}$ be the root vector of $G_2 \subset \mathfrak{g}$ associated with the positive root $2\alpha_2 + \alpha_3$. Then $e^+ := [X, [X, e_1]]$ is the highest weight vector of $\mathfrak{g}_1$ as a $\mathfrak{g}_0$-module. Since $\mathfrak{g}_2$ is one-dimensional, it must be spanned by $E = [e_1, e^+]$.

If $e_\beta \in \mathfrak{g}_1$ is a weigh vector not proportional to e_1 or e^+, both $[e_\beta, e_1]$ and $[e_\beta, e^+]$ vanish since they lie in the image of $L(2\alpha_1) \subset \wedge_s^2(\mathfrak{g}_1)$ under the Lie superbracket. Hence $[e_\beta, E] = 0$. We also have $[e^+, e^+] = 0$, and the Serre relation $[e_1, e_1] = 0$. Thus $[e_1, E] = [e^+, E] = 0$. Therefore, $[\mathfrak{g}_1, E] = 0$, which implies $\mathfrak{g}_k = \{0\}$, for all $k \geq 3$.

4.1.8 The case of $D(2,1;\alpha)$

We have $\mathfrak{g}_0 = \mathfrak{sl}_2 \oplus \mathfrak{sl}_2 \oplus \mathfrak{gl}_1$, and $\mathfrak{g}_1 \cong \mathbb{C}^2 \otimes \mathbb{C}^2$. The tensor $\wedge_s^2(\mathfrak{g}_1)$ decomposes into the direct sum of two irreducible $\mathfrak{g}_0$-submodules,

$$\wedge_s^2(\mathfrak{g}_1) = L_{(2;2)} \oplus L_{(1^2;1^2)}, \quad L_{(2;2)} = L_{(2)} \otimes L_{(2)}, \quad L_{(1^2;1^2)} = L_{(1^2)} \otimes L_{(1^2)}.$$

The notation here only reflects the $\mathfrak{sl}_2 \oplus \mathfrak{sl}_2$-module structure, as there is no need to specify the $\mathfrak{gl}_1$-action explicitly (see Remark 4.4 below).

We have $\dim L_{(2;2)} = 9$ and $\dim L_{(1^2;1^2)} = 1$. The lowest weight vector for $L_{(2;2)}$ is $v(2) = e_1 \otimes e_1$. Let

$$e_{--} = e_1, \quad e_{+-} = [e_1, e_2], \quad e_{-+} = [e_1, e_3], \quad e_{++} = [e_{+-}, e_3],$$
$$v(1^2) = e_{--} \otimes e_{++} + e_{++} \otimes e_{--} - e_{+-} \otimes e_{-+} - e_{-+} \otimes e_{+-}.$$

Then the vector $v(1^2)$ spans $L_{(1^2;1^2)}$.

The Lie superbracket maps $L_{(2;2)}$ to zero because $[e_1, e_1] = 0$. Note that the element $[e_{--}, e_{++}] + [e_{+-}, e_{-+}]$ belongs to the image of $L_{(2;2)}$, thus is zero. Hence $\mathfrak{g}_2$ is spanned by $E = [e_{--}, e_{++}]$. Now it is easy to show that $[E, \mathfrak{g}_1] = 0$.

Remark 4.4 This proof is essentially the same as that in the case of $D(2,1)$, except for that the $\mathfrak{gl}_1$ subalgebra of $\mathfrak{g}_0$ acts on $\mathfrak{g}_1$ by different scalars in the two cases. However, this scalar is not important in the proof of Lemma 3.3, and that is the reason why we did not specify it explicitly.

4.2 Comments on the proof

Let us recapitulate the proof of Lemma 3.3 in the distinguished root systems.

1. By Lemma 3.4, the proof of Lemma 3.3 is reduced to showing that the parabolic subalgebras $\mathfrak{g}(A)_{\geq 0}$ and $L(A)_{\geq 0}$ are the same.
2. The elements $\{h_s\} \cup \{h_i, e_i, f_i \mid i \neq s\}$ and those defining relations of $\mathfrak{g}(A)$ obeyed by them give a Serre presentation for the reductive Lie algebra $\mathfrak{g}_0$. Then it essentially follows from Serre's theorem that $\mathfrak{g}_0 = L_0$, see Lemma 4.1.
3. Given item (2), it suffices to show that $\mathfrak{g}(A)_{>0} = \oplus_{k>0}\mathfrak{g}_k$ and $L(A)_{>0} = \oplus_{k>0}L_k$ are isomorphic as $\mathfrak{g}_0$-modules.
4. Equation (4.1) gives the necessary and sufficient conditions for $\mathfrak{g}_1$ to be a finite dimensional irreducible $\mathfrak{g}_0$-module with lowest weight α_s, hence $\mathfrak{g}_1 = L_1$ as $\mathfrak{g}_0$-modules.
5. The standard and higher order Serre relations involving e_s are conditions imposed on $\mathfrak{g}_0$-lowest weight vectors of $[\mathfrak{g}_1, \mathfrak{g}_1]$, which are the necessary and sufficient to guarantee that $\mathfrak{g}_2 = L_2$.
6. The fact that $\mathfrak{g}_3 = 0$ follows (trivially in the type I case) from the result on $\mathfrak{g}_2$ and graded skew symmetry of the Lie superbracket, thus no additional relations are required. The vanishing of $\mathfrak{g}_3$ implies that for all $k \geq 3$, $\mathfrak{g}_k = 0$, and hence $\mathfrak{g}_k = L_k$.

In non-distinguished root systems, one can still prove Lemma 3.3 by following a similar strategy, as we shall see in the next section. However, there are important differences in several aspects.

There are many such $\mathbb{Z}$-gradings as defined in Sect. 3.2.2 for the Lie superalgebras $\mathfrak{g}(A,\Theta)$ and $L(A,\Theta)$. This works to our advantage.

Given any such $\mathbb{Z}$-grading $\mathfrak{g}(A,\Theta) = \oplus_{k\in\mathbb{Z}}\mathfrak{g}_k$, the degree zero subspace $\mathfrak{g}_0$ forms a Lie superalgebra, which is not an ordinary Lie in general. Thus the requirement that $\mathfrak{g}_1$ be an irreducible $\mathfrak{g}_0$-module is much more difficult to implement, and usually leads to unfamiliar higher order Serre relations.

In general $\mathfrak{g}_3 \neq 0$. In order for $\mathfrak{g}_k$ to be equal to L_k for $k \geq 3$, higher order Serre relations are needed at degree $k \geq 3$.

5 Proof of key lemma for non-distinguished root systems

In this section we prove Lemma 3.3 in non-distinguished root systems by following a similar strategy as that in Sect. 4. In particular, Lemma 3.4 will be used in an essential way.

Assume that the Cartan matrix A is of size $r \times r$. Fix a positive integer $d \leq r$, we consider the corresponding $\mathbb{Z}$-gradings for $\mathfrak{g}(A,\Theta)$ and $L(A,\Theta)$ defined in Sect. 3.2.2. We shall first establish that $\mathfrak{g}_0 = L_0$. Since the roots of $L(A,\Theta)$ are known explicitly (see Appendix A.1), we have a complete understanding of the $\mathfrak{g}_0$-module structure of every L_k. Thus once we have a description of the weight spaces of each $\mathfrak{g}_k$ as $\mathfrak{g}_0$-module for all $k > 0$, an easy comparison with the root spaces of L_k will enable us to prove the key lemma.

Remark 5.1 In the proof of Lemma 3.3 given below, we shall only describe the weight spaces of $\mathfrak{g}_k$ $(k > 0)$, and leave out the easy step of comparing them with those of L_k in most cases.

For convenience, we introduce the parity map $p : \{1,2,\dots,r\} \longrightarrow \{0,1\}$ such that $p(i) = 1$ if $i \in \Theta$ and $p(i) = 0$ otherwise. Then e_i and f_i are odd if $p(i) = 1$, and even if $p(i) = 0$.

5.1 Proof in type A

We use induction on the rank r together with the help of Lemma 3.4 to prove Lemma 3.3 and Theorem 3.3.

If $r = 2$, the Dynkin diagram in the non-distinguished root system has two grey nodes. In this case, there exists no relation between e_1 and e_2, and $[e_1,e_2]$ is another positive root vector. Note that $[e_1,[e_1,e_2]] = 0$ and $[e_2,[e_1,e_2]] = 0$ by the graded skew symmetry of the Lie superbracket. Thus Lemma 3.3 is valid and $\mathfrak{g}(A,\Theta) = L(A,\Theta)$

When $r > 2$, we take $d = r$. Then $\mathfrak{g}_0' = [\mathfrak{g}_0,\mathfrak{g}_0]$ is a special linear superalgebra of rank $r-1$ by the induction hypothesis, and thus $\mathfrak{g}_0$ is a general linear superalgebra.

Define the following elements of $\mathfrak{g}_0$:

$$X_{ij} = ad_{e_i} \cdots ad_{e_{j-2}}(e_{j-1}), \quad i < j \leq r, \tag{5.1}$$

where $X_{j,j+1} = e_j$. In view of the general linear superalgebra structure of $\mathfrak{g}_0$, we conclude that $\mathfrak{g}_1$ is isomorphic to the irreducible $\mathfrak{g}_0$-module with lowest weight α_r (which is in fact the natural module possibly upon a parity change) if and only if

$$[X_{ik},[X_{jr},e_r]] = 0, \quad j \neq k.$$

By using the $\mathfrak{g}_0$-action, we can show that these conditions are equivalent to the relation

$$[e_{r-1},[[e_{r-2},e_{r-1}],e_r]] = 0 \tag{5.2}$$

and the relevant relations in (3.1). For $p(r-1) = 1$, (5.2) is a higher order Serre relation associated with the sub-diagram 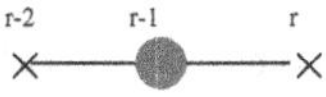

with $sgn_{r-2,r-1} = -sgn_{r-1,r}$. If $p(r-1) = 0$, it can be derived from

$$[e_{r-1},[e_{r-1},e_r]] = 0, \tag{5.3}$$

which is a standard Serre relation.

Consider $\wedge_s^2\mathfrak{g}_1$, which is an irreducible $\mathfrak{g}_0$-module. The lowest weight vector is

$$e_r \otimes e_r, \quad \text{if } p(r)=1, \quad \text{or}$$
$$e_r \otimes [e_{r-1}, e_r] - [e_{r-1}, e_r] \otimes e_r, \quad \text{if } p(r)=0.$$

Thus $\mathfrak{g}_2 = 0$ if and only if

$$\begin{aligned} &[e_r, e_r] = 0, \quad \text{if } p(r)=1, \quad \text{or} \\ &[e_r, [e_r, e_{r-1}]] = 0, \quad \text{if } p(r)=0, \end{aligned} \tag{5.4}$$

both of which are standard Serre relations. This proves that Lemma 3.3, and hence Theorem 3.2, are valid at rank r.

Remark 5.2 The proof presented here includes an alternative proof for Serre's theorem in the case of $\mathfrak{sl}_n$. This can be generalised to all finite dimensional simple Lie algebras. In particular, the proof for the other classical Lie algebras can be extracted from the next two sections.

5.2 Proof in type B

Consider the first Dynkin diagram of type B in Table 2, where the last (that is, r-th) node is white, and take $d = r$. In this case, $\mathfrak{g}_0$ is a general linear superalgebra, and we have already obtained a Serre presentation for it in Sect. 5.1.

We require $\mathfrak{g}_1$ be isomorphic to the irreducible $\mathfrak{g}_0$-module with lowest weight α_r, which is in fact the natural module for $\mathfrak{g}_0$. This is achieved by relations formally the same as (5.2) or (5.3).

As $\mathfrak{g}_0$-module, $\mathfrak{g}_2$ is isomorphic to $\wedge_s^2\mathfrak{g}_1$, which is irreducible with the lowest weight vector $E := [e_r, [e_r, e_{r-1}]]$. Now $\mathfrak{g}_3 = 0$ if and only if $[E, \mathfrak{g}_1] = 0$. This in particular requires that

$$(ad_{e_r})^3 (e_{r-1}) = 0. \tag{5.5}$$

We shall show that this in fact is the necessary and sufficient condition.

If $p(r-1) = 1$, then $[E, [e_{r-1}, e_r]] = 0$ trivially since $[e_{r-1}, e_{r-1}] = 0$ in $\mathfrak{g}_0$. For $K = [e_{r-2}, [e_{r-1}, e_r]]$, we also have $[K, E] = 0$. This follows from $[K, e_{r-1}] = 0$, which is one of the higher order Serre relations associated with a sub-diagram of type A. Applying ad_{e_r} to it twice and using (5.5), we obtain the desired relation. These relations imply that $[X, E] = 0$ for all $X \in \mathfrak{g}_1$ in this case. If $p(r-1) = 0$, the fact that $[X, E] = 0$, for all $X \in \mathfrak{g}_1$, follows from

$$[[e_{r-1}, e_r], [[e_{r-1}, e_r], e_r]] = 0,$$

which can be derived from (5.5).

The other Dynkin diagram (where the last node is black) can be treated in essentially the same way. We omit the details.

5.3 *Proof in types C and D*

The Dynkin diagrams of type C formally have the same forms as two of the Dynkin diagrams of D. The only difference is in the numbers of grey nodes, see Remark Appendix A.1. This enables us to treat both types of Lie superalgebras simultaneously.

5.3.1 Case 1

Consider the Dynkin diagram

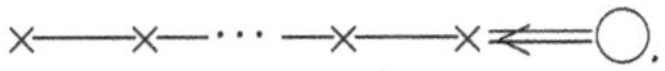

We label the nodes from left to right, thus the r-th node is the one at the right end. Set $d = r$, then $\mathfrak{g}_0$ is a general linear superalgebra.

As a $\mathfrak{g}_0$-module, $\mathfrak{g}_1$ is generated by e_r. We require it be isomorphic to the irreducible module $\overline{L}_{\alpha_r}$ with lowest weight α_r. Appendix B.2 describes the structure of the generalised Verma module $\overline{V}_{\alpha_r}$ with lowest weight α_r and the irreducible quotient $\overline{L}_{\alpha_r}$. We immediately see that the relevant relations in (3.1) and the relations

$$[X_{ir}, [X_{jr}, [X_{kr}, e_r]]] = 0, \quad \forall i \le j \le k \le r-1, \tag{5.6}$$

are necessary and sufficient conditions to guarantee that $\mathfrak{g}_1 \cong \overline{L}_{\alpha_r}$. Here X_{ir} are elements of $\mathfrak{g}_0$ defined by (5.1). The conditions (5.6) are equivalent to

$$\begin{aligned} &[e_{r-1}, [e_{r-1}, [e_{r-1}, e_r]]] = 0, && \text{if } e_{r-1} \text{ is even}, \\ &[X_{r-2,r}, [X_{r-2,r}, [e_{r-1}, e_r]]] = 0, && \text{if } e_{r-1}, e_{r-2} \text{ are both odd}, \\ &[X_{r-3,r}, [X_{r-2,r}, [e_{r-1}, e_r]]] = 0, && \text{if } e_{r-1} \text{ is odd}, e_{r-2} \text{ is even} \end{aligned} \tag{5.7}$$

because of the $\mathfrak{g}_0$-action. Here Remark Appendix B.1 is also in force.

Note that the different situations where the relations apply are mutually exclusive. The first relation is a standard Serre relation. The second and third are higher order Serre relations respectively associated with the sub-diagrams

or

Recall that $\mathfrak{g}_2$ is the image of $\wedge_s^2 \mathfrak{g}_1$ under the Lie superbracket. As $\mathfrak{g}_0$-module, $\wedge_s^2 \mathfrak{g}_1$ is irreducible with the lowest weight vector $e_r \otimes [e_{r-1}, e_r] - [e_{r-1}, e_r] \otimes e_r$. Thus $\mathfrak{g}_2 = 0$ if and only if

$$[e_r, [e_r, e_{r-1}]] = 0. \tag{5.8}$$

This is again a standard Serre relation.

5.3.2 Case 2

Now we consider the case with the Dynkin diagram

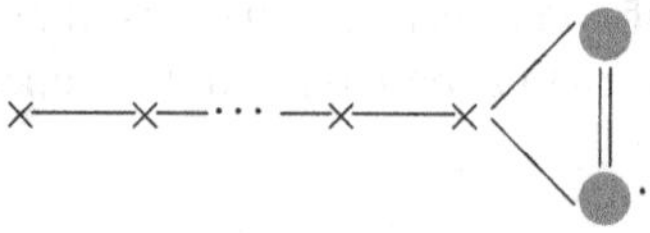

Let us first assume that $r = 3$. We have the Dynkin diagram of $\mathfrak{osp}_{2|4}$ (resp. $\mathfrak{osp}_{4|2}$) if $p(1) = 0$ (resp. $p(1) = 1$). Label by 1 the node marked by $\times$, and take $d = 1$. The diagram obtained by deleting this node is

This is a non-standard diagram of $\mathfrak{osp}_{2|2} \cong \mathfrak{sl}_{2|1}$. Equation (3.1) by itself suffices to define this Lie superalgebra.

Now $\mathfrak{g}_0 = \mathfrak{osp}_{2|2} \oplus \mathfrak{gl}_1$ (isomorphic to $\mathfrak{gl}_{2|1}$). Let $\bar{\mathfrak{b}}_0$ be the Borel subalgebra of $\mathfrak{g}_0$ generated by f_2, f_3 and all h_i, and define the lowest weight Verma module $\overline{V}_{\alpha_1} := U(\mathfrak{g}_0) \otimes_{U(\mathfrak{b}_0^-)} \mathbb{C}_{\alpha_1}$ for $\mathfrak{g}_0$, where $\mathbb{C}_{\alpha_1}$ is the irreducible $\bar{\mathfrak{b}}_0$-module with lowest weight α_1. Direct computations show that the maximal submodule M_{α_1} is generated by the vector $(e_2e_3 - e_3e_2) \otimes 1$. The irreducible quotient $\overline{L}_{\alpha_1}$ is four dimensional, with a basis consisting of the images of $1 \otimes 1$, $e_2 \otimes 1$, $e_3 \otimes 1$, and $[e_2, e_3] \otimes 1$. Its restriction to $\mathfrak{osp}_{2|2}$ is the natural module.

We need $\mathfrak{g}_1 \cong \overline{L}_{\alpha_1}$, possibly up to a parity change depending on the parity of e_1. From the description of $\overline{V}_{\alpha_1}$ and M_{α_1} above, we see that the necessary and sufficient conditions are the relevant quadratic relations involving e_1 in (3.1), and

$$[e_2, [e_3, e_1]] - [e_3, [e_2, e_1]] = 0. \tag{5.9}$$

Note that this is a higher order Serre relation associated with the sub-diagram (6) given in Theorem 3.3.

To proceed further, we need to specify the parity of e_1.

If e_1 is even, the Lie superalgebra $L(A, \Theta)$ is $\mathfrak{osp}_{2|4}$. Now $\wedge_s^2 \mathfrak{g}_1$ is the direct sum of a seven dimensional indecomposable $\mathfrak{g}_0$-submodule and a one dimensional $\mathfrak{g}_0$-submodule. The seven dimensional submodule is generated by the two lowest weight vectors

$$e_1 \otimes [e_2, e_1] - [e_2, e_1] \otimes e_1, \quad e_1 \otimes [e_3, e_1] - [e_3, e_1] \otimes e_1,$$

and the one dimensional submodule by

$$[e_2, e_1] \otimes [e_3, e_1] + [e_3, e_1] \otimes [e_2, e_1] + e_1 \otimes [[e_2, e_3], e_1] - [[e_2, e_3], e_1] \otimes e_1.$$

In this case, we need $\mathfrak{g}_2$ to be isomorphic to a one dimensional $\mathfrak{g}_0$-module with weight $2\alpha_1 + \alpha_2 + \alpha_3$. Thus the seven dimensional indecomposable submodule of $\wedge_s^2 \mathfrak{g}_1$ is sent to zero by the Lie superbracket, or equivalently,

$$[e_1, [e_1, e_2]] = 0, \quad [e_1, [e_1, e_3]] = 0, \tag{5.10}$$

which are standard Serre relations. The image of the one dimensional submodule is $\mathfrak{g}_2$, which is spanned by

$$[[e_1,e_2],[e_1,e_3]] - [e_1,[e_1,[e_2,e_3]]] = -[[e_1,e_2],[e_1,e_3]],$$

where (5.10) is used to obtain the identity. By using (5.9) and (5.10), one can easily show that $[\mathfrak{g}_2,\mathfrak{g}_1] = 0$, and hence $\mathfrak{g}_3 = 0$.

If e_1 is odd, the Lie superalgebra $L(A,\Theta)$ is $\mathfrak{osp}_{4|2}$. By dimension counting, we need $\mathfrak{g}_2 = 0$. Now $\wedge_s^2\mathfrak{g}_1$ is also a direct sum of a seven dimensional indecomposable $\mathfrak{g}_0$-submodule and a one dimensional submodule. Given the condition $[e_1,e_1] = 0$, the seven dimensional submodule vanishes automatically under the Lie superbracket, and the image of the one dimensional submodule is spanned by $[[e_1,e_2],[e_1,e_3]]$. Taking the Lie superbraket of e_1 with both sides of (5.9), we obtain $[[e_1,e_2],[e_1,e_3]] = 0$. Hence $\mathfrak{g}_2 = 0$.

Now assume $r \geq 4$. We take $d = r-3$, then $\mathfrak{g}_0$ is the direct sum of a general linear superalgebra and $\mathfrak{osp}_{4|2}$ or $\mathfrak{osp}_{2|4}$.

If e_{r-2} is even, the condition that $\mathfrak{g}_1$ is an irreducible $\mathfrak{g}_0$-module of lowest weight α_{r-3} is given by the relevant relations in (3.1),

$$[e_{r-2},[e_{r-2},e_{r-3}]] = 0,$$

and also

$$\begin{aligned} &[e_{r-4},[e_{r-4},e_{r-3}]] = 0, && \text{if } p(r-4) = 0, \\ &[e_{r-4},[e_{r-5},[e_{r-4},e_{r-3}]]] = 0, && \text{if } p(r-4) = 1. \end{aligned} \tag{5.11}$$

As $\mathfrak{g}_0$-module, $\wedge_s^2\mathfrak{g}_1$ is the direct sum of three irreducibles. The $\mathfrak{osp}_{2|4}$ subalgebra of $\mathfrak{g}_0$ acts trivially on one of the irreducible submodules, and $\mathfrak{g}_2$ is isomorphic to it. The necessary and sufficient conditions for the Lie superbracket to annihilate the other two irreducible submodules are

$$\begin{aligned} &[e_{r-3},[e_{r-3},e_{r-2}]] = 0, \quad [e_{r-3},[e_{r-3},e_{r-4}]] = 0, && \text{if } p(r-3) = 0, \\ &[e_{r-3},e_{r-3}] = 0, \quad [e_{r-3},[e_{r-4},[e_{r-3},e_{r-2}]]] = 0, && \text{if } p(r-3) = 1, \end{aligned} \tag{5.12}$$

as can be shown by examining lowest weight vectors of the submodules.

Remark 5.3 Let $E = [[e_{r-3},e_{r-2}],e_{r-1}]$ and $E' = [[e_{r-3},e_{r-2}],e_r]$. Then at least one of the vectors $[X,E]$ and $[X,E']$ vanishes for any $X \in \mathfrak{g}_1$.

Let v denote a lowest weight vector of $\mathfrak{g}_2$. We can take $v = [E,E']$ if e_{r-3} is even, and $v = [[e_{r-4},E],E']$ if e_{r-3} is odd. Then by Remark 5.3, we have $[v,X] = 0$ for any $X \in \mathfrak{g}_1$. Hence $\mathfrak{g}_3 = 0$.

If e_{r-2} is odd, the condition that $\mathfrak{g}_1$ is an irreducible $\mathfrak{g}_0$-module of lowest weight α_{r-3} translates into the relations (5.11),

$$\begin{aligned} &[e_{r-2},[[e_{r-2},e_{r-1}],e_{r-3}]] = 0, \\ &[e_{r-2},[[e_{r-2},e_r],e_{r-3}]] = 0, \end{aligned}$$

plus the relevant relations in (3.1). Here we have used some facts about generalised Verma modules for $\mathfrak{osp}_{4|2}$.

As $\mathfrak{g}_0$-module, $\wedge_s^2 \mathfrak{g}_1$ is again a direct sum of three irreducibles. One of them restricts to a direct sum of one dimensional $\mathfrak{osp}_{4|2}$-modules, and $\mathfrak{g}_2$ is isomorphic to it. The other two irreducibles are both mapped to zero by the Lie superbracket. The necessary and sufficient condition for this to happen is still (5.12).

Note that Remark 5.3 remains valid in the present case if we define E and E' in the same way. Let $v = [E, E']$ if e_{r-3} is odd, and $v = [[e_{r-4}, E], E']$ if e_{r-3} is even. Then v is a nonzero lowest weight vector of $\mathfrak{g}_2$. It follows from Remark 5.3 that $[v, X] = 0$ for any $X \in \mathfrak{g}_1$. Hence $\mathfrak{g}_3 = 0$.

5.3.3 Case 3

Finally we consider the Dynkin diagram

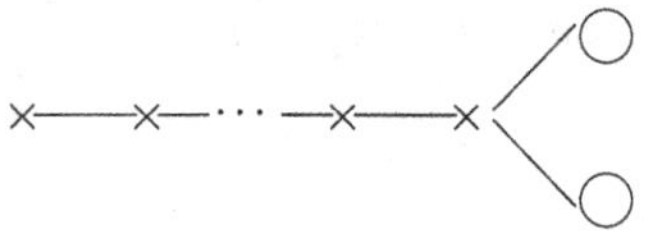

assuming that there are at least two grey nodes (as otherwise this would correspond to the distinguished root system of type D). This forces $r \geq 4$.

This case is quite easy, thus we shall be brief. We choose d to be the largest integer such that $p(d) = 1$. Then $\mathfrak{g}_0$ is the direct sum of a general linear superalgebra and an even dimensional orthogonal Lie algebra.

From Sect. 5.1, we see that the necessary and sufficient conditions for e_d (which must be odd) to generate an irreducible $\mathfrak{g}_0$-module are the relevant relations in (3.1) and the higher order Serre relation involving e_d associated with the following sub-diagram 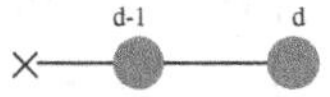

of the Dynkin diagram if $p(d-1) = 1$. Note that if $d = 2$, this becomes vacuous.

As $\mathfrak{g}_0$-module, $\mathfrak{g}_1$ is the tensor product of the natural modules V_A and V_D respectively for the general linear superalgebra and orthogonal algebra contained in $\mathfrak{g}_0$. Here V_D is purely even, and the grading of V_A gives rise to the grading of $\mathfrak{g}_1$.

Now $\wedge_s^2 \mathfrak{g}_1 \cong \wedge_s^2(V_A) \otimes (S^2(V_D)/\mathbb{C}) \oplus S_s^2(V_A) \otimes \wedge^2(V_D) \oplus \wedge_s^2(V_A) \otimes \mathbb{C}$ as $\mathfrak{g}_0$-module. The images of the first two irreducibles under the Lie superbracket are set to zero by the relation $[e_d, e_d] = 0$ and the higher order Serre relation(s) associated with the sub-diagram(s) of the form 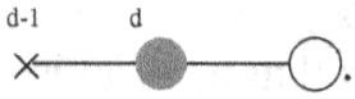

. Note that if $d < r-2$, there is only one such diagram, but there are two if $d = r-2$, as the last node can be $(r-1)$ or r. We have $\mathfrak{g}_2 \cong \wedge_s^2(V_A) \otimes \mathbb{C}$.

One can show that $[\mathfrak{g}_2, \mathfrak{g}_1] = 0$ by using the same arguments as those in Sect. 4.1.3 and Sect. 4.1.4, thus $\mathfrak{g}_3 = 0$.

5.4 Proof in type $F(4)$

Now we turn to $F(4)$, which is considerably more complicated than the other types of Lie superalgebras.

5.4.1 Case 1

Consider first the root system corresponding to the Dynkin diagram

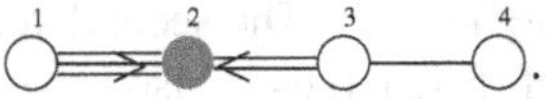

We take $d = 2$. Then $\mathfrak{g}_0 = \mathfrak{sl}_2 \oplus \mathfrak{gl}_3$. The standard Serre relations plus the relevant relations in (3.1) are the necessary and sufficient conditions rendering the $\mathfrak{g}_0$-module $\mathfrak{g}_1$ irreducible. We have $\mathfrak{g}_1 \cong \mathbb{C}^2 \otimes \mathbb{C}^3$ up to a parity change.

As $\mathfrak{g}_0$-module, $\wedge_s^2 \mathfrak{g}_1$ is a direct sum of two irreducibles. The condition $[e_2, e_2] = 0$ forces one of the irreducibles to be in the kernel of the map $\wedge_s^2 \mathfrak{g}_1 \longrightarrow \mathfrak{g}_2$. Thus $\mathfrak{g}_2$ is an irreducible $\mathfrak{g}_0$-module generated by the lowest weight vector $E = [[e_1, e_2], [e_2, e_3]]$. We have $\mathfrak{g}_2 = \mathbb{C} \otimes \wedge^2(\mathbb{C}^3)$.

Now $\mathfrak{g}_3 = [\mathfrak{g}_2, \mathfrak{g}_1] \cong \mathbb{C}^2 \otimes \mathbb{C}$ with a basis consisting of vectors $[E, [e_2, [e_3, e_4]]]$ and $[E, [E', e_4]]$, where $E' = [e_1, [e_2, e_3]]$. One immediately sees that

$$[\mathfrak{g}_3, e_2] = \mathbb{C}[E, [E, e_4]],$$

which generates $\mathfrak{g}_4 = \mathbb{C} \otimes \mathbb{C}^3$.

To consider $\mathfrak{g}_5$, we only need to look at $[\mathfrak{g}_4, \mathfrak{g}_1]$. If $X \in \mathfrak{g}_1$ is any lowest weight vector for $\mathfrak{sl}_2 \subset \mathfrak{g}_0$, the higher order Serre relation associated with the Dynkin diagram (see diagram (7) in Theorem 3.3) renders $[\mathfrak{g}_4, X] = 0$. Since the $\mathfrak{sl}_2$ subalgebra of $\mathfrak{g}_0$ acts trivially on $\mathfrak{g}_4$, it follows that $[\mathfrak{g}_4, \mathfrak{g}_1] = 0$, that is, $\mathfrak{g}_5 = 0$.

5.4.2 Case 2

For the Dynkin diagram

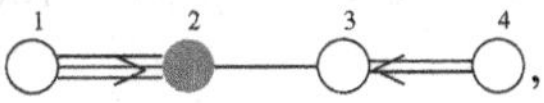

we also take $d = 2$ as in the previous case. Then $\mathfrak{g}_0 = \mathfrak{gl}_2 \oplus \mathfrak{sp}_4$. The relevant relations in (3.1) and standard Serre relations guarantee that e_2 generates an irreducible $\mathfrak{g}_0$-module, which is isomorphic to the tensor product $\mathbb{C}^2 \otimes \mathbb{C}^4$ of the natural modules for $\mathfrak{gl}_2$ and $\mathfrak{sp}_4$ up to a parity change.

Now $\wedge_s^2 \mathfrak{g}_1$ decomposes into the direct sum of three irreducible $\mathfrak{g}_0$-modules, which are respectively isomorphic to $S^2(\mathbb{C}^2) \otimes S^2(\mathbb{C}^4)$, $\wedge^2(\mathbb{C}^2) \otimes (\wedge^2(\mathbb{C}^2)/\mathbb{C})$ and $\wedge^2(\mathbb{C}^2) \otimes \mathbb{C}$. The necessary and sufficient conditions for the Lie superbracket to map the first and the third submodules to zero are $[e_2, e_2] = 0$ and the higher order Serre relation

$$[[e_1, e_2], [[e_2, e_3], [e_3, e_4]] - [[e_2, e_3], [[e_1, e_2], [e_3, e_4]] = 0 \qquad (5.13)$$

associated with the Dynkin diagram (see diagram (8) in Theorem 3.3). Now $\mathfrak{g}_2$ is isomorphic to $\wedge^2(\mathbb{C}^2) \otimes \frac{\wedge^2(\mathbb{C}^2)}{\mathbb{C}}$ with lowest weight vector

$$E = [[e_1, e_2], [e_2, e_3]].$$

Formally $[\mathfrak{g}_2, \mathfrak{g}_1]$ decomposes into the direct sum of two irreducibles, respectively having lowest weight vectors

$$[E, e_2], \quad [e_2, [E, [e_3, e_4]]].$$

The first vector vanishes by $[e_2, e_2] = 0$. The second vector is the supercommutator of e_2 with the left hand side of (5.13), thus is also zero. This shows that $\mathfrak{g}_3 = 0$.

5.4.3 Case 3

Consider the Dynkin diagram

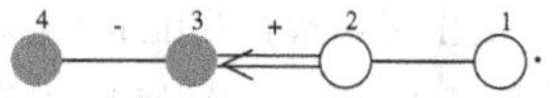

We take $d = 4$, and delete the 4-th node from the diagram to obtain

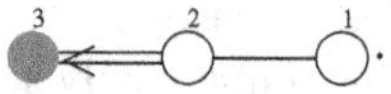

This is a non-standard diagram for $\mathfrak{sl}_{1|3}$, where the double edges can be got rid of by a normalisation of the bilinear form on the weight space thus are immaterial. The presentation for $\mathfrak{sl}_{1|3}$ involves no higher order Serre relation. We have $\mathfrak{g}_0 = \mathfrak{gl}_{1|3}$.

Let $\overline{\mathfrak{p}}$ be the lower triangular maximal parabolic subalgebra of $\mathfrak{g}_0$ with Levi subalgebra $\mathfrak{l} := \mathfrak{gl}_3 \oplus \mathfrak{gl}_1$. Let $\overline{L}^0_{\alpha_4} = \mathbb{C}v_0$ be the 1-dimensional $\overline{\mathfrak{p}}$-module with lowest weight α_4, which is assume to be a purely odd superspace. Since α_4 is a typical $\mathfrak{g}_0$ weight, the generalised Verma module $\overline{V}_{\alpha_4} = U(\mathfrak{g}_0) \otimes_{U(\overline{\mathfrak{p}})} \overline{L}^0_{\alpha_4}$ is irreducible, i.e., $\overline{L}_{\alpha_4} = \overline{V}_{\alpha_4}$. It is multiplicity free, and the set of weights is given by

$$\Delta^+ \backslash \{\Delta^+(\mathfrak{g}_0) \cup \Delta_2^+\}, \tag{5.14}$$

where Δ^+ is the set of the positive roots of $F(4)$ relative to the Borel subalgebra under consideration, $\Delta^+(\mathfrak{g}_0)$ is the set of the positive roots of the subalgebra $\mathfrak{g}_0$, and

$$\Delta_2^+ = \left\{ \frac{1}{2}(\delta + \varepsilon_1 + \varepsilon_2 + \varepsilon_3),\ \varepsilon_i + \varepsilon_j, i \neq j \right\}. \tag{5.15}$$

The $\mathfrak{g}_0$-module $\wedge_s^2 \overline{L}_{\alpha_4}$ is not semi-simple. To avoid the laborious task of determining the indecomposable submodules, we simply examine the $\mathfrak{l}$ lowest weight vectors in $\wedge_s^2 \overline{L}_{\alpha_4}$. Of particular importance to us are the vectors

$$\begin{aligned}
z_1 :=& v_0 \otimes v_0;\\
z_2 :=& e_3 v_0 \otimes [e_2, e_3] e_3 v_0 - [e_2, e_3] e_3 v_0 \otimes e_3 v_0;\\
z_3 :=& v_0 \otimes e_3 v_0 - e_3 v_0 \otimes v_0;\\
w_1 :=& v_0 \otimes [e_2, e_3] e_3 v_0 + [e_2, e_3] e_3 v_0 \otimes v_0;\\
w_2 :=& v_0 \otimes [e_1, [e_2, e_3]][e_2, e_3] e_3 v_0 - [e_1, [e_2, e_3]][e_2, e_3] e_3 v_0 \otimes v_0.
\end{aligned}$$

The space of $\mathfrak{l}$-lowest weight vectors of $\wedge_s^2 \overline{L}_{\alpha_4}$ is spanned by w_1, w_2 and the $\mathfrak{l}$-lowest weight vectors in the $\mathfrak{g}_0$-submodule M generated by z_1 and z_2. It is important to observe that w_1 and w_2 are not in M, but $w_1 \in U(\mathfrak{g}_0)w_2$. Furthermore, one can verify that $\wedge_s^2 \overline{L}_{\alpha_4}/M$ is multiplicity free with the set of weights Δ_2^+.

Now we take $v_0 = e_4$ and require $\overline{\mathfrak{p}}$ act on it by the adjoint action. Then $\mathfrak{g}_1 = \overline{L}_{\alpha_4}$. We require that the Lie superbracket maps z_1 and z_2 to zero. This leads to the following relations:

$$\begin{aligned} &[e_4, e_4] = 0; \\ &[[e_3, e_4], [[e_3, e_4], [e_2, e_3]]] = 0. \end{aligned} \tag{5.16}$$

Under the first condition, the Lie superbracket automatically maps z_3 to zero. Note that the second relation in equation (5.16) is the desired higher order Serre relation associated with the sub-diagram

The vectors w_1 and w_2 have non-zero images under the Lie superbracket, and we have $\mathfrak{g}_2 \cong \wedge_s^2 \overline{L}_{\alpha_4}/M$. By considering the possible $\mathfrak{l}$-lowest weight vectors, we can show that $[\mathfrak{g}_1, \mathfrak{g}_2] = 0$, thus $\mathfrak{g}_3 = 0$.

Now the proof of Lemma 3.3 in this case is completed by comparing the weights in (5.14) and (5.15) with the roots in L_1 and L_2.

5.4.4 Case 4

Consider the Dynkin diagram

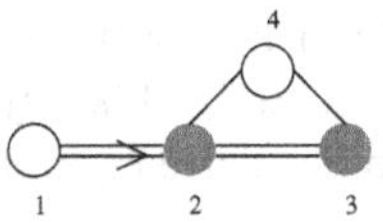

Take $d = 1$, then $\mathfrak{g}_0 = \mathfrak{osp}_{2|4} \oplus \mathfrak{gl}_1$. The presentation of $\mathfrak{osp}_{2|4}$ relative to the Dynkin diagram

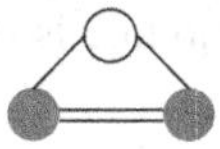

has been constructed, thus the defining relations among e_i, f_i, h_i for $i > 1$ are all known. The parabolic subalgebra of $\mathfrak{osp}_{2|4}$ defined in Appendix B.1 together with the ideal $\mathfrak{gl}_1$ form a parabolic of $\mathfrak{g}_0$. Then e_1 spans a 1-dimensional module for this parabolic, which induces a generalised Verma module $\overline{V}_{\alpha_1}$ of lowest weight type for $\mathfrak{g}_0$. The structure of $\overline{V}_{\alpha_1}$ can be understood by using results of Sect. B.1. In particular, imposing the condition (B.1.1), which in the present case reads

$$[e_2, [[e_2, e_3], e_1]] = 0, \tag{5.17}$$

sends $\overline{V}_{\alpha_1}$ to the irreducible quotient, which is $\mathfrak{g}_1$. Note that (5.17) is a higher order Serre relation associated with diagram (9) in Theorem 3.3. It is a non-standard diagram of $\mathfrak{sl}_{1|3}$.

Now $\mathfrak{g}_1$ is 10-dimensional. A basis for it can be deduced from Sect. B.1. For every vector b in this basis, we have $[b,e_1]=0$. This holds trivially for most basis vectors, but for $b=[e_2,[[[e_1,e_2],e_3],e_4]]$, we have

$$\begin{aligned}[e_1,b] &= [[[e_1,e_2],e_3],[[e_1,e_2],e_4]] \\ &= \frac{1}{2}(ad_{e_1})^2[[e_2,e_3],[e_2,e_4]].\end{aligned}$$

One can deduce from the defining relations for $\mathfrak{osp}_{2|4}$ that $[[e_2,e_3],[e_2,e_4]]=0$, hence $[b,e_1]=0$. This implies that the commutators of e_1 with all the remaining basis vectors are zero. Therefore, $\mathfrak{g}_2=[\mathfrak{g}_1,\mathfrak{g}_1]=0$.

5.4.5 Case 5

In the case of the Dynkin diagram

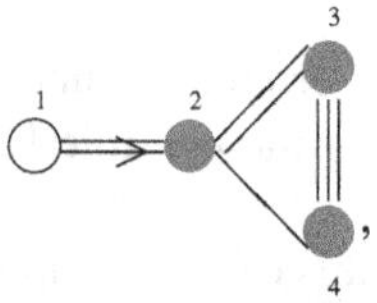

we take $d=4$ and delete the 4-th node to obtain the diagram

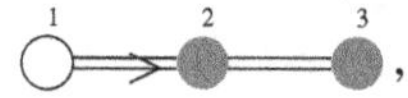

which is a non-standard diagram of $\mathfrak{sl}_{1|3}$. Thus we have a relation formally the same as (5.17).

Now $\mathfrak{g}_0=\mathfrak{gl}_{1|3}$. The Verma module of lowest weight type for $\mathfrak{g}_0$ generated by e_4 contains the primitive vector $2[e_2,[e_3,e_4]]-3[[e_2,e_3],e_4]$, which generates the maximal submodule. Thus the higher order Serre relation

$$2[e_2,[e_3,e_4]]-3[[e_3,e_2],e_4]=0, \tag{5.18}$$

associated with diagram (10) in Theorem 3.3, is all that is needed to guarantee that $\mathfrak{g}_1$ is an irreducible $\mathfrak{g}_0$-module. This module is typical relative to the distinguished Borel subalgebra, and has dimension 8.

Restricted to a module for $\mathfrak{gl}_3\subset\mathfrak{g}_0$, the even subspace of $\mathfrak{g}_1$ is the direct sum of the natural $\mathfrak{gl}_3$-module and a 1-dimensional module, while the odd subspace is the direct sum of the dual natural module (twisted by a scalar) and a 1-dimensional module.

Now consider $[\mathfrak{g}_1,\mathfrak{g}_1]$. We can easily work out its decomposition into irreducible $\mathfrak{gl}_3$-submodules. The corresponding $\mathfrak{gl}_3$ lowest weight vectors can be worked out, which include the following vectors:

$$[e_2,e_2], \quad \left(ad_{[e_2,e_4]}\right)^2[e_2,e_1], \quad [[e_2,e_4],[e_3,e_4]].$$

It follows from the higher order Serre relation (5.18) that

$$[[e_2,e_4],[e_3,e_4]]=0.$$

Now we impose the relations

$$[e_2, e_2] = 0, \quad \left(ad_{[e_2,e_4]}\right)^2 [e_2, e_1] = 0,$$

where the first is a standard Serre relation, and the second is a higher order Serre relations associated with

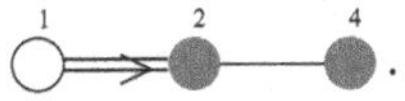

Under these conditions, all other $\mathfrak{gl}_3$ lowest vectors in $[\mathfrak{g}_1, \mathfrak{g}_1]$ vanish, except

$$\left(ad_{[e_2,e_4]}\right)^2 e_1, \quad [[e_3, e_4], [[e_1, e_2], [e_2, e_4]]],$$

where the first one is actually a $\mathfrak{g}_0$ lowest weight vector. It generates a 4-dimensional irreducible $\mathfrak{g}_0$-module containing the second vector. This module is isomorphic to the dual of the natural $\mathfrak{g}_0$-module twisted by a scalar. This gives us $\mathfrak{g}_2 = [\mathfrak{g}_1, \mathfrak{g}_1]$. We can further show that $[\mathfrak{g}_2, e_4] = 0$, hence $\mathfrak{g}_3 = 0$.

5.5 Proof in type $G(3)$

5.5.1 Case 1

Consider the Dynkin diagram

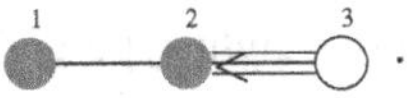

We take $d = 3$, then $\mathfrak{g}_0 = \mathfrak{gl}_{2|1}$. Let $\overline{V}_{\alpha_3}$ be the lowest weight Verma module for $\mathfrak{g}_0 = \mathfrak{gl}_{2|1}$ with lowest weight α_3. Denote by v_0 the lowest weight vector, which is assumed to be even. Then the maximal submodule of $\overline{V}_{\alpha_3}$ is generated by $e_1 v_0$ and $e_2[e_1, e_2]^3 v_0$. The irreducible quotient $\overline{L}_{\alpha_3}$ is multiplicity free and has weights

$$\begin{aligned} &\alpha_3 + k(\alpha_1 + \alpha_2), && k = 0, 1, 2, 3, \\ &\alpha_3 + p(\alpha_1 + \alpha_2) + \alpha_2, && p = 0, 1, 2. \end{aligned}$$

In fact $\overline{L}_{\alpha_3}$ is isomorphic to the third $\mathbb{Z}_2$-graded symmetric power of the natural module for $\mathfrak{g}_0$ tensored with a 1-dimensional module. Thus $\wedge_s^2 \overline{L}_{\alpha_3}$ is completely reducible; it is the direct sum of two irreducibles.

Now we take v_0 to be e_3, and let $\mathfrak{g}_0$ act on it by the adjoint action. Then the generators of the maximal submodule of $\overline{V}_{\alpha_3}$ in this case are $\left(ad_{[e_1,e_2]}\right)^3 [e_2, e_3]$ and $[e_1, e_3]$. Thus $[e_1, e_3] = 0$ and the higher order Serre relation

$$\left(ad_{[e_1,e_2]}\right)^3 [e_2, e_3] = 0 \tag{5.19}$$

(associated with diagram (11) in Theorem 3.3) render $\mathfrak{g}_1 = \overline{L}_{\alpha_3}$.

One of the irreducible submodules of $\wedge_s^2 \mathfrak{g}_1$ has a lowest weight vector of the form $e_3 \otimes [e_2, e_3] - [e_2, e_3] \otimes e_2$. We require that this submodule be in the kernel of the Lie superbracket. This leads to the standard Serre relation $[e_3, [e_3, e_2]] = 0$.

The other irreducible submodule of $\wedge_s^2 \mathfrak{g}_1$ is mapped surjectively onto $\mathfrak{g}_2$. A lowest weight vector of $\mathfrak{g}_2$ is given by $\overline{X} := ad_{e_3}\left(ad_{[e_1,e_2]}\right)^2 [e_2,e_3]$. This irreducible module is 4-dimensional and has weights

$$-(2\varepsilon_3 - \varepsilon_1 - \varepsilon_2),\ \delta + \varepsilon_2 - \varepsilon_3,\ \delta + \varepsilon_1 - \varepsilon_3,\ 2\delta,$$

in the notation explained in Appendix A.1. It is easy to see that $[X, \overline{X}] = 0$ for all $X \in \mathfrak{g}_1$. Thus $\mathfrak{g}_3 = 0$.

By examining the weights of $\mathfrak{g}_1$ and $\mathfrak{g}_2$, we see that Lemma 3.3 holds.

5.5.2 Case 2

Consider the Dynkin diagram

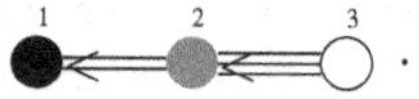

We take $d = 1$, and delete the first node from the Dynkin diagram to obtain

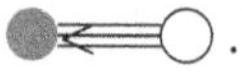

This is a nonstandard diagram for $\mathfrak{sl}_{1|2}$, which can be cast into the usual Dynkin diagram of $\mathfrak{sl}_{1|2}$ in the distinguished root system by normalising the bilinear form on the weight space. Note that no higher order Serre relations are required to present this Lie superalgebra. We have $\mathfrak{g}_0 = \mathfrak{gl}_{1|2}$.

Now the $\mathfrak{g}_0$ Kac module of lowest weight type generated by e_1 is typical thus irreducible, hence $\mathfrak{g}_1 \cong \overline{L}_{\alpha_1}$ with basis

$$e_1, \quad [e_2,e_1], \quad [[e_2,e_3],e_1], \quad [[e_2,e_3],[e_2,e_1]].$$

As $\mathfrak{g}_0$-module $\wedge_s^2 \mathfrak{g}_1$ is the direct sum of two irreducible typical submodules, respectively generated by the lowest weight vectors $e_1 \otimes e_1$ and $v - \frac{1}{2}v'$, where

$$v = e_1 \otimes [[e_2,e_3],[e_2,e_1]] + [[e_2,e_3],[e_2,e_1]] \otimes e_1,$$
$$v' = [e_2,e_1] \otimes [e_3,[e_2,e_1]] - [e_3,[e_2,e_1]] \otimes [e_2,e_1].$$

We require that $v - \frac{1}{2}v'$ and thus the $\mathfrak{g}_0$-submodule generated by it be mapped to zero by the Lie superbracket. This leads to

$$[[e_2,e_1],[e_3,[e_2,e_1]]] - [[e_2,e_3],[[e_1,e_1],e_2]] = 0,$$

which is one of the higher order Serre relations associated with the Dynkin diagram (see diagram (12) in Theorem 3.3). Therefore, $\mathfrak{g}_2 \cong \overline{L}_{2\alpha_1}$ and has a basis

$$[e_1,e_1], \quad [[e_1,e_1],e_2], \quad [e_3,[[e_1,e_1],e_2]], \quad [[e_2,e_3],[[e_1,e_1],e_2]].$$

Now we consider $[\mathfrak{g}_2, \mathfrak{g}_1]$. One can easily see that $(ad_{e_1})^3 e_2$ is a $\mathfrak{g}_0$ lowest weight vector. We require that the $\mathfrak{g}_0$-submodule generated by it be zero, hence we have the standard Serre relation

$$(ad_{e_1})^3 e_2 = 0.$$

This leaves $\mathfrak{g}_3 = [\mathfrak{g}_2, \mathfrak{g}_1]$ to be an indecomposable $\mathfrak{g}_0$-module cyclically generated by the lowest weight vector $[[e_1,[e_2,e_3]],[[e_1,e_1],e_2]]$, which is 7-dimensional and multiplicity free. One can easily write down a basis for this module. We should remark that no $\mathfrak{g}_0$ lowest weight vector in $\mathfrak{g}_3$ is annihilated by all f_i for $i = 1,2,3$.

One can show by direct computations that $[e_1, \mathfrak{g}_3] = 0$ and $[[e_1,e_1],\mathfrak{g}_2] = 0$. Hence $\mathfrak{g}_4 = 0$.

An inspection of the weight spaces of $\mathfrak{g}_i$ for $1 \leq i \leq 3$ shows that they agree with those of L_i for $1 \leq i \leq 3$. This completes the proof in this case.

5.5.3 Case 3

The final case of $G(3)$ is the diagram

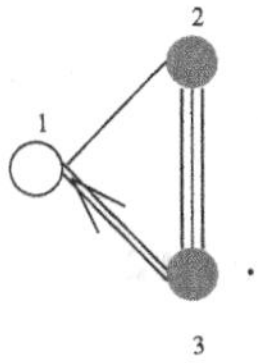

We take $d = 3$, then $\mathfrak{g}_0 = \mathfrak{gl}_{2|1}$. The $\mathfrak{g}_0$ Kac module of lowest weight generated by e_3 is atypical. We set the primitive vector to zero to obtain

$$2[[e_1,e_2],e_3] - [e_2,[e_1,e_3]] = 0,$$

which is a higher order Serre relation in the present case. Then $\mathfrak{g}_1$ is an irreducible $\mathfrak{g}_0$-module with lowest weight α_3, which is isomorphic to the third $\mathbb{Z}_2$-graded symmetric power of the natural module for $\mathfrak{g}_0$ twisted by a scalar. It has 3 odd and 4 even dimensions. A basis for $\mathfrak{g}_1$ is given by

$$e_3, \quad [e_1,e_3], \quad [e_1,[e_1,e_3]], \quad [e_2,e_3],$$
$$[[e_1,e_2],e_3], \quad [[e_1,e_2],[e_1,e_3]], \quad [[e_1,e_2],[e_1,[e_1,e_3]]].$$

The rest of the analysis is similar to Sect. 5.5.1. Now $\wedge_s^2 \mathfrak{g}_1$ is the direct sum of two irreducible $\mathfrak{g}_0$-submodules. The images of theirs lowest weight vectors in $[\mathfrak{g}_1, \mathfrak{g}_1]$ are repectively $[e_3,e_3]$ and $E = [[e_1,e_3],[e_1,e_3]]$. Both generate typical $\mathfrak{g}_0$-submodules, which respectively have dimensions 20 and 4. The standard Serre relation $[e_3,e_3] = 0$ removes the 20-dimensional submodule, thus $\mathfrak{g}_2$ is the 4-dimensional irreducible $\mathfrak{g}_0$-module generated by E.

We can also show that $\mathfrak{g}_3 = 0$ without imposing further relations. Inspecting the weights of $\mathfrak{g}_1$ and $\mathfrak{g}_2$, we see that the claim of Lemma 3.3 indeed holds.

5.6 Proof in type $D(2,1;\alpha)$

The Dynkin diagrams having only one grey node can be treated in exactly the same way as for the distinguished root system, thus we shall consider only the diagram with three gray nodes here. Set $d = 3$, then $\mathfrak{g}_0 = \mathfrak{gl}_{2|1}$. The $\mathfrak{g}_0$ Verma module of lowest weight type generated by e_3 contains the primitive vector

$$\alpha[e_1,[e_2,e_3]] + (1+\alpha)[e_2,[e_1,e_3]],$$

which in fact generates the maximal submodule. The higher order Serre relation requires this vector to be zero. This is equivalent to taking the irreducible quotient of the Verma module, and we obtain $\mathfrak{g}_1$. A basis for $\mathfrak{g}_1$ is

$$e_3, \quad [e_1,e_3], \quad [e_2,e_3], \quad [e_1,[e_2,e_3]].$$

An easy computation using the higher order Serre relation shows that $[\mathfrak{g}_1, e_3] = 0$. Hence $\mathfrak{g}_2 = 0$. A quick inspection on the weights of $\mathfrak{g}_1$ shows that Lemma 3.3 indeed holds in this case.

6 Remarks on affine Lie superalgebras

We wish to mention that the generalisation of the method to affine Lie superalgebras is in principle straightforward conceptually. Consider, for example, the untwisted affine superalgebra $\hat{\mathfrak{g}}$ of a contragredient Lie superalgebra $\mathfrak{g}$. We want to present $\hat{\mathfrak{g}}$ with the standard generators e_i, f_i, h_i with $0 \le i \le r$ and relations. Here the generators e_i, f_i, h_i with $1 \le i \le r$ are those for $\mathfrak{g}$. By results of earlier sections, we may assume that all the Serre relations and higher order ones obeyed by e_i and f_i with $1 \le i \le r$ are given.

We introduce the standard $\mathbb{Z}$-grading of $\hat{\mathfrak{g}}$ by decreeing that all h_j and e_i, f_i with $1 \le i \le r$ have degree 0, but e_0 and f_0 have degrees 1 and -1 respectively. Then $\hat{\mathfrak{g}} = \oplus_{k\in\mathbb{Z}} \hat{\mathfrak{g}}_k$, with $\hat{\mathfrak{g}}_0 = \mathfrak{g} \oplus \mathfrak{gl}_1$. Now we require that as $\hat{\mathfrak{g}}_0$-modules, all $\hat{\mathfrak{g}}_k$ are isomorphic to $\mathfrak{g}$. The (necessary and sufficient) conditions meeting this requirement give rise to the defining relations of $\hat{\mathfrak{g}}$.

To illustrate how this may work, we consider the untwisted affine algebra $\hat{\mathfrak{g}} = \hat{\mathfrak{sl}}_{r+1}$. The relations

$$[e_1,[e_1,e_0]] = 0, \quad [e_r,[e_r,e_0]] = 0, \quad [e_i,e_0] = 0, \; i \ne 1, r$$

arise from the requirement that $\hat{\mathfrak{g}}_1$ be an irreducible $\hat{\mathfrak{g}}_0$-module. In $[\hat{\mathfrak{g}}_1, \hat{\mathfrak{g}}_1]$, there are $\hat{\mathfrak{g}}_0$ lowest weight vectors $[[e_1,e_0],e_0]$ and $[[e_r,e_0],e_0]$, which have weights different from any roots of $\mathfrak{g} = \mathfrak{sl}_{r+1}$. Thus the condition that $\hat{\mathfrak{g}}_2$ is isomorphic to $\mathfrak{g}$ as $\mathfrak{g}_0$-module requires

$$[[e_1,e_0],e_0] = 0, \quad [[e_r,e_0],e_0] = 0.$$

Now we have derived all the Serre relations needed for e_0, and those for f_0 can be similarly obtained. Together with relations defining $\mathfrak{g}$, these relations define $\hat{\mathfrak{g}}$.

We hope to treat the affine superalgebras on another occasion.

Appendix A Dynkin diagrams

We describe the Dynkin diagrams for both the distinguished and non-distinguished root systems in this Appendix. The roots of all the simple contragedient Lie superalgebras will also be listed [12, 13].

A.1 Roots

Let ε_i ($i = 1,2,\ldots,k$) and δ_j ($j = 1,2,\ldots,l$) be a basis of a real vector space $E(k,l)$ equipped with a non-degenerate symmetric bilinear form. Then for each simple contragredient Lie superalgebra $\mathfrak{g}$, the dual space $\mathfrak{h}^*$ of the Cartan subalgebra is either $\mathbb{C}\otimes_{\mathbb{R}} E(k,l)$ for appropriate k,l or a subspace thereof, which inherits a non-degenerate bilinear form that is Weyl group invariant.

For the series A, B, C or D, the bilinear form is defined by

$$(\varepsilon_i, \varepsilon_{i'}) = \delta_{ii'}, \quad (\delta_j, \delta_{j'}) = -\delta_{jj'}, \quad (\varepsilon_i, \delta_j) = 0, \quad \forall i, i', j, j'.$$

The roots of the simple contragredient Lie superalgebras can be described as follows.

$A(m|n)$:

$$\begin{aligned}
\Delta_0 &= \{\varepsilon_i - \varepsilon_{i'} \mid i,i' \in [1,m+1], i \neq i'\} \cup \{\delta_j - \delta_{j'} \mid j,j' \in [1,n+1], j \neq j'\},\\
\Delta_1 &= \{\pm(\varepsilon_i - \delta_j) \mid i \in [1,m+1], j \in [1,n+1]\},
\end{aligned}$$

where $[1,N]$ denotes $\{1,\ldots,N\}$ for any positive integer N.

$B(0,n)$:

$$\begin{aligned}
\Delta_0 &= \{\pm\delta_j \pm \delta_{j'},\ \pm 2\delta_j \mid j,j' \in [1,n], j \neq j'\},\\
\Delta_1 &= \{\pm\delta_j \ \mid j \in [1,n]\}.
\end{aligned}$$

$B(m,n)$, $m > 1$:

$$\begin{aligned}
\Delta_0 &= \{\pm\varepsilon_i \pm \varepsilon_{i'},\ \pm\varepsilon_i \mid i,i' \in [1,m], i \neq i'\}\\
&\quad \cup \{\pm\delta_j \pm \delta_{j'},\ \pm 2\delta_j \mid j,j' \in [1,n], j \neq j'\},\\
\Delta_1 &= \{\pm\varepsilon_i \pm \delta_j, \pm\delta_j \ \mid i \in [1,m], j \in [1,n]\},
\end{aligned}$$

$C(n+1)$:

$$\begin{aligned}
\Delta_0 &= \{\pm\delta_j \pm \delta_{j'},\ \pm 2\delta_j \mid j,j' \in [1,n], j \neq j'\},\\
\Delta_1 &= \{\pm\varepsilon_1 \pm \delta_j \mid j \in [1,n]\}.
\end{aligned}$$

$D(m,n)$, $m > 1$:

$$\begin{aligned}
\Delta_0 =& \{\pm\varepsilon_i \pm \varepsilon_{i'} \mid i,i' \in [1,m], i \neq i'\}\\
&\cup \{\pm\delta_j \pm \delta_{j'},\ \pm 2\delta_j \mid j,j' \in [1,n], j \neq j'\},\\
\Delta_1 =& \{\pm\varepsilon_i \pm \delta_j \mid i \in [1,m], j \in [1,n]\}.
\end{aligned}$$

$F(4)$:

$$\Delta_0 = \{\pm\varepsilon_i \pm \varepsilon_j,\ \pm\varepsilon_i \mid i,j = 1,2,3,\ i \neq j\} \cup \{\pm\delta\},$$
$$\Delta_1 = \left\{\frac{1}{2}(\pm\varepsilon_1 \pm \varepsilon_2 \pm \varepsilon_3 \pm \delta)\right\},$$
$$(\delta,\delta) = -6, \quad (\varepsilon_i,\varepsilon_j) = 2\delta_{ij}, \quad (\varepsilon_i,\delta) = 0, \quad \forall i,j = 1,2,3.$$

$G(3)$:

$$\Delta_0 = \{\varepsilon_i - \varepsilon_j, \pm(2\varepsilon_k - \varepsilon_i - \varepsilon_j) \mid 1 \leq i,j,k \leq 3, \text{pairwise distinct}\} \cup \{\pm 2\delta\},$$
$$\Delta_1 = \{\pm\delta + (\varepsilon_i - \varepsilon_j),\ \pm\delta \mid i \neq j\},$$
$$(\delta,\delta) = -2,\ (\varepsilon_i,\varepsilon_,) = \delta_{ij},\ (\varepsilon_i,\delta) = 0,\ \forall i,j = 1,2,3.$$

$D(2,1;\alpha)$, $\alpha \in \mathbb{C}\backslash\{0,-1\}$:

$$\Delta_0 = \{\pm 2\varepsilon_i \mid i = 1,2\} \cup \{\pm 2\delta\},$$
$$\Delta_1 = \{\pm\delta \pm \varepsilon_1 \pm \varepsilon_2\},$$
$$(\varepsilon_1,\varepsilon_1) = 1, \quad (\varepsilon_2,\varepsilon_2) = \alpha, \quad (\delta,\delta) = -(1+\alpha), \quad (\varepsilon_i,\delta) = 0,\ \forall i.$$

Denote by $\Pi = \{\alpha_1, \ldots, \alpha_r\}$ the set of simple roots of $\mathfrak{g}$ elative to the distinguished Borel subalgebra. We have

$A(m|n)$: $\Pi = \{\varepsilon_1 - \varepsilon_2, \ldots, \varepsilon_m - \varepsilon_{m+1}, \varepsilon_{m+1} - \delta_1, \delta_1 - \delta_2, \ldots, \delta_n - \delta_{n+1}\}$;
$B(0,n)$: $\Pi = \{\delta_1 - \delta_2, \ldots, \delta_{n-1} - \delta_n, \delta_n\}$;
$B(m,n), m > 1$:
$\Pi = \{\delta_1 - \delta_2, \ldots, \delta_{n-1} - \delta_n, \delta_n - \varepsilon_1, \varepsilon_1 - \varepsilon_2, \ldots, \varepsilon_{m-1} - \varepsilon_m,\ \varepsilon_m\}$;
$C(n+1)$: $\Pi = \{\varepsilon_1 - \delta_1,\ \delta_1 - \delta_2, \ldots, \delta_{n-1} - \delta_n, 2\delta_n\}$;
$D(m,n), m > 1$:
$\Pi = \{\delta_1 - \delta_2, \ldots, \delta_{n-1} - \delta_n, \delta_n - \varepsilon_1,\ \varepsilon_1 - \varepsilon_2, \varepsilon_2 - \varepsilon_3, \ldots, \varepsilon_{m-1} - \varepsilon_m, \varepsilon_{m-1} + \varepsilon_m\}$;
$F(4)$: $\Pi = \left\{\frac{1}{2}(\varepsilon_1 + \varepsilon_2 + \varepsilon_3 + \delta),\ -\varepsilon_1,\ \varepsilon_1 - \varepsilon_2,\ \varepsilon_2 - \varepsilon_3\right\}$;
$G(3)$: $\Pi = \{\delta - \varepsilon_1 + \varepsilon_3,\ \varepsilon_1 - \varepsilon_2,\ 2\varepsilon_2 - \varepsilon_1 - \varepsilon_3\}$;
$D(2,1;\alpha), \alpha \in \mathbb{C}\backslash\{0,-1\}$: $\Pi = \{\delta - \varepsilon_1 - \varepsilon_2,\ 2\varepsilon_1,\ 2\varepsilon_2\}$.

Note that there is a unique simple root, which we denote by α_s, in each Π. Thus $\Theta = \{s\}$.

The simple roots relative to other Borel subalgebras can be obtained by using odd reflections [24]. Let $\Pi_{\mathfrak{b}} = \{\alpha_1, \ldots, \alpha_r\}$ be the set of simple roots relative to a given Borel subalgebra $\mathfrak{b} \subset \mathfrak{g}$. Take any isotropic odd simple root $\alpha_t \in \Pi_{\mathfrak{b}}$, and define the odd reflection s_t by

$$\begin{aligned} s_t(\alpha_t) &= -\alpha_t, & & \\ s_t(\alpha_i) &= \alpha_i + \alpha_t, & &\text{if } i \neq t \text{ and } a_{it} \neq 0, \\ s_t(\alpha_i) &= \alpha_i, & &\text{if } i \neq t \text{ and } a_{it} = 0. \end{aligned}$$

Then $s_t(\Pi_{\mathfrak{b}}) = \{s_t(\alpha_1), \ldots, s_t(\alpha_r)\}$ is the set of simple roots relative to another Borel subalgebra, which is not Weyl group conjugate to $\mathfrak{b}$. Further odd reflections can be defined with respect to isotropic roots in $s_t(\Pi_{\mathfrak{b}})$, which turn $s_t(\Pi_{\mathfrak{b}})$ into sets of simple roots relative to other Borel subalgebras. All the distinct sets obtained this way correspond bijectively to the conjugacy classes of Borel subalgebras.

A.2 Dynkin diagrams

A.2.1 Dynkin diagrams in distinguished root systems

The Dynkin diagrams in the distinguished root systems are listed in Table 1 below, where r is the number of nodes and s is the element of Θ. Note that the form of Dynkin diagrams in the distinguished root systems is quite uniform in the literature. Table 1 is essentially the corresponding table in [12] with a slight modification in the Dynkin diagram for $D(2,1;\alpha)$.

Table 1 Dynkin diagrams in distinguished root systems

Lie superalgebra	Dynkin Diagram	r	s
$A(m,n)$		m+n+1	m+1
$B(m,n)$ $m>0$		m+n	n
$B(0,n)$	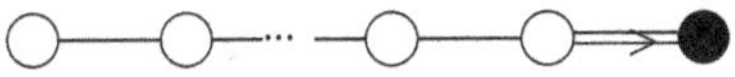	n	n
$C(n)$ $n>2$	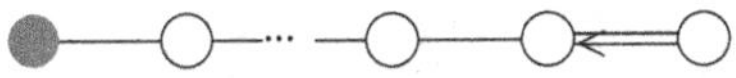	n	1
$D(m,n)$ $m>1$	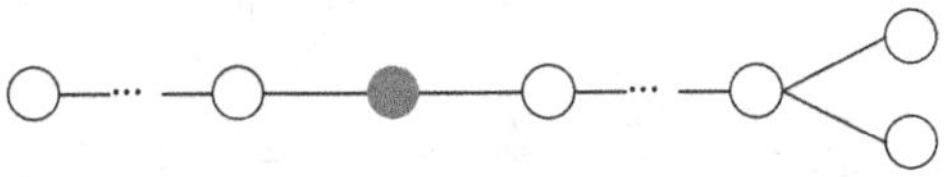	m+n	n
$F(4)$		4	1
$G(3)$	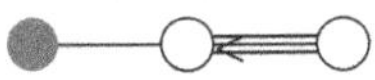	3	1
$D(2,1;\alpha)$	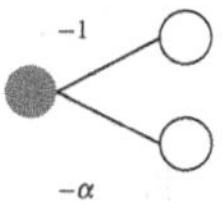	3	1

A.2.2 Dynkin diagrams in non-distinguished root systems

Table 2 gives the Dynkin diagrams of the non-distinguished root systems. A nice graphical explanation can be found in [3, §4] (see also [6]) on how to obtain the Dynkin diagrams in Table 2 by applying odd reflections to those in Table 1.

Table 2 Dynkin diagrams in non-distinguished root systems

Lie superalgebra	Dynkin Diagram
$A(m,n)$	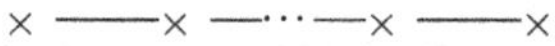
$B(m,n)$ $m>0$	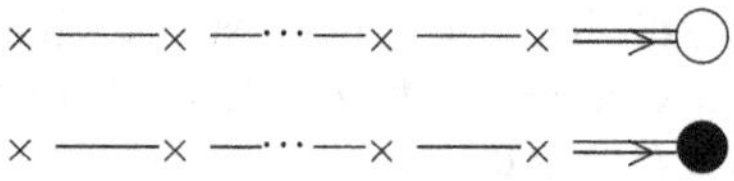
$C(n)$	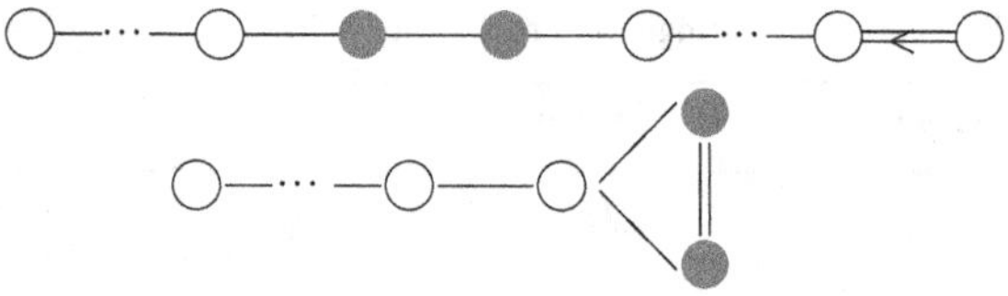
$D(m,n)$ $m>1$	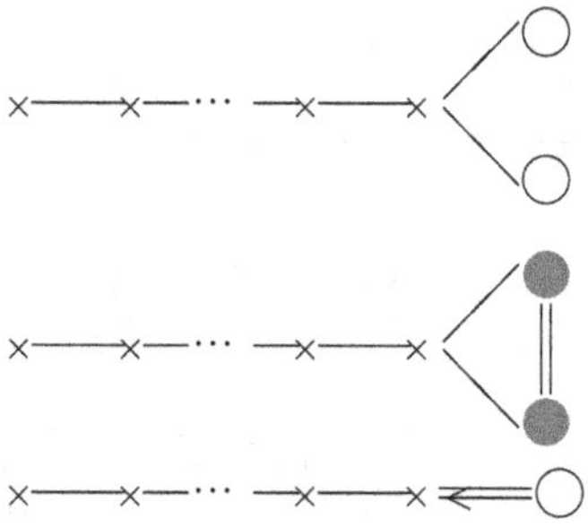
$F(4)$	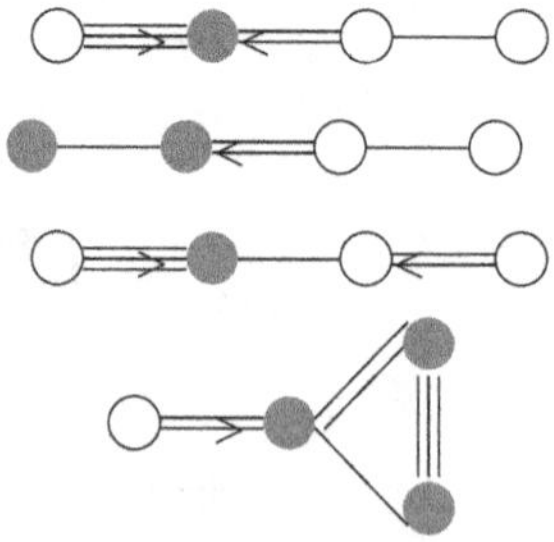

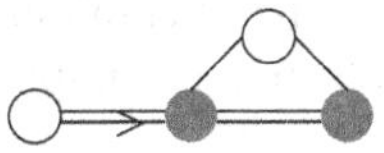

$G(3)$

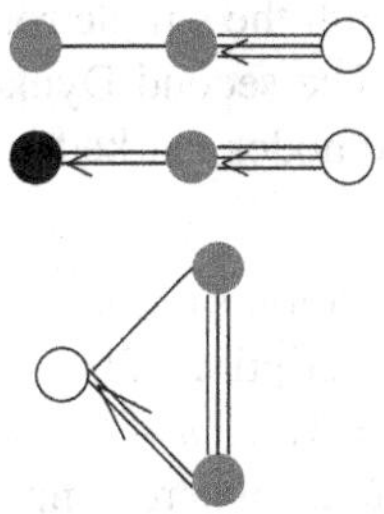

$D(2,1;\alpha)$

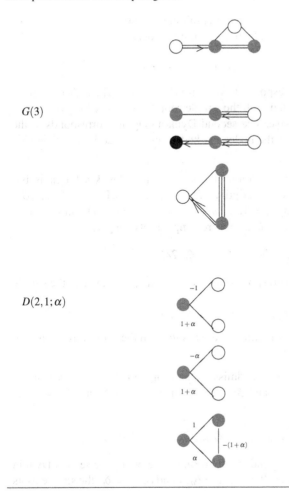

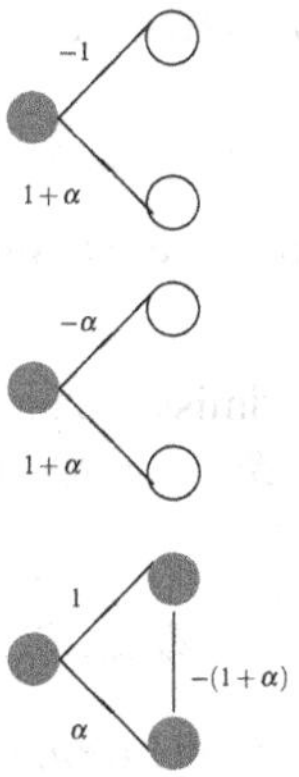

In the diagrams in Table 2, a node marked with $\times$ can be white or grey. However, the precise rule for assigning colours requires the knowledge of the simple roots, which are described below.

$A(m,n)$. An ordering $(\mathscr{E}_1, \mathscr{E}_2, \ldots, \mathscr{E}_{m+n+2})$ of ε_i and δ_j is called admissible if ε_i appears before ε_{i+1} for all i and δ_j before δ_{j+1} for all j. Each admissible ordering corresponds to one Weyl group conjugate class of Borel subalgebras, with the associated simple roots given by $\mathscr{E}_a - \mathscr{E}_{a+1}$ $(1 \le a \le m+n+1)$. In particular, the distinguished Borel corresponds to the admissible ordering such that all the ε_i appear before the δ_j. Let us define $[\mathscr{E}_a]$ $(a = 1, 2, \ldots, m+n+2)$ by $[\mathscr{E}_a] = 0$ (resp. $[\mathscr{E}_a] = 1$) if $\mathscr{E}_a$ is some ε_i (resp. δ_j). The a-th node from the left in the Dynkin diagram is associated with the simple root $\mathscr{E}_a - \mathscr{E}_{a+1}$, which is white if $[\mathscr{E}_a] = [\mathscr{E}_{a+1}]$ and grey otherwise.

$B(m,n)$, $m > 0$. Let $(\mathscr{E}_1, \mathscr{E}_2, \ldots, \mathscr{E}_{m+n})$ be an admissible ordering of ε_i $(i = 1, \ldots, m)$ and δ_j $(j = 1, \ldots, n)$. Then the corresponding simple roots are

$$\mathscr{E}_1 - \mathscr{E}_2, \ldots, \mathscr{E}_{m+n-1} - \mathscr{E}_{m+n}, \mathscr{E}_{m+n}.$$

The first Dynkin diagram corresponds to the case $\mathscr{E}_{m+n} = \varepsilon_m$. The a-th node $(a < m+n)$ from the left is associated with the simple root $\mathscr{E}_a - \mathscr{E}_{a+1}$, which is white if $[\mathscr{E}_a] = [\mathscr{E}_{a+1}]$ and grey otherwise. The second Dynkin diagram corresponds to the case $\mathscr{E}_{m+n} = \delta_n$. The colours of the nodes marked $\times$ are assigned in the same way as in type A.

$C(n)$. We have already specified the colours of the nodes in the Dynkin diagrams, but it is still useful to have an explicit description of the simple roots. Let $(\mathscr{E}_1, \mathscr{E}_2, \ldots, \mathscr{E}_n)$ be an admissible ordering of δ_j $(j = 1, \ldots, n-1)$ and ε_1. The first Dynkin diagram corresponds to the case with $\mathscr{E}_n = \delta_{n-1}$, where simple roots are given by

$$\mathscr{E}_1 - \mathscr{E}_2, \ldots, \mathscr{E}_{n-1} - \mathscr{E}_n, 2\mathscr{E}_n.$$

The second Dynkin diagram corresponds to the case with $\mathscr{E}_n = \varepsilon_1$, where the simple roots are given by

$$\mathscr{E}_1 - \mathscr{E}_2, \ldots, \mathscr{E}_{n-1} - \mathscr{E}_n, \mathscr{E}_{n-1} + \mathscr{E}_n.$$

The colours of the nodes marked with $\times$'s are assigned in the same way as in type A and type B.

$D(m,n)$. Let $(\mathscr{E}_1, \mathscr{E}_2, \ldots, \mathscr{E}_{m+n})$ be an admissible ordering of ε_i $(i = 1, \ldots, m)$ and δ_j $(j = 1, \ldots, n)$. If $\mathscr{E}_{m+n-1} = \varepsilon_{m-1}$ and $\mathscr{E}_{m+n} = \varepsilon_m$, or $\mathscr{E}_{m+n-1} = \delta_n$ and $\mathscr{E}_{m+n} = \varepsilon_m$, the simple roots are given by

$$\mathscr{E}_1 - \mathscr{E}_2, \ldots, \mathscr{E}_{m+n-1} - \mathscr{E}_{m+n}, \mathscr{E}_{m+n-1} + \mathscr{E}_{m+n}.$$

The first Dynkin diagram corresponds to the former case, while the second Dynkin diagram corresponds to the latter. If $\mathscr{E}_{m+n-1} = \delta_{n-1}$ and $\mathscr{E}_{m+n} = \delta_n$, the simple roots are given by

$$\mathscr{E}_1 - \mathscr{E}_2, \ldots, \mathscr{E}_{m+n-1} - \mathscr{E}_{m+n}, 2\mathscr{E}_{m+n}.$$

The third Dynkin diagram corresponds to this case.

We assign colours to the nodes marked with $\times$ in the same way as in the other cases.

Remark Appendix A.1 There are at least three grey nodes in the Dynkin diagrams of type $D(m,n)$ in Table 2, but in each of the Dynkin diagrams of type $C(n)$, there are only two grey nodes which are always next to each other.

Appendix B Presentations of irreducible modules

In general it is hard to give an explicit description of a finite dimensional irreducible module for a Lie superalgebra as the quotient of a (generalised) Verma module in a form similar to [10, Theorem 21.4] in the context of ordinary semi-simple Lie algebras. However, this is possible in some special cases, e.g., the natural module for $\mathfrak{gl}_{m|n}$ in arbitrary root systems as discussed in Sect. 5.1. Here are two further cases, which are used in the proof of Lemma 3.3.

B.1 An irreducible $\mathfrak{osp}_{2|4}$-module

Let $\mathfrak{g}$ be the Lie superalgebra $\mathfrak{osp}_{2|4}$ with the choice of Borel subalgebra corresponding to the Dynkin diagram

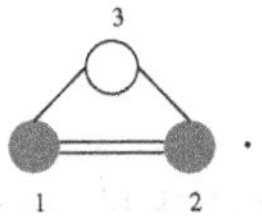

We present $\mathfrak{g}$ in the standard fashion using Chevalley generators e_i, f_i, h_i $(i = 1,2,3)$ and relations with the higher order Serre relations being those associated with diagram (6) in Theorem 3.3. To be specific, we denote by α_i the simple roots and take

$$(\alpha_1, \alpha_3) = (\alpha_2, \alpha_3) = -1, \quad (\alpha_1, \alpha_2) = 2, \quad (\alpha_3, \alpha_3) = 2.$$

Let $\overline{\mathfrak{p}}$ be the parabolic subalgebra generated by all the generators but e_1. Then $\overline{\mathfrak{p}} = \mathfrak{l} \oplus \overline{\mathfrak{u}}$ with $\mathfrak{l} = \mathfrak{gl}_{2|1}$ and $\overline{\mathfrak{u}}$ spanned by

$$\begin{array}{lll} \zeta_1 := e_1, & \zeta_2 := [e_1, e_3], & \\ X_1 := [e_1, e_2], & X_2 := [[e_1, e_2], e_3], & X_3 := [[[e_1, e_2], e_3], e_3]. \end{array}$$

Given the irreducible $\overline{\mathfrak{p}}$-module $\overline{L}^0_\lambda = \mathbb{C}v_0$ with lowest weight λ such that

$$(\lambda, \alpha_2) = 0, \quad (\lambda, \alpha_3) = 0, \quad (\lambda, \alpha_1) = -2,$$

we construct the generalised Verma module $\overline{V}_\lambda = U(\mathfrak{g}) \otimes_{U(\overline{\mathfrak{p}})} \overline{L}^0_\lambda$. Then the maximal submodule M_λ of $\overline{V}_\lambda$ is given by

$$M_\lambda = U(\mathfrak{g})\zeta_1 X_1 v_0. \tag{B.1.1}$$

The irreducible quotient $\overline{L}_\lambda = \overline{V}_\lambda / M_\lambda$ is 10-dimensional with a basis

$$\begin{array}{lllll} v_0, & X_1 v_0, & X_2 v_0, & X_3 v_0, & X_1 X_3 v_0, \\ \zeta_1 v_0, & \zeta_1 X_2 v_0, & \zeta_1 X_3 v_0, & \zeta_1 X_1 X_3 v_0, & \zeta_1 \zeta_2 v_0. \end{array}$$

B.2 Graded symmetric tensor for $\mathfrak{gl}_{m|n}$

Let $\mathfrak{g} = \mathfrak{gl}_{m|n}$ and set $r = m + n - 1$. Choose an arbitrary homogeneous basis for the natural module $\mathbb{C}^{m|n}$ with the last element being odd. We regard $\mathfrak{g}$ as consisting of matrices relative to this basis. Take the subalgebra consisting of the upper triangular matrices as the Borel subalgebra, which corresponds to an admissible ordering $(\mathscr{E}_1, \mathscr{E}_2, \ldots, \mathscr{E}_{m+n})$ of ε_i $(1 \le i \le m)$ and δ_j $(1 \le j \le n)$ with $\mathscr{E}_{m+n} = \delta_n$. See Appendix A.2 for more details.

Let $\mathfrak{l}$, $\mathfrak{u}$ and $\bar{\mathfrak{u}}$ be subalgebras respectively spanned by matrix units $e_{r+1,r+1}$ and e_{ij} with $1 \le i, j \le r$, by $e_{i,r+1}$ with $1 \le r$, and by $e_{r+1,i}$ with $1 \le r$. Set $\bar{\mathfrak{p}} = \mathfrak{l} \oplus \bar{\mathfrak{u}}$, which is a parabolic subalgebra, and $\mathfrak{g} = \bar{\mathfrak{p}} \oplus \mathfrak{u}$.

For $\lambda = 2\delta_n$, we consider the generalised Verma module $\overline{V}_\lambda := U(\mathfrak{g}) \otimes_{U(\bar{\mathfrak{p}})} \mathbb{C}_\lambda$ of lowest weight type, where $\mathbb{C}_\lambda$ denotes the irreducible $\bar{\mathfrak{p}}$-module with lowest weight λ. Let v_0 denote a generator of $\mathbb{C}_\lambda$, then

$$\begin{aligned} &f_r v_0 = 0, \\ &e_i v_0 = 0, \quad f_i v_0 = 0, \qquad 1 \le i \le r-1, \\ &e_{jj} v_0 = 2\delta_{j,r+1} v_0, \qquad 1 \le j \le r+1, \end{aligned} \tag{B.2.1}$$

where $e_i = e_{i,i+1}$ and $f_i = e_{i+1,i}$.

Now $\overline{V}_\lambda \cong U(\mathfrak{u}) \otimes \mathbb{C}_\lambda$ as $\mathfrak{l}$-module, where $U(\mathfrak{u}) = S_s(\mathfrak{u})$, the $\mathbb{Z}_2$-graded symmetric algebra of $\mathfrak{u}$. This superalgebra has a $\mathbb{Z}$-grading with $\mathfrak{u}$ having degree 1. It induces a natural $\mathbb{Z}$-grading on $\overline{V}_\lambda$. The unique maximal submodule M_λ of $\overline{V}_\lambda$ is the direct sum of the homogeneous subspaces of degrees greater than or equal to 3, which is generated by $U(\mathfrak{u})_3 \otimes \mathbb{C}_\lambda$, the homogeneous subspace of degree 3. The irreducible quotient $\overline{L}_\lambda$ of $\overline{V}_\lambda$ is isomorphic to the $\mathbb{Z}_2$-graded symmetric tensor of the natural $\mathfrak{g}$-module at rank 2.

The natural $\mathfrak{l}$ action on $U(\mathfrak{u})$ (obtained by generalising the adjoint action) respects the $\mathbb{Z}$-grading. In the present case, each homogeneous component is in fact an irreducible submodule. We are interested in $U(\mathfrak{u})_3$. If u_3 is a nonzero lowest weight vector of $U(\mathfrak{u})_3$, then M_λ is generated over $\mathfrak{g}$ by $u_3 \otimes \mathbb{C}_\lambda$. The form of u_3 depends on the ordering of the basis for $\mathbb{C}^{m|n}$. Denote by $E_{ij} \in U(\mathfrak{g})$ the image of $e_{ij} \in \mathfrak{g}$ under the natural embedding. The u_3 can be expressed as follows:

$$\begin{aligned} &u_3 = E^3_{r,r+1}, && \text{if } E_{r,r+1} \text{ is even;} \\ &u_3 = E^2_{r-1,r+1} E_{r,r+1}, && \text{if both } E_{r,r+1} \text{ and } E_{r-1,r} \text{ are odd;} \\ &u_3 = E_{r-2,r+1} E_{r-1,r+1} E_{r,r+1}, && \text{if } E_{r,r+1} \text{ is odd but } E_{r-1,r} \text{ is even.} \end{aligned} \tag{B.2.2}$$

Remark Appendix B.1 The third case becomes vacuous if $r = 2$; and both the second and third cases are vacuous if $r = 1$.

The irreducible quotient $\overline{L}_\lambda = \overline{V}_\lambda / M_\lambda$ is isomorphic to the graded skew symmetric rank two tensor $\wedge^2_s(\mathbb{C}^{m|n})$ of the natural $\mathfrak{g}$-module.

Acknowledgements I wish to thank Professor Dimitry Leites for helpful suggestions and Professor Victor Kac for historical remarks. This work was supported by the Australian Research Council.

References

1. Bouarroudj, S., Grozman, P., Lebedev, A., Leites, D.: Divided power (co)homology. Presentations of simple finite dimensional modular Lie superalgebras with Cartan matrix. Homology, Homotopy Appl. **12**(1), 237–278 (2010)
2. Bracken, A.J., Gould, M.D., Zhang, R.B.: Quantum supergroups and solutions of the Yang-Baxter equation. Modern Phys. Lett. **A 5**(11), 831–840 (1990)
3. Chapovalov, D., Chapovalov, M., Lebedev, A., Leites, D.: The classification of almost affine (hyperbolic) Lie superalgebras. Journal of Nonlinear Mathematical Physics, **1**(1), 1–55 (2009)
4. Drinfeld, V.G.: Quantum groups. In: Proceedings of the International Congress of Mathematicians, Vol. 1, 2 (Berkeley, Calif., 1986), pp. 798–820, Amer. Math. Soc., Providence, RI (1987)
5. Floreanini, R., Leites, D.A., Vinet, L.: On the defining relations of quantum superalgebras. Lett. Math. Phys. **23**(2), 127–131 (1991)
6. Frappat, L., Sciarrino, A., Sorba, P.: Structure of basic Lie superalgebras and of their affine extensions. Comm. Math. Phys. **121**(3), 457–500 (1989)
7. Gabber, O., Kac, V.G.: On defining relations of certain infinite-dimensional Lie algebras. Bull. Amer. Math. Soc. (N.S.) **5**(2), 185–189 (1981)
8. Grozman, P., Leites, D.: Defining relations for Lie superalgebras with Cartan matrix. Czechoslovak J. Phys. **51**(1), 1–21 (2001)
9. Grozman, P., Leites, D., Poletaeva, E.: Defining relations for classical Lie superalgebras without Cartan matrices. The Roos Festschrift, Vol. 2. Homology Homotopy Appl. **4**(2), part 2, 259–275 (2002)
10. Humphreys, J.E.: Introduction to Lie algebras and representation theory. Graduate Texts in Mathematics, Vol. 9. Springer-Verlag, Berlin New York (1972)
11. Jimbo, M.: A q-difference analogue of $U_q(\mathfrak{g})$ and the Yang-Baxter equation. Lett. Math. Phys. **10**(1), 63–69 (1985)
12. Kac, V.G.: Lie superalgebra. Advances in Math. **26**(1), 8–96 (1977)
13. Kac, V.G.: Representations of classical Lie superalgebras. Differential geometrical methods in mathematical physics, II. In: Proc. Conf., Univ. Bonn, Bonn, 1977, pp. 597–626. Lecture Notes in Math., Vol. 676. Springer-Verlag, Berlin Heidelberg (1978)
14. Kac, V.G.: Infinite-dimensional Lie algebras, 3rd edn. Cambridge University Press, Cambridge (1990)
15. Kaplansky, I.: Superalgebras. Pacific J. Math. **86**(1), 93–98 (1980).
16. Kaplansky, I.: Afterthought: superalgebras. In: Selected papers and other writings. With an introduction by Hyman Bass, p. 225. SpringerVerlag, New York (1995)
17. Lanzmann, E.: The Zhang transformation and $U_q(osp(1,2l))$-Verma modules annihilators. Algebr. Represent. Theory **5**(3), 235–258 (2002)
18. Leites, D., Serganova, V.: Defining relations for classical Lie superalgebras. I. Superalgebras with Cartan matrix or Dynkin-type diagram. Topological and geometrical methods in field theory (Turku, 1991), pp. 194–201. World Science Publ., River Edge, NJ (1992)
19. Links, J.R., Gould, M.D., Zhang, R.B.: Quantum supergroups, link polynomials and representation of the braid generator. Rev. Math. Phys. 5(2), 345–361 (1993)
20. Lusztig, G.: Introduction to quantum groups. Progress in Mathematics, Vol. 110. Birkhäuser, Boston, MA (1993)
21. Manin, Yu.I.: Quantum groups and noncommutative geometry. Université de Montréal, Centre de Recherches Mathématiques, Montreal, QC (1988)
22. Scheunert, M.: The theory of Lie superalgebras. An introduction. Lecture Notes in Mathematics, Vol. 716. Springer-Verlag, Berlin Heidelberg (1979)
23. Scheunert, M.: The presentation and q deformation of special linear Lie superalgebras. J. Math. Phys. **34**(8), 3780–3808 (1993)
24. Serganova, V.: Automorphisms of simple Lie superalgebras. Izv. Akad. Nauk SSSR Ser. Mat. **48**(3), 585–598 (1984), in Russian. English translation: Math. USSR-Izv. **24**(3), 539–551 (1985)

25. Yamane, H.: Quantized enveloping algebras associated with simple Lie superalgebras and their universal R-matrices. Publ. Res. Inst. Math. Sci. **30**(1), 15–87 (1994)
26. Yamane, H.: On defining relations of affine Lie superalgebras and affine quantized universal enveloping superalgebras. Publ. Res. Inst. Math. Sci. **35**(3), 321–390 (1999)
27. Zhang, R.B.: Braid group representations arising from quantum supergroups with arbitrary q and link polynomials. J. Math. Phys. **33**(11), 3918–3930 (1992)
28. Zhang, R.B.: Finite-dimensional representations of $U_q(\mathrm{osp}(1/2n))$ and its connection with quantum $\mathrm{so}(2n+1)$. Lett. Math. Phys. **25**(4), 317–325 (1992)
29. Zhang, R.B.: Finite-dimensional irreducible representations of the quantum supergroup $\mathrm{U}_q(\mathrm{gl}(m/n))$. J. Math. Phys. **34**(3), 1236–1254 (1993)
30. Zhang, R.B.: Quantum supergroups and topological invariants of three-manifolds. Rev. Math. Phys. **7**(5), 809–831 (1995)
31. Zhang, R.B.: Structure and representations of the quantum general linear supergroup. Comm. Math. Phys. **195**(3), 525–547 (1998)
32. Zhang, R.B. Quantum superalgebra representations on cohomology groups of non-commutative bundles. J. Pure Appl. Algebra **191**(3), 285–314 (2004)
33. Zhang, R.B., Bracken, A.J., Gould, M.D.: Solution of the graded Yang-Baxter equation associated with the vector representation of $\mathrm{U}_q(\mathrm{osp}(M/2n))$. Phys. Lett. B **257**, 133–139 (1991)
34. Zhang, R.B., Gould, M.D., Bracken, A.J.: Solutions of the graded classical Yang-Baxter equation and integrable systems. J. Physics A **4**, 1185–1197 (1991)

Printed in the United States
By Bookmasters